METHODS FOR CATCHING BEETLES

Carlos Aguilar Julio

NATURALIA
SCIENTIFIC COLLECTION
Edited by
Jorge Barrett Viedma
MONTEVIDEO - ASUNCION 2010

Naturalia Scientific Collection
Montevideo - Asunción
ISBN 978-9974-98-133-1
PRINTED IN URUGUAY

CONTENT

Acknowledgements 5
Contributors 6
Reviewers 8
Translators 9
Pictures contributors 9
Prologue 10
Introduction 11

Chapter I
Collection plan 17
Planning a trip 17

Chapter II
Environments 21
Deserts and other xerophytic areas 21
The Gran Chaco 28
Car-catcher 31
Dunes and beaches 33
Collecting in aquatic and semiaquatic environments 35
Mangroves 54
Forest and humid forests 58
Collection of underground beetles 70
Collecting in the mountains 72
Beetles associated with caves and other underground habitats 79
Other places to search beetles 88

Chapter III
Sampling methods and traps 89
Night collection 89
Xilophagous beetles 96
Traps for xilophagous beetles 101
Rearing larvae living in wood 106
Beetles in fungi 108
Beating Method 115
Sweeping Method 118
Fogging trees 119
Vacuum sampling in shrubs and trees 122
Beetles associated with vertebrates 123
Beetles associated with invertebrates 127
Collecting in insect nests 129
About traps 130
Pitfall traps 131
Pitfall traps made of small plastic containers 131
Traps with fruits 139
Purple traps 141
Concave colored pan traps 142
Flight intercep traps 144

Some device to collect beetles 148

Chapter IV
Where do they live... 151
Content and Taxonomic list of the Beetles families 151
Where do they live? What do they eat? How to collect the different Beetles families? 154
Suborder Archostemmata 154
Suborder Myxophaga 155
Suborder Adephaga 157
Suborder Polyphaga, Series Staphyliniformia, Superfamily Hydrophiloidea (incl. Histeroidea) 165
Superfamily Staphylinoidea 171
Series Scarabaeiformia, Superfamily Scarabaeoidea 179
Series Elateriformia, Superfamily Scirtoidea 184
Superfamily Dascilloidea 186
Superfamily Buprestoidea 187
Superfamily Byrrhoidea (incl. Dryopoidea, Psephanoidea) 188
Superfamily Elateroidea (incl. Cantharoidea, Artematopodoidea) 191
Series Derodontiformia, Superfamily Derodontoidea 200
Series Bostrichioformia, Superfamily Bostrichoidea (incl. Dermestoidea) 201
Series Cucujiformia, Superfamily Lymexyloniodea 205
Superfamily Cleroidea (By John M. Leavengood, Jr.) 205
Superfamily Cucujoidea 210
Superfamily Tenebrionoidea 227
Superfamily Chrysomeloidea 241
Superfamily Curculionoidea 243
Order: Stresiptera 248

Chapter V
Killing, Temporary storage, Preservation, Labelling, Formulas, Rearing 249
How to kill, store and labelling the collected beetles 249
Basic collection kit 253
Formulas 256
Glossary 260
Useful books for study 264
References 264

To Vidalina, my wife, for his inexhaustible patience

ACKNOWLEDGEMENTS

I wish to send my deep gratitude to the contributors, reviewers, translators and those who offered some pictures for this book since they have made possible to complete this work.

Because they gave me part of their experience in collecting beetles I am indebted to:

Albert Allen (USA)
Edgard Benitez (Paraguay),
Eduardo Faúndez (Chile),
George Hangay (Australia),
Joel R. Cantero (Paraguay)
Kipling Will (*)
Paolo Gandini (Italy),
Peter Cate (Austria),
Reinhard Foerster (Argentina),
Ulf Drechsel (Paraguay)
Wolfgang Barries. Wien (Vienna), Österreich (Austria).

My gratitude also for:

Jorge Barrett Viedma, editor, who worked tirelessly day after day;
Luci Aparecida Roberto (Brazil) who allowed us to use her home and computers;
Rafael Barrett Viedma, also editor, who handled the graphic design, texts and other aspects concerning the preparation of this book;
Helena Julio, my mother, who came from USA and whose presence I could not enjoy for many days while I was engaged in the final papers of this book.

* http://ecnr.berkeley.edu/facPage/dispFP.php?I=643

CONTRIBUTORS

John A.Kochalka, Ms.C.
Head of Dept. Invertebrados, Microbiología y Paleontología,
Museo Nacional de Historia Natural del Paraguay.
Km. 10,5 Ruta Mcal. Estigarribia, Suc. De Correos N° 1, CP
2169. San Lorenzo – Paraguay, Tel: (595) 21.585.208. http://
sites.google.com/site/ibnpmuseo/home/inventario
"PROLOGUE", Page 10,
Contributor, Reviewer, Translator and Pict. Contributor.

John M. Leavengood, Jr. Ph,D.
University of Kentucky, Department of Entomology
S-225 Agricultural Science Center Building – North
Lexington, KY 40546-0091
Email: tokay@ufl.edu, tokaytokay@uky.edu
« SUPERFAMILY CLEROIDEA », Page 205,
Contributor and Reviewer.

Michael D. Ulyshen Ph. D.
Michael D. Ulyshen, Department of Entomology,
Ohio Agricultural Research and Development Center,
The Ohio State University, Wooster, OH 44691, USA.
E-mail: mulyshen@hotmail.com
"SILPHIDAE", Page 174.
Contributor and reviewer.

Prof. Dr. Andreas Herrmann
Bremervörder Strasse 123 - 21682 Stade-
Germany (Deutschland)
Email: herrmann@coleopterologie.de
Web: http://www.dermestidae.com
"DERMESTIDAE", Page 201.
Contributor, Reviewer, and Pict. Contributor.

Rolf Oberprieler, Ph.D.
Zimmerman Fellow, CSIRO Entomology,
GPO Box 1700, Canberra, ACT 2601-AUSTRALIA.
Tel. (02) 6246 4271- Fax (02) 6246 4264-
Zimmerman Weevil Research Laboratory
Tel. (02) 6246 4234- Email: rolf.oberprieler@csiro.au
http://www.ento.csiro.au/
He provided most of the information on
SUPERFAMILY CURCULIONOIDEA, Page 243

Josef Jelínek, Ph. D.
Department of Entomology, National Museum,
CZ-148 00 Praha 4 – Kunratice 1, Czech Republic;
E-mail: JJ.Nitidula@seznam.cz
"MONOTOMIDAE", Page 213
"KATERETIDAE", Page 218
"NITIDULIDAE", Page 219
Contributor and Reviewer.

Eric Chapman Ph. D.
University of Kentucky
S-225 Agricultural Science Center North,
Lexington, KY 40546 - (859) 257-2759
Email: ericgchapman@gmail.com
COLLECTING IN AQUATIC AND SEMIAQUATIC ENVIRONMENTS, Page 35.
Contributor and Reviewer.

Stewart B. Peck, Ph. D.
Dept. Biology, Carleton University,
1125 Colonel by Drive, Ottawa, ON K1S 5B6,
Canada office: 4640 CTTC Building,
Tel. 613 520 2600, ext. 3860 (mornings);
Lab. 4690 CTTC Building, ext. 3868;
Fax 613 520-3539, located in Dept. Office, Nesbitt Building.
"BEETLES ASSOCIATED WITH INVERTEBRATES", Page 127,
Contributor and Reviewer.

Sergio D. Ríos Díaz
Section of Invertebrates.
Museo Nacional de Historia Natural del Paraguay.
Km. 10,5 Ruta Mcal. Estigarribia, Suc. De Correos N° 1, CP 2169
San Lorenzo – Paraguay, Phone: (595) 21.585.208;
Email: sergiord40@gmail.com
"COLLECTING IN AQUATIC AND SEMIAQUATIC ENVIRONMENTS", Page 35,
FOREST AND WET FORESTS, Page 58.
Contributor and Reviewer.

Guillermo González.
Agricultural Engineer.
Email: willogonzalez@yahoo.com
Santiago – La Reina – Chile.
www.coccinellidae.cl
COCCINELLIDAE, Page 224,
Contributor.

REVIEWERS

Albert Allen. 9235 West Wenatchee Court, Boise, Idaho 83709, USA, Phone 208 362 1699.

Alex Delobel, Dr. Laboratoire Itnorstom de Protection des Stocks, B.P. 1386, Dakar, Sénégal., Muséum National d'Histoire Naturelle, 45 rue Buffon, 75005 Paris, France.

Andreas Herrmann. Bremervörder Strasse 123 - 21682 Stade-, Germany (Deutschland), http.//www.dermestidae.com

Bolívar Garcete-Barrett. Curitiba, Brazil.

David W. Edmonds. http.//www.unl.edu/museum/research/entomology/workers/DEdmonds.htm; P.O. Box 426, Marfa, Texas 79843.

Eric Chapman, University of Kentucky, S-225 Agricultural Science Center N Lexington, KY 40546 (859) 257-2759 Email. ericgchapman@gmail.com.

Gene Hall, Invertebrate Zoology - UCB 265 CU, Museum of Natural History - University of Colorado, Boulder, CO 80309-0265, Phone. 303.735.5262, CU Zoology. http.//cumuseum.colorado.edu/Research/Zoology/index.html, http.//tolweb.org/Ptiliidae.

Hiroyuki Yoshitomi. Entomological Laboratory, Faculty of Agriculture, Ehime University, Tarumi 3-5-7, Matsuyama, 790-8566 Japan TEL. 089-946-9898 http.//book.geocities.jp/maruhananomi/index.html.

Jiří Háva. Branická 13, CZ-14700 Praha 4, Czech Republic.

John A. Kochalka. Museo Nacional de Historia Natural de Paraguay, http.//www.esnips.com/user/johnkochalka.

John Leavengood Jr. University of Kentucky, Department of Entomology, S-225 Agricultural Science Center Building – North, Lexington, KY 40546-0091 Email. tokay@ufl.edu, tokaytokay@uky.edu.

Josef Jelínek. Department of Entomology, National Museum, CZ-148 00 Praha 4 – Kunratice 1, Czech Republic.

Karina Atkinson. Scotland, http.//www.paralatierra.org.

Keith Philips. Laboratory Department of Biology Western Kentucky University 1906 College Heights Blvd. Bowling Green, KY 42101 Office. (270) 745-3419 Lab. (270) 745- 2394 Fax. (270) 745-6856. http.//bioweb.wku.edu/faculty/Philips/.

Michael D. Ulyshen. Department of Entomology, Ohio Agricultural Research and Development Center, The Ohio State University, Wooster, OH 44691, USA.

Michael S. Caterino. Santa Barbara Museum of Natural History 2559 Puesta del Sol Rd. Santa Barbara, CA 93105-2998, USA.

Mizell,Russell F,III. Entomology and Nematology Department, Cooperative Extension Service, Institute of Food and Agricultural Sciences, University of Florida, Quincy, 32351, www.ufinsect.ifas.ufl.edu/hoover_sampling.htm.

Olivier Oboilly. Musée d'Histoire Naturelle de Lille19 rue de Bruxelles59000 LILLE-France E-mail. oboilly@voila.fr, Phone (805) 682-4711 Ext. 151 Fax (805) 563-0574.

Richard A. B. Leschen. New Zealand Arthropod Collection, Landcare Research Invertebrate Systematics, Private Bag 92 170, Mt. Albert Road Mt. Albert, Auckland, New Zealand, Tel.. +64 9 815 4200 ext. 7031 (work) +64 9 849 6323 (home), Fax. 6498497093. http.//www.landcareresearch.co.nz/research/biodiversity/antarctica/html/beetles.html.

Richard Cunningham. 3889 Walnut Avenue. Chino, CA, U.S.A.

Robert Otto. City, state, country. Suring WI USA- http.//bugguide.net/node/view/63946.

Rolf Oberprieler. Zimmerman Fellow, CSIRO Entomology, GPO Box 1700, Canberra, ACT 2601-Australia., Tel. (02) 6246 4271- Fax (02) 6246 4264-, Zimmerman Weevil Research Laboratory, Tel. (02) 6246 4234- http.//www.ento.csiro.au/.

Sergio Ríos. Asunción. Azahares 406., Barrio Loma Pytá, Email. sergiord40@gmail.com.
Sonja Ukwuoma, Prof. London, England.
Stewart B. Peck. Dept. Biology, Carleton University, 1125 Colonel By Drive, Ottawa, ON K1S 5B6, Canada office. 4640 CTTC Building, tel. 613 520 2600, ext. 3860 (mornings); lab. 4690 CTTC Building, ext. 3868; fax 613 520-3539, located in Dept. Office, Nesbitt Building.
William Wyatt Hoback. Department of Biology Bruner Hall of Science 905 West 25th Street University of Nebraska at Kearney Kearney, NE 68849 phone (308) 865-8602 fax (308) 865-8045.
Warren Steiner. Smithsonian Institution, National Museum of Natural History, 10th & Constitution NW, Washington, DC 20560-0187, http.//entomology.si.edu/StaffPages/SteinerW.html.
William D. Shepard. 201 Wellman Hall #3112, University of California, Berkeley, CA, 94720-3112, (510) 643-0804, USA. http.//essig.berkeley.edu/about/shepard.
Wolfgang H. Rücker. SuVon-Ebner-Eschenbach-Str. 12 D-56567, Neuwied, Germany. http.//www.latridiidae.de/; coleoptera@latridiidae.de.

TRANSLATORS

Alain Aguilar J., New Jersey, USA
Alicia Revello, Ciudad de la Costa, Uruguay.
Bolívar Garcete-Barrett: Curitiba, Brazil.
Carlos Bruyn, Miami Florida, USA.
John A. Kochalka, M.S.: Asunción, Paraguay.
Maruzella Aguilar Julio, Utah, USA.
Sergio Ríos, Asunción., Paraguay, sergiord40@gmail.com.

PICTURES CONTRIBUTORS

Andreas Herrmann, Germany.
BioQuip Products, USA, www.bioquip.com.
Bolívar Garcete-Barrett, Curitiba, Brazil.
Carlos Aguilar Julio, Asunción, paraguay.
Danilo Carriel ramírez, Santiago, Chile.
Dr. Enrique Martínez , Montevideo, Uruguay.
Dr. Eric Chapman, Lexington, KY, USA.
Frank Rodríguez, Caracas, Estado Miranda, Venezuela.
Gonçalo Lopes, São Domingos de Rana, Portugal.
Jack Owens, Ontario, Canada.
John A. Kochalka, M.S.: Asunción, Paraguay.
Jorge Barrett Viedma, El Pinar, Uruguay, barrett.george@gmail.com.
Olivier Oboilly, France.
Pablo Marano, Buenos Aires, Argentina, pablo_marano@hotmail.com.
Russell F. Mizell III, Florida, USA, www.ufinsect.ifas.ufl.edu/hoover_sampling.htm.
Sergio García, Buenos Aires, Capital Federal, Argentina.

PROLOGUE

The World was not created for Man alone. It could argueably be stated that Coleoptera are the dominant life form on this planet. It was 61 years ago, in 1959, when J.B.S. Halsane expressed his much-quoted opinion that "God has an inordinate fondness for beetles." With 300,000 described species, according to Wilson (1992) or 350,000 according to Liebherr and McHugh (2003), and many more waiting to be discovered, their diversity surpasses that of all other insects. It also surpasses that of all other non-insect animals. Even the higher plants, with 248,000 known species, fall short of this number. All known and described species of lower plants, algae, protozoans, fungi, bacteria, and virusus taken together, do not approach the biodiversity of Coleoptera.

Great discoveries are being made in the biodiversity of microorganisms, based on molecular and genetic techniques, and soon scientists may very well discover that God also has an inordinate fondness for microbes.

True, very little is known about microbes, but neither do we know how many species of Coleoptera there are in the world. The 350,000 described species are only those that have been scientifically discovered and published, and this number may be only a relatively small percentage of the total diversity of beetles in the world. For example, Nielsen and Mound (2000) estimate that there could be 850,000 species, and according to Simon (1996) and Grove & Stork (2000) there could be between 2,4 and 4,7 million species of beetles. There is much work to be done.

You can take part in this discovery. Once you start reading this book, you will learn very quickly that beetles can be found in virtually all habitats and micro/habitats on Earth, and that they occupy virtually all ecological niches imaginable, and even some that you may never have imagined. Furthermore, you will learn how to collect them, by direct observation and with a myriad of specialized techniques described in this book.

You can start looking right now, in you house, in your backyard, in a forest by a stream, on a mountain, at a friends house or while on vacation. As you get better and better at collecting beetles, all the best beetle experts in world will know that you are the best, and – who knows? You might even get invited to participate in an "All Taxon Biological Inventory" or in some other kind of international multidisciplinary scientific research project.

You must start now, and never, never, stop looking for new species.

John A. Kochalka - 2010
Dept. of Invertebrates
Museo Nacional de Historia Natural del Paraguay

"Every time a species becomes extinct we are more alone on this planet."

INTRODUCTION

On many occasions I have seen several insect collections with many specimens in poor condition, lacking legs or antennae. And many times, I asked the collection's owner where a particular specimen was collected, and he was not able to tell me due to the fact that he neither registered nor kept the data properly.

On other occasions, some students (also some professionals) had been collecting in the same place I did before, and seeing the specimens that I collected, they wanted to know how I got them, because in spite of a lot of work on the same site, they could not catch them. From time to time a student or colleague will come to me inquiring as why he did not collect the same beetles that I had at the same locality despite all of their efforts.

These reasons prompted me to write this book. At first, I thought it was an absurd task since on the internet and in the literature in general there is a large amount of data concerning the collection of beetles. But when I was looking for material, I noticed that data were widely scattered throughout the world in thousands of books and websites, which made their access very difficult to those who needed them. Finally, the necessity of a basic order for all this material, the collection and compilation of the most important data, the addition of my personal experience and that of some of the world's most renowned scientists, were the facts that determined and shaped the contents of this book.

I do not intend in any way to say that this book contains all data that I have gathered. It would be impossible to cover in a book of this size such an overwhelming amount of information.

I have worked at the National Museum of Natural History of Paraguay, specifically the Department of Entomology, for 18 years as a volunteer officer. This has allowed me to continue studying the Neotropical Scarabaeidae.

Some visitors to the Museum, neophytes in terms of entomology (and even some junior biologists) have asked me some questions, such as the ones I quote below. Questions, dear reader that you may also be asking yourself:

- Why are there so many insects gathered in museums if forests are now disappearing? Wouldn't it cause their extinction? Why not consider studying what has already been collected?.

Those questions seem logical, and to answer them I ask you to read the following.

In 1994, the **"27 Committees of Systematics Agenda 2000"** published two reports, called "Systematics Agenda 2000: Charting the Biosphere" (SA2000, 1994), with the results of their calculations about the possibility of an inventory of all living beings in the world within a time frame of 25 years. It was estimated in that meeting that the total cost of the project -something too big to be considered a single project- was 3 billion dollars per year, during 25 years, which totaled an approximate of US$ 75.000.000.000. The toughest task was to get the money to fund such a project, but they thought that it should come from somewhere because the task in question is truly important, and therefore an economic and healthy investment. However, 16 years have passed since 1994 and still it has not been possible to create an inventory all living beings. Currently, approximately 1,8 million species have been described worldwide and according to "Systematics Agenda 2000" between 10 to 100 million species have not been described yet (I. Inv. Re B. A. H. 1998). In many parts of the world no faunal inventory has been made, and in many others the local fauna has been poorly studied at the taxonomic level, so a large percentage of their biodiversity is still unknown.

The trouble is that most wild habitats are systematically disappearing worldwide. The lack of studies hinders efficient and sustainable conservation of these habitats (García and Chacón 2005).

Taxonomy currently faces several challenges, including the small number of specialists capable of identifying and naming the various species. Taxonomists are aware that the biggest problem is that the process of species description and dissemination of information takes place very slowly in relation to the rate of destruction of ecosystems, leading to the extinction of thousands of species before they are even described[(1)]. To save a species, first it must be collected and described, and that is the moment when the study of its biology and behavior has just begun. Hence the importance of taxonomists, who are responsible for describing new species, is inherent.

Among all animal groups on earth insects are the most diverse, occupying most of the habitats and ecosystems on the planet and among insects, beetles (Coleoptera) comprise more species than any other group.

Comparing extinct and living species has helped to understand the different evolutionary processes, hereditary traits, morphological adaptations, and the ecological relationships these animals have with the environment. Also, due to their large number of species and their high degree of evolution [adaptiveness], the knowledge of current species distribution will reveal with more precision the geographic changes of the earth's surface. And this will allow our understanding how the fauna of a particular region was created.

Ribera & Foster (1992) propose to use beetles to make ecological assessments to determine the degree of habitat conservation. In fact, Coleoptera have already been used as indicators of environmental changes due to their different characteristics and ecological requirements (Wettstein and Schmid 1999).

Bournaud et al. (1980) argue that information obtained from aquatic beetle research is sufficiently useful to characterize the different environments of a river.

Moron (1997) is another entomologist who proposes beetles as excellent environmental indicators of terrestrial habitats.

In a single micro-habitat, large numbers of small species with large populations can coexist, so when using traps, hundreds and sometimes thousands of specimens are caught, which must be further processed and identified. However, according to Moron (1997),

1. Several alternatives have been proposed to solve the serious problem of lack of qualified taxonomists who could describe new species, which would help in some way to save many of them before they disappear forever. Some alternatives are the use of internet which has spread and promoted taxonomy. For example, in the E-magazine Zootaxa (http://www.mapress.com/zootaxa) more than 10,000 animal species have been described from 2001 to date. Never before have so, many species had been described in only one decade. Also a website called "Encyclopedia of Life" or "EOL" (http://www.eol.org) is including the description, general information, and photographs of each of the approximately 1,8 million species known worldwide.

although it involves hard work for the taxonomist who uses traps, this peculiarity of the insects allows one to focus on the detailed study of a complex community within a confined space.

Access too many parts of the world is difficult and generally collecting trips are short. Consequently, the samples are relatively small and sporadic and do not allowproper examination of the site. For example, in any place you can collect the adult of a species found in the month of January, but species in the larval stage will not be collected since they will emerge in winter. One way to solve this problem is preparing parataxonomists(2) to be supervised by a biologist, working in such inventories. With their help we will have a good representation of the species present throughout the year in the studied area. This option is economically favorable because less money is spent with parataxonomists living on the site and also saves time moving faster in the preliminary separation, labeling, assembly and semifinal separation of morpho-species(3).

Thanks to parataxonomists, the specialist can dedicate more time to his specific job, which is the study of collected species.

Many times, these technicians conduct studies on their own time, offering important observations and contributing to expand scientific knowledge of species.

Almost all biologists agree that scientific collections do not affect the insect populations causing their extinction, although some others think the opposite. But all of them at the end agree that habitat destruction almost always causes species extinction.

Márquez & Asiain (2000) argue that scientific collections represent the raw material for the generation of biological knowledge in different fields; they are part of the cultural heritage of mankind, the germplasm of life, the memory of nature. It is therefore very important to promote their development and conservation. Already in 1974 Alfredo Barrera, a Mexican entomologist, argued that to assign names to species, first it is necessary to identify them taxonomically, and this is always based on information obtained directly from organisms deposited in one or more collections.

What is said above supports the idea that it is necessary to make biological inventories and even enhance the existing insect collections because, due to the information they generate, they contribute to the conservation of animals and their environment.

Several years ago I was preparing a project which sought to obtain financial support to inventory Paraguayan butterflies and I asked my friend John A. Kochalka: "*how can I justify scientific research?*" and he answered: *"Scientific research always brings surprises. It is not necessary to investigate if you know in advance what will be the result. The reason to research is to find out something that previously was not known, and it is impossible to know something before you experience it. The information received through investigation is always useful. Sometimes we discover just what is wanted to find, and its usefulness is immediate. In other cases the new discovery becomes useful after several years,*

2. The parataxonomists may be people who help at the laboratory or field, of course with the assistance of a professional. These parataxonomists are selected by nomination and highly competitive interviews. People who apply to this work may be ordinary citizens with a minimum basic education and after being selected, they are prepared with an intensive course of about 5 or 6 months. This covers what a person needs to know in order to be a collector, trainer and connoisseur of arthropods, independent and full-time, under rural conditions in a tropical country (Jansen, 1991). Parataxonomists may be people who live in the place where the inventory is done, or may be people who have time to make trips at different times of the year.

3. "Morpho-species" is a temporary identification and not always accurate, though approximate, of a particular species, pending definite identification.

when other investigations put it in a new context and, although directly or indirectly, information always has economic value, and beyond monetary value, its value lies on the fact that it helps the survival of human beings and all life on this planet" .

The experience of collecting insects allows much more learning than simply reading books. To see live insects in their natural habitat, eating, parasitizing and mimicking each other, reproducing, emerging and growing is an experience that allows the observer to learn much about the ecology of this planet. The young biologist learns a lot more when he collects the insect and takes it to a microscope to identify and know what it is. The direct comparison of a live insect in the field with the same specimen dead, being carefully studied under a microscope, is an unparalleled experience. This experience helps the scientist and the young biologist to understand the true importance of these small and seemingly "insignificant" creatures which dominate this world.

There are two ways to collect insects: The first one by direct manual collection in their natural habitats. It is true that by gathering in this way you can learn a lot about the ecology and behavior of these insects, but generally most species caught this way are the most commonly found during the day. Furthermore, nocturnal species or those who hide among the vegetation or in the soil are rarely seen. Therefore, it is convenient to use the second method that involves the use of traps to catch the most elusive species. With these traps you can find species that are normally hidden in their habitats or those leaving in search of food at any time of day or night.

In the course of history, since people were interested in knowing the world of insects, many types of traps have been designed to catch them (this can be checked by reviewing the extensive bibliography in this book). Thanks to the progress in collection methodology, developed by biologists and applied in biological inventories of wildlife at local or regional level, in recent decades the discovery of new species has been more accelerated.

With this book, I also pursue the goal of discovering new species because many of the methods and traps presented here can be used and standardized for use in biological inventories and rapid ecological assessments at any location on earth.

I think that the methodology presented in this book will help you to become a good collector of beetles.

STRUCTURE OF THE BOOK

In Chapter I, **"Collection plan"** as the name implies, you will find details of the equipment and important preventive actions to be taken into account for successful sampling.

In Chapter II, "**Environments**" I attempt to cover the main environments in the world, with some of their physical and ecological characteristics. I also include some examples of beetle species which can be found in certain places, and what methods and traps are the most appropriate to catch them. I could not include all environments in the world, since that is impossible, but I believe that the ecosystems described in this work are present in many countries around the world. So when you visit a desert or a mangrove or a simple pool of water, you will know how to collect there.

In Chapter III, **"Sampling methods and traps"** I have tried to include as many homemade traps as possible, at low cost, which can be used to obtain the greatest number of Coleoptera species, almost anywhere in the world. Some of them can be purchased at specialized shops. The section, "Systematization" explains the proper way to kill, prepare, label, and transport the collected specimens to be studied later at the laboratory. Also in this chapter I explain how rearing the immature stages of Coleoptera. I propose some useful chemical compounds to be used in traps and preserve the specimens collected.

Chapter IV, **"Where do they live..."** in this chapter are all the known families of Coleoptera. I included basic information about where they live, how they feed and how each known family of Coleoptera is trapped, being the essential details to know how to collect them.

It is impossible to establish a complete list of species found in each habitat. For this reason, in the chapters dealing with the environment, I cited only the most characteristic families. But you do not need a list if you study this fourth chapter thoroughly, because through the knowledge that you will find here, you will be able to infer the species to be found anywhere you go, no matter what corner of the world you might be. Thus, the fact that only the main environments existing in the world are detailed will not be an obstacle for you to define exactly what methods you will use to make a sample. It will be sufficient to study what is visible and what is known in the environment, for example what kind of animals live in the site, what kind of vegetation, if there are streams or lakes, if the place is damp, warm, dry, cold, etc. For example, local inhabitants could tell you that there is some herbivorous mammal, and then you will be sure to find dung beetles.

Taxonomic description is not included in this chapter. I would have liked to do it, but this is a book about collection, not taxonomy. Anyway, to be able to make a basic identification of the different Coleoptera families, or at least observe their pictures, you can search on Internet. There you will find extensive literature that will help you to identify the beetles that you collect, at least to family level, some perhaps even to genus and/or species.

I am including a glossary which provides the explanation of some terms, which I could not avoid when writing this book.

I got the information contained in this work by consulting a lot of literature over four years. You will find this bibliography at the end of the book. But a lot of information presented here is based on my own experience in collections, and on data kindly provided by many friends who are famous entomologists who have collected beetles, and still do, in many parts of the world.

This is an essay, so I prefer to call it, despite of having more than 300 pages in which, as explained above, I could not quite cover all methods and types of traps to collect beetles However; I hope this book will be useful to any professor of entomology, facilitating the arduous task of teaching this interesting branch of zoology and, of course, also that it will help any student interested in getting many and interesting specimens of beetles.

C.A.J.

CHAPTER I
COLLECTION PLAN

PLANNING A TRIP

Before embarking on a search for beetles and to be able to succeed in the collection, it is desirable to develop a good plan.

As the collection grows, the interest in exploring new areas awakens, motivating the desire to obtain more scientific knowledge of each site, resulting in new discoveries of species. Visiting these new areas is crucial for the spread of knowledge about the abundance and distribution of various species and subspecies already known. It is not essential to visit remote areas, since there are interesting species to be discovered everywhere.

The best way to find interesting places to collect is by reading and studying what others have written on the subject; several books, brochures, checklist and lists of research projects written by entomologists are available. There you will find useful information such as dates of appearance of certain species, specific collection sites, etc. Those papers have been published in several magazines and books around the world, which can be obtained in a good library and on the Internet.

It is also useful to consult other entomologist's collections, either private, museums or universities. This information is also available in many clubs and entomological societies.

Road maps might guide you as you travel to the site, but they will have no details on the areas you intend to study, so it would be advisable to get a topographical map and specialized maps of national parks, departments or provinces you want to visit. Very detailed maps are often expensive and difficult to obtain.

It is recommended to search for a set of maps known as "An Atlas state and/or provincial". Each insect collector should be familiar with these maps, because they are of great help to locate good sites for collection. These useful books and their maps will help you get to remote areas, secondary roads, and roads which are not usually traveled.

The Google Earth program on the Internet is very useful as well because the satellite images can be printed, becoming aerial photographs of the place you choose.

It is also important to draft a tentative schedule of activities to be undertaken at the site of collection, including mealtime and rest. This will maximize the time efficiency and will help you get the best results.

To ensure the collection success it is very important to know in advance the dates on which the moon will be visible or not, since insects do not come to light traps when the moon is full. On the article "Night collection" (Chapter III), there is a detailed explanation on this issue.

Besides it is important to have on hand a list of materials and equipment that will be taken to the site of collection (see "Basic collection kit" and others of the Chapter V).

Below there is a list of items to consider before or during the trip. Only the basic is recommended here. Each place of collection should be evaluated in advance and the necessary adjustments should be made according to the characteristics and requirements of area. For example, if you are going to carry out an aquatic collection, use the proper clothing (plastic boots to enter the water, etc). If you are going to explore a mountain, it will require appropriate equipment for cold environments.

1. In order to know what kind of traps and materials will be taken to a collection site you need to know what you are going to collect. This will also determine the particular habitat we are going to put emphasis on.

2. Determining the collecting season according to the country to be visited (see below), taking into account the lunar calendar.

3. Become familiar with local roads and other accesses with the help of maps.

4. The whole place must be very familiar to the collector. It is convenient to walk

Monday - ACTIVITIES	
06:00 hrs.	Breakfast
07:00 hrs.	Putting the pitfall traps in the forest trail
08:30 hrs.	Beating in the forest
12:30 hrs.	Lunch
13:30 hrs.	Putting Malaise traps and other pitfall traps in forest clearing
14:10 hrs.	Collecting Cicindelidae
14:50 hrs.	Looking for beetles among fungi
17:00 hrs.	Sweeping in the pasture
18:00 hrs.	Prepare and place the light trap
18:30 hrs.	Prepare the specimens collected during the day
19:00 hrs.	Turn on the light trap and visit the nearby lights
20:00 hrs.	Night collection
	Choose the appropriate time for dinner

Table: Tentative schedule for the first day of activities in the field

around the area and familiarize with it by inquiring the local people.

5. Take into account the special characteristics of the area and habitats, and locate each one on a map.

6. If you are looking for aquatic beetles, you need to find the available environments with many micro-habitats, such as water in the holes of tree trunks, in stumps, in fallen bamboo, in bromeliads etc., and records all this information in a notebook for later visits. Furthermore, you must detect the micro-habitats for the frequent or sequential sampling, specifically those with waterfalls, springs, etc. and see how accessible they are; we must also detect water filtrations, springs, wetlands and other special places.

7. Design a schedule to visit the traps, and the samplings. They can be quite detailed, but at least they should contain information on daily activities, for example, what will be done every day in the morning and time for each activity (see table).

Always be flexible when designing a program as opportunities are just as important to consider when they are presented. The sampling of different habitats, at different times and conditions, is more important than a strict maintenance of the program. Note that you must seize the opportunities to collect rare or uncommon beetles. You must take all the time necessary to do the best job, not the fastest. The quality of the collection -not the quantity- and the accuracy of data are most important.

The collection at a particular site only ends when you have collected in its entire area. However, when you consider that different environmental conditions and times during the year bring different species which come and go, you can conclude that almost no place in the world has been entomologically fully evaluated.

It is essential to know the best time to collect in the region to be visited. For example, Chile is a country with temperate climate; winters are very cold and rainy. Therefore, the insects in that country spend winter in immature stages and become adults in the spring and summer, which begins in September and lasts until late autumn (May). In other latitudes as the tundra, summer lasts only two months; the rest of the year is very cold to collect insects. In almost any temperate country the best time of collection begins in mid spring and extends until late summer or early fall.

You should also have some previous and basic information on the groups of beetles you want to collect. For example, in the Southern cone of South America, Carabidae are abundant all year long, but adults of Scarabaeidae, Lucanidae and Cerambycidae are present mainly during the spring and summer and spend the rest of the year as eggs, larvae or pupae. Some adult species however, are found almost throughout the year.

With regard to aquatic beetles -at least in the Neotropical region- although they are present everywhere and throughout the year, it is advisable to collect them during the dry season. In those months, streams and water bodies are generally smaller and a larger amount of aquatic beetles is concentrated per square meter. On the contrary, during the rainy season, when water bodies are larger, beetles are much more dispersed (pers. comm. Bill Shepard, 2005).

When planning your trip is necessary to remember to get information in the respective embassies, or in the Internet, about the political situation in the country you will visit and what places may be more or less dangerous with regard to local diseases and poisonous animals, in case you need some vaccination one, etc.

Coming up with a good collection plan will not assure you success. Sometimes there can be difficulties, such as bad weather or bad road conditions, but it will ensure a large percentage of success.

SOME WORLD-WIDE METEOROLOGY LINKS

http://www.wunderground.com/global/SA_ST_Index.html
http://www.nrlmry.navy.mil/sat-bin/global.cgi

OCEANIA

http://www.bom.gov.au/announcements/media_releases/climate/change/20070103.shtml
http://www.weatherzone.com.au/satellite.jsp?state=AUS

SOUTH AMERICA

http://faculty.smu.edu/bakewell/BAKEWELL/images/rainfall.jpg
http://www.seedman.com/zonemap.htm

CENTRAL AMERICA
http://unosat.web.cern.ch/unosat/asp/prod_free.asp?pid=1003
http://bostongringo.com/haiti/images/map-central_america.jpg
NORTH AMERICA
http://wwwghcc.msfc.nasa.gov/GOES/
http://www.weather.gov/view/national.php?thumbs=on
AFRICA
http://www.bushtracks.com/large_maps/south-africa-map.html
http://www.bushtracks.com/destinations/eastern-africa/temperature-and-rainfall
ASIA
http://www.naturalhistoryonthenet.com/Continents/asia.htm
http://www.artphotoasia.net/weather_map_southeast_asia.php
EUROPE
http://www.ecmwf.int/
http://www.metoffice.gov.uk/

CHAPTER II
ENVIRONMENTS

DESERTS AND OTHER XEROPHYTIC AREAS[1]

Typical deserts

The deserts of the world (Fig. 1) vary significantly in their annual rainfall and drought seasonality. Annual precipitation can vary from 0 mm to over 500 mm. The range of temperatures may go from below 0°C to over 50°C (Ward 2009). The cold deserts have hot summers, compensated by cold winters (Flegg 1993).

Deserts are not fully dry, although places with water may be separated by large distances. These water sources produced by poor rains are usually separated from each other by hundreds of kilometers. The water, in the saturated part of the aquifer, flows along the course slope until it joins the desert surface; this is called an oasis, which can also be saline. Only a few herbaceous plants can survive in the wetland ecosystem (marshes), except the reeds which can be present. Narrow oases can also take form along the rivers, such as the Nile River in Africa and the Rio Grande and Colorado River in North America.

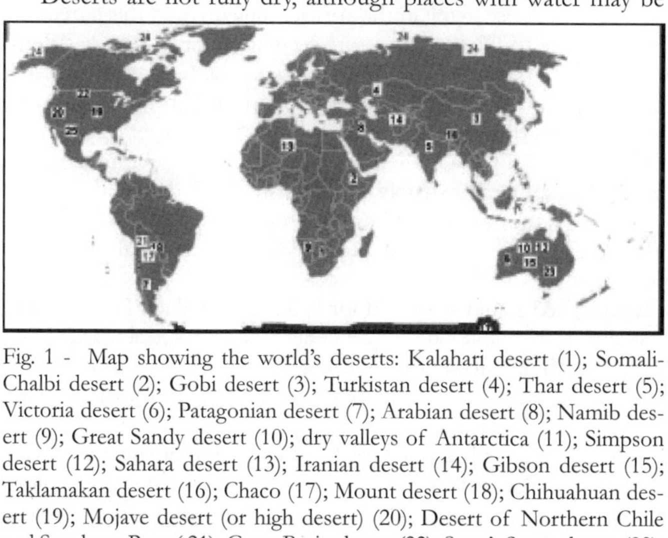

Fig. 1 - Map showing the world's deserts: Kalahari desert (1); Somali-Chalbi desert (2); Gobi desert (3); Turkistan desert (4); Thar desert (5); Victoria desert (6); Patagonian desert (7); Arabian desert (8); Namib desert (9); Great Sandy desert (10); dry valleys of Antarctica (11); Simpson desert (12); Sahara desert (13); Iranian desert (14); Gibson desert (15); Taklamakan desert (16); Chaco (17); Mount desert (18); Chihuahuan desert (19); Mojave desert (or high desert) (20); Desert of Northern Chile and Southern Peru (21); Great Basin desert (22); Sturt's Stony desert (23); Arctic Tundra (24); Sonoran Desert (25) (Modified from Ward 2009).

1. Due to the fact that xerophytic regions are scattered throughout the world and each one has a broad range of habitats and thousands of beetle species, it is beyond the limits of this book to explain and describe thoroughly all xerophytic regions of the world. Consequently, this article only provides an introduction to the most outstanding worldwide, giving just a few examples of beetle's collection in these regions. Techniques and methods explained above can be used safely in all arid or xerophytic environments.

In South America there are different xeric habitats: the Atacama and Peru deserts in the Pacific coast, Mount Desert in central Argentina, the Chaco, Patagonian Steppe, the Andean Puna, the Brazilian Cerrado and Caatinga.

The Atacama Desert in Chile is considered the driest of the world; no rain had fallen since 1570 and recently it rained in 1971 (Flegg 1993). However, the regions farther north in the Chilean desert, as well as in Southern Peru, present some oases (Azapa Valley, Lunahuaná Valley, etc.) with many endemic species of Buprestidae, Meloidae, Scarabaeidae, Tenebrionidae, Anthicidae, Mordellidae, Chrysomelidae, Cerambycidae, Curculionidae, etc., which appear only in certain years, together with a dampening fog (Camanchaca), which forms the phenomenon known as " desert blooming ". This can happen every 7 years or more. In drier years, the eggs of these beetles will remain in diapause, while the optimal conditions of humidity and temperature are not present to be able to emerge. The same phenomenon manifests itself in a similar way, at intervals of some years, in other deserts like the desert of Namibia, the Sahara desert and Baja California. Although rainfall is very limited, the fog from the sea is enough to sustain life in those deserts. The humid fogs brought in from the sea provide an excellent habitat for the growth of lichens (Rundel 1978), and lodge interesting beetles.

In arid and semiarid neotropical ecosystems the mesquites *(Prosopis flexuosa* De Candolle*)* produce micro-habitats under their treetops, with micro-climatic conditions and with very different communities of those living in the surrounding area, so these trees are considered "islands of fertility" due to the high moisture content, organic material, detritus and essential minerals they lodge (Flores et al. 2004). In Argentina, Gustavo Flores and his colleagues collected 23 families of Coleoptera in the leaf litter beneath these trees. So it is interesting to examine these micro-habitats, where one can collect several species of beetles. Also under the litter of *Acacia* lives a numerous and varied beetle community.

Most beetles living in deserts are active mainly at night and hide during the day to prevent water loss through evaporation. Other species excrete wax to cover their body and thus prevent desiccation (Hadley 1970; McClain et al. 1985). The wax is also useful to collect water run-off to drink.

In deserts, the most represented family on the floor is Tenebrionidae. For example, Louw (1984) found in two sites in Namibia's desert 29 genera with 45 species, and 13 of which were represented by the abundant genus *Onymacris* (Ward 2009). However, according to Louw, in the soil of Kalahari Desert Carabidae are more abundant than other beetles, outnumbering even the Tenebrionidae.

The tenebrionid subfamily Pimeliinae has three tribes, Nycteliini, Praocini and Physogasterini, with 470 species endemic to South America and widely distributed in arid regions (Roig-Juñent and Flores 2001). Other genera of wide distribution in South American deserts are: *Epipedonota, Callyntra, Auladera* and *Mitragenius.*

Most darkling beetles are cryptic or black, thus minimizing the effects of ultraviolet radiation and prevent drying by heat. Most species have long legs, which allow them to run very fast on the sand. Many run with the body raised above the ground, and other species have a white abdomen to prevent dehydration. As a curiosity, these species have specialized in the deserts more than in coastal regions (Ward and Seely 1996a).

Tenebrionidae can be caught with pitfall traps using almost any bait, since they feed on a huge range of organic materials and obtain water from their own food. During the day one should look under rocks and between roots and trunks of shrubs, trees and grass, where one may also find: subfamily Lagriinae, Meloidae, Alleculidae, Carabidae, Buprestidae, Elateridae, Cerambycidae, and other families. Under rocks near streams, dry or damp, there are many Carabidae, Tenebrionidae, Anthicidae, Chrysomelidae, Curculionidae, etc. Some Carabidae (Scaritini) hide in the cracks formed in the hard ground.

Collecting in salt lakes

In many arid lowlands there is high salt concentration. Generally this happens in the channels or mouths of Endorheic rivers[(2)]. So, water is lost by evaporation and salt remains. These lands may look like lakes in the rainy seasons but they can also be drier than a desert in other seasons. Good examples of these habitats are:

1) North Africa: especially in Algeria, Tunisia and Egypt.

2) Asia: Middle East, Caspian Sea, Aral Lake, Mongolia, etc.

3) North America: Great Basin Desert, Mojave Desert, Chihuahuan Desert, Sonoran Desert. Additionally, there are smaller deserts within the Columbia Plateau/Columbia Basin, the Snake River Plain, and the Colorado Plateau regions.

4) Canada: A few small "deserts" such as the Nk'mip Desert, which is actually shrub steppe; another "desert" near Carcross (Yukon) is commonly called a desert, but it is not a true desert, rather an area of northern sand dunes.

5) South America: Salt lakes and Chilean Desert, the Peruvian Coast and the Chaco.

6) Much of mainland Australia.

In many other places there are smaller salt deposits (salt lakes in the small islands of Cyprus and other Mediterranean islands, also in Anatolia, etc.). In biological terms, the salt lakes are usually considered as extreme environments with low biodiversity.

Very poor vegetation grows in these habitats. The few plants present in these places are able to separate the salt from the salt water (which is usually stored in the leaves). Since these plants have a high salt concentration in their tissues, only a few animals are able to eat them (among the large mammals, only the camels). However, some species of Chrysomelidae and Curculionidae can live in these salt plants. Therefore, some of their Carabidae predators are also attracted.

Fig. 2 – " Laguna Capitan" (Salt water), Chaco, Paraguay. *(Photo: Jorge Barrett).*

Generally in South America Cicindelinae *(Brasiella* spp. and *Cylindera* spp.*)* are seen during the day running on salt, especially if the floor is wet at night it is possible find (depending on the region) *Tetracha* and other genera. In saline formations there are isolated pockets of vegetation and among their twigs or leaves some Buprestidae, Mordellidae, Curculionidae, Chrysomelidae, Melo-

2. In the deserts there are two types of rivers: the Exorheic Rivers which flow into the sea or into a river running into the sea, and the Endorheic rivers which disappear in the desert. Along these rivers there are forests called "gallery forests". Often these forests lodge interesting species of beetles because of their geographical isolation.

idae, Cantharidae, etc., can be found. And on the floor there could be Carabidae, Tenebrionidae, Scarabaeidae, Anthicidae (*Notoxus* spp.), etc. However, most of the beetles living in these places are nocturnal.

The main methods of collection are the same as those used in deserts: Pitfall traps with barriers, intercept traps such as Malaise traps, etc. Colored pan traps are particularly effective on sunny days, and light traps at night.

Collecting in the flowers

There are many beetles in flowers, as they are a rich source of protein. The beetles feed on pollen, nectar and/or floral structures. In addition, flowers of different plant species are attractive to different types of beetles. Therefore, flowers should not be overlooked, and each species should be checked at different stages of development. During the day one can find Scarabaeidae, Meloidae, Cantharidae, Lampyridae, Mordellidae, Cerambycidae, Elateridae, Chrysomelidae, etc. After they have been checked, a sweep net may be passed over them. Yellow or white flowers attract different species of Buprestidae, Meloidae, Scarabaeidae and other beetle families. At night one should repeat this activity, which can be performed with the aid of a headlight/headlamp, but plugging the focus with yellow cellophane (to avoid the attraction of too many insects which may distract the collector). During the day cactus flowers should be checked to find Buprestidae and other families. We must also look at the plants of Asteraceae, in which there are usually Buprestidae, Cerambycidae, Scarabaeidae, Meloidae, Mordellidae, Cleridae, Melyridae, Cantharidae, etc. Flowers can be reviewed and collected by hand or can be shaken on a net or beating sheet.

The collector should be fast and agile when he finds flower beetles, since most of these insects are very fast fliers.

It is also effective to locate a Malaise trap in places full of flowers or flowering shrubs (see "Flight intercept traps" in Chapter III).

Other places to collect

During the day, walking in the soil among vegetation one may encounter: Tenebrionidae, Scarabaeidae, Trogidae, Glaphyridae, Carabidae, Meloidae, Silphidae, Staphylinidae, Curculionidae, Chrysomelidae, Scydmenidae, Anthicidae, etc.

Pitfall traps baited with vinegar mixed with water or ethanol are useful to catch many species of Carabidae (and Cicindelinae), Ptiliidae, Scydmaenidae, Anthicidae, Staphylinidae, etc.

It is very productive turning the rocks, logs, branches, fallen cactus and any other object lying on the ground to be able to find Tenebrionidae, Carabidae, etc. Remove the plants from the earth in search of insects among the roots, stems and leaves: there are Curculionidae, Chrysomelidae, Tenebrionidae, Scarabaeidae, etc.

Artic tundra [3]

Together with the boreal coniferous forest or taiga, the Arctic Tundra is the nearby biomes to the. In the Northern Hemisphere it is a continuous band encircling the Arctic Ocean, distributed across Northern Alaska, Canada, Scandinavia, Siberia and the Arctic

3. The subjects related to the alpine tundra and the Paramo, are explained in the article about the mountains.

Islands (in Greenland the Arctic Tundra belongs only to coastal areas, since the interior is occupied by glaciers). The Southern boundary is connected to the coniferous forest. Due to this distribution it is called arctic tundra.

The Arctic tundra has a smooth relief, resulting from the passage of glaciers. The winters are long and cold, with short, cool summers which do not last more than two months, and the temperatures do not exceed 10° C. There is almost no rain; however, during the summer, wetlands and flooded soils are abundant because there is little evaporation and snowmelt hardly infiltrates the underground, which is a permanently frozen soil called the permafrost layer (French 2007). The floor and the surface layer only thaw during the summer. Permafrost is present throughout the region and its thickness can reach over 300 meters. It decreases with distance from the Arctic tundra and becomes discontinuous (as in the taiga, located farther south).

Fig. 3 – Denali National Park, Alaska *(Picture courtesy of Jack Owens, Canada).*

In the colder and drier areas of Arctic islands, which are above 75° North latitude, the Arctic deserts appear (French 2007), where vegetation does not exceed 5 cm in height and consists almost exclusively of mosses, lichens and some creeping willows. But in less extreme climates, more typical of the arctic tundra, there are heath-type shrubs and flooded areas with mosses, sedges, grasses and rocks rich in lichens and mosses (Quinn 2008). Birches are present in valleys and areas near the sea. Vegetation is always more scarce on the slopes and ridges exposed to wind. Southbound higher trees develop. Birches with floors covered with lichens are typical of this area. Finally, farther South, the coniferous boreal forest or taiga is well defined.

The melting ranges from 30 to 100 cm deep, which makes plants, have very small roots, avoiding the development of trees in these regions. The few seed plants are very small and require low energy reserves. However, there are about 400 species of flowering plants and abundant mosses and lichens. The soils are very poor in humus (peat type), which is present only in humid areas with more vegetation and it is permanently frozen (Quinn 2008). In these regions beetles are only active in limited ranges of temperature and humidity, so it is best to look for them in sunny days and where vegetation is more abundant. Sometimes, rotten wood can be found which usually contains some beetles and/or larvae inside. According to French (2007) many species of beetles are found in Arctic and Subarctic environments. Also, in a private message my friend John Leavengood told me the following: *"I have a friend who works on Dytiscidae and has collected throughout Canada. He explained to me that Canada was great for collecting them. The current literature in the family shows that many species only occur in arctic zones".*

Some species of Tenebrionidae and Carabidae can be hidden in vegetation, which is mostly of low altitude and stunted or creeping growth. Predator species as Carabidae (even spiders) are active at night, dedicated to attacking insects immobilized by cold (Quinn 2008). According to Quinn, the moss has a rich variety of Staphylinidae, but the sparse undergrowth supports few insects.

According to Chernov et al. *(2000):"90 carabids have been recorded from tundra (In Eurasia) and about 170 from the tundra and forest-tundra combined. The potential carabid fauna of the whole Arctic in the broadest limits is estimated to comprise 240–300 species, which constitutes 0.6–0.8% of the total number of species in this family."* Also there are Dytiscidae, Staphylinidae, Chrysomelidae, Curculionidae, Catopidae, etc. (Chernov et al. 2000). Regarding the previous comment from Leavengood, the Dytiscidae are particularly interesting because (unlike the other families mentioned in this sentence) they are rather speciose in the Arctic (John Leavengood, pers. comm. 2010).

In the *Arctic desert (Arctic Tundra)* of North America there are almost no records of Coleoptera (McAlpine 1964), nor were there many records in the Polar deserts of Eurasia (Korotkevich 1972). However, Makarova and his colleagues (2007) have collected a few small wingless beetles in the archipelago of Severnaya Zemlya such as *Micralymma brevilingue* Schiodt (Staphylinidae), *Chrysolina subsulcata* Mannerheim, *Chrysolina septentrionalis (*Ménétriés*)* (Chrysomelidae), *Dienerella filum* (Aube) (Latridiidae), and a few other species.

Chernov et al. (2001) cited 25 to 27 genera of Arctic ground beetle living in the Arctic tundra, belonging to nearly 20 tribes. Most tribes are represented by only one or two genera: Carabini *(Carabus),* Notiophilini *(Notiophilus),* Bembidiini *(Bembidion),* Agonini *(Agonum),* Pterostichini *(Poecilus* and *Pterostichus)* Nebriini *(Nebria* and *Pelophila,)* Zabrini *(Amara* and *Curtonotus),* and only the tribe Elaphrini comprises three genera *(Blethisa, Diacheila* and *Elaphrus).* The Carabini tribe is represented in the tundra by 5 - 7 species of *Carabus.* According to Chernov et al. (2001), Chrysomelidae and Curculionidae represent the main part of phytophagous species and the most advanced groups of beetles among the arctic fauna.

Although insects are scarce, there are a few species of mosquitoes and biting flies that breed during the summer, unmercifully attacking both animals and people, especially in flooded areas. Experiments have shown that the attacks of mosquitoes can extract up to 115 g of blood in one animal in one day (Quinn 2008).

MacLean and Pitelka (1971) trapped few carabids beetles on sticky boards in the Arctic Alaskan tundra. According to Norment (1987), to collect carabids and many other beetles in the Tundra one can use pitfall traps.

Placing corrugated cardboard traps in low vegetation or shrubs may be also useful.

Savannas

Savannas are located in tropical climates where the vegetation forms a continuous herbaceous stratum composed of perennial grasses, with the occasional tree, shrub or bush individually or in small groups of less than 10 m., typically, the savannas are transition zones between forests and steppes. Savannas are located between 10° to 30° latitude N and S. These environments sometimes are mixed with other ecosystems (bush, chaparral). Savannas are hot with temperatures of 20°C to 30°C. The rains are frequent, with two rainy periods separated by a dry season. The soil is reddish with ferrous tones; the relief is flat.

There are several types of savannas

Savannas of the intertropical zone (the Serengeti in Africa, Llanos of Venezuela, etc.).

Temperate grassland savannas: an example is the Pampas of Argentina, which has a temperate climate (semi-arid and humid) and in a large portion of Patagonia there is a warm season, and usually another cold season in winter. The dominant vegetation is grasses, reeds and other pasture plants.

Mediterranean savannas are located in the middle latitudes of the five continents with a Mediterranean climate.

Mountain meadows include the prairies and altitude scrubs (montane, subalpine and alpine) from around the world. Mountain meadows are characterized by annual vegetation composed of grasses and flowering plants. In the Northern hemisphere there are great snowy forests.

In savannas and grasslands sweeping is advisable, since insects are very active on the branches and flowers of various plants, both day and night (Eduardo Faúndez, pers. comm. 2010).

In the dung and carrion one may find many Scarabaeidae, Hydrophilidae, Histeridae, Staphylinidae, Silphidae, Ptiliidae, Geotrupidae, Trogidae, Scaphidiidae, Dermestidae, Cleridae, etc.

A wide range of traps can be used, from flight intercept traps, colored pan traps and pitfall traps, using any type of bait derived from organic material. For example, dog food attracts many species of beetles. Other bait: several dead Meloidae in a pitfall trap attract many species of *Notoxus* (Anthicidae), which is a genus present in most Neotropical deserts. Here it is also useful to use corrugated cardboard traps in low vegetation.

The canopy of xerophytic savannas and grassland is more open and less complex than that of forests, and with fewer plant species. However, methods to collect beetles are usually the same as those used in forests.

When examining savanna trees, pay attention to groups of leaves and flowers in the outer ends of the branches. Beetles tend to fly around the trees and land on these leaves and/or flowers.

Dead trees should be examined closely. Most of beetles are in the shadows rather than exposed to direct sunlight. The wood of dead trees is quickly dried by long exposure to the sun's rays. This creates very high temperatures and the beetles must hide to avoid death due to the intense heat. However, there are micro-species that prefer hot and dry habitats. In such exposed sites one can collect some species, but the diversity is much lower than that of micro-habitats protected from heat.

Beetles living in exposed sites differ from those of protected sites. If the sampling sites are subjected to extreme fluctuations in humidity, it is advisable to collect beetles inhabiting fallen plant debris after the rainy season, when humidity is not so high.

In this book, only some of these ecosystems and how to collect the beetles in these environments are described in detail. However, the methods used to collect beetles in all the savannas of the world are very similar.

Patagonia

Patagonia is a geographical region located in the Southernmost part of South America. It covers areas in Argentina and Chile, separated mainly by the Andes.

In large part the area is covered with dense forest and evergreen coniferous forests and cold deciduous Fagaceae and Magellanic temperate forest (Bridges 1952). Extra-Andean Patagonia has oriental and semi-arid climate, with soils poor in organic matter. In wetter areas near the Andes, and the region between the Santa Cruz River basin and the Estrecho de Magallanes, the land is covered by xerophytic grass steppes. The rest of the territory is mainly covered by tough grassesof one to two meters in height (Walker 1955). In the Mediterranean areas is "paramo" and "vegas" or small fertile areas, located at the bottom of valleys and canyons.

Fig. 4 - Patagonia Argentina 2007 *(Picture courtesy Danilo Carriel Ramirez, Chile).*

In Patagonia there are many endemic insects closely related to other insects in the far-southern extremes of the world (Spinelli et al. 2009). Among them are many species of Tenebrionidae in the subfamily Pimeliinae (Flores 2000, 2001), Scarabaeidae (*Taurocerastes patagonicus* Philippi) and many other beetles (Carabidae, Staphylinidae, etc.).

The methods for collecting beetles in Patagonia are almost the same as those used in other xerophytic regions: use a Sweeping bag, light traps, pitfall traps, underground traps, etc., but the baited traps in the trees are not very productive. Also, one must consider that the wind is very strong for most of the year, and that limits the use of intercept traps and light traps (Eduardo Faúndez, pers. comm. 2010). During the day it is helpful to look for beetles in the foliage and the roots of plants, although, among the flowers usually there are many beetles (Eduardo Faúndez, pers. comm. 2010). Moreover, in Patagonia the spring and summer are very short, and the winter is long. There is a wide range of temperatures between summer and winter. There is also a very marked difference in temperature between day and night. The average winter temperature is -2°C and summer maximum temperatures can reach 40°C in the Alto Valle de Río Negro and Neuquén. Tierra del Fuego reaches 30°C. The Andean Patagonia is humid; the rainfall exceeds 2000 mm per year and eastward it is increasingly dry up the Atlantic zone, where the average annual rainfall is only 200mm. In Tierra del Fuego, there are approximately 800 mm per year. Then, the best time to collect beetles in Patagonia runs from December to early February.

THE GRAN CHACO[4]

The Gran Chaco, located approximately between 17° and 33° south latitude and 60° and 65° west longitude, is a vast plain that is located in Northern Argentina, Southeastern Bolivia, Northwestern Paraguay and a small area of Southwestern Brazil. It extends approximately 1500 km from North to South and about 700 km from East to West, without the presence of any physical barrier (Riveros, no date). Its rainfall decreases from 1.100

4. In other similar biomes or habitats such at Cerrado, the Caatinga, etc., one can use the same methods and traps discussed in this article. Daly and Mitchell (2000) give an account of most South American ecosystems this paper is available online (see bibliographic references) information of many ecosystems in: Biomes of the world in: http://science-class.net/Ecology/Ecosystems_Biomes/biomes_intro.pdf

mm per year in Formosa to 300 mm per year in the Western.

The predominant vegetation consists of xerophytic forests of "Quebracho colorado" *(Schinopsis* spp.*)*, "Quebracho blanco" *(Aspidosperma quebracho-blanco (*Schltdl.*))* and mesquites *(Prosopis* spp.*)*. The Chaco is one of the largest woody grassland areas of central South America (Riveros, no date).

In Paraguay, the Chaco is divided into two parts: the Humid Chaco (or Lower Chaco) and the Dry Chaco or Boreal. In the Lower Chaco is a great plain covered mostly with palm groves *(Copernicia alba* Morong*)*, which cover thousands of hectares. Among the palms there are some trees and shrubs. These areas remain flooded for most of the year.

Lower Chaco should be visited during the spring, but the reader should know that the environment in this great plain is very unpleasant. During the day there are almost no shadows anywhere. Although there are some trees, their foliage is too sparse to provide significant shade. Generally, the shrubs in this region are thorny. Therefore, the very high temperatures and excessive humidity, along with hordes of mosquitoes, make it very difficult to collect in this ecosystem. However, if the collector can stoically endure these environmental conditions, he will be able to find several species of beetles. For example, in rotting trunks of palms there are large Chrysomelidae and Curculionidae attracted by the scent of decaying wood. Also one should check all the flowers since they usually have Buprestidae, Cerambycidae, Cantharidae, Scarabaeidae, etc.

Among the herbs, sweeping and/or pan traps can be used to attract Buprestidae and other families. As the soil generally has plenty of water, there are numerous abundant species of aquatic beetles. They can be collected with an aquatic net or with light traps at night, which also attract other interesting species of non-aquatic beetles. However, in the wet Chaco, ground beetles are more difficult to find because there is water everywhere, except in higher places, where Cicindelinae and other Carabidae may be found.

In the Wet Chaco there are some rivers accompanied throughout their course by gallery forests. Within these coastal forests almost every collection method explained in this book can be used.

The Boreal Chaco (or Dry Chaco) (Fig. 5) does not present the palm grove like the Lower Chaco. There, palms are replaced by a wide variety of plants adapted to drought, such as Cactaceae, Bombacaceae, *Prosopis*, *Acacia* and many leguminosae, etc. In this region water is absent in most of the year. So, it is advisable to visit the Dry Chaco of Paraguay at the beginning or during the rainy season. It is difficult to know exactly when it will rain, since in this region the rains are highly variable. The closer to the Paraguay River (Humid Chaco or Lower Chaco) (Fig. 6) rains appear earlier in the spring (November or December). Going west, towards the proximities of Bolivia, the rains appear after February.

Fig. 5 – Dry Chaco, Paraguay *(Photo: Jorge Barrett)*.

Cicindisini is a tribe of Carabidae found only in the Argentine Chaco (in very dry and saline soils) and Iran (Kavanaugh and Erwin 1991).

It should be taken very seriously when it rains in the Chaco; roads are absolutely impassable, even operating vehicles with four-wheel drive. Almost flat terrain and low absorption clay soil makes the flooding very extensive and totally isolating, sometimes for months, for populations and entire regions. The villagers plan their annual labor knowing and anticipating these long periods of isolation in their properties. Another point to consider is that the roads become so slippery that only the few maintained by the villagers are passable. Airplane travel is limited to landing strips with paved tracks. Helicopters would be appropriate, but are expensive and rarely accessible.

Fig. 6 – Humid Chaco or Lower Chaco, Paraguay *(photo: Jorge Barrett)*

Except in the rainy season, water, and especially fresh water, is very scarce in the Chaco and that makes it impossible to collect insects. Rains usually occur after February, but there are extremely dry years and plants bloom only after the rains. Then we can find the following beetles: Cerambycidae, Buprestidae, Scarabaeidae *(Cyclocephala, Pelidnota)* and many Melolonthinae and *Gymnetis.* At night, using light traps, we can collect: Scarabaeidae *(Megasoma, Strategus, Heterogomphus, Coelosis,* etc.), Carabidae *(Calosoma, Galerita, Brachygnathus, Brachynus, Cnemalobus,* etc.), Cicindelinae *(Tetracha, Phaeoxantha, Brasiella, etc)* or Meloidae *(Lytta, Epicauta,* etc.). During the day, on dirt roads or along them, in moist areas (near the pools of water) we can collect: Cicindelinae such *as Odontocheila, Brasiella, Cicindela, Cylindera, Pentacomia,* etc. With pitfall is possible to collect Scarabaeidae (Scarabaeinae) such as: *Coprophanaeus, Sulcophanaues, Deltochilum, Dichotomius, Canthon, Ontherus, Canthidium, Ateuchus,* etc. Also, as in most arid regions, there are Tenebrionidae and many other families.

When using pitfall traps in the Chaco these should be checked every day, because the baits dry in a few hours, rendering them unusable. To prevent some baits from drying out (especially feces), put them in cloth bags, wet them and suspend them over the pitfall trap. Moreover, some salt water can be put within the trap which helps to keep moisture in the bait. However, to catch Troginae, it is advisable to put dry meat in the pitfall trap, or look under dried remains of animals, skins, skeletons with traces of dried meat, feathers, etc.

Moreover, being that water is very scarce in the Dry Chaco, traps imitating water reflections can attract several species of aquatic beetles (see article "Aquatic and semi-aquatic environments", Chapter II). Malaise traps are a good alternative also.

In the salt lakes or salt pits *(Fig. 2)* there are plants adapted to live in those environments saturated with salt, just as several species of Chrysomelidae and Curculionidae and some Cantharidae. On the floor Cicindelinae and others Carabidae are running around which may be captured with pitfall traps, barrier traps or a hand net. Light traps, either ultraviolet or mercury, attract thousands of specimens in the evening (Hydrophilidae, Dytiscidae, Heteroceridae, etc.).

Colored pan attract many species of Buprestidae and some other flying families. The colors that attract more beetles to traps are purple, blue and white, and while yellow and red are less attractive to these beetles, it is better to try with all these colors. Try to locate at least 50 colored dishes in very sunny and flourished places, preferably near streams, ponds or lagoons Bottle traps in the trees are also advisable.

It is interesting to check the seeds of *Acacia* spp., *Prosopis* spp. and other legumes, or any plant, shrub or tree with hard seeds and pods. They usually have several species of Bruchidae and Curculionidae.

John Leavengood in a private message informed me: *"I have Collected pupae of the mesquite borer (*Cerambycidae: *Callon rimosa (*Buquet*)) by hatcheting the root crowns of mesquite in Southern Texas. I would find the oviposition site on the root crown by looking for discarded wood shavings and frass. Then, starting at the entrance, hatchet gradually lower and lower, revealing the tunnel, until reaching the bottom of the gallery where i would reach the pupa. Later I rear it to adulthood in tupperware."*

During day and night, carefully check the trunks of trees and shrubs seeking the beetles walking on them. The beating sheet and the sweep bag can also be used; this method allows catching some Buprestidae, Elateridae, Mordellidae, etc., which are very agile with intense sunlight. A good method to prevent these beetles (Buprestidae, Rhipiphoridae and Mordellidae) from escaping is to stand between the sun and the insect so there is a shadow on the beating sheet. Most Buprestidae have a bad habit of running through the branches or thorny stems. If the thorns are not very long and hard, Wolfgang Barries (pers. comm.2009) suggests using a thick cloth padded with cotton and, with a quick push of the hands, press the beetle against the branch or trunk, so it is trapped between the cotton and the thorny branch.

In the Chaco, carabids abound during the day and one may collect different species of *Cicindela* spp., *Odontocheila* spp., *Brasiella* spp. and other genera. Under the sticks and sometimes under clods *Tetracha* or *Phaeoxanta* are hidden along with families of Coleoptera. In the evening one may encounter *Calosoma* spp., *Lebia* spp. *Galerita* spp., *Chlenius* spp., *Brachygnathus* spp. and many others. Also Tenebrionidae abound during the day: *Entomoderes* spp., *Epipedonota* spp., *Mitragenius* spp., *Psectrascelis* spp, *Scotobius* spp., and *Cicindis horni* Bruch. There are also some Dynastinae: *Megasoma.joergenseni. penyai* Nagai, *Megasoma janus* Felsche, *Dyscinetus, Eutheola, Dynastes,* etc., and many different species of Melolonthinae which are, together with Carabidae, one of the dominant beetle groups in species abundance in the Chaco. Different species of Meloidae and other minor families also abound. Moreover, one of the best methods of collection in the Chaco is the light traps which allow catching thousands of specimens each night. Beating sheets and sweep bags must be used.

CAR-CATCHER

This method uses cars with large collection nets, which are attached to vehicles or they are employed by the companion of the driver

According to Peck and Davies (1980) most beetles fly at 1.5 m or less above ground level. Dr Peck sought to take advantage of this fact and developed a lightweight and easily assembled vehicle-mounted net to capture beetles flying at this height.

According to Peck and Cook (1992) the net is mounted on a bicycle and the car mounted models were developed by Masner and Goulet in the early 1980s. Andreas Herrmann

(2001), who kindly provided an example, described this interesting collection method that is currently fashionable among German entomologists for collecting beetles in savannas or prairies: *"An Auto-catcher basically consists of a metal loop, which is fixed on the roof of the vehicle. This loop is in the form of a half circle with a diameter of about one meter and 20cm. Fixed to this metal loop is a bag of fabric with a very fine mesh, the frontal aperture being formed by the metal loop. This bag tapers gradually towards the back and ends in a smaller collection bag, which usually is attached to the larger bag with Velcro which makes changing the collection bag quick and easy."*

Fig. 7- Collecting beetles with Car-catcher. *(Picture courtesy of Andreas Herrmann, Germany).*

Peck and Cook (1992) suggested putting the net in front of the vehicle, because the great advantage is that here there is a boundary layer in the air (a cushion-like layer) that cushions impact of the insects with the net, without churning and breaking them. Also, Peck suggests painting the net with an insecticide that prevents the insects from escaping when the vehicle stops.

Andreas Herrmann goes on to explain in his paper: *"It is very important to choose the right material as on the one hand even small beetles (like* Ptiliidae*) need to be caught, but on the other hand the bag must be sufficiently permeable to air in order to avoid air turbulence in front or the bag which would prevent beetles from entering (one mesh that has been found satisfactory has a thread thickness of about 0.1mm and a spacing between threads of about 0.2mm). The length of the bag is usually such that, when parking the car or during tuning maneuvers the bag does not touch the ground and instead hangs freely across the back of the vehicle."*

According to Peck and Cook (1992) the collection speed should be slow, and never as much as 20 to 30 km/hr on narrow roads in forests.

In a personal message (2010) Andreas Herrmann explained: *"Several species are known only from the car-catcher and never have been collected in another way. This is a problem, because you cannot say anything about their biology and furthermore you cannot tell colleagues the exact place where they have been caught. You drive e.g. 3 hours at 40 km/h and then empty the fabric bag. In this time you drive a pretty big distance, whilst many scientists want to know exactly the part of the site or woods where the interesting beetle has entered the catcher."*

Also, Peck and Cook (1992) explained that Rutanen and Muona (1982) had used a car-net to collect 607 beetle species in 47 families (especially Staphylinidae, Ptiliidae, Cryptophagidae, Leiodidae, Nitidulidae, Scolytinae, and Lathridiidae) in Finland. And Dr. Peck (1992) has collected 45 different families of beetles by evening car-netting on forest roads in Eastern and Northern Ontario.

For more details on this method of collection, I recommend reading the work of Andreas Herrmann (2001) and Peck and Cook (1992), both papers are cited in the references of this book.

DUNES AND BEACHES

Dunes

Dunes are located in four different types of landscapes: shores of lakes, river valleys, in arid regions and near the sea (Fig. 8).

They are called "coastal dunes" to those found above the high water mark in the sand of the beach.

Dunes are present both in the Northern and Southern hemispheres, from the Arctic to Antarctica.

According to Anwar (2009), dunes of the sea coasts are ecologically similar to those of the deserts; they are common in temperate climates, but less common on tropical and subtropical coasts. These are also common near the mouths of rivers where the sand is deposited after being transported by the river (Carter et al. 1990b).

In the most extreme deserts, plants can not survive in the dunes, except in a wet valley or oasis, where most of them are xerophytic.

Despite not being very rich in biodiversity, the dunes are home to an interesting fauna of beetles. Living among the grasses that grow on sand dunes, there are Chrysomelidae, Curculionidae, Dynastinae (*Cyclocephala* spp.) Glaphyridae, Anthicidae, Limnichidae, etc.

Fig. 8 - Dunes near a beach in Canelones, Uruguay. *(Photo: Jorge Barrett).*

Buprestidae and Cerambycidae larvae feed on the roots of various grasses growing in the same habitats. Also in the dunes one may find Geotrupidae, Trogidae and Scarabaeidae feeding on carrion of dead animals and debris left by the waves on the shores of the beaches.

On some beaches in Mexico, Troginae may be found among the remains of turtle eggs and / or dead hatchlings. On plants and their roots, adults can be found sometimes, and more often larvae of several Dynastinae *(Eutheola* spp., *Dyscinetus* spp.*)*, Melolonthinae *(Serica* spp., *Phyllophaga* spp., *Macrodactylus* spp. etc.) and Rutelinae *(Anomala* spp., *Leucothyreus* spp., *Bolax* spp.*)* and some larvae of Buprestidae, Meloidae, Tenebrionidae and other beetles.

Among the vegetation of dunes one may find rare species of Cicindelinae (*Cicindela* spp.). Among the dead remains of invertebrates on the sea shore, there are Tenebrionidae, Carabidae, Staphylinidae, Histeridae, Hydrophilidae, etc.

Using a kerosene lamp during the night, one may look for the footprints left by beetles on the sand, and follow them to fine the beetle, if you are lucky.

The corrugated cardboard traps and folded spiral are an effective method for collecting beetles, which use these traps as a shelter. The traps should be attached in the vegetation (with a strong rope) to prevent the wind push them far away and to prevent them be lost under the sand.

On the dunes, colored pan traps can be placed on the ground (yellow, white, purple, red or blue), filled with soapy water, and the beetles will fly into them. Remove any object lying on the ground, natural or not, (which can be made) of any material such as wood, plastic, metal, glass, etc., since many beetles are hidden there: Carabidae, (Cicindelinae and others), Tenebrionidae, Staphylinidae, Chrysomelidae, Histeridae, etc. It is convenient to sift the sand that lies between the roots of grasses and other herbs and also the sand under organic debris at the sea shore, but not the very wet sand which is not good to shelter the beetles because of its high salt content.

On the dunes, and even in deserts, metal bright sheet that mimic water attract a variety of Dytyscidae, Noteridae, Gyrinidae and other families of aquatic beetles.

Where there are patches of trees or shrubs, cut the branches and collect them into separated groups to attract Cerambycidae, Buprestidae, Anthribidae, etc., and check after one week or more.

The dunes are a very restricted habitat, usually small and very fragile. Already there have been many studies which have shown the effect caused by vehicles that run on the dunes around the world. So, to avoid damage to the fauna of the dunes, will be healthier for these (and for you too) that you explore them on foot, so, you will learn much more about this fragile ecosystem, which if you were driving a car over them.

Beaches

To collect on the beaches (Fig. 9) a sweep net must be passed through the vegetation to (be able to) catch Cerambycidae, Pedilinae, Scarabaeidae, Buprestidae, Meloidae, Mordellidae, Rhipiphoridae, Cleridae, Cantharidae, Meloidae, etc. It is also convenient to use pitfall traps for coprophagous and scavenger species. Dead shrimp, fish or ethylene glycol-based in antifreeze are good bait. Use vinegar for nocturnal Cicindelinae and others Carabidae, or just look for them with a flashlight or headlamp on the ground where they walk at night and catch them by hand collecting. Diurnal Cicindelinae should be collected with aerial nets. Diurnal carabids, running on the sand, can be taken by hand.

Fig. 9 - Habitat for several species of Scarabaeidae, Tenebrionidae, Chrysomelidae and Curculionidae (beach in Argentina). *(Picture courtesy of Pablo Marano, Argentina).*

Turn over any algae or animal remains lying on the sand. Many beetles, such as Tenebrionidae, Staphylinidae, Histeridae, Hydraenidae, Phycosecidae, etc. feed on these remains left by the sea (marine algae, dead fish, etc.). The remains of algae can be put into a bucket with salt water which will make the beetles float and then may be collected by hand or with a fine sieve. You can also put the seaweed on a wire mesh and place it over a bowl. This device is placed in the sun. The insects will descend to avoid the heat, and will fall to the basin, where they can be caught by hand. The remains of algae and other plants can also be put inside a Winkler/Moczarski eclector. On the beaches pour some water on the dry soil or sand, so the underground beetles will come out.

Saturated soils, with mosses and plant remains found along the margins of water, often contain a mix of terrestrial and aquatic beetles. Most aquatic beetles pupate (they become pupae) in terrestrial sites, near the water's edge. Adults can be found under stones and logs by examining leaf litter or by spilling the topsoil on a beating sheet.

On the white sand beaches of some rivers, such as most rivers of Paraguay, Brazil, Bolivia and Argentina, there are many species of Cicindelinae wandering at night on the sand *(Tetracha bilunata (*Klug*), Tetracha cruciata (*Brulle*), Tetracha brasiliensis* Kirby*), Tetracha fulgida (*Klug*),* etc.). In some areas of the Pilcomayo River in the Chaco of Western Paraguay, Northern Argentina and Southeastern Bolivia, you can find the rare *Phaeoxantha bucephala* (Horn) running on the ground near the river in the evenings from October to November. When these beetles are observed here in their natural habitat, they can be easily collected by hand at night, aided with a flashlight.

Light traps work very well on the beaches and the banks of rivers. There you will find Scarabaeidae, Anthicidae, Cerambycidae, Silphidae, Carabidae, Geotrupidae, Elateridae, Throscidae, Meloidae and many smaller families of beetles.

COLLECTING IN AQUATIC AND SEMIAQUATIC ENVIRONMENTS

By Carlos Aguilar J., Eric Chapman[1], Sergio Ríos[2]

Aquatic beetles include many different families of Coleoptera: crawling water beetles (Haliplidae), water scavenger beetles (Hydrophilidae), predaceous diving beetles (Dytiscidae), water penny beetles (Psephenidae), riffle beetles (Elmidae), whirligig beetles (Gyrinidae), minute moss beetles (Hydraenidae), etc. Some aquatic beetles are substrate dwellers (e.g., Elmidae, Psephenidae), while others are free swimming (e.g., Dytiscidae, Gyrinidae, Haliplidae, Hydrophilidae).

Aquatic beetles are present in all available habitats: ponds, lakes, marshes, swamps, streams, rivers, seepages (i.e., spring heads), holes in trees, cavities or cracks that can hold water, etc. Thanks to their small size and high capacity of adaptation, beetles can live in almost any aquatic habitat available, and are only absent in the middle of the oceans (although a few species live in the intertidal zone).

Beetles occupy most non-marine aquatic habitats, both lotic and lentic. Over 80% of aquatic species live in freshwater lentic systems (Pond Action 1994). These systems include temporary pools and streams, which are an important habitat for Coleoptera. Beetles are present throughout the water phase and many species have specific ecological requirements, which help biologists to detect and prevent alterations in ecosystem quality and/ or simply help them to characterize habitats. Also, the general characteristics of the water cycle and the extension of the drought period in some cases can be detected with great accuracy using aquatic beetles as indicators, as demonstrated by Wiggins and his colleagues in 1980 (quoted in Correia 1999), therefore the characteristics of the immediate history of the habitat can be determined.

As for environmental assessments, information obtained from aquatic beetles provides a better method of evaluation of different river environments than the one of other

1. Eric Chapman: Department of Entomology, University of Kentucky, S-225 Agricultural Science Center N Lexington, KY 40546 (859) 257-2759- Email: ericgchapman@gmail.com
2. Sergio Ríos: Asunción. Azahares 406, Barrio Loma Pytá Email: sergiord40@gmail.com

groups traditionally used (Bournaud et al. 1980; Kaesler et al. 1978). Among them are included species of superfamily Bhyrroidea (Shepard & Aguilar 2010). Aquatic beetles are good water quality indicators, reflecting environmental conditions. The presence of indicator taxa can help regulators make priority decisions on areas to be preserved. However, aquatic beetles are generally poorly represented in most samples of statistical studies of the water fauna. The reasons vary: - Many species have low population densities

- Most aquatic beetles and their larvae are often found in places of difficult access, such as:

  * Dense vegetation
  * Debris in or on the water surface
  * Holes in trees
  * Cracks in rocks, etc.

To obtain increased representation of aquatic or semi-aquatic Coleoptera it is best to visit and examine as many habitats as possible. This article will serve as the basis for determining where to sample in different environments.

To learn where to find specific groups of beetles the reader can see the article "Where do they live ...?", Chapter IV.

It is not a fundamental requirement to review and collect a large area, which eventually becomes exhausting. However, studying a large area means: sampling of a relatively large area and employment of techniques useful to separate the beetles from the different substrates. On the other hand, Eyre et al. (1986) and Eyre & Foster (1989) indicate that most species of aquatic Coleoptera can be collected in any season and any time of day. Nevertheless, when the courses or bodies of water are scarce or become smaller, there is greater concentration of aquatic species than in the wettest period in which water is abundant (William Shepard, pers. comm. 2010).

Before going out to collect aquatic or semi-aquatic beetles, it is good to have some idea about the microhabitats in which they live. This helps the researcher decide which habitat(s) is/are more appropriate to conduct the study. Moreover, this information should be recorded on the labels that accompany each sample of specimens. For that reason, a brief overview of aquatic and semi-aquatic environments where these beetles are abundant, and methods and tools that can be used to catch them is presented.

Aquatic habitats are classified into lotic (flowing water) and lentic (standing water):

Coleoptera are present in different terrestrial, aquatic and semi-aquatic habitats and each species occupy specific niches within each ecosystem[(3)]. Therefore, for a better understanding of aquatic and semi-aquatic ecosystems and to know where to find the different families in these habitats, the following classification is presented (based on Rosenbauer 1999):

Lotic habitats

Composed of rivers streams and springs, these habitats are characterized by the constant flow of their waters, even though some are temporarily dry.

Rivers and streams have two types of conformation of their beds that are important

3. However, some competing terrestrial species can coexist on the same resource with no observable evidence of niche differentiation. For example, some Chrysomelidae of the subfamily Hispinae, which eat the same food and occupy the same habitat, can coexist without any evidence of segregation or exclusion (Strong 1982).

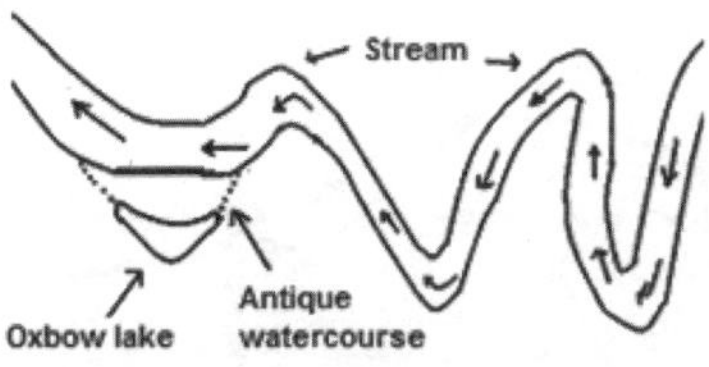

Fig. 10 - Oxbow (or small lagoon), it is formed when the watercourse has cut a curve of a stream or river.

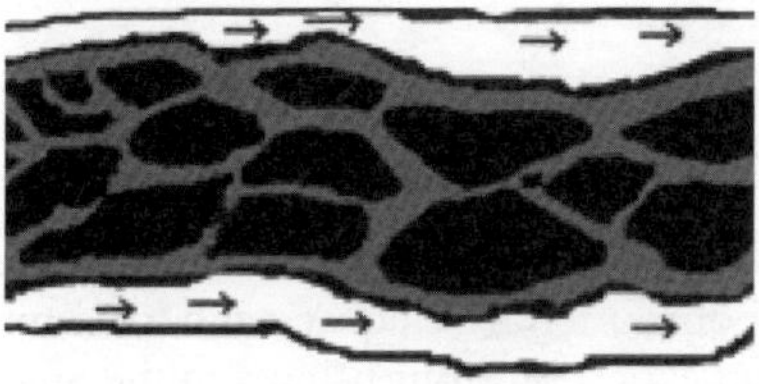

Fig. 11 - Braided stream. *(Drawing of Carlos Aguilar).*

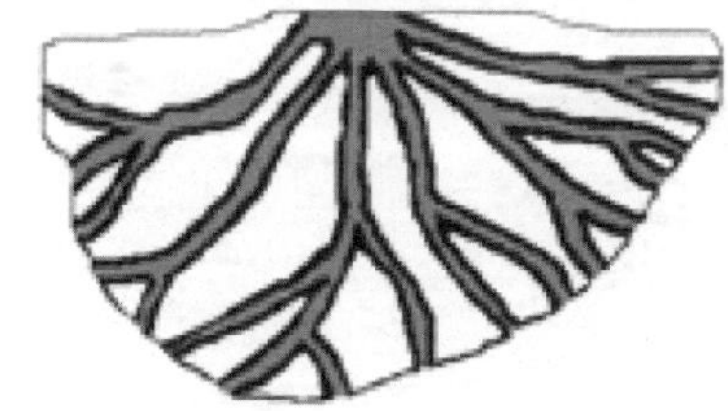

Fig. 12 – Fan shaped streams. (Drawing of Carlos Aguilar)

for Coleoptera collection: pools and riffles. Also important are stream margins, especially where the root masses of terrestrial plants hang into the water and undercut banks where you can get large numbers of dytiscids and gyrinids, especially as winter approaches, because many species overwinter in undercut banks.

Different stream habitat types: A stream is a body of water with a current, confined within a bed and stream banks. In some countries or communities a stream may be defined by its size. The biological habitat in the immediate vicinity of a stream is called a riparian zone.

There are five types of Streams:

a) Meandering Streams: An unmodified or natural stream usually consists of successive meanders (curves). Meandering streams develop in relatively flat areas, such as a floodplain, and where sediment consists primarily of fine sands, silts, and mud. They form in lowland areas with low overall gradients, either where sediment is not very available or where the stream has lost so much of its sediment behind as its gradient has declined that it no longer has enough to carry to accomplish its work. Over time the stream is cut into bends, forming small lagoon-shaped gaps. In these meanders there are usually copious amounts of aquatic vegetation which serve as habitat for many species of beetles, especially Hydrophiloidea.

b) Braided Streams: These streams sometimes carry loads water that exceed their normal capacity (Fig. 11). During such spate events, sediments are rapidly added to the channel by external erosion processes or where discharge (capacity) fluctuates drastically. Many aquatic plants grow in these sediment deposition areas which create a variety of aquatic beetle habitats.

c) Fan-Shaped Streams: Where canyons open out onto flatter and more open land, large amounts of sediment is carried by the stream and deposited in the lower basin stream in a fan-shape. Much of the sediment, particularly the coarser parts, may be lost to produce an alluvial fan (Fig.12). Here you can find beetles; however, it can be difficult to collect in these places because the heavy clay can make footing difficult and clog collecting devices.

d) Riffles: Riffles occur in areas of streams where larger rocks and stones have been deposited where the bed is elevated in relation to areas immediately up and downstream. Such areas have much faster flow than deeper adjacent areas and can be very violent and impetuous. When the bed has gravel, stones or very small waterfalls, and the water drains rapidly between the stones, beetles of the families Elmidae, Dryopidae and Lutrochidae can be found in the interstices.

e) Stream Online: The erosion that brings the upstream part of the river down to its equilibrium level will be down cutting erosion creating a deep canyon and a straight channel (Fig. 13). The pools which are formed in these streams, discussed above, are habitats for many beetles. It is formed by an input, where the stream enters, and also has a tongue, which is the place where the stream begins to narrow. On each side of the tongue, at the end of the input, there is a turbulence zone. There, the current weakens forming the backwaters along the sides, where the water has very little flow. Along the banks, the pools often have aquatic and/or semi-aquatic vegetation. According to Wladimir (2010), an expert on trout fishing, in the junction there is a centrifugal force in which insects are trapped. For this reason, large trout congregate here to feed on insects trapped in the swirling current. However, Eric Bachman (pers. comm. 2010) stated that most of beetles hide among root mass to avoid predation by fish. In the area of the stream pool, sheltered under stones, are found mainly Elmidae and Dryopidae. Finally, there is the pool exit, where the stream enters the next part of the river. There, under rocks and emerging branches, or hanging over the water surface, water beetles can also be found (Elmidae, Dryopidae, Psephenidae larvae, Lutrochidae).

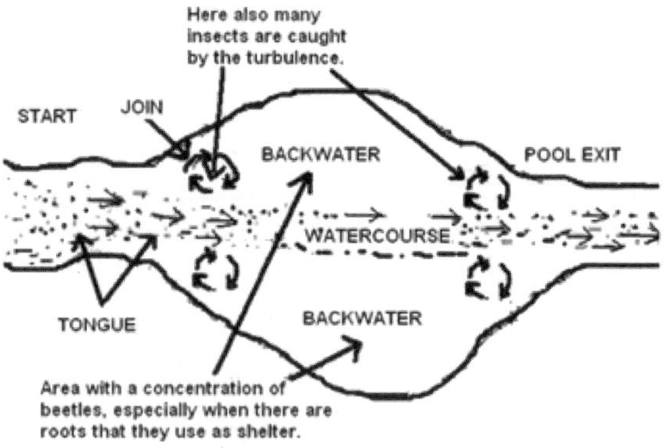

Fig. 13 – Stream online *(drawing based and modified from Wladimir, 2010).*

f) Stream curve (Fig. 14): these are shaped like an "L", but there are pools where the curve is more gradual. In the elbow, the current flows against one of the river banks, causing erosion. Waste produced by erosion is dragged to the back of the elbow, producing an area full of mud and almost barren, in which some beetles living in the mud can be collected (e.g., Heteroceridae).

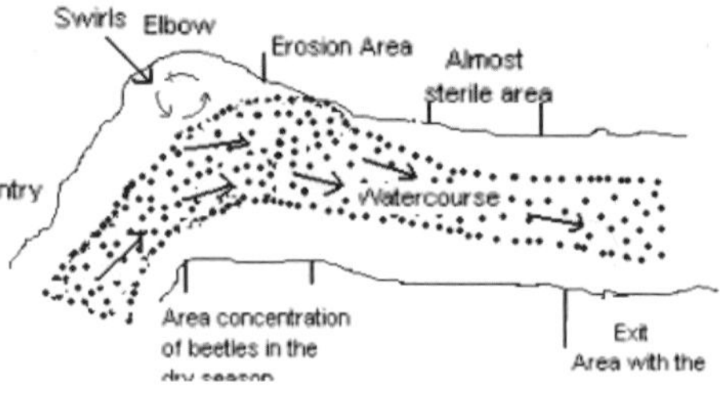

Fig. 14 - physical characteristics of a curve pool (*drawing based and modified from Wladimir 2010).*

In the angle of the elbow the water is slow and turns, producing a whirlpool. The speed of rotation will be higher or lower, depending on the river characteristics. At the exit of the pool, where water depth is low, there is lighter and greater oxygenation, the greatest diversity of beetles is found. However, in the dry season, when the rivers have little water, the beetles are concentrated in the inner curve.

g) Springs: These are areas where groundwater flows upward to the surface. The range in size is variable, from tiny seep holes with only enough water to form a small puddle to large spring streams flowing at a rate of 1,000 gallons per minute (Helfrich and Parkhurst 2009).

Lentic habitats[4]

These are environments where there is no evidence of water flow in the past or present. These are classified according to the water permanence time:

I) Permanent Lentic Environments have presence of water throughout the year. They include lakes, ponds, marshes, and swamps, etc.:

a) A **marsh** is a standing water habitat dominated by herbaceous vegetation. Salt marshes occur along of many coastal areas. They are dominated by salt-tolerant plants (halophytes) like cord grass and are greatly influenced by tides (Helfrich and Parkhurst 2009).

b) A **swamp** is a standing water habitat dominated by woody vegetation. Swamps are sometimes wet in part of the year and dry the remainder of the year, others are permanently wet. Many families of Coleoptera occur in these environments such as Hydrophilidae, Dytiscidae, Dryopidae, Noteridae, Haliplidae, Gyrinidae, Hydraenidae, Carabidae, Chrysomelidae, Curculionidae, and Cerambycideae.

c) A **lake** is larger and deeper than a pond. Lakes can range in size from small (10 acres) to very large (e.g. the Caspian Sea, the Great Lakes, Lake Tanganyika) (Helfrich and Parkhurst 2009), and they are present in continental areas of different countries of the world. There are freshwater lakes and saltwater lakes (Fig. 15). The latter are dominated by salt-tolerant plants and are found in deserts and mountainous areas (e.g., the Andes). There are almost no beetles in the water of saltwater lakes, but their beaches can provide habitat for many species of Tenebrionidae and Carabidae and on the plants Chrysomelidae, Curculionidae, Scarabaeidae, Bruchidae, Cantharidae, Mordellidae and other families can be found. Some sampling methods to collect beetles in the salt marshes are explained in the article on the Deserts, Chapter II.

Fig. 15 – "Laguna Capitán." It is a salt lake in the Paraguayan Chaco. *(Photo of Jorge Barrett).*

II) Temporary Lentic Environments: These areas contain water only during the rainy season (Fig. 16). They include small ponds or puddles of water, roadside ditches, etc. Temporary ponds are environments with dry periods of varying duration (Williams 1997). Due to their cyclical nature, these environments impose different stresses on their inhabitants than those with permanent water (Fontanarrosa et al. 2004).

Some species of temporary waters specialize in such habitats and are therefore not likely to be found in other habitats, whereas other species are facultative and can be found in both temporary and permanent bodies of water. The existence of insects in this type of environment requires that they survive dry periods in situ, or migrate to other types of

4. Beetles may be collected in these aquatic habitats using just a very fine mesh strainer. With this method of manual collection, one can learn a lot about their habitats, but many specimens are not obtained using only a hand strainer. In addition, many beetles live in inaccessible places (cracks or crevices, or very deep water current, etc.). Therefore, the sampling methods explained in these pages should also be used.

aquatic environments, and/or to be re-colonized after the restructuring of the place (Wiggins et al. 1980). Wiggins et al. (1980) point out that the fauna of temporary ponds is poor due to their ephemeral nature. However, in the last decade it was shown that temporary environments can have similar species richness, as compared to permanent bodies of water (Boix & Sala 2002). Beetles are well represented and are the most varied in number of species in almost any temporary aquatic environment. For example, in several temporary ponds in the city of Buenos Aires, Fontanarrosa et al. (2004) found 36 taxa belonging to 26 genera from 7 families (Carabidae, Gyrinidae, Haliplidae, Noteridae, Dytiscidae, Hydrophilidae and Scirtidae). Moreover, these authors suggest that the wealth of insects (with 58 taxa) in temporary ponds was lower than that of the lagoons, where they found 73 species. Other studies in temporary environments showed a similar number of species to that of permanent environments (Williams 1996). But the diversity of aquatic insects is typically lower in temporary aquatic environments than in permanent ones. Temporary aquatic environments depend on drought periods imposing strict conditions, which restrict the number of species that can survive in them (Wiggins et al. 1980).

Fig. 16 – A small pool of water which has many Hydrophilidae. (Dep. Central, Paraguay) *(Photo of Jorge Barrett).*

The collection of beetles in standing or stagnant water, such as swamps, shallow ditches and edges of small ponds, can be very productive. But the size of beetle populations depends largely on the abundance of vegetation, fish, substrate type and the size and depth of the water body. The presence of small predatory fishes significantly reduce the mean abundance and species richness of beetles via habitat selection behavior alone. Therefore, aquatic beetles have clear habitat preferences for fishless ponds (Binckley & Resetarits 2005).

Dytiscidae, Gyrinidae, Noteridae, Haliplidae, and Hydrophilidae are the dominant insects in temporary environments (Colinson et al. 1995). Cicindelinae and others Carabidae inhabit their banks.

Lakes and Ponds: A pond is a body of standing water, either natural or man-made, that is usually smaller than a lake. The technical distinction between a pond and a lake has not been universally standardized. Biologists have proposed formal definitions for a pond, in part to include 'bodies of water where light penetrates to the bottom of the water body, 'bodies of water shallow enough for rooted water plants to grow throughout,' and 'bodies of water which lack wave action on the shoreline' (Wikipedia, the free encyclopedia). Each of these definitions have met with resistance or disapproval, as the defining characteristics are each difficult to measure or verify. But, accordingly, some organizations and researchers have settled on technical definitions of pond and lake which rely on size alone (Biggs et al. 2005). Sarmiento (2001) defines a pond as "small expanse of water, where the coastal zone is very large in percentage and limnetic or deep regions are absent."

There is no universally recognized standard for the maximum size of a pond. The Ramsar Convention of Wetlands International sets the maximum size of a pond at 8 acres.

However, biologists, have not universally adopted any of these conventions. Researchers from the organization "British charity Pond Conservation" defined a pond as a body of water made by man or a natural water body measuring between 2-5 ha, and that holds water for four months of the year or more (Céréghino et al. 2008). Other European biologists have set other sizes. Sarmiento (2001) defines a lake as a continental body of water of large size which is located in depressions in the ground, and will eventually fill with sediment.

According to Helfrich and Parkhurst (2009) "A pond is a small (usually less than ten surface acres in size), shallow depression in the earth filled with water from rain or snowmelt runoff, springs, or groundwater. Ponds can lose water through seepage, evaporation, or transpiration (plant respiration)."

Equipment and collection methods

Aquatic Net: The most frequently used instrument for collecting aquatic beetles is the aquatic net.

One can learn a lot about the ecology of aquatic beetles, while sitting in front of a pool of water. They can be collected by hand, or using a tea strainer. However, many species of beetles are difficult to catch with this method, such as beetles living in a lagoon, at the bottom of a pond, in a river or stream, or in all water sources where the water is deep and/or flowing very quickly. In such cases, it is better to use an aquatic net.

Fig. 17 - Heavy Duty Aquatic Nets. *(Photo of BioQuip, www.bioquip.com)*

A good net consists of a strong and tough triangular (or rectangular) frame in a "D" or "V" shape. Each face of the frame will measure approximately 30 cm long. The net must be designed so that the bag hangs on the frame, but it does not bend completely over it. The bag for the net must be made of a thick tulle with many small holes and needs to have, a thick canvas material on the rim for protection of the tulle and a heavy aluminum or wood handle, about 1-1.5 m long (Fig. 17). These features allow the net to be strongly pushed through areas with vegetation or abrasive rocks.

How to use the aquatic net: When collecting with the aquatic net is a good idea to wear long boots (e.g., hip or chest waders). But such boots can be uncomfortable in hot weather.

For species that live at the bottom of rivers or streams, the best technique is to use the aquatic net vertically, but it will work better if the mud or sand of the bottom of the water course is removed with the feet before (in front of the net). In places with vegetation, the aquatic net must be passed several times back and forth in the same place, moving the water (with the net) and trying to get close to the shore as much as possible. Repeated passes through the same area eventually collect beetles that became dislodged but not captured in earlier passes with the net.

Beetles living on habitats inundated with clay, mud or stones, are more difficult to sample with aquatic nets, but the net can be pulled softly on the surface of the water, trying not to take too much mud or sand into the net. To use the aquatic net in these places, removing the mud with the hands or feet is very useful. In some places with abundant vegetation, it may be necessary to splash the water by jumping into the water and catching the splash with the net (most easily acomplished by two people).

Other habitats, such as tiny pools associated with spring seeps, may require passing a very small net (e.g., an aquarium net) through the substrate once it has been stepped on, to create a temporary depression through which to sample.

In streams it is a good idea to sweep the branches hanging above the water surface, and also among those that rise above the water, where you can collect several interesting beetle families, like Dryopidae, Elmidae and Lutrochidae.

In small waterfalls, other families can be captured (e.g. Amphizoidae, Gyrinidae, Pelobiidae, Dytiscidae and Hydrophilidae subfamilies: Hydrochinae and Spercheinae), in these places the aquatic net should be used with energy (moving it vigorously from the bottom up to the surface) especially in places with dense vegetation.

In many streams there are places where the water flows very quickly (especially in streams that have stones on the bottom; e.g., riffles) These area can be kick-sampled: the mouth of the net is placed facing upstream and the rocks that are just opposite of the aquatic net are disturbed with one's heel. The beetles that are hidden under the rocks (e.g. many Byrrhoidea) will be dislodged and swept into the net by the force of the water.

Sweeping the net through floating vegetation can result in the capture of many species of Chrysomelidae and Curculionidae and even species that feed on the flowers of floating plants which, although they aren´t aquatic beetles, are also interesting (e.g., Cerambycidae).

The beetles that are trapped in the aquatic net can be taken by forceps or with an aspirator; in the latter case you can use the following method:

The material collected with the net is emptied on a white tray with a 1 cm wire mesh, placed over the tray. The contents of the net (debris and plant material) are placed on the mesh where the beetles trying to escape will fall through the mesh into the tray where they are more easily collected. This makes sifting through the debris much easier.

Shovel with grid: In stream beds, there is often debris or very abrasive small stones between which can be inhabited by many species of beetles and their larvae. A shovel with grid can be used to collect them. This consists of a metal frame shovel, fitted with a plastic grid or solid wire with approximately 4 mm mesh. At the back there is a small cap. The handle is solid, 1 m long or a little more (Fig. 18). The shovel is dragged through the stream bed and its contents can be poured over a white tray from which the beetles can be extracted.

Fig. 18 - Shovel to collect aquatic beetles. (Drawing of Carlos Aguilar).

Screen barrier net: Many beetles can be found in shallow and fast-flowing streams, under rocks and other debris. When lifting the stones, the beetles can escape very quickly, being carried away by the current. To collect them, a screen barrier approximately 1 m^2 can be used, which can be manufactured with a flexible wire net or a strong plastic net, with very small mesh size (Fig. 19).

The screen barrier should have two bars (as handles), one on each side. The same technique as with the D-frame net is employed: the screen is placed a meter or so downstream of the stones or sticks to be disturbed, maintaining the grid firmly with both hands, somewhat diagonally with respect to the bed of the stream or river. It should be noted that the smaller elmid species are between 1-2 mm in length (with a cross-section < 1 mm) and can go right through nets if the mesh size is not small enough

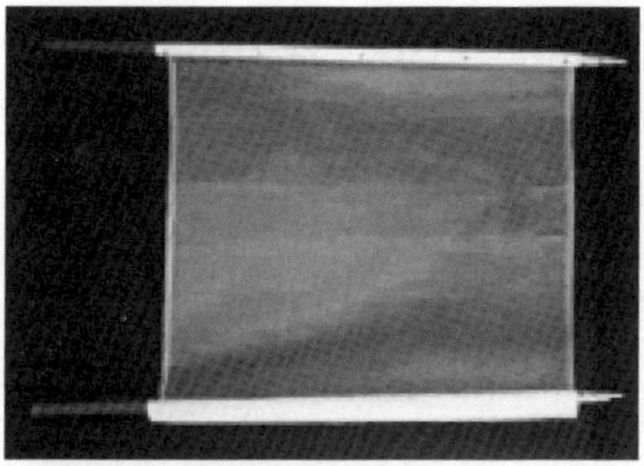

Fig. 19 - Screen barrier (Photo: BioQip, www.bioquip.com).

The stones and sticks under water are turned over and rubbed together to dislodge the insects by another person situated 1 m upstream (Fig. 20). The insects are washed downstream onto the grid where they can be captured. The contents are emptied on one of the devices displayed on these pages, or the screen can be laid flat in the sun on the bank where the slower-moving beetles (e.g., Elmidae) will begin to crawl as the screen dries - they can be tough to see when the sitting still on a wet screen, as they can become encrusted with minerals and appear like small stones until they begin to crawl.

Fig. 20 - Collecting aquatic beetles with a grid. *(Photo: Carlos Aguilar)*

Grids for draining: Metal grids with different mesh sizes are useful for collecting larvae and pupae of Donaciinae (Chrysomelidae) and aquatic weevils that occur on the roots or at the juncture of the roots and stems of aquatic plants. An aquatic plant is unearthed with its roots and is placed on the grid to drain the contents onto a tray or a plastic cloth. These grids are also useful for sifting moss and plant litter from dry or wet soil.

Fig. 21 - Washing aquatic vegetation *(Photo: Carlos Aguilar)*.

Strainer with tray and grid: Beetles caught in a net can be collected manually with forceps. However, a separation device can facilitate quicker sorting. A solid grid with mesh size between 5 to 10 mm is placed above a white bucket. Vegetable debris and other wastes obtained with the aquatic net are dumped onto the grid. Water is then poured over the debris constantly for a few minutes washing the beetles into the container (Fig. 21).

Once the bucket is nearly full of water, the beetles can be collected by pouring the contents of the bucket back through the net and can be removed with forceps or an aspirator. This method circumvents manually searching through mud and dead leaves for each specimen. Many species of beetles are found in large clumps of algae and other debris deposited on beaches and other shorelines by waves or by flooding of waterways. This separation device can be applied to these samples as well.

It should be noted that that some of the smallest Coleoptera (e.g., some Hydrophilidae), are less active, and will stay among the leaves and algae. To extract them, the debris or algae is placed on the grid and the device placed in the sun. These less active beetles and larvae will eventually migrate to the bottom and fall into the tray or bucket where they can be caught by hand, strainer or aspirator. If water accumulates in the tray, it can be poured over a fine sieve to recover the insects. Aquarium nets are useful for screening small Gyrinidae and other aquatic beetles.

Bucket with water: Adults belonging to some groups of Hydraenidae, small Hydrophilidae, Elmidae, Dryopidae, etc. are not good swimmers; therefore, when these substrates are placed in water and agitated, the beetles come to the surface. To facilitate their collection, the substrates can be shaken in a bucket with water. When floating, the beetles can easily be captured manually or with a suitable strainer.

Fig. 22 - Aquatic Light Trap. *(Photo: BioQuip, www.bioquip.com).*

Aquatic bottle trap and aquatic light trap: Some specialized shops offer aquatic traps as shown in figure 22. However, similar traps can be made at home from plastic 2 liter bottles. The bottles should have a wide enough mouth to accommodate large dytiscids, and 2 liter bottles have such a mouth. A good design is to remove the top from one bottle and insert it into the bottom of another (Fig. 23).

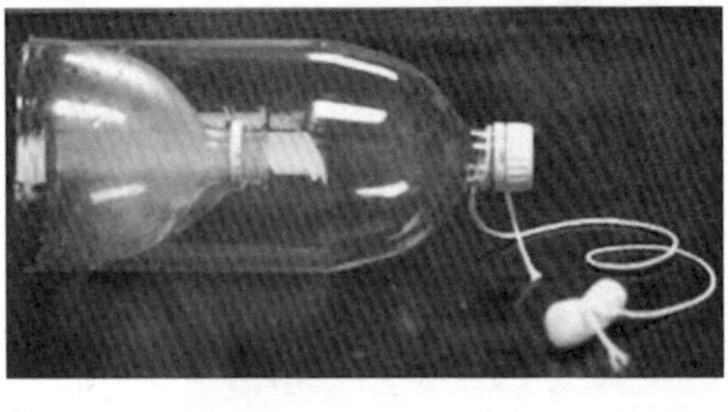

Fig. 23 - Aquatic bottle trap. *(Photo: Eric Chapman)*

The advantage of leaving the top on the bottle is that when the trap is collected, the cap can be removed and the contents poured through an aquarium net or small sieve. A 4" piece of rebar can be put in the trap to act as a weight, which will fall into the net as the contents are poured out (too big of a weight will plug the top). A piece of string tied to the neck with a piece of Styrofoam on the other end lets the user find it quickly, especially when placed in murky habitats where the traps are not so obvious from shore. It is also important to remove all air bubbles when placing the trap, otherwise, the beetles will not drown and may feed on other beetles in the trap, damaging the specimens.

Bottle traps can be very ineffective in streams, as any flow can cause sediments to build up in the trap, plugging the funnel and diminishing the catch. Furthermore, care must be taken not to place traps in pools where amphibians are developing, as the traps can be found completely full of drowned tadpoles. Thus, these traps can be used in swimming pools, artificial ponds, lakes and other lentic environments. They can collect large Dytiscidae (e.g., *Megadytes* spp, *Dytiscus* spp. etc.) as well as many Hydrophilidae (e.g., *Stethoxus* spp. etc.). These traps can be used without baits, or baited with liver or fish. A waterproof flashlight can also be placed within the bottle, functioning as an aquatic light trap.

Floating bottle trap: This is a modification of the trap of Aiken and Roughley (1985), which is similar to the trap described above, but this is placed on the surface of the water in pools, ponds and lakes. This trap can be made with plastic 2 liter bottles.

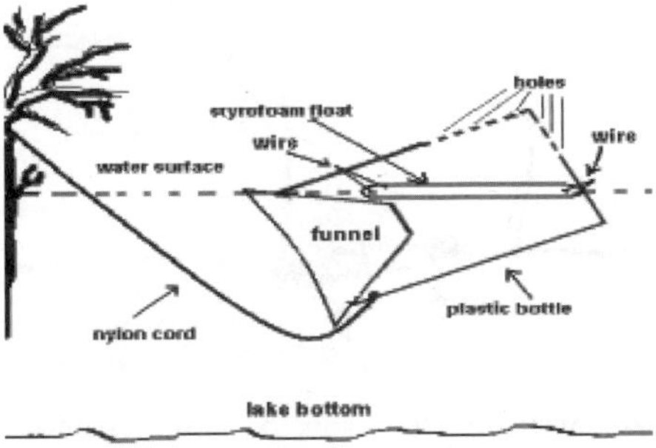

Fig. 24 - Floating bottle trap tied to a tree trunk *(Drawing of Carlos Aguilar)*

The posterior side has several small holes and a float with wires attached to the inside of the bottle. One cone with a 2 cm opening at the apex is inserted into the bottle. The base of the cone is attached over the mouth of the bottle and it is attached at the mouth with wires. The cord is attached at one end to a branch or tree or a brick (or any heavy object) near the water's edge (Fig. 24**)**. A piece of liver inside the trap can be used as bait. This type of trap is extremely successful in trapping large numbers of aquatic beetles.

Aquatic bucket trap

Andreas Herrmann (pers. comm. 2010) explains a very effective trap for collect aquatic beetles:

Figure 25a-b shows a small pond in a swamp with a white 5 or 6 liter plastic container on the bottom of the pond. A piece of pork liver (bait) and a stone is placed in the container, so that it is heavy and can again be plunged into the pond bottom. A round hole is made in the center of the lid with a knife (3 or 4 cm in diameter) for the beetles to enter the container where the liver can be fed upon. When finished feeding beetles have a hard time finding their way out of the bucket. Alternatively, a plastic pipe or bottle neck can be cut and pasted to the opening of the container, so the beetles will find it even more difficult to escape. The beetles will eventually drown because of lack of air, therefore it is important to remove all air bubbles when placing the trap. After the beetles die, their bodies will attract more beetles into the container. When the bin is emptied after a week, it should have many dead beetles. The diagram in figure 26 shows the composition of this trap.

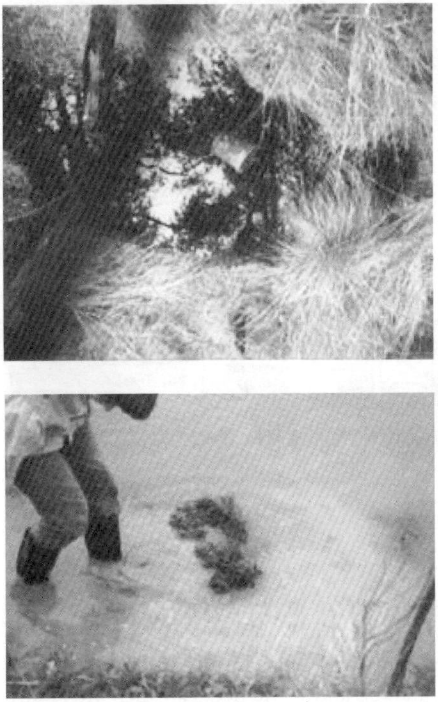

Fig. 25a-b - Aquatic beetles bucket trap submerged in a pool of water *(both photos of Andreas Herrmann)*.

Aquatic traps can be left in the habitat for a number of days. During colder months when the water temperature is near freezing, the traps can be serviced weekly, whereas during warmer weather, the traps should be serviced every 3 days or so to prevent the earliest arriving specimens from rotting inside the trap.

Bright foil traps: Small pools of water often form on surfaces such as paved roads, or on the concrete of ruinous buildings. Such pools often attract many species of aquatic beetles, flying as they move from one temporary aquatic habitat to another. Certain wavelengths of light are reflected from the surface of water bodies which beetles use to detect the presence of an aquatic habitat.

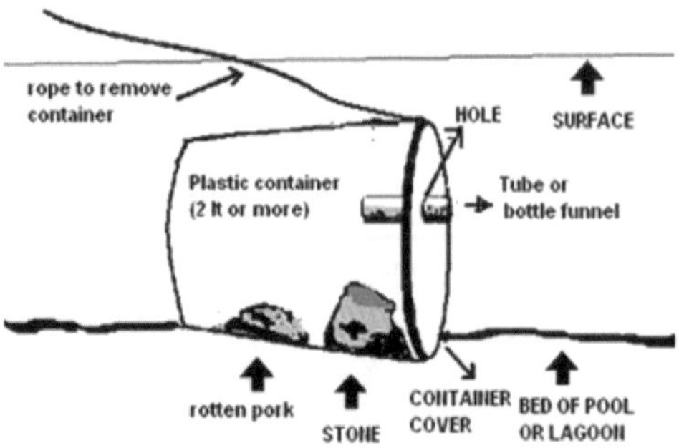

Fig. 26 - Structure and location of an aquatic bucket trap. *(Drawing of Carlos Aguilar).*

The reflection of such water accumulations can be imitated with aluminum foil or Mylar sheets (Fig. 27). This lamina works like fly paper or the "purple trap" (on the top surface transparent non-toxic glue is sprayed), but in this case, the sheet extends on the ground where the beetles will become trapped on the sticky foil. This foil mimics an aquatic environment and is commercially available (Eclipse miler 4 mil: www.mylarstoreonline.com). Ideal glue for such traps is Tangle-Trap Insect Trap Coating Spray (The Tanglefoot Company, Grand Rapids, Michigan, USA). Placing these sheets on the ground in different environments simulates pools of water to the eyes of aquatic beetles, to which they are attracted.

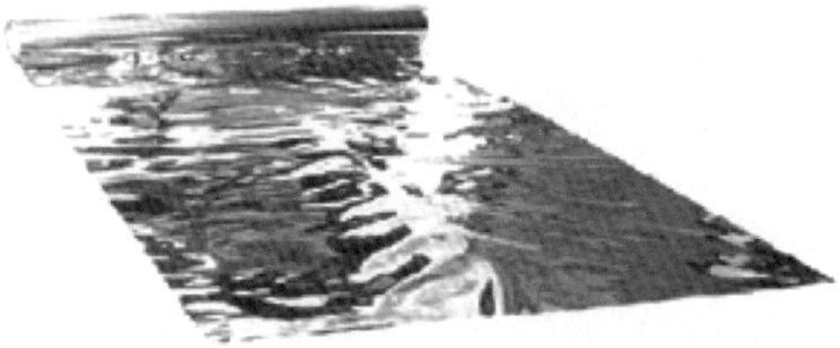

Fig. 27 - This lamina mimics a water source that attracts many aquatic beetles. *(Picture Jack Owen, Canada).*

Pot trap with light: This collecting tool is based on the fact that some species of beetles are phototropic (attracted to light). The pot has a lateral hole in which a transparent bag is inserted, which acts as a bridge (in the form of tunnel) between the pot and a transparent outer container (Fig. 28). A light source (60 watt) is located in front of the jar.

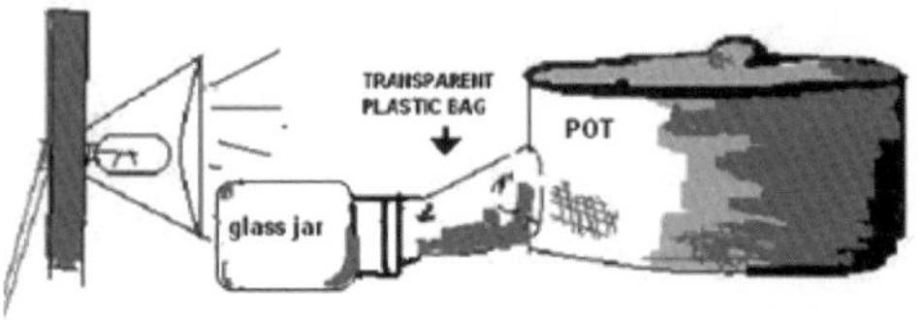

Fig. 28 - Pot trap with light *(Drawing of Carlos Aguilar).*

Plant material such as humus, algae, mosses, etc., is placed into the pot such that some parts at the bottom of the pot remain free of organic material. This is done so that insects can walk easily and go into the exterior bottle. Larvae and some beetles which are among the plant debris will be attracted to the light and will stay in the bag. Others will walk into the pot.

Collecting in stream ravines and banks: On the banks and ravines of a stream, river or any watercourse, it is usually possible to find several species of Coleoptera, mainly semi-aquatic. To collect them, water can be poured on the bank or at its foot. Beetles buried below the surface of the bank will try to escape and can be caught by hand or with a small strainer. Some of the earth can be placed into a bucket of water and the insects will float to the surface. These comments also apply to other land ravines, i.e., those dug in the earth by heavy machinery.

Some ravines, especially near a stream, quite often have some water seeps, or have small waterfalls produced by the rain.

If the ravines have some vegetation (mold, moss or algae), it can be gathered in bunches and placed into a bucket with water; the beetles will float and can be caught by hand or aquarium net. Ravines can also be sprayed with pyrethroids and the beetles that are hiding in the crevices of the cliff or plants can be evicted with a brush (Fig. 29).

Fig. 29 - Spraying a bank . *(Photo: Carlos Aguilar).*

Fig. 30. Brushing a rock wall after spraying. *(Photo: Carlos Aguilar J.*

Fogging: In the crevices of rocks, beetles can be so well hidden that unless they move, they are virtually invisible. A fogging technique can be used to force them out of their hiding places using a pyrethrum insecticide spray. The principle is similar to that used in the fogging of a tropical canopy. A small cloud of poisonous gas is applied to rocks with moss or other vegetation, forcing hidden beetles to come out, facilitating their collection. This technique can also be applied to mosses and other small plants in areas where water splashes or occasionally flows such as the edge of moist gullies, the edges of waterfalls in small streams and over exposed rocks of rivers.

The fogging method does not work very well on smooth, polished surfaces, since very few beetles inhabit such surfaces. Beetles usually inhabit small drops of mud, behind the mosses or liverworts small plants, even in crevices.

When spraying small habitats, an aerial net (with very fine mesh) or a beating sheet can be used. The device is located beneath the site to be fumigated, so insects stunned or killed by the poison will fall onto the net/sheet. Along with the net, a camel's hair or other soft bristled brush can be used to scour the sprayed habitat to remove the beetles between the cracks and crevices (Fig. 30). Larger habitats can also be fumigated, for example, the very slowly flowing and shallow backwater areas of streams. To do so, a net is placed between the rocks, so that it catches the drifting beetles. After spraying moss or other vegetation and waiting 5 to 10 minutes, the area is then sprinkled with water either by splashing with hands or using a spray bottle. Many small aquatic or semi-aquatic beetles and/or their larvae can be collected in this way. To collect beetles floating on the water surface with the fumigation method, it is best to use a natural biodegradable pyrethroid spray. Synthetic pyrethroids such as permethrin can also be used, but with caution, as they can be damaging to the environment.

Collection of beetles among cane or grasses that grow in moist soils: According to Andreas Herrmann (pers. comm. 2010) in Germany: beetles occur in bunches of canes or grass growing on a wet ground, mostly near a pond or a lake. In winter, beetles move to the bottom of such bunches looking for protection against the cold, ice and snow. This can result in a dense concentration of nearly frozen or very cold (but living) beetles inside the bunch. The bunch can be cut with a knife near the base of the bunch and shaken vigorously over a fabric bag into which the beetles will fall (together with some pieces of earth and small parts of grass). When the bag is emptied in the warmth of a lab, the beetles begin to move; they are not typically killed by the cold.

This method delivers only species which can be caught also in the summer by other methods, but in this way a large number of specimens can be collected and it is an easy method to collect beetles during the winter.

Fig. 31 - Pouring water on a beach (*Photo: Carlos Aguilar*).

Spilling water: Stewart Peck (2006) who studied beetles of the Galapagos Islands explains the following method of collection: "Netting of aquatic and shore-line species. Several species of aquatic beetles occur in coastal saline lagoons, and temporary and permanent freshwater pools in the uplands. They are easily caught with a small net. Several beetles live in burrows or litter at the margin of freshwater and saline pools (hydraenids and *Bledius* staphylinids). Splashing water on the shore causes these to come out of hiding to where they can be collected from the shore, or in the water."

Fig. 32 - Many species of beetles live on the beaches of streams and they can be collected by pouring water on the dry gravel beaches. *(Photo: Carlos Aguilar).*

By pouring water directly onto beaches, on the banks of streams, lakes, lagoons and ponds, often semi-aquatic and ground beetles are collected which emerge rapidly trying to escape the water. Sphaeriusidae, Carabidae, Staphylinidae, Limnichidae, Heteroceridae, Lepiceridae, etc. can be caught this way (Fig. 31). When water is poured on the gravel that lies on the banks of streams (Fig. 32), small species of Elateridae can sometimes be collected while they try to escape, climbing among the rocks and then flying away.

Fig. 33 - Removing the floating vegetation on the edge of a river. *(Photo: Jorge Barrett).*

Removing and collecting from floating vegetation: A good method is to push with both hands and remove vegetation, mud or sand at the bottom, on the edge of a stream, lake or swamp, so that the substrate is altered and beetles are dislodged (Fig. 33). Some beetles which are poor swimmers, become trapped floating on the surface and can be caught with a small mesh tea strainer or aquarium net. Also, Andreas Herrmann (pers. comm. 2010) explained that in Germany:

Fig. 34 - Accumulation of water in a bromeliad, which is a habitat available for aquatic beetles. *(Photo Jorge Barrett).*

"In the summer we used to trample with our feet the earth of river - or seaside into the water. We also transported some plants together with their roots to the water and let them dive. After a while the beetles appeared on the surface of the water and could be collected with a sieve. A variation of this method is to carry the water to ruderal places in the surrounding and splash it to the ground, then the beetles came out off the earth."

Water accumulated in plants: Some plants form cups that collect water where many species of beetles can be found (Fig. 34). Insects that prefer small aquatic habitats occur in these waters, called phytotelmata "plantheld waters" (Varga 1928). The best method to remove water that accumulates in plants, together with the insects that live there, is to remove the plant and pour the water that is contained in a bucket. Afterwards, the plant is washed several times with more water until it is clean and free of most insects.

In case that the plant should not be extracted, a noninvasive method is to collect beetles by inserting a small hose into the phytotelmata and extracting the water via suction and pouring it through a sieve or fine-mesh net to remove the beetles. More water can then be added and this procedure repeated to remove any beetles that evaded collection the first time.

Collecting in natural tree holes

Natural cavities in the trunks of trees can collect rainwater (Kitching 1971a). These water-filled cavities occur in hardwood forests throughout the world (Fish 1983; Kitching 2000) and such systems are most abundant in tropical forests. These holes are varied in form and size. They are classified into several types according to the morphology of the opening hole and the ratio of water volume to the surface area (Fincke 1992a). Some holes occur in the tops of the trees and may contain more than 50 liters of water (Fincke1992a; Yanoviak 1999a and 1999b), while most are much smaller, and many occur below 2 meters, and are more readily accessible. According to Yanoviak (2001) a variety of beetles use tree holes as breeding sites (Dytiscidae: *Copelatus*, *Laccophilus*), Ptilodactylidae larvae, and some Scirtidae (*Prionocyphon* and *Scirtes*), and many species breed exclusively in these habitats. Aquatic insects dominate the macro- fauna in tree hole cavities.

Leather et al. (2005) suggested that sampling at the bottom of a hole can be summarized as a five-step process:

1) First use a flashlight to see that insects are inside the hole and then measure the chemical constituents of water (if this is important to the study).

2) Debris and sediment is then removed (see below);

3) All water is removed from the hole (see below).

4) Clean water is added to the hole, removed and the washing is repeated several times.

5) The inner walls of the empty hole are then inspected with a flashlight.

Researchers have devised a variety of techniques to accomplish this task, but reasonably complete samples are obtained by removing the debris and water and sifting macroorganisms (Jenkins & Carpenter 1946; Walker & Merritt 1988; Barrera 1996). The most common collection method is the removal of the contents of the hole into a white font, pan or tray. In small holes a simple method is to use a fine tea strainer to remove the debris and insects, placing them in a white tray and removing the beetles with an aspirator. Larger holes can be emptied by using a flexible garden hose to siphon the water into a pail. Detritus is then removed by hand (or by using a wooden ruler or trowel to lift small packs of leaves), the tree hole rinsed and the water and set aside. Rather than completely refilling the hole with a large quantity of water, rinsing the walls with a few liters of clean water is an effective way to dislodge remaining insects (Leather et al. 2005).

Pan trap to mimic natural tree holes: This method utilizes a bucket or pan which is placed between the axils of the trunks of trees (Fig. 35). The inner walls of the bucket are lined with a dark thick cloth bag with some holes. Water is poured into the bucket until almost full. A bit of litter is added to the container along with a small piece of wood, bark or small stone (so that half of this is out of the water). A single fruit fall can result in a pulse of superabundant nutrients (Fincke et al. 1997). According to Leather et al. (2005) to keep nutrients above some minimum, it may be necessary to add small amounts of litter or a substitute nutrient (e.g. fish food or yeast). Furthermore, according to Leather et al. (2005) it is best to put this trap inside a wire cage to prevent oviposition by odonates, because they are the most common top predators in these holes (Fincke 1998) and can severely reduce beetle populations.

Fig. 35 – Caged bucket trap to mimic natural tree holes. *(Photo: Carlos Aguilar).*

The bucket with its contents is removed two to three weeks later. The contents are poured through a net or sieve to remove the beetles, and the contents emptied onto a white pan or tray to facilitate collection. A similar trap is made with a piece of Bamboo that is of 50 cm long and 15-20 cm diameter. The Bamboo is tied to a tree trunk and filled with water and a cage must be put around the top of the trap to prevent odonate oviposition also. The trap is serviced as above once every month.

Collection in Hygropetric habitats: Broadly, these correspond to rocks or wet rock walls which are constantly exposed to water action. It is quite difficult to collect beetles in this type of habitat. For example, looking at two cliffs, one may not have beetles, but the other despite being nearly equal in size, can lodge many species.

Rock walls which are always wet are ideal habitats. A rock face which is only wet when it rains will not lodge water beetles.

Some species of beetles can hide in cracks and moss of water falls whereas they cannot inhabit fast streams where they would be washed away. Therefore, they hang on either side of the waterfall, where water splashes, or in areas where there is a smooth film of water.

Some methods have been devised for collecting in these hygropetric environments, which are detailed below:

Collecting inmature stages

Saturated soils, moss and plant debris found along the margins of water, usually have a mix of terrestrial and aquatic beetles. Most larvae of aquatic beetles pupate in terrestrial sites near the water's edge. To find adults, the stones and logs should be lifted and carefully examined. The leaf litter should also be checked since, both adults and immature stages of these beetles can occur there (immatures can be reared at the laboratory (See below). For this task a knife (or shovel) can be used which serves to remove the topsoil, which may be loose.

Rearing aquatic larvae: Aquatic Coleoptera larvae are best kept in a split terrarium/aquarium because many aquatic beetles pupate out of water, and offering both habitats gives beetle larvae choices of pupation sites. It is best to put one specimen in each container to prevent the chance of cannibalism or predation. If one knows that the species at hand will pupate in water, a simple rearing chamber is a plastic beverage cup with a loose fitting lid. More elaborate trays with mesh sided cages and aerators to provide oxygen and current to the larvae can increase the success rate. Each species of aquatic beetles requires different rearing conditions to successfully complete its life cycle and finding these conditions can be tricky, as some species or groups have not been successfully reared (e.g., many Haliplidae). It may be necessary to feed the larvae, and other invertebrates collected in the same habitat make good candidates. With luck the larva will develop to maturity and the adult insect will emerge. For further reading on rearing aquatic beetles, see Brown (1973); White & Jennings (1973); Glaister (1985); Inoda & Kamimura (2004) Fernandez & Campos (2005); Inoda et al. (2009).To preserve the immature stages that are dead: very large larvae should be dipped in hot water for a few moments (this prevents disintegration of the internal tissues thereby maintaining the integrity of their bodies), and then preserved in alcohol; small larvae can be put directly into 75% alcohol; if there are many hundreds, it is preferable to keep them in alcohol at 85% or more, and should be labeled for later identification at the laboratory. A good rule of thumb is to have at least 5-10 times the amount of alcohol as there is total volume of specimen(s).

Flood debris

Collecting beetles in buildup of organic materials washed away by floods: Part of the waste (detritus) carried by stream waters form floating masses of organic detritus which accumulates in the banks, pools and corners of watercourses.

Fig. 36 - Collecting beetles in flood debris in Germany *(Photo courtesy of Andreas Herrmann).*

Flood debris (Fig. 36) can trap arthropods that accidentally end up in the water (Irurzun & Moreno 2003). An example is the delta formed by the mouths of two or more rivers where, after a heavy rain, the rising water carries many cubic meters of floating vegetation and other organic waste from upstream areas. Often there are plants which do not correspond to the local vegetation; and sometimes this vegetation does not even belong to the country where it is deposited. Reaching the delta, this debris is pushed toward the shore by the currents.

The same applies to streams and ditches. Vegetable waste is not only found where two streams converge, it also occurs against fallen trees in creeks, against large emergent rocks, and particularly in backwaters.

During the course of the year, the size and composition of flood debris varies greatly (Irurzun & Moreno 2003). Sometimes vertebrate carcasses are found, such as lizards, fishes, frogs, birds, rodents and even larger animals. Occasionally, even living animals can be found. This debris sometimes can gather hundreds or even thousands of beetles of all sizes, and from different families (e.g., Dytiscidae, Hydrophilidae, Gyrinidae, Carabidae, Scarabaeidae, Curculionidae, Dryopidae, Chrysomelidae, Buprestidae, Elateridae, Anthicidae, etc.; Schaefer 1949; Cooter 1991; Bordy 2000; Southwood 1987).

Importance of collection in flood debris: According to Irurzun and Moreno (2003), the number of insects trapped by water in each period is a reflection, although somewhat distorted, of the number of currently active insects. Therefore, continuous collection and data accumulation can provide interesting phenological and faunal information in a particular place, especially in spring and summer, which coincides with maximum vegetative growth and the flowering of most plants.

Also, according to Irurzun and Moreno (2003), insects carried by water are studied both for their large numbers and the large number of species; therefore, it is important to perform quantitative studies during short periods. The combination of short and long term studies show the beetles' rhythms of activity: annual, inter-annual, seasonal and daily.

Finally, in reference to these collection methods, Irurzun and Moreno (2003), explain:

"Since walking and jumping species, together with others of widespread appearance and more or less awkward flight, seem more likely to be victims of the currents, and the buoyancy of different species is distinct and makes them more or less likely to be dragged to the bottom, the limitations of this sampling method and study should be taken into account. However, you must properly assess the potential of this type of insect collection since, despite their limitations and drawbacks, properly made they can become an interesting tool for entomological studies."

How to collect in flood debris: Old debris that is in poor condition is not good for collecting beetles. Furthermore, decomposing vertebrates are often found in these places, so the task may not be so pleasant (Irurzun & Moreno 2003).

Also, it is important to follow the obvious hygiene standards. Improper handling of these materials can be unhealthy; therefore the use of rubber gloves, tweezers or sticks is advisable when attempting to collect beetles from such substrata.

The removal of debris for examination is not always simple and depends on the mass, density and the sizeof its components. Occasionally it may be necessary to use your hands to remove large portions of debris, but this can also be done with a large metal sieve, which is attached with tape or bolted to the end of a stick. This tool can be used on other occasions.

The best time to collect in debris is immediately after heavy rain, when rivers overflow and considerable amounts of debris are deposited on their margins (Cooter 1991).

In summer, floating debris can be very compact; therefore it must be separated it before removing it from the water. To do so, a gardener's rake can be used. This yields from a few grams to several kilograms of material and its examination requires time and patience. The remains removed from water can be spread on a beach or on the ground, and the beetles can be collected as they begin to move. To collect them, the debris is placed on a 1 cm wire mesh over a white tray (As mentioned in the **"How to use the aquatic net"** section). The beetles can be collected once they fall/drop into the tray. Placing the tray in the sun might aid this process as the beetles will try to move away from the sun's heat.

Smaller insects are less visible in the debris, which is usually very dirty due to foam and other items. So these remains should be washed in a bucket with grid, pouring water with a bottle. To remove the insects that have fallen into the bucket the water can be strained with a fine strainer. After the first inspection of debris left on the grid, these wastes are turned back on the same grid and the bonded portions are studied in more detail. Once the debris has been examined, it can be spread it in a thin layer over a clear plastic cloth until nearly dry. Then the debris can be put in a cloth bag and taken to the laboratory and placed in an emergence box (a dark box with a clear bottle attached such that the beetles can crawl into it). The beetles will leave the debris and will be attracted by light to the transparent bottle. The box should be large enough to allow debris to form a thin layer, making it easier for insects to get out and come to the light.

If the layer is thick, many insects will die and material in poor condition will accumulate at the bottom. This system is ineffective when there is a large amount of debris. It is important to note that it is not possible to collect all the beetles living among the debris. When sifting through large amounts of debris, it is best to take note of or collect only those of interest.

Variations in the collection process: Andreas Herrmann (pers. comm. 2010) presents a similar method of collection, but with interesting variations in the selection process:

"In spring, when the snow melts in the mountains, rivers increase their flow. A similar situation occurs at harvest time, when the North Sea and its storms push water against the mouths of rivers. Beetles trying to survive climb quickly to the plants and all that is floating around. These floating objects (mainly wood chips, twigs, leaves and trash) loaded with insects, concentrate on the water surface with less flow, and [get] stuck in the banks.

With a strainer or net, we remove the material from the water allowing it to drip for a few seconds and place it in a pillowcase or a large cloth bag. At home, the pillowcase or cloth bag with the material inside is put in a spin dryer at 3000 rpm. It is surprising that none of the insects die or suffer some damage with this procedure. Then, some of the bag contents is spread out in a thin layer at the bottom of a large plastic container. Of course, this material could also be spread out on a desk or a towel, but most of the beetles are extremely fast and the container's wall will prevent them from escaping.

With this method you can collect some beetles which are difficult to find with conventional methods, for example, those living within the earth".

"In spring, when the snow melts in the mountains, rivers increase their flow. A similar situation occurs at harvest time, when the North Sea and its storms push water against the mouths of rivers. Beetles trying to survive climb quickly to the plants and all that is floating around. These floating objects (mainly wood chips, twigs, leaves and trash) loaded with insects, concentrate on the water surface with less flow, and get stuck in the banks.

With a strainer or net, we remove the material from the water allowing it to drip for a few seconds and place it in a pillowcase or a large cloth bag. At home, the pillowcase or cloth bag with the material inside is put in a spin dryer at 3000 rpm. It is surprising that none of the insects die or suffer some damage with this procedure. Then, some of the bag contents is spread out in a thin layer at the bottom of a large plastic container. Of course, this material could also be spread out on a desk or a towel, but most of the beetles are extremely fast and the container's wall will prevent them from escaping.

With this method you can collect some beetles which are difficult to find with conventional methods, for example, those living within the earth".

Killing and conservation

Beetles collected in the flood debris are killed in the usual way, with ethyl acetate, which stops the decomposition of insects. At the laboratory, they are carefully washed with water or soapy water, while keeping them in a colander. Once the water is drained, they can be dried with a paper towell or toilet paper and placed in clean jars with ethyl acetate, 70-80% alcohol or some other preservative. It should be noted that to preserve insects for DNA work, >95 % non-denatured alcohol (ethanol or isopropyl) is preferable. Ethyl acetate destroys the DNA and denatured alcohols can have components that do the same (denaturing agents are variable, depending on the brand). The best preservation method for DNA work involves storing the specimens in >95 % ethanol at -20°C.

MANGROVES

The mangrove association is a type of ecosystem formed by salt-tolerant trees that occupy the intertidal zone near the mouths of freshwater streams, along the coasts of tropical latitudes. Mangrove areas include estuaries and coastal zones. They have an enormous biodiversity with high productivity, containing large amounts of birds, fishes, crustaceans, molluscs, insects, etc (Hogarth, 2007).

Mangroves vegetation is composed of several families and genera of dicotyledonous woody shrubs or trees that are virtually confined to the tropics. They often form dense forests that dominate intertidal muddy shores, frequently consisting of virtually monospecific patches or bands (Hogarth, 2007).

The word Mangrove has a Caribbean origin and refers to the features that mangroves have in common, so the word "mangrove" is not taxonomic, but rather refers to physiological attributes (Kricher, 2008).

Distribution in the world

Mangroves are distributed in the West African region and both coasts of the Americas and the Caribbean, including East Africa, Southern Asia and the Pacific, from Oceania to Australia, where the greatest diversity of mangroves is concentrated. In the Americas and the Caribbean there are mangroves in all coastal countries except Chile, Argentina and Uruguay (Carrera and Lugo, 1978).

Mangrove Characteristics

All species of mangroves are woody plants that tolerate high internal levels of salt, a consequence of frequent immersion in high salinity waters (Kricher, 2008). Their aerial roots, in the form of descending stilts, allow them to be anchored in unstable soil; their floating seeds allow a greater dispersion and special upward pointing root structures (pneumatophores) help in gas exchange in anaerobic soils.

The community of organisms living in, on or around a mangrove is heterogeneous and depends on the mangrove to fulfill their biological cycles. The animals living there may be

Fig. 37 - Mangrove of Brazil/Northeast/Bahia *(Picture courtesy of Gonçalo Lopes , Pt. 2010).*

permanent residents, or can occupy the mangrove temporarily, either seasonally or for only part of their life cycles. They may be unique to mangrove habitats, or they live in any humid tropical environment (Balasubramanyan, et al., no date).

Mangroves virtually lack any understory vegetation (Hogarth, 2007), but there is a variety of other plants: the climbers include vine and *Caesalpinia Derris* (Leguminosae), lianas, mangrove palms (*Erinaceus and Calamus),* orchids, ferns, and others whose roots are in the soil.. Some plants are truly epiphytes and grow exclusively on the mangrove trees, with no direct connection with the ground. They include lichens growing on the bark of trunks and branches, and vascular plants (Hogarth, 2007) such as several species of ferns, and orchids, and bromeliads in the New World.

Mangrove and beetles

The beetle fauna differs between different species of mangrove. The most obvious presence of beetles appears in the leaves of mangroves, which show the damage caused by those beetles at the edges of the leaves, holes in leaves or mines (Kathiresan, 1992). Among flowers and fruits may be found Curculionidae (including Scolytinae), Mordellidae, Cerambycidae, Buprestidae etc. Chrysomelidae and Lampyridae are among the most common families in the mangroves.

Fig. 38 - Aerial branche trap (cloth). Mangrove of Brazil. *(Picture courtesy of Gonçalo Lopes, Pt. 2010).*

Moreover, little is known of other insects in mangrove habitats. The insect fauna of the mangrove canopy is similar to that of terrestrial forests and many species are common to both (Kathiresan, 1992).

Insects live in different ecological niches within mangrove forests. They may be permanent residents or temporary visitors only. However, most of the beetles present in the mangroves are temporary visitors and are present in many other habitats (Balasubramanyan, et al., no date).

According to Kathiresan (1992) and Murphy (1990) Lepidoptera and Coleoptera are the most important phytophagous insects living in mangrove forests.

Many of these beetles are scavengers, like Scarabaeidae *(Heliocopris, Onthophagus).* Among Lampyridae, *Photuris* and *Photinus* can be found, which live under rocks in depressions of the soil and under leaf litter.

Although insects are the most abundant group, the tropical heat and lack of fresh water create a hostile environment for them. Only a few species are active during the day (Kathiresan, 1992). Most beetles avoid sunlight and prefer to feed at night or live entirely within the plants. As a result, traditional methods of collection traps, baits, or vegetation spraying with pesticides, have a low efficiency, giving poor results. However, by dissecting parts of the plants and collection of larvae we can realize that the insect fauna of the mangrove is much more diverse and ecologically important than one might suspect (Kathiresan, 1992). For example, in the living branches of several red mangrove species, specialized wood beetles are found (Scolitinae, Platypodinae, Bostrichidae, Anobiidae, etc.). Many beetle larval stages feed inside the hollow branches where there are also some of their predators (Histeridae, Trogossitidae, Chaetosomatidae, etc.), which use these sites for hunting, nesting and refuge from the relentless sun.

How to collect in the mangroves

To collect in mangroves one can use some of the collection methods utilized in forests. One of them is a trap consisting of a perforated bag (or cloth) full of leaves and branches, which is suspended in the air between the mangroves and is checked after several days (or more weeks) (fig 2).

Light traps are effective in the mangroves and the ideal time for night collection is just after a light rain on a dark night.

Although scavenger beetles are present in mangrove forests, pitfall traps are not very productive, since these habitats are flooded with the tides. However, pitfall traps can be used in places where the ground is drier. By hanging traps in trees, such as aerial Malaise trap, bottles baited with fruits, etc., can be capture several species of beetles.

Bundles of branch cuttings

Also in the mangroves (and in all types of forests) a good method to collect species of wood-feeding beetles (e.g. Cerambycidae, Brentidae, Curculionidae, etc.) Is to cut branches of trees or shrubs, and tie them into a bundle. This bundle of twigs is suspended from the lower parts of trees or shrubs. Some days later, the beetles attracted by these branches are collected by beating the bundle above a beating sheet. This method attracts generalist and specialist (host specific) wood-boring beetles, which may be active for only short periods of time, or throughout the rainy or dry season (Peck, 2006).The bundles can also be placed into emergence cages to see what adult beetles emerge from immature developing in the branches.

Colored Purple traps and pan traps also are useful for collecting beetles that live in mangroves.

An extra show ...

One of the most spectacular images of mangroves in Southeast Asia is the synchronization of firefly flashing (*Pteroptyx* spp). According to Berenguer (2007) the Selangor River, on the west of peninsular Malaysia, is surrounded by mangrove trees and at nightfall, most of these forests are alive with thousands of tiny flashes of light. The regularity of the flashes is amazing. All flashes appear simultaneously in a tree, and the lights in the surrounding trees fall almost in syncronism. The phenomenon seems to occur only in Southern Asia and Western Pacific, east of India, through Thailand, Malaysia and Indonesia to Philippines and Papua New Guinea. Different flashes are produced by different species of Lampyridae which have different wave lengths (colors) of light.

To collect Lampyiridae flashlights with Led bulbs can be used, which imitate the natural frequencies of the insects flashing lights (see "Where do they live...", Chapter IV). Additionally, the signal lights of adults and larvae of these beetles can be impressively abundant on mats of floating vegetation of still bodies of water in interior South America (Peck Stewart, pers, comm. 2010).

Warning!!

In the mangroves, mosquitoes often are incredibly numerous and their degree of abundance is exceptional (Mavnea, 1968). The populations are often dense. Thangam and Kathiersean (1993) reported that the Pichavaram[1] mangrove are 18 species of mosquitoes, these authors also reported the presence of filariasis, malaria and encephalitis. The high incidence of filariasis is because of the high population of the filaria vector (*Culex quinqufasciatus*). In the Solomon Islands, the mosquito (*Aedes pangiforceps*) was captured while biting a species of Periophthalmus[2] (Sloof and Marbs, 1965). Also, biting flies of the family Tabanidae[3], are abundant in the Mangroves.

1. **Pichavaram mangrove:** is the second largest mangrove forest in the world, located near Chidambaram in the Southern India.
2. **Tabanidae:** family of biting flies, commonly known as horse flies or deer flies. (Tábanos in Spanish). The name sand flies usually refer to bitting flies in the family Ceratopogonidae.
3. **Periophthalmus:** mudskippers are members of the subfamily Oxudercinae (tribe Periophthalmin), within the family Gobiidae (Gobies). They are completely amphibious fish; they can use their pectoral fins to walk on land.

FOREST AND WET FORESTS

By Carlos Aguilar J. and Sergio D. Ríos D.

There are many different types of tropical forests (Fig. 39). Among the most famous are the equatorial rainforests, found in a belt around the Equator line, into three main regions of the world: South America, Africa and Southeast Asia. These forests contain the greatest diversity of plants and animals, the largest biomass of living material per unit area and maintain a pattern of growth uninterrupted throughout the year (Moore 2008).

Fig. 39 - Mbaracayú forest (Dpto Canindeyú, Paraguay). *(Photo: Carlos Aguilar).*

The tops of forest trees are evergreen, have a great variety of flowers, fruits and billions of leaves and branches which

help to cover the needs of myriad of vertebrates, insects and microbes living there. This is the forest known as "the jungle". "Jungle" is a word that derives from the ancient Sanskrit language of India, and means "a desert". This word was adopted by the English language and distorted in their pronunciation and meaning, and came to be applied specifically to tropical forests. Particularly to the impenetrable growth of vegetation associated with forest edges (Moore 2008).

The richest forests in species are the Primary forests, which are virgin forests with a underbrush free of weeds, litter on the floor, stratificated tall trees, broad leaf, with a much larger number of species than individuals per species. They have no human interference, there are a full coverage of foliage and are equivalent to mature forests (Sarmiento 2001).

To give an arbitrary distinction between forests and the more open conditions of woodlands, ecologists have given the value of 40% canopy cover as the maximum for the woods. If the canopy covers more than 40% of the ground, then the vegetation can be defined as forest rather than woodland (Moore 2008). The woodland is an oligo-specific tree planting made for the purpose of commercial exploitation or decorative (Sarmiento 2001).

The forest structure

According to Brokaw and Lent (1999); Newton (2007) and Steeter et al. (no date) cited in: Sarmiento (2001), the structure of forest habitat is divided into several layers and is classified as:

Vertical stratification:

Belowground

Aboveground which includes:

the Creeping, the Herbaceous and the Understory

The logs (Overstory)

The treetops (Canopy).

Emergent layer, which includes trees that stand out above the general canopy.

Forb layer, of epiphytes and/or aroids.

Horizontal stratification refers to the existence of different vegetation types from the center to the boundaries of the ecosystem. These variations are due to climate and local factors of the substrate.

However, until now, the layers of primary forests have not been clearly defined (Kricher 2008 and Richard 1952).

Departing from the equator to the South or North, the weather gets drier. This means

that forests are becoming less dense in structure and evergreen trees are replaced by deciduous.

Finally, forests are replaced by savannas with thorny vegetation and dry seasons are longer, producing longer droughts. In some tropical locations, such as Northern India, heavy rains occur in a single period of the year (Monsoon) and as a result a distinctive monsoon forest appears. In high mountain tropics, forests are subjected to different environmental conditions as they acquire other aspect. The tropical mountain forests are sometimes called "elfin forest" because the small stature of the vegetation and lichen cover gives them a miniature look.

The following are some important places to take into account when collecting beetles in a tropical forest:

Clearings

Occasionally some very strong winds hit the top of the forest and the tallest trees, and the tallest trees pushed by the wind fall slumped to the ground, dragging in its way the smaller trees (Fig. 40). This fall removes several square meters inside the forest, producing what is known as "clearing". Sometimes the man produces these "clearings", cutting down a swath of jungle. Furthermore, it should be noted that the gaps are essential to the existence of the jungle, since the smaller plants and other shrubs, including the lowest level of the forest (underbrush), have the opportunity to grow receiving the sunlight. Thus, vegetation may change as time passes (Kricher 2008).

Fig. 40 - A clearing in the jungle, produced after the fall of a great tree, the tree was removed by farmers (Canindeyú, Paraguay). *(Photo: Carlos Aguilar).*

If the clearings did not exist, then the amount of plant species within the forest would be minimal, because only trees that could reach great heights in order to carry out photosynthesis would be able to prosper. The lower strata simply would perish, either by lack of light or completely devoured by hordes of herbivores and indirect herbivores (ants, for example). When collecting within the forest it is always more productive to look for clearings and inspect the wood with sunlight, in them; many species of beetles and sometimes in large numbers can be found in these habitats. Among the leaves and flowers of the shrubs in the underbrush, or in old clearings, especially when they have flowers or fruit, there are many Buprestidae, Cerambycidae, Scarabaeidae, Chrysomelidae, Nitidulidae, Curculionidae, Carabidae, Elateridae, Dascillidae, Cantharidae, Cleridae, Melyridae, some Meloidae, Oedemeridae, etc. An aerial net with a long handle is useful. The sunlit logs and trunks usually have Buprestidae and Cerambycidae. On the ground, among or under leaf litter, in fungi, roots and fallen sticks, live many species of Scarabaeidae, Pselaphidae, Staphylinidae, Carabidae, Ptiliidae, Tenebrionidae, Histeridae, Anthicidae and many other families. Inside the rotten trunks there are Passalidae, Scarabaeidae, Elateridae, Bostrichidae, Scolitynae, Platypodinae and larvae of several different families.

In these places ligth traps are very productive and better if the trees collapsed less than a month ago because the smell of dying wood attracts many beetles which lay their eggs in it and the lights will attract these beetles. The lights can also be used in the wide roads recently opened in the forests. Also, in the clearings it is very interesting to use Malaise traps, but it is better to choose places without people or domestic animals, who often destroy or remove the traps. If there are no gaps inside the forest, it is best to collect on the edge because the beetles within the forest are well hidden and those which become visible are generally common species. But if you check the humus and litter fallen in the forest, you will have many different species of Pselaphidae, Anthicidae, Carabidae, Tenebrionidae, Ptiliidae, Staphylinidae and many other families of small size.

The Canopy

Different groups of beetle species live in different levels of the forest habitat, ranging from the forest floor to the highest level of canopy (Fig. 41).

Fig. 41 - Canopy in the forest of Misiones, Argentina *(Photo: Jorge Barrett).*

An expert on Carabidae, Terry Erwin (1983), conducted a study in the Peruvian jungle, clearing the base of a tree and setting it free from shrubs and other plants. Then he spread some clothes beneath and proceeded to fumigate. He caught all invertebrates that fell into the sheet (most from the canopy) capturing more than 300 species of beetles on a single tree.

Erwin repeated the same study in another tree, where a similar number of insects were caught. By doing some calculations comparing the amount of Coleoptera found in each tree, with the number of species of known trees across America, he found that probably the number of Coleoptera still waiting to be discovered can be about several million species. At present about 400,000 species are known worldwide. Erwin's calculations may seem exaggerated, but you must bear in mind that only now the Canopy is beginning to be studied. Moreover, this is a very incipient study which has been done only in some tropical countries. It is necessary to do these studies in the rest of the planet. The Canopy is one of the last biological frontiers (along with the deep sea) which still remain largely unknown.

In addition, there are many regions in many countries, still unexplored. . It is important to collect in the litter and many other micro-habitats scattered throughout the world which remains very poorly known. Considering these facts, the number of Coleoptera that Terry Erwin believes are still to be discovered does not seem so absurd.

In recent decades there have been several studies to examine the stratification of insects in a forest (Sutton & Hudson 1980; Sutton 1983; Davis et al. 1997; Bruhl et al. 1998), and for this purpose different sampling techniques have been used. The methods to access the canopy were discussed by Lowman et al. (1993) (interesting work available online, see

bibliography) and Moffett (1993), while sampling in the canopy was deeply discussed by Basset et al. (1997) and Leather et al. (2005). The means to access the treetops have been an obstacle for those who are studying the canopy, and the access to it has developed in the tropics. One of the first attempts to climb the canopy was performed by means of ladders and pulley systems used by scientists in an expedition to Guyana in 1929 by the University of Oxford. The few studies conducted before 1970 used fixed systems such as multiple towers, platforms, walkways and stairs. In 1978, Donald Perry used a single cable where he tied a box, using it to climb the trees in mountain forests. It was the same technique used by speleologists to climb the vertical walls of the caves. In the next decade, new methods were tested which allowed access to the canopy, including the canopy raft and accompanying sledge. The canopy is also studied with satellites. In December 1999 four missions with canopy raft were completed and four canopy cranes were used regularly in the tropics. In an interesting article available on the Internet, Basset, Yves et al. (no date, see references) described the key scientific findings contributed by the International Canopy Crane Network (ICCN) since 1990.

Coleoptera are collected by taking samples of the habitat (soil, leaf litter, or moss) in the canopy and then removing the animals in the laboratory either by active extraction (e.g. Winkler extraction, Berlese funnel) or by washing (Behan-Pelletier et al. 1996; Leather et al. 2005).

In any case, to collect in the canopy it is advisable to fogging the foliage (discussed in Chapter III of this book) and the use of different aerial pitfall traps, bottles with bait, malaise traps, others intercept traps and also if possible, must be climb trees and beating the foliage. The epiphytes bromeliads must also be checked. Light traps can also be used very effectively in the tree tops to collect actively flying insects (Ozanne 2005).

Lowman and Moffett (1993) in their work (available online, see bibliography) discusses the ecology of the canopy.

The Ecotone

The ecotone is the transition zone between two ecological systems and has a unique set of characteristics, defined by time, space and the strength of interaction.

The ecotone generally hasmore species than nearby ecosystems, because apart from the species present on both sides, the typical species are generally restricted to the ecotone. This trend toward diversity and intensity increased at the juncture of communities is known as "edge effect" and is ex-

Fig. 42 – Hill in Central Department, Paraguay. *(Photo: Carlos Aguilar).*

plained by adjacent hybridization of the genomes involved (Sarmiento 2001). That is a reason to practice beating and sweeping in vegetation on the edge of the forest (ecotone) you can practice beating and sweeping; there are many species of Elateridae, Carabidae, Cantharidae, Coccinellidae, Meloidae, Nitidulidae, Mordellidae, Chrysomelidae, Curculionidae, Bruchidae, and other families of beetles. At the edge of a forest a ligth trap is very helpful, this must be located a few meters outside the forest. Also in these, the traps on the trees baited with fruit are successful, attracting a wide variety of Cerambycidae, Elateridae, Cetoniinae and other Scarabaeidae. These traps should be hung in trees at different heights, to have a representation of all the different stratifications.

Forest Fragments (Forest Patches)

It is hard to decide where to collect when you have to choose between multiple pieces of wood. With respect to these fragmented forests, Brazil is conducting a study to decide what happens to the wildlife that lives there. Kricher (2008) gives a good overview of the project, which is known as "Biological Dynamics Project in Forest Fragments" (BDFFP).

The intent of this project is to evaluate the ecological effects generated by changes in the size and degree of isolation in forest fragments. The project is designed to serve as a model for the theory of island biogeography (MacArthur and Wilson 1963 and 1967).

This theory describes a hypothetical mathematical relationship between the biodiversity of the island, its size (total area) and its distance to the source of colonization (land).

With this model it follows that at some point all the islands reach a balance with regard to the richness of species, where species immigration rate equals the rate of extinctions. In balance, the large islands near the mainland will have a larger richness of species than small islands and more distant from the mainland. The important thing is that this model predicts the difference in richness of species between different islands with different conditions.

Fig. 43 - Aerial view of fragmented forests, Southern Brazil.

The model is quantitative and can be tested. As fragmented forests are like islands, it was recognized that the model had implications for the structuring conservation of parks and reserves (Diamond 1976).

The project researchers worked together with owners of cattle, to whom the Brazilian government required to protect 50% of the forests on their land. They were asked to cut down their forests leaving some fragments of different sizes and different distances from a protected area of 1,000 hectares (source of colonization). The area of each fragment was thus: 5 fragments (1 ha. each); 2 fragments (10 ha. each); 2 fragments (100 ha. each) and 1 fragment (200 ha).

Studies of BDFFP are still being pursued, but some interesting results were already obtained. It has been shown that all species studied have varying degrees of sensitivity to the effects of area and distance. For example, underbrush birds decreased dramatically after 7 months compared to the number of species that existed before fragmentation, thus part of the biodiversity plummeted, an effect known as "Collapse of fauna" (Kricher 2008). Moreover, Euglossinae bees were reluctant to fly to fragments that were more than 80 meters from other forests.

The decline of wildlife did not occur in all cases, for example, populations of small mammals in fragments of 1 hectare increased, compared to 10 ha fragments and continuous forest. Some Scarabaeidae and other scavengers decreased in small fragments, but that was probably due to lack of dung and carrion which disappeared when larger animals migrate.

According to the author's experience collecting in fragmented forests of varying sizes, the species of Scarabaeidae collected in small islands, scavengers and coprophagous, are linked to bovine cattle living in nearby areas.

When a large forest is cut, the beetles are very abundant during the first two years. Thereafter, their populations drastically reduce, and some species disappear. Probably a 10 hectares forest may consistently contain several species of interesting beetles. As an example, in central Paraguay (Department Paraguarí) the author and others have collected many rare species of Coleoptera families, including the Subfamily Cicindelinae *(Oxycheila pochoni* Mandl, *Brasiella minarum* (Putzeys), etc), Cerambycidae *(Navosoma luctuosa* (Schoenherr), several *Callipogon,* and many rare Lamiinae), Scarabaeidae *(Strigidia, Pelidnota, Dichotomius, Dendropaemon, Eurysternus* and Canthonina, *Coprophanaeus, Megasoma,* Phileurini, etc.) and many other families. These forests today are highly fragmented; some of them have only a few hectares and are very disturbed. However, according to Laurance and Lovejoy (2002), in the course of time, opportunistic and generalist species are those that dominate the fragmented forests.

A study conducted in Europe on fragmented forests suggests that reducing the fragment size, adversely affects the capture of Carabidae and its richness in species.

Besides, urbanization affects also the populations of Carabidae. For example, when two forest fragments are separated by a road (common in South America), these beetles rarely move from one fragment to another. Therefore, the roads are dispersive barriers for forest Carabidae (Matti and Hendrik 2005).

When choosing a small forest fragment, it is important to have a great variety of trees, much undergrowth, and trees of considerable size (pers. obs.).

Kotze and Samways (1999) conducted a comparative study of Staphylinidae in several forest fragments in Colombia. They concluded that small forest fragments (3 hectares to 6 hectares), have few species of Staphylinidae, and medium-sized fragments (more than 6 hectares) have more species. To get that forest fragments have variety of species, it is important that they are not too far from each other. Still better if they are connected to each other, or connected somehow between themselves or to a larger one, even with some planting of introduced trees containing underbrush.

Apparently, the presence or absence of some beetles is determined by the intensity of light entering into the forest and also by humidity and other environmental factors (presence of humus, canopy, etc.), which change dramatically when clearing a forest or the larger trees are removed.

In some savannas and natural grasslands there is a kind of small forest called "island of forest", with a few hectares in size which have not nearly been modified or at least have not been logged. In fact its main degradation has been the effect of grazing by the cattle.

Many of these forests are flooded the majority of the year; however there may be found some interesting beetles.

To get forest fragments having a great have variety of species, it is important that they are not too far from each other, as insects and other animals use these fragments to move from one forest to another, as in the gallery forests, which follow along water courses such as the Scarabaeinae *Sulcophanaeus faunus* Fabricius (Edmonds 2000). Still better if they are connected to each other, or connected or to a much larger one, even if part of the vegetation is introduced (see below).

With regard to the collection, at the edges of these forest fragments it is advisable to use beating-sheet, pitfall traps and ligth traps. Bottle traps in trees are not very productive except in certain fragments connected to larger ones.

Forest Islands

The forest islands (Fig. 44), which are in the midst of some natural savannas or within a pasture or natural grassland, often have been formed from the remains of a forest that was cleared, or it is what remained of a great forest that led to a natural savanna. This is a process of advance and retreat of the savannas, which sometimes takes thousands of years. Many of these forest islands have soils that used to be flooded, sometimes with the accumulated water of the rains, others by a stream or creek flowing through that produces seasonal flooding. Some of these small forests are composed of a ground devoid of grass, but with many terrestrial bromeliads and epiphytes in trees. There are also trees with half-naked trunks and with many lianas.

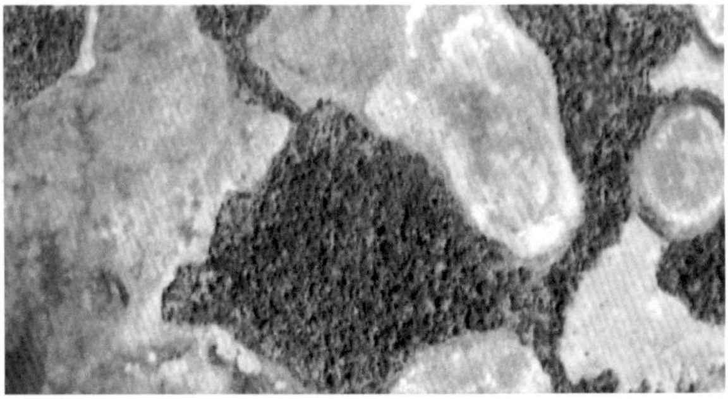

Fig. 44 - Forest island surrounded by floodplain

In this type of small size forest there are not many species of beetles, except for some Scarabaeidae, Cerambycidae, Bruchidae, Tenebrionidae, Carabidae, etc. which may be in dry soil, or partly humid. To collect many species of beetles examine the branches and trunks and even the tops of the trees, especially when they have mature flowers or fruits. It is important to note that, in the open areas surrounding the forest islands, in floodplains, there are several species of Scarabaeinae. But the traps must be buried when these places are not flooded.

Forests of introduced trees

In South America there are many pine and eucalyptus forests (not native, originally from Australia, see fig. 45). In general, these are not good places to collect beetles. However, if they have plenty of native underbrush they may contain interesting species of beetles. For example, in Chile was made a comparative study of the epigeal Coleoptera in a reserve, including the surrounding native forest fragments and surrounding pine plantations. The averages of beetles collected were slightly lower in pine plantations than in the native forest, but these differences were not significant (Grez et al. 2003). The study concluded that

Fig. 45 - Eucalyptus forest in Pando stream, Uruguay *(Photo: Jorge Barrett)*.

if the pine forest has rich native underbrush, then it might still harbor an interesting fauna of beetles.

Most beetle species such as Cerambycidae and Scarabaeidae (Rutelinae, Cetoniinae and Dynastinae) and other families live almost exclusively in native trees that are not present in introduced forests (though these have native underbrush). In any case, when collecting in these environments, light traps, sweep net, beating-sheet, pitfall traps, malaise traps, etc. can be used. Bottle traps in trees are not very productive for collecting Cetonidae, Scarabaeidae and Cerambycidae since introduced trees do not have many interesting species of beetles. However, traps can be used for other xilophagous beetles such as Scolytinae, Platypodinae, Bostrichidae or Anobiinae as many of these species attack the plantations.

Forests in gallery

The "forests in gallery" are a type of riparian forest, the term Riparian comes from the Latin word "Riparius", meaning "of or pertaining to the bank of a river" (Webster's Revised Unabridged Dictionary in 2010). According to Peck (2007) the riparian, refers to biotic communities that occupy the banks of rivers and streams. Riparian areas are the transition zone between the terrestrial environment and rivers.

Fig. 46 - Margins of a stream into a forest (Naranjo, Department Cordillera, Paraguay. *(Photo: Carlos Aguilar)*.

The "forests in gallery" are common in Uruguay, Paraguay and many other countries worldwide. Its structure is somewhat different from the "jungle" and includes forests that follow the banks of streams, where vegetation is set on a very wet soil and occasionaly flooded in certain months of the year, and sometimes permanently. They are characterized by deciduous tree species in climates with summer drought. In these environments branches near the ground are somewhat bare of leaves and often prickly. The riparian vegetation of rivers is relatively independent of the territory general climate. In areas of high rainfall, differences with the dominant vegetation in the vicinity may not be very pronounced. In contrast, in areas of dry climates, or even semi-arid Mediterranean, "riparian" vegetation appears as a tree or shrub leafy system, in linear distribution, or along the paths of water showing enough contrast to the surrounding landscape; for this reason this type of foresthas been called "riparian forest" (Aguilella and Russell 2004).

In the fallen trunks in gallery forests, beetles are not frequent, although sometimes, inside rotting logs, Passalidae, Lucanidae and other families can be found. In these habitats, it is better to use bottles of intercept traps in trees and logs. Although gallery forests do not have such a variety of beetle species as the most extensive jungles, these forests are an important biological corridor because they connect two or more forests, where the rivers with "riparian forests" pass through. This allows finding many rare and interesting species of beetles. For example, in the Amazon basin is widely distributed a set of coprophagous Scarabaeinae (*Sulcophanaeus* spp.) and among them, until a few years ago, the nocturnal species *Sulcophanaeus faunus* (Fabricius). This species was known only in the Amazon basin. The author of this book found it in a much more Southern forest (with pitfall traps baited with human feces) in the biological reserve of Mbaracayú (in Northern Paraguay). Probably, this beetle arrived in Paraguay from the ancient gallery forests that connected the Paraguay River with a large area of Mato Grosso, and consequently, to the Amazon basin (Edmonds 2000).

In the gallery forests, after flooding, the soil should be examined both day and night, since there can be found walking Cicindelinae, others Carabidae, Tenebrionidae, Anthicidae, etc.

In the meanders, or small pools of muddy water formed by the flooding of rivers, Hydrophilidae and other aquatic beetles can be found. Also at the edges of these habitats there are many Carabidae, Anthicidae, etc. Places with land snails (which are common in these habitats), must be found because there can locate some interesting Carabidae (*Brachygnathus* spp., Carabus spp.) that feed on these snails. On the tree branches Buprestidae are common (especially in sunnier sites). On the trunks and between them (and branches) Anthribidae, Cerambycidae, Tenebrionidae can be collected (including Alleculinae, Lagriinae, etc.). Under the bark there are Scolytinae, Bostrichidae, Anobiinae, Platypodinae and other xilophagous beetles. Nitidulidae abound in fruit fallen to the ground.

In this type of forest (as in the patches), only the pitfall traps should be used during the period in which there is no flooding. The use of malaise traps and intercept traps gives good results, like the light traps. In any case, if you intend to use pitfall traps, it is convenient baited them with fish or shrimp and/or feces, but they should be placed preferably on the higher mounds, avoiding the traps to be filled with water when it rains.

Occasionally "Gallery forests" are not very wide, either by deforestation or because these forests are located in rivers running through a savanna or natural prairie.

The sampling methods in the forest

In the big forests, most of the beetles live near the tops of trees, in flowers, fruits, leaves and branches. The main families living among the foliage, in both the canopy and the underbrush are: Scarabaeidae (Cetoniinae, Rutelinae, Trichiinae, Dynastinae, Melolonthinae, etc), Lucanidae (some small species have been found on the flowers) Glaphyridae, Cerambycidae, Buprestidae, Throscidae, Lampyridae, Lycidae, Coccinellidae, Chrysomelidae, Curculionidae, Carabidae, etc. So, to collect the widest possible range of beetles within the forest it is appropriate to review:

Collecting at night: Walk into the jungle to find the beetles during the day is not a very good system, however it is convenient to do it at night with a flashlight or lantern. In this time of the day, it can be found on the branches or trunks specially some Tenebrionidae, Anthribidae, Brentidae, Cerambycidae, Lucanidae, Erotylidae, etc., and between beetles walking in the soil can be found Carabidae, especially Calosoma spp, Galerita spp, Brachynus spp, Ceroglossus spp, and some Cicindelinae Subfamily species.. Many species hide during the day and just take a walk at night, so their collection is difficult.

To catch them, different methods can be used: beating the foliage, strain the leaves and humus and even the earth below the humus, check the fungi. To do so, must be used Berlese funnels or "Winkler/Moczarski" eclectors, and different types of traps, like pitfall traps or bottles in the trunks and other baited traps. Intercept traps and light traps are also very productive.

Collecting in flowers and fruits: Flowers are very attractive to many beetles, so, it is important review them constantly. Fruits, especially those which are rotting and/or those that are fermenting are an excellent food source for beetles. The fruits attract various species of beetles, such as Elateridae, Nitidulidae, Helotidae, Silvanidae, Cerambycidae, some Scarabaeinae and other families. The fruits in a very advanced stage of putrefaction attract only Nitidulidae, Silvanidae, some Histeridae and Staphylinidae. Fruits also can be used as bait in the traps. The accumulations of petals fallen on the floor may have varied species of Carabidae and other families.

Collecting in decomposed organic material: Odoriferous particles of dead bodies, feces and other decaying substances are more easily detected by the beetles which live at a certain height, than those sitting on the floor. Some Scarabaeidae perch on shrub leaves, either a few centimeters above the soil or even some meters height. The olfactory helps to detect feces and/or decomposed bodies and also to keep track of certain mammals (monkeys, koatis, tapirs, etc.). This is an energysaving method because they do not have to fly constantly in search of food. This behavior has been recorded in almost all the world's forests. *Onthophagus semiaureus, Onthophagus semicupreus, Sisyphus thoracicus* (Hanski 1989), to name a few, have been found in Malaysia, perched on the leaves. Although these species live primarily in the soil, Davis et al. (1997) revealed that some species of *Onthophagus* are arboreal. They have been found a few meters above the forest floor, even in the upper canopy. These beetles feed on primate dung living in the Canopy (Hanski 1989). In Paraguay, the author has collected many times *Canthon virens* Mannerheim, some C*anthidium* spp. and *Chalcocopris hesperus* Olivier, perched on leaves in the forest at low altitude.

According to Howden and Young (1981): *"Canthon angustatus* Harold *and another congeneric species (Canthon subhyalinus* Harold*), have evolved the ultimate behavior pattern for locating food in the threedimensional space of a tropical rain forest. Howler monkey faeces, placed at the top of a 23 m tower was approached first at a height of only 15 to 60 cm above the ground, and then, from the base of the tower, by vertical flight up to the top of the tower. These two species are able to form and roll primate faeces adhering to leaf surfaces. The beetle remains attached to the ball as it falls to the forest floor, managing to land on top of it."* So when pitfall traps are used, it is convenient to review the underbrush foliage in the vicinity of these traps.

Collecting in plants or shrubs that have leaves chewed: In these cases the culprits must be sought, which usually are Chrysomelidae and other beetle's larvae. Furthermore, some species of Othnidae can be found running on the foliage present in the wettest parts of the forest.

Beetles walking on the trunks: The following families are usually found walking on dead or alive trunks: Tenebrionidae, Alleculidae, Brentidae, Anthribidae, Curculionidae, Cerambycidae, etc. and under the bark can be found: Cucujidae, Lucanidae, Buprestidae, Chelonariidae, etc.

Collecting in stumps, logs and dead branches: Examining the dead wood is one of the most productive methods for obtaining a wide diversity of beetles. This is because many beetles and their larvae use wood as a source of food and feed on other insects living there, algae, mosses, vascular plants, fungi, slime molds and even colonies of bacteria. Should look bellow the trunk and/or on the bark and break it to be able to find: Carabidae, Lucanidae, Histeridae, Elateridae, Bostrichidae, Anobiinae, Cleridae, Sphindidae, Cucujdae, Colydiidae, Tenebrionidae, Cerambycidae, Anthribidae, Curculionidae and many others. Also must be hit the bark with a stick to make the smaller specimens fall on a white cloth. Different species of beetles can be found in different ages of the wood. The freshly cut wood will attract different species than those which are attracted when it starts to rot.

Other species are attracted to even older wood. Must be carefully examined each part of a trunk. After reviewing over the bark it should be removed carefully so that beetles fall to the ground. It is better to lift the bark sections, helped with a small knife or machete and place a white cloth or beating-sheet under the trunk, so it will be easier to locate some beetles when they fall. It is also useful to examine the holes and galleries made by other insects inside the trunks and branches; if there are small brown elliptic pellets, usually they are Cetoniidae droppings (or other Scarabaeoidea), so must be carefully review them. To find the species hidden among the dead roots and trunks, a small hatchet or knife can be used. In very old and rotten wood you can find many Tenebrionidae and Carabidae, and also some Cerambycidae, Lucanidae, Passalidae, etc. There are larvae or pupae of many rare species protected by a brown shell, like a small bird's egg, covered in their droppings.

Collecting in natural tree holes: Many trees have natural cavities and these are filled with rainwater, sometimes these are very large. Many beetles (and other animals) used these cavities as habitat. In the article "Collecting in aquatic and semiaquatic environments" (Chapter II), is explained in more detail the ecology and methods to collect the beetles that live within these cavities.

Collecting in fungi: (See "Beetles in fungi", Chapter III).

In fungi can be found Myxophaga, Endomichidae, Mycetophagidae, some Geotrupidae, Staphylinidae, Tenebrionidae, and many other small beetles.

Collecting in dung(1)**:** Search for Scarabaeidae (Scarabaeinae; Phanaeina, Canthonina, Dichotomina, also Onthopagini, Aphodiinae, etc.), Geotrupidae, Trogidae, Hybosoridae, Histeridae, Staphylinidae, some Hydrophilidae, etc.

Collecting in carrion: Here there are Silphidae and Scarabaeidae (Scarabaeinae; Phanaeina, Canthonina, Dichotomina, also Onthopagini etc. and Aphodiinae), Hybosoridae, Trogidae, Geotrupidae and some Hydrophilidae, Histeridae, Staphylinidae, Dermestidae, Cleridae, etc.

---

1. In mountainous forests, humidity is permanent throughout the year and necrophagous and coprophagous appear in the warmer months of spring and summer. In humid forests just as in some tropical savannas beetles are abundant all year. When there is a light wind, necrophagous and coprophagous use to be more active than in heavy wind or rain.

During prolonged droughts beetles are hardly found, except in dry regions. In these regions, beetles are much more abundant after a rainfall.

Beetles living in association with other animals: Some beetles are associated with the nests of ants, termites, birds, mammals and reptiles which use the dead logs (see: beetles associated with others animals, chapter III).

Using the pitfall traps: To collect Ground beetles pitfall traps must be used with which you can catch mainly Carabidae, Tenebrionidae, Staphylinidae, Histeridae and Scarabaeidae. Locate these traps on the floor, even better if it is near a tree, a fallen tree or other natural barrier.

Using traps in the trees: The traps in the trees are very good to collect Cetoniinae, Scarabaeidae, Cerambycidae, Lucanidae, etc. The bottles with holes and stuffed with sawdust, can be hung on trees to attract some Cerambycidae, Scolytinae, Platipodinae, Anthribidae, Brentidae, etc. (See chapter III).

Using sticky traps: With sticky traps on logs, stumps, branches, soil, etc. many species of beetles that use to walk on them can be caught.

Using traps for underground beetles: The underground traps like tubes with holes, filled with sawdust, are useful to collectmany small subterranean species (Staphylinidae, Pselaphidae, Carabidae, various larvae, etc.). Another method consists of using a wire grid (5 mm mesh), for draining the rotted leaves and leaf litter on a white cloth. Collect the beetles walking on the cloth. The leaves and twigs lying on the grid are discarded, but the remains of humus, which are smaller and lying on the white cloth, are stored in cloth bags (with all data collection) and later will be separated with a Berlese funnel or with a Winkler/Morczarski eclector.

Beating method: A very effective method to collect many species is the use of an beating sheet. The roads within the forests are advisable, especially where they have been recently opened.

Bottle traps in trees, light traps, intercept traps, pitfall traps, etc. are also used.

Fogging: Find a tree (or fallen log) that is off the ground by one end, or leaning against another tree (see "Fogging trees", Chapter III).

Below ... Must lift up anything that is on the ground: stems, bark, stones, piles of branches or fallen leaves, etc. Many species are hidden there, especially under or near logs where the highest soil moisture is concentrated, such as: Carabidae, Erotylidae, Endomichidae, Tenebrionidae, Staphylinidae, Corylophidae and many other families.

Rearing larvae to get adults: In the rainforest some pieces of wood (not too old) can be picked up, to be ableto raise the larvae of many borer species (Cerambycidae, Tenebrionidae, Buprestidae, Bostrichidae, Mordellidae, etc.) and their predators (Cleridae, Elateridae, etc.).

Warning: In the "Riparian forests" and "patches" in general, mosquitoes and other biting insects abound, and almost all of them have a strong bite, so you should enter into these forests with a good dose of repellent. When visiting a forest where it is suspected that many mosquitoes are present (which are almost always there haunting their victims), it is best to dress in light clothing covering most of the body; shirts and pants in "jean" are fresh and useful for this purpose. It is not advisable at all going into those habitats in short pants or short sleeve shirts.

COLLECTION OF UNDERGROUND BEETLES

Several rare species of underground (soil dwelling) Coleoptera can be collected by burying a perforated trap bottle. Examples are in families such as Staphylinidae, Pselaphidae, Ptiliidae, Leiodidae (Cholevinae) and some blind beetles. The trap (Fig. 47) consists of a plastic bottle with one liter capacity, with several perforations of different diameters.

Inside the bottle can be placed damp non-acidic sawdust (but not too wet), or the sawdust can be mixed with small pieces of meat, or with some kind of black molasses or sugar, or some humus. One can experiment with pieces of Bananas (or other fruit), but avoid citrus fruits. The bottle is buried beneath the soil surface, preferably under the litter of a forest. The buried bottles should be covered with the surrounding soil and humus. The organic material, inside of the bottles, will have a strong odor as it decomposes and will attract many species of beetles and other invertebrates (worms, mites, etc.) will also approach. It is best to prepare and bury several of these bottles.

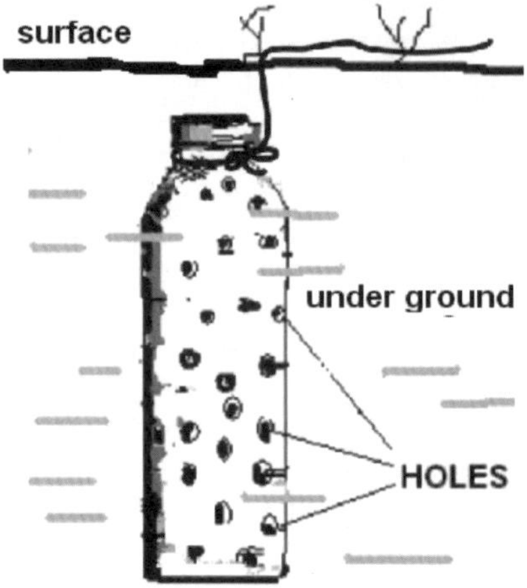

Fig. 47 - Perforated bottle trap, baited and buried.

After burial of the bottles, it is important to mark a tree nearby, or leave a permanent and visible sign to remember where the bottle was buried. It is better to tie it to a strong and non-degradable nylon or plastic string, leaving a piece about 30 cm jutting out from the ground, so you just pull the rope to lift up the trap. After 1 month or more, remove the traps with the material contained in it and extract the insects mixed with the garbage.

This bottle can also be placed lying on the surface of the ground to attract beetles that live on the soil surface (Fig. 48), but in this case is wise to dig a small ditch where the bottle will be half buried.

After a month or more you dig up the trap, and the beetles that are inside (which came through the holes), are relatively visible among the sawdust; they can then be collected and labeled.

Fig. 48 - Bottle with sawdust lying on the ground.

A variant of this trap is similar to López and Oromí (2010): in this case using two plastic bottles (2 liters), cut the two bottles on the top and these are attached to each with a wire. At the top and the bottom of trap is left an area without holes. The device loaded with propylene glycol and baited glass container with meet, fruits, etc. (Fig. 49 and 50). The whole system is covered with soil and litter to hide and protect the trap. Is installed inside the hole such that the perforations of the trap stay at the ground level, and the top remains a few centimeters below the ground surface. A stone or a conspicuous marker is used to indicate the traps location. At the end of a sampling interval, the trap is removed from the ground taking care to avoid soil falling inside and its contents poured for transport to the laboratory. This trap can also be used to collect the beetles that live among the cavities and crevices of the lava deposits (see Article "Mountains", pag. 72).

Many species of Scarabaeidae, some of them of a very large size (*Strategus*, *Heterogomphus*, *Enema* and other genera) live among the roots of trees, shrubs or plants. On certain days of spring, summer or early fall emerge from the soil as adults by hundreds.

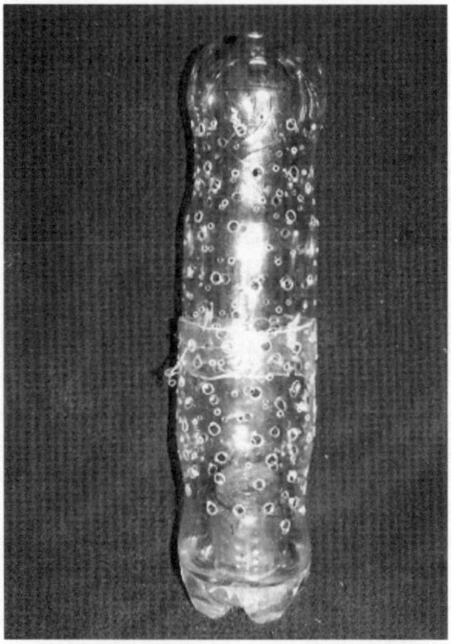

Fig. 49 - Baited perforated bottle. *(Photo: Carlos Aguilar)*

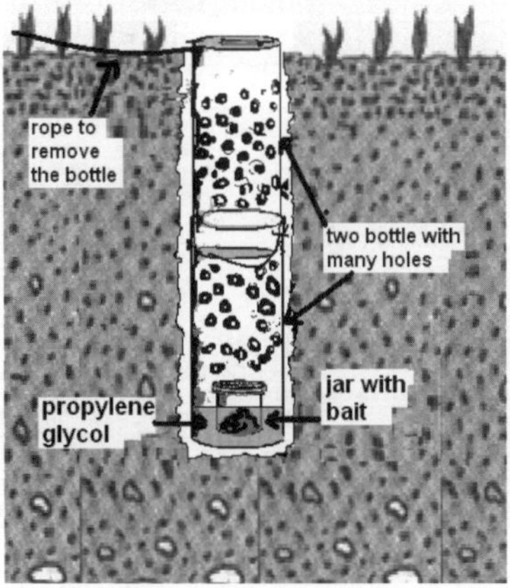

Fig. - 50 Buried perforated bottle with bait.

To collect in roots of various plants, first they must be examined for holes in the ground. Dig into these holes with a small machete or shovel since beetles are often found there. Before digging, a twig should be inserted to the bottom of the hole. This will serve as a guide to dig into the gallery and not to destroy the insects inside .with the shovel or other tool.

Sometimes there are two or more species together in one burrow, as some found for *Athyreus* and *Neoathyreus* genera etc of the Geotrupinae.of. These beetles dig small burrows under the ground and can easily be collected using this method. Leiodidae (Leiodinae) are sometimes found in the burrows of Geotrupinae, probably feeding on the fungi the Geotrupinae have buried (Peck Stewart, pers. comm. 2010).

Sometimes, the litter around the roots must be removed and, with a 50 x 50 cm sieve or sifter, sift the soil or humus onto a cloth or white plastic for insect detection. They must be captured by hand, with soft blunt forceps or an aspirator. Then carefully remove the roots of the complete plant, without shaking it, and put it above the strainer or sifter with a white cloth spread on the ground. Shake the roots on the sieve-sifter and spread the residue on the white cloth. In all these cases the residue can be carried to the laboratory and separated with a Berlese funnel.

In grasslands various species of Coleoptera can be found among the grass roots. Use here the same method as explained above.

Inside a forest, when you find a place with litter and humus, remove 1 m^2 of litter and a bit of humus. Put this material in labeled bags, take them to the lab and put them into a Berlese funnel. After removing the litter from the delimited area, dig into the root mat and mineral soil to a depth of 15 to 20 cm. Depositing the removed soil into a large bucket of water and stir this carefully. Wait a minute or so for the insects to float to the top of the water, along with other floating debris. Then with a fine hand strainer (a small tea strainer can be used), remove what is floating on the water and discard the sediments at the bottom of the bucket. Drain the floating sample on a screen. Make small mounds of a half cup or so. Wait until the water drains, and then introduce the drained remains into a Winkler/Moczarsky eclector. This can be done at the camp. It is better to let the sample drain more using cloth bags, but do not put them into the sun. Reaching the laboratory proceed without delay to put the samples into Winkler/Moczarsky eclector or a Berlese funnel.

Another method consists of putting the soil obtained under the litter into a funnel-shaped plastic bag. Water is added into the bag. Then take what floats and place it in another similar bag, eliminating what settled to the bottom of the first bag. Put water in this new bag and discard again what is left at the bottom, taking care to save what is floating. This process can be repeated several times using the same bags. Finally, the floating material from these samples is deposited in a Winkler/Moczarsky eclector or Berlese funnel, after they have been drained on a screen.

OBS: By stirring the water in the bucket, the heavier materials (sand, rocks, some branches and dirt) settle at the bottom, while insects float on the water surface. Salt can be added to the water to make it denser and the insects may float more easily. Keep in mind that not many specimens are obtained with this method, but certainly some species of Carabidae, Staphylinidae and other very rare specimens, sometimes blind, can be obtained.

COLLECTING IN THE MOUNTAINS

Below are discussed several methods of collection of beetles, but before, will be explained some interesting characteristics of mountain regions, in order to know something about the biology of the beetles living in the mountains.

Alpine tundra

The Alpine tundra is present in the high mountains of the world, and among its characteristics is the presence of dwarf trees, shrubs, grasslands and many other plants that are similar to those found in the Artic tundra (Fig. 51). The Alpine Tundra lacks the permafrost of the tipical Arctic tundra, so the water drainage is more developed (Quinn, 2008), and since the seasons and daylight hours are longer, the plants growth period lasts for several months. Nighttime temperatures are almost always below 0 °C, but during the day it gets warmer.

Fig.51 - Alpine tundra from North of Europe. *(Picture courtesy of Enrique Martinez, Uruguay).*

The Alpine Tundra is higher in the mountains close to the Equator line. In North America, the largest expanses of alpine tundra occur in the Northern Rocky Mountains, and in South America in the Andes. In Asia, the Tundra is manifested in the Himalayas, Pamir, the mountains of interior Siberia and the Caucasus. Europe: the Alps and Pyrenees. Oceania: the mountains of New Zealand.

The Tropical alpine tundra, as it is called when it is near the Equator line, also occurs in Africa as the Afro-alpine Belt, which expands from Ethiopia and Uganda to Kenya, Tanzania and South Africa. In Asia there is a smaller portion of this tropical alpine tundra, in parts of New Guinea and Indonesia.

In the Equatorial East African Mountains the insects are not well inventoried, but most of the more conspicuous species are known (Quinn 2008). According to Quinn (2008) in these mountains there are many specialized beetles restricted to the environment of mole rat burrows. Some species are apparently limited to one mountain or habitat only, but there has been little study of afro-alpine invertebrates to confirm this. Also, according to Quinn (2008) the Ethiopian plateau (moorland) is a land area of over 6,500 feet (2,000 m) high, with mountain ranges and peaks over 9,850 feet (3,000 m) high. In contrast to the mountains of East Africa, which are surrounded by wetter climates, the Ethiopian highlands are in drier areas. The mountain climates are largely unknown, but weather changes according to the height (Quinn 2008).

The lower part of the alpine zone (in Africa) is a mountainous forest, and there are also several lakes and grasslands disturbed by grazing. It is common to collect brachypterous or apterous specimens, a condition that adversely affects these insects dispersion (Brandmayr 1991).

The percentage of species having small wings, or a complete apterous condition, increases with elevation. For example in the Himalayas about 60% of beetles living above 13,200 ft (4,000 m) have small wings, and several are unable to fly at all, compared with lowland species with normal wings (Quinn 2008). For this reason, in the mountains is helpful to use pitfall traps to collect various species which are walking on the surface of the ground.

Usually, beetles are small in the mountains and hide in the crevices of rocks. To collect them, the area must be fumigated with pyrethroids (the use of turpentine is a very effective method) and then after those holes should be examined and use the brush to get those beetles hiding there, especially when there are lichens, mold or moss on the rocks. Beetles can be found in the soil or under rocks in very sunny and not so windy days.

The Andes

In central Andes, the "Altiplano" (or "Puna") is the largest area of tropical alpine habitat in the world. It extends from the central and Southern Peru and Bolivia to Northern Chile and Argentina, corresponding to a vast plain located at an average altitude of 3,600 meters above sea level. In East Africa this kind of habitat is called Afro-alpine Zone, and in Indonesia and New Guinea is called Tropical Alpine[1].

The elevated region of the Andes, in central Argentina and parts of Chile, at 27° - 39° S, can be considered as alpine tundra or grasslands, as it contains elements of both. Elevations reach from 4,250 m in the North to about 5,500 (1,700 m) in the South.

The morphology of the Puna consists of major volcanic reliefs, with detrital surfaces with salt pan, salt marshes, beaches, and lakes or dry mud (Alonso 1998). Its climate is arid

1. The highlands of Hawaiian volcanoes are higher than the trees growth line. The lower limit of that zone is difficult to define because it can be a no-reçognizable tree line. The alpine zone is above the forest or cloudy forest and the tree line may consist of groups of shrubs or stunted trees covered with epiphytes and surrounded by grasslands. In the dry hillsides where there is no forest, shrubs and succulent plants become the typical vegetation in the Alps (Quinn 2008).

subtropical with continental basis, reaching harsh desert conditions (Locascio et al. 2005).

Temperatures can be situated below 0° C, up to +20° C, with intense sunlight and strong winds. In January (summer) is 55° F (13° C). These temperatures are warmer than in most alpine regions, but these elevations are covered with snow for several months in winter. Vegetation consists mainly of grasses. Under the soil of these grasses the use of bottle traps will be very useful for collecting beetles underground, sometimes also can be collected some Scarabaeinae with pitfall traps. The high mountain wetlands (3,000 and 5,000 m) are interesting environments like islands and the variability of interacting factors determining unique characteristics, with a rare biota (Locascio et al. 2005). In these places is interesting to look for burrows of mammals also, especially those of rodents that are often common in these mountain habitats and trapping them to collect the beetles that live as hosts of these mammals.

According to Quinn (2008), a line of tall shrubs becomes in small shrubs, clumps of grasses and cushion plants. In the subnival zone the small herbaceous and rosettes dominate, but the cushions are common too. Some shrubs are endemic: *Pinifolia, Adesmia, Chuquiraga ruscifolia* D.Don and *Chuquiraga echegarayi* Hieron. Endemic cushions include *Adesmia subterranea* Clos, *Laretia acaulis* (Cav.) Gillies & Hook, *Azorella cryptantha* (Clos) and *Oxychloe bisexualis* Kuntze (Quinn 2008). Many of these plants have beetles adapted to live in them. During the day mainly Chrysomelidae and Curculionidae can be found.

Epipedonota spp. (Tenebrionidae) and other genera may be collected walking on the ground. The corrugated cardboard traps are also very effective in these places as elsewhere where there are shrubs or trees, because these traps could serve as a refuge for the beetles that live in these environments. To collect in these habitats beating and sweeping vegetation is a good method. Also, here is productive to review the roots of plants.

In the Andes, from medium to high altitudes, from Venezuela to Bolivia and in the mountains of Panama and Costa Rica, lives several species of Cicindelinae (eg. *Pseudoxycheila* spp.), for example, in Colombia there are two species of this genus, at high elevations on the slopes of bare clay soils (Cassola 1997).

In the Puna and Altiplano more species of beetles are captured with Malaise traps than with pitfall traps. But, as reported by Sackmann (2006), both methods are highly complementary because the species obtained with these traps are quite different. This demonstrates the advantage of using different sampling methods to evaluate a community in a complete way.

Paramo and moorland or fells

The Paramo (Fig. 52) is a mountain ecosystem, formed by large areas covered with grass and microphilous shrubs. It lies between the upper forest (3400-3600 m) and the zone of perpetual snow (Moret 2006). Shrubs and grasses dominate the Paramo vegetation. The relative height and composition of the heath community vary with elevation. There are very strong winds and frequent fog or drizzle.

The ground and litter on the floor of the paramo is always wet due to inclement weather. Most beetles are found in protected sites among the leaf litter, beneath the ground at the foot of plants, under rocks and among the clumps of grass.

In the Paramo you can found many Carabidae of the tribes Migadopini, Bembidiini, Trechini, Platynini, Pterostichini, Dercylini, Harpalini and Lebiini. Most of the Coleoptera

Fig. 52 – Paramo in Tolima, Colombia *(Picture courtesy of Enrique Martinez).*

living in the Paramo are endemic species whose distribution is limited to mountainous areas (Moret Pierre 2006). For example, representatives of *Bradycellus* genus live there, of which only 12 species are known in the South American mountains (Moret 2001), however, undoubtedly more species will be discovered with more collecting in different parts of the Andes, as this genus is distributed in many mountains in other parts of the world (Moret 2001).

Platynini are a well represented tribe in the mountains around the world. Most species have a uniform external appearance but some of them are slightly modified, like some *Platynus* which are flat and are found living under bark or in bromeliads (Martinez and Ball 2003).

Another genus is *Dyscolus*, which has adaptations to survive in extreme cold and dry weather (Somme et al. 1996).

As in the Tundra, in the Paramo, the Moorlands and others similar places, the climatic conditions are very harsh: it is very cold, with strong winds or raining. That is why it is hard to find suitable conditions in order to collect insects in the paramo, since many beetles only fly when weather conditions are good. Thus, the sweeping and beating in the bushes is essential for obtaining many species of beetles. Moreover, light traps are not very effective in such environments, but it could be very interesting to use lights if weather conditions are good. The use of corrugated cardboard traps is a good alternative; these traps can be placed both on the ground and in the bushes.

Another effective sampling method to collect beetles in the paramo, is sifting the soil humus on a white cloth and use a Berlese funnel or Winkler/Moczarski eclector. Remember that in mountain it is appropriate to lift fallen sticks and the stones to look for beetles hiding there.

Alpine mountains from Europe

According to Kerstin (2009) the Alpine region in Europe (Fig. 53) comprise the Alps, the Apennines (that run down the spine of Italy), the Pyrenees (on the border between Spain and France), the Scandes (which straddle Sweden, Finland and Norway), the Carpathians (that extend in an arch from Slovakia to Romania), the Balkan mountains, and the Rhodopes.

Also, according to Kerstin (2009) independently of their geographical location, all Alpine mountains are characterized by a relatively cold and harsh climate, high altitude and often complex and varied topography. Forests and semi-natural grasslands cover the lower slopes but, as the altitude increases and the temperature decreases, trees begin to dwindle giving way to alpine pasture, rocky hills and bushes; in the highest parts of the mountains, between the rocks and snow, the vegetation is reduced to only a handful of plants well adapted to extreme conditions.

Fig. 53 - Alpine mountain from Europe *(Picture courtesy of Enrique Martinez).*

The Alpine region in Europe is also home to an immense diversity of beetles (Kerstin 2009) and many wood-eating beetles (Scolytinae, Platypodinae, Cerambycidae, Bostrichidae, Lyctinae, Buprestidae, Anthribidae, etc) are associated with the mountain forests, which can be collected in the same way as explained in article "Xylophagous beetles", Chapter III, however, some are now so rare that they are included in the "Habitats Directive" (European species of protected fauna) such as some *Carabus* and some Cerambycidae which live in dead wood or holes in trees. On the other hand, In the Alps, 29 different insect families were observed visiting edelweiss (Quinn 2008).

Snowy mountains and glaciers

There are very few species of beetles living on high mountain peaks and glaciers (Fig. 54). However, some species are relatively common under rocks or logs near the snow (Tenebrionidae, Carabidae, etc). For example, the author collected some rare Cerambycidae in the Andes ("El Volcán", Chile), in great heights and just at the edge where the snow begins.

Fig. 54 - Glacier Perito Moreno, Argentina *(Picture courtesy of Enrique Martinez).*

Some beetles almost frozen on the ground, right where the snow starts, apparently adult life of several species of beetles (mainly species of larger size) that live at high altitudes is generally not very long.

According to Quinn (2008) in the Arctic and Northern mountains some insects live in or adjacent to unfrozen lakes and rivers where the temperature is close to freezing point and snow cover protects them from cold. Many rely on super-cooling so that the temperature at which they freeze is decreased. An insect's freezing point of 17.5 grades F (-8 grades C) in the summer may adjust as temperatures drop in the fall, reaching 30 grades F (-35 grades C) in mid-winter. This process is reversed in the spring when the temperature is higher (Quinn 2008).

It is interesting to examine the surface of snow and glaciers since very rare Tenebrionidae, Scarabaeidae, Carabidae, etc. may appear. Dead insects in perfect condition are often found on the snow, which have been swept away by the spring wind from the lower regions to the highest.

Collecting in lava deposits

Some beetles (Tenebrionidae, Carabidae, Staphylinidae, Melyrtinae, Malachinae,) have been collected in the lava deposits found on volcanoes (Fig. 55) with pitfall traps. The trap was composed with Turquin's liquid (which also attracts and preserves beetles). In the crevices of rocks the beetles are so well hidden that unless they move, they are virtually invisible. Recent lava flows are similar to deserts, high mountains and Polar Regions in their lack of plants (Myrtle and Ashmote 1988). These areas seem to have no life, but in reality are often populated by diverse communities of animals. Several kinds of beetles dominate the caves fauna of lava deposits (away from the entrances) (Myrtle and Philips 1988). Jeannel & Leleup (1952) provided excellent examples for preadaptated ancestors of Pselaphidae after studying high altitude (2000-2900 m) beetles on Mount Kivu in Congo (Oana & Géza 2007). At this level, the species are exclusively humic inhabitants, deepened at few centimeters in humus, where proper conditions, such as constant temperature (10°C) and high humidity, are fulfilled (Jeannel & Leleup 1952; Oana & Géza 2007). There are also species of beetles that live among the patches of lichens. Also in these places, pitfall traps must be buried, baited with Turquin's liquid or cheese, dung or meat.

Fig. 55 - Lava deposits from Hawaii *(Picture courtesy of Enrique Martinez).*

Light traps in the hills and mountains

The summits of the hills are very interesting for light trapping. Most species found in these places get blown by the wind from the valleys and hillsides. If the top of the mountain is occupied by a military or commercial installation, with restricted access (radio or television towers), try to obtain a permit to collect, addressing those responsible for these sites.

When a collect is performed in the mountains or hills, it is important to note that the wind direction and temperature vary with height and time of day. The following details should be considered to define time and site to collect beetles with different methods or to place light traps.

In the mountains and hills, winds are moving in different directions depending on day or night. The size, shape and orientation of topographic features have influence on the air movement (Bullard and Peterson 1982).

There are many combinations between the mountainous areas, among them: a valley between mountains, a single mountain on flat terrain, mountains with deep valleys inside, or even a valley on flat ground. As shown in figures 56 and 57, the air tends to rise over an obstacle that comes in its way and try to advance through the different sides. If a change in high temperature (warm air over cold air) covers the higher elevation, then the air will try to find its way by the sides of the mountain (Bullard and Peterson 1982). When airflow is blocked, there is an entrapment or air recirculation (Whiteman 2000). At night, hills and mountains produce descending wind movements because the air is colder at high elevations (Monti 2002).

The mountains and valleys are heated unevenly (Fig. 58). To collect beetles during the day (by beating, sweeping or by hand-collecting) it is important to develop this activity on the slopes most exposed to sunlight. Very early in the morning, the sun heats and illuminates one side of a mountain or valley and the other side is still dark and cold. The air rises over the illuminated side and descends on the dark one (Fig. 58). At noon, "falls" on both

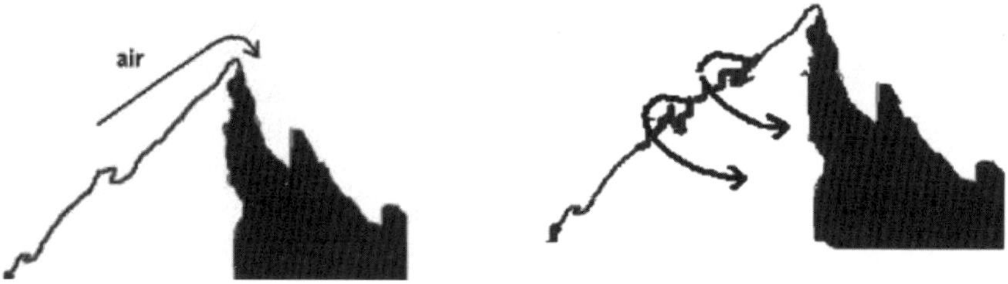

Fig. 56-57 Circulation of hot and cold air of a mountain

sides and heat them. By late afternoon, the situation is similar to that of the morning. After dark, as the air cools due to radial cooling, air descends into the valley from the highest hills (Bullard and Peterson 1982).

Figure 58 show the descending and ascending winds that occur during the day and night. In the case of a valley, the descendent winds can occur in the slopes of the valley, this causes that the cold air is accumulated in the valley. This cold air can get into the valley and this movement of air is due to cold air drainage. As the cold air descends to the valley floor, the air in altitudes becomes warmer. This results in a temperature inversion restricting the vertical transport of air (Bullard and Peterson 1982). The valley winds are chained because of its shape. They blow predominantly toward the top or toward the bottom of the valley. This can lead to high concentrations of air in the soil, because the

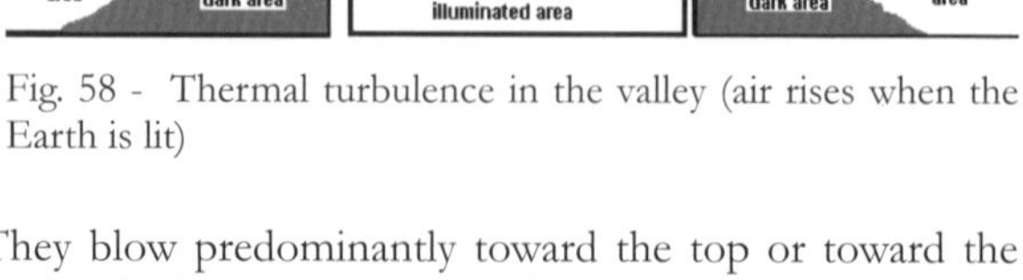

Fig. 58 - Thermal turbulence in the valley (air rises when the Earth is lit)

geometry of the valley restricts variations in wind directions (Bullard and Peterson 1982). Circulation within valleys may also be modified by solar heating of the valley slopes. Winds form during the day when hillside slopes are heated more than the valley floor. The differential heating of contact air causes air to flow upslope. Furthermore, according to Singh (2006) the air above the mountains being rises upwards is warmer, this is followed by the air from the valleys, which also moves upwards towards the higher mountains to replace it. This produces valley winds. At night, the more exposed upper mountain slopes are cool more rapidly and the air resting over them becomes cold. The colder air being heavier flows down the slopes into the valleys in the form of mountain winds. The light trap must be placed into the slope of the mountain (but not far from the valley) because the insects tend to flying following the currents of warm air rising from the valleys, but if the wind is very strong it is not worth using this trap.

In examples given in figures 15-16, the best places for collecting beetles will be: during the morning in the sunny slope and, at noon, both will be good (if there is sunlight).

Finally, be noted that in the mountain forests (for example, in the Yungas Forest of Argentina and Bolivia) it will be use the same sampling methods used in the jungles of lower regions.

BEETLES ASSOCIATED WITH CAVES AND OTHER UNDERGROUND HABITATS

Caves, artificial cavities, fissures, lava tunnels, lava flows, interstitial fillings and microspaces in talus slopes in valleys, are suitable places for unique subterranean species[1], terrestrial cave habitats are typically cool, moist and dark, supporting simple detritus-based foodwebs (Juberthie et Decu 1994; Carabajal et al. 1996).

With few food sources available, caves and other underground habitats have relatively few species and individuals. Beetles living there have very strange and interesting extreme adaptations. According to Gigli (1998) most species found in subterranean habitats are typically small, with long legs, antennae and palpi, while their eyes are often vestigial or absent. Their abdomen is usually very large in order to house a very long intestine, usefull to digest any available food and most are testaceous or entirely white in color. Dark colors (intended to protect them from UV rays) are not needed and bright colors that could be used to send messages to their peers are useless in the dark.

Caves and other underground habitats are not very abundant in terms of species and individuals because food sources are much reduced. Beetles living there have very strange, interesting and extreme adaptations. Most species found in subterranean habitats are very small, with long legs, antennae and palpi, while their eyes are very small to completely absent. Their abdomen is usually very large and has a very long intestine, useful to digest any available food or contain a large fat body as a reserve. Many of them are depigmented and testaceous in color. Dark colors —protecting them from UV rays— are not adaptive, and bright colors that could be used to send visual messages to their peers are non-functional in the dark.

Subterranean beetles represent an excellent object of study for historical biogeography because they are the group with the best representation in the subterranean domain

1. More information in the article "Underground beetles" (Chapter II).

(Moldovan & Rajka 2007; Oana and Rajka 2007).

1. They are the best represented group in the subterranean karst environment or domain, with many species inhabiting caves and the mesovoid shallow substratum (Juberthie et al. *1980*).

2. Most species are terrestrial and therefore have reduced mobility; and while they are not limited to limestone/karst areas, most taxa inhabit caves.

3. Their representatives display different degrees of adaptations to life in caves. Some of the clades have endogean and hypogean species, of which the last is more or less adapted to subterranean environments

Fig. 59 - Vallemi Cave in Northern Paraguay. *(Photo: Bolívar Garcete B., Brazil)*

4. They are a well known group, with many specialists studying various aspects of their biology, including their taxonomy, adaptations, behavior and molecular phylogeny.

Colonization of the caves

Bellés (1991) enumerates three adaptive pressures for cave colonization:

1. Survival, when external stress pressures animals to find refuge in caves;

2. Opportunism by colonization of an empty space;

3. Convenience or escape from surface competition which uses the same trophic resources.

Classification of the underground habitats

Racovitza (1907) is one of the first authors who intended a classification of the underground habitats:

Caves accessible to man

Cracks and fissures inaccessible to man

Phreatic layers with circulating water

Microcaves: constructed or excavated by animals (vertebrates or invertebrates)

Artificial caves: mines, tunnels and other man-made cavities

Howarth (1983) considered that the size of the cavities was a very important factor. He made the following classification:

Macrocaves: cavities larger than 20 cm

Mesocaves: spaces from 1 to 20 cm with a favorable microclimate

Microcaves: spaces less than 1cm.

Juberthie, Belay & Bouillon (1980) described what they called the superficial subterranean environment, which is formed by the natural breakup and fragmentation of the bedrock. The resulting debris accumulates on the slopes, forming colluvial deposits and protecting some empty spaces. These deposits, with the passage of time, become covered with humus and later with soil, isolating the microspaces from the external medium. A subterranean environment with its own microclimate thus develops there. These microspaces have sizes ranging from 1 mm to 10 cm and differ from endogeous by its porosity (Carabajal et al. 2006).

This discovery greatly expanded the boundaries and extent of subterranean environments, confirming the existence of troglobitic faunas in every place where the rocks crack or break appart, with the subsequent formation of interconnected spaces which allow the propagation and movement of underground animals

(Carabajal et al. 1996).

Cavernicolous animals groups

Christiansen (1962) divided cavernicolous animals into four groups:

1. Trogloxens: accidentally or temporarily cavernicolous.

2. Epigeomorphs: They live and reproduce in caves but do not show anatomical adaptations to a cavernicolous way of life.

3. Ambimorphs: with some morphological adaptations for a cavernicolous life, but keeping most of their epigeic features.

4. Troglomorphs: their bodies have completely adapted for life in caves, so they are totally different from non- cavernicolous (epigaeic) animals.

The habitat-based criteria are usefull but insufficient to classify cavernicolous beetles since biogeographical knowledge of these groups is limited.

Galan (2004) summarizes cavernicolous species into three categories:

1. Troglobitic:

Troglobitic are those species whose life cycle takes place entirely in the subterranean environment, without ever venturing out. Knowledge of their morphology alone does not clearly separate troglobitic beetles from edaphobitic animals (Arbea & Baena 2003). Troglomorphic features such as depigmentation, anophthalmia and other structural adaptations give clues, but are not enough to characterize a troglobitic species, since similar features are usualy found also in soil dweling (euedaphic) species (Thibaud & Massoud 1973; Arbea & Baena 2003). Some special features present in troglobitic species include:

- They complete their development in and are adapted to mesocaves and the deep environment of caves

- Some species can facultatively enter intermediate or transitional environments

- They have a high degree of troglomorphism

2. Troglophiles:

Troglophiles live and reproduce within caves, but they are able to leave them. There are also sub-troglophiles, which are species that spend part of their life cycle in the subterranean environment. Only a precise knowledge of their biology and ecophysiology allow classifying them as troglophiles (Thibaud & Vannier 1986):

- They complete their development and are adapted to the intermediate environment of caves

- They can also live in transitional habitats and sometimes can enter the deep cave anvironment, but do not complete their life cycle in it.

- They have a low degree of troglomorphism when compared with troglobitic species, but it is developed when compared with related epigeic species.

3. Trogloxenes:

Trogloxenes penetrate the subterranean environment in a casual or accidental manner:

- They live in the surface environment, but can also enter seasonally or temporarily into the intermediate environment
- They visit caves and can complete there only part of their life cycle
- They can also be found in moist and dark epigeic environments
- Troglomorphism is undeveloped, but they can be adapted to move in the dark
- Some species can maintain populations in the surface environment

Caves and rock crevices are not the only environments that host subterranean beetle faunas. Many other species live in the forests subsoil, as happens with many hypogaeic staphylinids that spend their entire lives underground. For this reason, some researchers have proposed a classification system for the interstitial fauna —stygobious, stygophilous and stygoxens— (Mezquita 2000) in addition to the classification of the cavities.

Cavernicolous beetles are found in all kinds of subterranean environments (Aldemaro 2009). The most typical cave beetles are in the families Leiodidae, Cholevinae (=Catopidae) and Carabidae, but there are also many staphylinids, histerids, etc. Most of the species live in caves inhabited by bats or other gregarious vertebrates because they are richer in nutrients. For instance, up to 29 species in 14 different families of Coleoptera have been found in a single cave in Venezuela, living on seeds and plant debris from *Steatornis* (oil bird) guano (Peck et al. 1989 and Herrera 1995). Some other species are aquatic and live in phreatic environments.

Cave beetles have a wide variety of eating habits, including predation, guanophagy, mycophagy, rhizophagy, xylophagy and scavenging and saprophytic opportunism. Many of them have typical troglomorphic adaptations: reduction or loss of eyes and ocelli, reduced pigmentation, thinning of the cuticle, wing reduction, body and appendage lengthening, reduced egg laying -even reduced to just a single but large egg per brood—, reduction of larval stages and phisiological adaptations that help a conservation of body fluids (Aldemaro 2009). For detailed information about feeding habits of cavernicolous beetles, must be study the chapter "Where do they live...". According to Decu & Juberthie (1998); Peck (1998); Newton (1998) and Encyclopedia of Caves (2009) the strictly subterranean beetle's species belong to 14 families and are widespread on all continents. However, the most they are concentrated mainly in the Atlantic-Mediterranean area and the Eastern Palearctic regions (81%), and in the South of the Nearctic region (13%). The troglophilic (mostly guanophilic species) and the subtroglophilic species are the most abundant in the tropical zone (Scarabaeidae, some Tenebrionidae, some Carabidae, Leiodidae Cholevinae, Histeridae, Colydiidae, Ptiliidae, Cryptophagidae, Hydrophilidae, etc.

Brief summary on the beetles families found in the caves of several countries

PTINIDAE: Some species of the subfamily Anobiinae are present in caves where animals nest or roost and where animal and plant debris occurs (Spilman 1968). Several ptinines have previously been reported from caves, mostly in Europe (*Niptus hololeucus* (Falderman) was found in European caves (Jeannel 1909) and *Niptus abditus* Brown was found in a North American cave (Brown 1959). Also, Spilman (1968) described two species (*Niptus abstrusus* Spilman and *Niptus absconditus* Spilman) found in caves in North America.

CARABIDAE: Cavernicolous carabids are often black and shiny. More than a half of the cavernicolous species in this family are troglomorphic and testaceous in color. Most of them are predators on other invertebrates. According to Moldovan & Rajka (2007) most of the world's subterranean carabid beetles are in the Trechinae. Some specialize on a particular kind of prey, (e.g.: eggs or larvae of specific insect species). They have been found on all continents except Antarctica. The following species (trogloxenes and troglophiles) have been cited from caves in Venezuela: *Colpodes combayensis* Goud, *Colpodes subviolaceus* (Chaudoir), *Pheropsopus aequinoctialis* Linnaeus *Selenophorus* sp., *Selenophorus* (or *Gynandropus*) sp., *Masoreus* sp., *Nototylus* sp., *Schizogenius* sp., *Agonia* spp., *Aephnidius* sp., *Pericalina* spp., *Brachinus* sp., and member of the tribes Clivinina (3 spp.) and Tachyina (Galan & Herrera 2006). According to Mitchell (1971a, b, and c) *Rhadine subterranea subterranea* (Van Dyke), occurring in Ranch Cave, Williamson County (Texas-USA) feeds almost exclusively on the eggs of cave crickets (*Ceuthophilus* spp.) buried in fine-grained, white, calcareous deposits. *Rhadine noctivaga* Barr, also has a similar behavior in such deposits in caves in Northern Williamson County (Reddell 1994). According to the authors mentioned above, the cave crickets are very important to the cave ecosystem.

CERYLONIDAE: *Euxestus erithaeus* Chevrolat has been collected in caves in Cuba (Peck et al. 1998) and from guano in caves in Jamaica (Peck 1992). It belongs to the first-level troglophile category and is a scavenger. It is widespread but uncommon in Florida and the Greater Antilles.

CURCULIONIDAE: The hypogean forms usually have elongated faces and legs and the eyes may be rudimentary or absent. About 30 species are known from caves (White 1983; Osella and Zuppa 2006; Aldemaro 2009).

DRYOPIDAE: Culver et al. (2000) mention cavernicolous species of *Stygoporus* and *Stygoparnus* from the United States (White 1983; Larson and Labonte 1994). According to Reddell (1994) in North America the adults of *Stygoparnus comalensis* Barr was collected in a spring mouth, because probably, there is a concentration of its food in this place. James Reddell believes that it is likely that their normal habitat is open caverns farther back in the system that feeds the springs.

DYTISCIDAE: *Haideporus texanus* Young & Longley has been found in the Ezell's Cave in San Marcos, Hays County and the famous artesian well at San Marcos in the United States (Reddell 1994; Culver et al. 2000). This is the only troglobitic predaceous diving beetle known in this country (Reddell, 1994). Some dytiscids have been found in caves in Cuba, but they were there apparently by accident (Peck 1998). Epigeic species of this family are carnivorous and good swimmers and fliers (White 1983). *Trogloguignotus concii* Sanfilippo is found in in the caves of Hueque (Sierra de San Luis, Falcon) in Venezuela, this species is depigmented, anophthalmic, with atrophied hind wings (Galan y Herrera 2006).

ELATERIDAE: A couple of species have been found in caves in Cuba: *Limonicus flavomarginatus* Knull and *Conoderus posticatus* var. *sticturus* Candeze This latter species is widespread across Cuba and its global distribution spans from Mexico to Brazil. It is a trogloxene predator and scavenger with larvae developing on guano. *Pyrophorus noctilucus* (Linnaeus) is a first-level troglophile, predator and scavenger. It is widespread across the West Indies, and from Mexico to Argentina. These luminescent beetles occur in caves where their larva feed on arthropods in guano. Larvae have been reported in the caves El Mudo and Cueva de los Majaes, Cuba (Peck et al. 1998). A population of *Limonius* sp. maintains itself in guano pile deep in a cave in Arkansas, USA.

ELMIDAE: Epigeic species feed on algae, moss and other plant materials found in small and crystalline streams with high oxygen concentrations (Aldemaro 2009). Almost nothing is known about cavernicolous species in this family (White 1983; Aldemaro 2009).

HISTERIDAE: Twelve cavernicolous species are known (Aldemaro 2009), which display different degrees of troglomorphisms and feed on small invertebrates in guano (White 1983; Peck 1998).

HYDROPHILIDAE: Species of the genus *Oosternum* have been found digging into wet guano in caves in Cuba (Aldemaro 2009). Species of this genus are often associated with caves, but their intimate association with them needs to be confirmed (Aldemaro 2009). Although most epigeic species in the family are aquatic, this member of the Sphaeridiini isusually found in wet guano, far away from water (Peck et al. 1998).

LATHRIDIIDAE: Hypogean species of this family have very small eyes. Some subterranean species are found in the Fiji Islands (Aldemaro 2009).

LEIODIDAE (mostly Cholevinae): They feed on decaying organic material and on "moonmilk" —cave clay containing algae, fungi and bacteria—. Over 700 cavernicolous species are known from Eurasia and the Americas together (Nicholas 1960; White 1983; Casale et al. 2004). Many of them live on guano (Aldemaro 2009). *Neotropospeonella decui* Pace, was described as the he only troglobitic, anophthalmic and depigmented Bathysciinae known from the Americas, in from Cueva del Guácharo, Venezuela (Galan & Herrera 2006), but it has been shown to be a mis-labeled European specimen! Cholevinae are detritivorous and saprophagous (Moldovan & Rajka 2007). According to Peck and Cook (2002) *Prionochaeta opaca* (Say) and *Catops gratiosus* (Blanchard) are frequent in caves in the Southeastern United States but occur frequently in forests to the North. No species in the Cholevini tribe in North America are cave limited.

NITIDULIDAE: cavernicolous nitidulids are saprophagous. *Stelidota guacharensis* Gillogly and other *Stelidota* spp. (Peck et al. 1989) are troglophilous species found in guano in Venezuelan caves.

NOTERIDAE: Many larvae feed on roots of submerged aquatic plants. According to Archangelsky (2001) the adults are predators. But, some of these are saprophagous or microphagous. Most species were discovered in caves, and springs, in areas between 45 degrees latitude North and South in both hemispheres.

PSELAPHIDAE: The cavernicolous species are usually orange-red colored. They either feed on fungi or are predators of other small invertebrates. About 150 hypogean species are known worldwide except for the Antarctica (White 1983; Peck 1998). According to Reddell (1994), in North America all are extremely rare and occur only on the underside of rocks buried in silt.

PTILIIDAE: The hypogean species are scavengers and probably feed on fungi in guano. They usually have very small eyes, whitish body and have no wings. Hypogean species of this family have been found in the Americas and Africa (White 1983; Peck et al. 1998).

SCARABEIDAE: Accidental cave inhabitants occur in karst regions throughout the world (Aldemaro, 2009). Some cave species can make balls of guano and lay their eggs on them (White 1983; Peck et al. 1998). The cetoniine *Gymnetis lanius sternalis* (Chevrolat) has been collected from caves in Cuba where they live in large numbers, depending on the guano as food for their larvae. These larvae occurred in balls of rather dry guano (Peck, Ruiz-Baliu and Gonzalez 1998). A species of *Onthophagus* is known only from guano in caves in the Ozark region of the central USA.

SCYDMAENIDAE: The hipogean New Caledonian species in the Southwest Pacific (*Scydmaenus aelleni* Besuchet) is eyeless wingless and depigmented (Besuchet 1981). A beetle

Fig. 60 - Collecting beetles in a cave. *(Picture: courtesy of Andreas Herrmann, Germany, 2010).*

belonging to this family was collected in rubbish below the mouth of San Isidro Cave, Merida, Mexico (Sperry and Banks 2007). Crowson (1981) also cited some species that live in caves.

STAPHYLINIDAE: About 30 hypogean species are reported from the Canary Islands, Northern Africa, Madeira, Southern Europe, Ascension Island, the Galapagos Island (White 1983; Peck 1998) and Cuba. According to Encyclopedia of Caves (2009) the approximately 27 known troglobitic species of this family are concentrated in the Canary Islands and Morocco, the rest occur in Algeria, Madeira, Spain, Italy, and Romania. The majority belong to the subfamilies Paederinae and Aleocharinae.

TENEBRIONIDAE: Hypogean species in this family have been described from Mexico, Cuba, Puerto Rico, Venezuela, Ascension Island and the Galapagos. The cavernicolous and epigeic species are scavengers (White 1983; Peck 1998). Peck at al. (1998) record several species found in cave guano, including *Alphitobius diaperinus* (Panzer). This species is found especially on dry guano, not only from caves, but also attics and chicken coops, and populations may number tens of thousands. *Alphitobius laevigatus* Fabricius found in Texas, Mexico, Cuba and the West Indies (Peck 1992), may group together in large numbers on bat guano in caves. *Orghidania torrei* Ardoin and *Trimytantron cavernicolous* Garrido and Gutiérrez are scavengers and first-level troglophiles, endemic to Cuba. *Zophobas rugipes* Kirsch is a first-level troglophile scavenger widespread from Mexico to Brazil, including Cuba to the Lesser Antilles. According to Reddell (1994) species of *Eleodes*, *Embaphion*, and other genera are frequently found in bat guano deposits in North America.

Additional families found in caves include Clambidae, Cantharidae, Limnichidae, Psephenidae, and Discolomidae (Galan and Herrera, 2006).

How to collect beetles in the caves

It is essential to have good flashlights when exploring a cave. According to Maree and Millar (2001): aim to collect as many species as possible, but not in large numbers, the selective nature of hand collecting makes it the most preferable method of colelect terrestrial cave fauna. The walls and floor must be carefully examined, as many small beetles (around 1 mm long) may be walking around. Silt banks and detritus deposits must be carefully examined, turning and putting back in place any rock, boulder, pieces wood and any other kind debris. Some individual may even be found away from the debris, looking for food. Pair of thin forceps is usefull to capture beetles between the cracks. A paintbrush damped with water or alcohol is a very good tool for catching small species that can then be easily transferred into alcohol in a glass vial. Examine wet walls very closely with the help of a good flashlight, especially considering all the nooks and crannies. This will help in locating some rare aquatic species.

Special care must be taken when examining tree roots hanging from the roof of some caves, because insects can easily fall down to the ground when disturbed and get out of sight. This is especially true in the case of large and dense root masses where insects may hide easily. Either a beating sheet put under the target roots or a white cloth spread on the ground, while gently shaking the roots, is always a good idea.

Pitffal traps are very effective for trapping invertebrates in caves, but they must not be left for long time, because they might affect the fragile population dinamics of some species. Also, traditional pitfall traps are ill-suited for many rocky substrates. Ramp pitfall traps can be used instead in such cases (Bouchard et al. 2000)

Baits produce different results compared to direct visual search (Poulson, Culver, 1968; Maree and Ian 2001). Many organisms that probably live hidden in inaccessible cracks are attracted to the baits and they may come from remote places deep in the cave, as the smell of the bait is easily spread in those closed spaces. Baits are to be placed safe from pathways and flooding, and the places should be carefully marked so they can be easily found later. Baited traps give better results if placed in varied situations and preferably under or close to shelters like rocks or other kind of debris which will provide refuge for the beetles attracted to them. This will also prevent the baits being eaten by vertebrates.

For example, in the Cave "del Yeso" (in Alemeria, Spain) Barranco et al. (2003) have used as an attractant Turquine solution (100 ml beer, 10 g chloral hydrate, 5 ml of acetic acid and 2 ml of formalin) in pitfall traps, but they changed it's to a solution of vinegar-beer-glycerine in the Cave "del Tesoro" and the "Cave of the Apas" (in Spain), as result of the extensive damage caused in both caves by small mammals.

Bottle traps, made from soft drink plastic bottles, and baited for aquatic invertebrates can also be used. The Fig. 61 a-b show a variation of an amended Mare and Millar trap (2001). A mesh covers the holes at bottom of the bottle, so water can drain freely through the trap. The inverted neck of the bottle forms a narrow tunnel, allowing easy access to invertebrates, but avoiding escape. The ribs of the bottle ensure the closure of the inverted neck in its position. The bottle can be tied and a stone put either inside or above the bottle to help keeping it in place. Nevertheless, this trap will hardly work if the water current is too strong, so rather a slowly flowing stream is recommended. Also consider that beetles can swim with or against the stream, so results are unpredictable and depend on testing. Small bits of shrimp, meat or cheese can be used as bait in the bottle. The bait should not be left adrift, it can be put inside a small mesh bag or piece of panty hose tied to the ballast stone. Be careful not to spread the bait when taking the insects out of the trap.

Some species of ptinids beetles are locally abundant and thousands of them have been collected in some caves such as Yeso (Almería, Spain). These beetles usually appear at the entrance, in the driest areas of the caves, or are associated with bat guano (Bellés 1976 and 1985; Decu et al. 1998). However, strictly cavernicolous species are unknown in this family. Those ptinids have been collected using pitfall traps with turquin solution (100 ml beer, 10 g chloral hydrate, 5 ml acetic acid and 2 ml formalin) in them as an attractant (Barranco et al. 2003). The same authors also used a solution of beer, vinegar and glycerin in another cave.

One must avoid over-collection of a single species. Caves support extremely fragile food webs and some evidence indicates that collecting can decimate entire populations of troglobitic species. Precise number is difficult to establish, but a sensible practice might be to limit a sample to a maximum of 6 to 10 specimens per species.

It is very important to label the samples as soon as they are collected, as it is easy to confuse and mix specimens from different locations or environments. Collecting data must

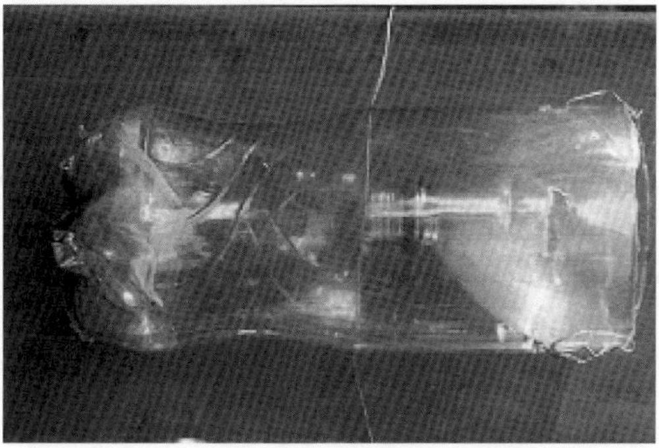

A

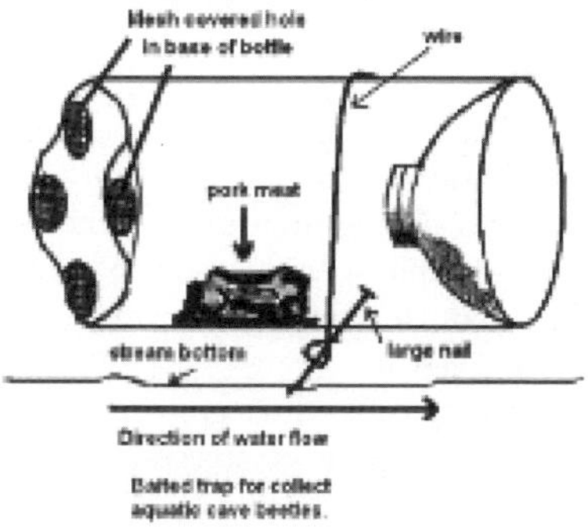

B

Fig. 61 a-b - Aquatic bottle trap to collect aquatic cave's beetles. (Photo and drawing: Carlos Aguilar)

include the name of the cave, as well as some other information such as where in the cave the specimens were found and the specific conditions of the place. Do not mix specimens from different caves. You should always have at hand several vials containing 80% ethanol. To transport living specimens to the laboratory predatory species must go in separate containers.

Fig. 62 - Head lamp. *(Picture courtesy of BioQuip, 2010 – www.bioquip.com)*

Materials

Protective helmet for the head
Light lamp
Head lantern (Fig.62)
Forceps
Paint brush
Aspirator
Vials with 80% ethanol
Glycerin
Labels
Rapidograph or pencil
Aquatic bottle trap
Baits (meat, feces, fruits, chemicals, etc)
Pitfall traps

OTHER PLACES TO SEARCH BEETLES

Mills

Wintertime is a good opportunity to examine mills. Some of the mills are frequently treated with toxic gas, so these places are not very interesting to look for beetles. But mills producing flour under "ecological" conditions are very interesting because a lot of synanthropic species lives inside and could be sieved in dark and damp parts of the mill. There are several species which are recorded only from mills (Andreas Herrmann, pers. comm. 2010).

The same is with e.g. storage houses in the harbour or similar buildings where it is possible collect pests of stored products.

Mills are interesting by the concentration of very old decaying rests of cereals in all the many dark corners and damp in the mill, together with the special micro climate, darkness, moderate temperature and living and dead animals (mouses, spiders, cockroaches) molds, cat's feces and the import's grain from other regions. These factors attract many species of beetles.

Sifting of garden, agricultural, and farm litter

These places are interesting to get beetles associated with agricultural activity. for example, Peck Stewart (2006), reports that in the Galapagos Islands: "This litter is a product of human activity and is composed of plant prunings, fruit and plant refuse, stored or discarded food stuffs, cow and horse manure, agricultural waste, and old hay from the tortoise pens at CDRS (Santa Cruz (CDRS, Playa Bacchas, and Tortuga Bay). These materials are most likely to contain stored product pests and species introduced by human activity."

Mounds of fruit or seeds

In cumulus of flower petals on the forest ground, there are many beetles, principally Carabidae family. The same applies to other plant debris, for example, in many farms there are mounds of seeds or fruits (such as coconut seeds) that farmers used as fertilizer or that they accumulate to make other products (oil, soap, etc.) (Fig.63), in these decaying seeds the author has found several beetles (Carabidae, Elateridae, Curculionidae, Nitidulidae, Staphylinidae, Scarabaeidae, etc.). These decaying seeds should be removed and with a aspirator can be collected beetles that are between these cumulus, but, by the author's experience, who in an occasion has spent many minutes under the hot sun and could catch some specimens only, so it is more advisable and productive to sift the seeds, this can be done with a grid (1 cm mesh) on a white cloth, so it is easier to catch the beetles which quickly escape.

Fig. 63. - The author collecting beetles in a mound of decaying coconut seeds. *(Photo: Jorge Barrett)*

CHAPTER III
SAMPLING METHODS AND TRAPS

NIGHT COLLECTION

The use of light traps is an effective method for collecting many species of nocturnal beetles that are attracted to different light spectra.

In any habitat where light is used you will quickly have an overview of many beetle families represented by large series and a significant diversity of species. You can use ultraviolet (UV) light, mercury vapor lamps, or combine both in front of a white bed sheet. Most species are attracted to both types of light, but some of them only come to UV and others only to mercury vapor light.

Because each collector has his own particular light trap, we will just outline some methods that work.

Fig. 64 - Night collection with UV light trap in Piribebuy, Paraguay. *(Photo: Bolívar Garcete B., Brazil)*

Optimum conditions for using light traps

- Just before or after a storm.
- Low atmospheric pressure.
- Hot and humid nights (perfect if it's cloudy).
- Night windless.
- Moonless sky (ref. moon rise / set) and from the third day before the new moon until the third day thereafter.

Negative conditions for using light traps

- During the rain.
- High atmospheric pressure.
- Cold nights.
- Nights with high moon visibility accompanied by clear skies.

What are the best nights to collect using light trap

Before making a trip where you intend to use light traps, it is imperative to obtain data on upcoming weather, about the phases of the moon and the moonrise and set (see explanatory text below).

This image (Fig. 65) illustrates appropriate dates and hours when a trap could be used in the month of December 2009, in Argentina. These time tables vary depending on the latitude where the collection site is located. Looking at this time table, the reader will notice that, from 5 until 19 December 2009 were the best days for light trapping in accordance with the phases of the moon. On Internet there are sources of this information (See "Planning a trip" , Chapter I)

It is also important to know the hours when the moon will be visible every night. For example, on Saturday the 5th there was time to prepare the light trap and make it work from dusk until 22:04 hrs. However on Sunday the 20th the moon was in its fourth quarter and was visible from 09:13 am until 22:34 pm; so from that Sunday onwards until the end of that month there were no good nights to collect with light traps. On the evening of Sunday 20th, after 22:34 hrs, we could have used a light trap, but the beetles attracted in that period would be less abundant, although sometimes interesting species fly in late hours of the night. However, collectors will usually be too tired to continue working into such late hours if results are not so promising.

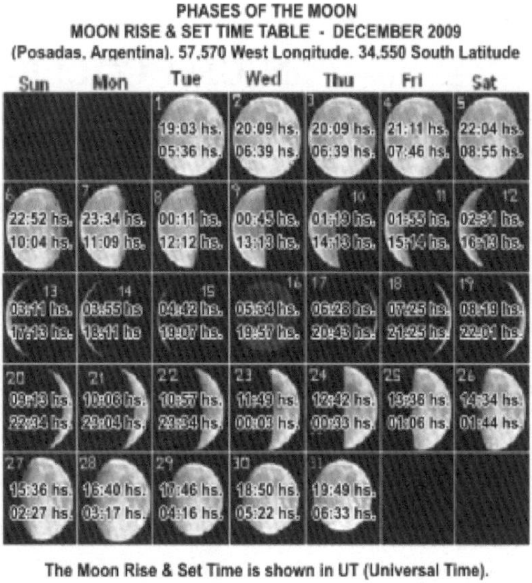

Fig. 65 - Example of moon rise and set time table.

Also bear in mind that if the night is cloudy (i.e., blocking the moon) a light trap can be used with better results.

Many species of beetles appear only on certain dates during the year, living as an adult for only a few days, like some species of Scarabaeidae *(Megasoma, Heterogomphus, Strategus,* etc.*)*, some Cerambycidae (*Macrodontia, Mallaspis, Callipogon,* etc.), and many Buprestidae and other families living in xerophytic regions. Therefore, if you want to collect these species, it is less important to know the phases of the moon and better to know the dates of these beetles' adult activity. Of course, reduced moon visibility would be preferable.

Traps using white bed sheets

Suspend a white bed sheet with a string and place a light source in front of it (but separated about 20 to 40 cm.), The light source can be a UV light (or black light) or mercury light (white light). If you want to get forest species, the light trap should be placed about 10 to 20 m from the forest treeline.

If you put the light inside a forest, do it in a clearing because it is important that the light can be seen by insects from afar, sometimes from several hundred meters.

Fig. 66 - White bed sheet light trap. *(Drawing: Jorge Barrett).*

The site chosen to locate the light trap is very important. The top of a hill is better than the bottom. Keep in mind that the hot air accumulated during the day in the lowlands rises in the evening and insects tend to follow those drafts.

It is also preferable that the front and back of the sheet is cleared, with nearby scrub or bushes at least 10 m away. This is so that the insects will not remain "trapped" among those bushes rather than your sheet. In addition, after weeds have been rooted out, an beating-sheet will be used to beat the cut vegetation.

Once you have chosen the place where the light trap will be located, look for two trees separated by at least 4 m. Tie the rope, on which the bed sheet will be hanging, to the two trees and in front of the light source. Start by tying the rope tightly to the first tree at a height of nearly 2 m. Pass the rope behind the second tree and go back to where it was tied around the first tree. Firmly tighten the rope and tie it again. The rope will form two parallel lines between the trees, one for the bed sheet and the other to suspend the light(s). Then the bed sheet is hung on the rope and is fixed with the help of clothespins.

On the ground lies another bed sheet that will be centered with respect to that which is hung, forming an inverted "T". The second bed sheet will be useful to allow the collector viewing the insects falling on the floor. It will also help to increase the area of light reflected onto the bed sheet and make the light observable from farther away. Using this sort of "drop sheet" is also useful for noticing insects which have fallen to the ground and flightless insects which are attracted to lights and walk over to them (e.g., the cerambycid *Romulus globosus* Knull in Florida).

Using a stone (like a paper weight) at each corner of the hanging sheet will prevent the wind from moving the sheet. Try to achieve a vertical surface as straight as possible, without wrinkles.

Hang the light bulb on the other rope, in front of the vertical bed sheet, separated about 20 or 40 cm. It is best to use a thin bed sheet so that the cloth, although illuminated on only one side, could attract insects from both sides. It is better to use a white, preferable somewhat shiny bed sheet.

Place the battery or electric generator near the bed sheet.

Because there are beetles that fly only at dusk or just before the sunset you should connect the bulb to the battery half an hour before sunset, or when the ambient light begins to fade. This allows diurnal species to be collected during their last diurnal flight. Leave the light on until 23:00 hr. Although in smaller quantities, there are species that appear after that time until sunrise.

Colder weather, or sudden drops in temperature, will discourage flight activity in most insects.

There should be minimal wind[1], a breeze at most, since few beetles fly in heavy winds. To find out if there is wind must have to look at the sheet. The trap must be located in a place that is sheltered from the wind, but that can be seen in the distance. If there is a slight breeze, then the light orientation is important because most insects fly directly against the wind, and when they come to the light they take refuge on objects where they still face the wind and can continue facing the light. That's why the light should be placed where the wind does not blow toward the light.-

If there are no trees available to install the light trap, but you have a vehicle, hang the bed sheet and one or two tubes of ultraviolet light behind the car, which are powered by the vehicle battery. If there is only one tree, your vehicle can substitute the second tree in order to tie the ropes and hang the bed sheet and lights. Remember to spread another bed sheet on the floor.

You can also put a basin of salt water on the hood of the vehicle with a plastic or metal grid with large holes above it. On the grid hangs the ultraviolet light tube, so larger insects that come to light will fall into the container with salt water, allowing the smallest beetles going through the grid. The largest ones (just as the moths) will remain on the grid and can be captured by hand. Adding soap to the salty water helps insects to break the surface tension and sink, if that is preferred.

It is important to note that the lights should have some type of roof to avoid getting wet in case of rain. You can use transparent glass screens so that light may still shine in all directions.

When you are catching the insects that come to the bed sheet do not cast a shadow standing in front of the bulb. To avoid this problem it is advisable to place the light above the collector's height, but always in front of the bed sheet.

It is very important to have on hand one or more spare light bulbs. It is also advisable that two people could participate in this activity since many beetles, including species rarely collected, stay away from the light source, often remaining at the edge of the luminous aura where darkness begins. Many beetles fly approaching up to a few meters and then walk toward the light; so one person should be devoted to collect a few meters away from the bed sheet. For example, many large Dynastinae such as *Megasoma, Heterogomphus, Dynastes,* etc., come only to the less lit areas in the vicinity of the bed sheet. It can also be practical to bury some shallow trays at ground level which are filled with salt water or other preservative liquid (See "Formules", Chapter V). The beetles approaching from the shadows or are attracted to the light but running on the ground (Cicindelinae and others Carabidae) they will fall in those trays and can be removed by hand or with a sieve.

Other species perch on branches close to the source of light, so the surroundings should be examined with a flashlight.

Rainy days are not very productive for light trap use, but many species of beetles fly only when it is raining, such as some species of Cerambycidae, Geotrupinae *(Athyreus, Neoathyreus, Parathyreus),* Hybosoridae, Ochodaeidae, Pleocomidae and Belohinidae. If you plan to collect these species place the light traps where they will be protected from rain.

---

1. Although it depends on the place, as with the experience of Andreas Herrmann (pers. comm. 2010), who informed me that on one occasion in the Cape Verde Islands, despite the strong winds prevailing in these islands and the difficulty in keeping the sheet in place, he was surprised by the large number of beetles that came to light. Presumably, these beetles have a different behavior due to the persistence of strong winds.

Light raps with cross-shaped vanes

This light trap consists of two cross-shaped vanes (of plastic, metal and even nylon) and at or near their intersection there is a source of ultraviolet light. Under the vanes there is a large funnel and below a large bowl, bucket or jar with preservative fluid. This light is hung from a tree, or from a rope between two trees, or from a vehicle. Leave the light running all night and check the next day. This trap catches thousands of specimens, but the problem is that beetles are mixed with moths -dropping in the bottle- and the scales from their wings pollute all specimens captured in the liquid. It is suggested that specimens are rinsed with water to remove most of these scales.

What kind of light is the best for catching beetles?

Fig. 67 - Light trap with cross-shaped vanes. *(Photo courtesy of Jack Owens, Canada)*

The ultraviolet light bulbs consume very little power, so they can be used with a rechargeable battery, at least for one night or a little longer before recharging. A mercury lamp consumes much more energy, so you need residential electric current or a generator. In Paraguay, these two kinds of lights attract different species, but in similar amounts. However, in some countries like the USA, both kinds of lights attract similar species and approximately in the same amounts. But the available literature does not show many comparative studies in this regard. The intensity of a 250 or 500 Watt mercury lamp will surely attract species which are farther away than those which might be attracted by a standard ultraviolet lamp of 15 watts This light trap consists of two cross-shaped vanes (of plastic, metal and even nylon) and at or near their intersection there is a source of ultraviolet light. Under the vanes there is a large funnel and below a large bowl or jar with preservative fluid. This light is hung from a tree, or from a rope between two trees, or from a vehicle. Leave the light running all night and check the next day. This trap catches thousands of specimens, but the problem is that beetles are mixed with moths -dropping in the bottle- and the scales from their wings pollute all specimens captured in the liquid. It is suggested that specimens are rinsed with water to remove most of these scales.

But in a dense forest there is not much difference. The difference could be noticed in a more open ecosystem such as a prairie.

In biological stations with permanent lights or street lighting, the lamps must be reviewed before dawn, as many species will go away as natural light increases. This happens with various species of Cerambycidae and Lucanidae arriving during the hours toward dawn, when they are active. Other rare species fly at dawn. If someone is willing to collect during those hours, be patient and wait for the beetles' appearance.

Beetles in various sizes come to the electric street lamps at biological stations and protected areas in general. During the first and even the second year of light bulbs inauguration, the species mixture attracted to them vary continuously, stabilizing from the second year, when they reach a maximum uniformity (asymptote). For example, large beetles, such as Scarabaeidae, Cerambycidae, Elateridae and others begin to be well represented in the spotlights in a minor period of time. And because of their size, these groups of beetles are often the most collected and subsequently the best studied. Therefore, collection in electric street lights, or light traps used in protected habitats such as national parks and nature reserves, should focus mainly on medium and small size beetles, and less on large beetles.

Large numbers of beetles and a huge diversity of other insects attracted by the white lights of protected areas, near a forest, roads and on the periphery of towns are eaten by other animals such as frogs, foxes, birds, lizards, etc. Even the artificial night lighting in natural areas negatively affects the behavior of many other animals (Rich and Longcore, 2006).The author thinks that it is unbelievable that even today white lights continue to be used on roads, in front of homes, within cities and elsewhere. They should be banned worldwide, in favor of environmental protection and conservation(2).

There is an alternative with the same price. Using slightly yellow lights will in no way hinder visibility for humans, improve visibility on foggy and/or rainy nights, and do not attract the vast majority of insects. Much is said and done for the protection of wildlife in terms of developing projects to improve or restore populations of mammals, reptiles, fishes, birds, amphibians and plants, but insects are the most attacked beings and, at the same time, the most neglected in conservation projects.

If yellow lights (preferably filtered) were used in biological stations and protected areas, more accurate statistical information would be produced when conducting rapid ecological assessments. Thus, using appropriate lights to attract insects such as UV light, mercury light, etc., an actual percentage of the insect population belonging to the place will be captured; the species and specimens that were previously diverted by white lights will be represented, resulting in the optimum evaluation of the sampling with more accurate representations of the population(s) studied. The final and most important benefit of installing permanent yellow lights, replacing the white lamps in public lightning and protected areas, will be the recovery of the original insect populations.

When you collect at street lights, we recommend a net with a telescopic handle because many specimens attracted by these lights are often out of reach.

Recommendations for using light traps

While traveling by car to the collection site take note of the lights at buildings and stores near the collection site since these will be good options for night collecting.

Two hours before sunset (if the site is new) take a walk practicing beating or sweeping, and choose the best places to locate the traps based on what is caught. Memorize the location of paths and dead or dying trees to facilitate navigation at night with a flashlight.

It is advisable to place two or three light traps in different environments, if you have the option.

Go back to the first light trap and arrange everything so that all lights are ready 15 minutes before sunset.

Visit light traps soon after sunset, as there are species that may arrive very early.

2. According to Rich and Longcore (2006) artificial night lighting should be considered by all ecological field researchers because of its effects on animal behavior, which in turn may influence the number of individuals counted by any given method. Replacing older high pressure mercury vapor lamps with high pressure sodium vapor lamps and combining them with full cutoff luminaries equipped with light-guiding mirror systems reduces energy waste (and associated CO_2 emissions) and insect mortality.

Take note

By looking directly at a source of ultraviolet light you may often feel a discomfort in your eyes, a feeling that other lights do not cause. To look directly into these lights may be dangerous to one's visual health.

The ultraviolet lamps contain phosphorus, which produces ultraviolet light when it burns; thus while the ultraviolet light sources are used phosphorus is slowly consumed until they produce no more ultraviolet light.

Fig. 68 - Precarious light trap, but with excellent results. At the right, a young assistant villager. *(Photo: Carlos Aguilar)*

A portable generator with lights allows exploring many different habitats. Placing a light in the clearings formed by fallen trees or logging, roads opened occasionally by animals, bridges over streams or rivers, etc., usually provides many different beetles.

Kerosene lamps are also useful to attract beetles if you do not have a generator. Flashlights are not very effective because the light emitted is not very strong and does not have a wide field of action.

When there are many people in front of the light trap they can produce shadows which decrease the attractiveness of the lights. Also do not allow people to move or handle the jars with specimens which are very delicate and are easily damaged by sudden movements.

Label with as much information as possible all specimens collected. When using two or more light traps in different environments, or those separated by great distances, keep several kill jars near each trap; never visit the lights with a single bottle in your hands! Keep a large jar with paper moistened with ethyl acetate at each trap. All this care avoids confusion when labeling the specimens.

Beetles, depending on their size, are taken directly by hand, with a vacuum/aspirator or a brush soaked in alcohol; larger species (Dynastinae and large Cerambycidae) may be injected with alcohol at 90% or ammonia , with a hypodermic needle (as those used for insulin). Take the insect with your hands and quickly plunge the needle above the pygidium and down the wings, reaching almost to the chest. Only a few drops are injected with a short and dry syringe grip. With this method the beetle is killed instantly.

It is advisable to have several kill jars at hand and never mix beetles of various sizes in a single bottle because they bite and destroy each other. That often happens with Cerambycidae and large Dynastinae. Inside a bottle only three to four specimens should be placed, and once dead, all the beetles collected must be moved to a larger flask, containing toilet paper or towel paper dampened with ethyl acetate. This liquid will ensure that the beetles are kept soft and do not dry, but the ethyl acetate cannot be used as a permanent preservative.

Materials for collecting with light in front of a bed seet

Numerous small kill jars with 75% alcohol or more

Large size kills jars with 75% alcohol or more.

Several bottles of various sizes with ethanol or preservative liquid to keep what is collected.

Camera (optional)

A tupperware with cotton or sawdust beds (to keep the insects collected)

Cardboard tubes with caps stuffed with sawdust (the cardboard absorbs moisture from the insects).

Tweezers (forceps)

hypodermic syringes

One bottle containing ammonia

Vacuums (aspirators)

Pencils and a notepad

Light trap equipment

(The types of lights and power sources are optional).

Two bed sheets (or more depending on the number of traps to be used); also consider the number of persons participating in the collection.

Strong Rope enough to hang the bed sheet and lights.

Thin rope to tie.

Electric cables to connect the electrical source with the light, and accessories ones (plugs, lamp socket, etc.).

Be sure to use something to protect the light bulb from the rain.

Clamps for hanging clothes.

2 or more mercury lamps.

2 UV lamps and accessories.

Electric generator and batteries.

12 Volt DC Battery Timer, See www.bioquip.com

12 Volt DC Battery Pack, See www.bioquip.com

Rigid Portable Heath Trap, See www.bioquip.com

XILOPHAGOUS BEETLES

Different beetle species require decaying wood of a different tree species as their food resource (Tero and Kotiaho 2004), for example, there are species found in living trees, others in fallen trees. Some beetles appear in the early days of the death of the tree, others arriving later when the wood is rotten. In Finland there are at least 4,000 species which are dependent on decaying wood only (Siitonen 2001).

Therefore, both living trees and those in their various states of putrefaction must be taken into account to gather the widest possible range of xylophagous beetles.

Collecting insects living inside trees, especially main branches and trunks, almost always involves some destruction. If the trees are already dead, then insects within can be fairly readily collected and counted. The problem is that this type of collection will destroy the habitats for these insects, and if, beetles are rare or endangered, then the very sampling procedure may make matters worse (Speight 2005).

Direct searching is probably the most effective sampling method when the aim is to find as many scarce beetles under bark or in timber as possible within a short time (Siitonen & Martikainen 1994). However, there are many types of traps and collection methods, in this article will show many sampling techniques for xilophagous beetles, that can be applied anywhere in the world with available habitat.

Fig.69 - In this case, sawdust reveals the presence of xilophagous beetles. *(Photo: Jorge Barrett)*

Fig 70 - Looking beetles under the bark of a log. *(Photo: Jorge Barrett)*

How to detect them

The presence of beetles is often noted through the traces they left, as loose sawdust in the place where they are active (whether adults or larvae), by the sounds they produce and/or the holes or galleries in the wood, which are sometimes covered by loose bark or sawdust (Fig. 69) and are not detected by the naked eye. The trunk, branch or stump must be carefully examined to confirm the existence of galleries in the wood. But first check if there are fungi in the wood (see Article "Beetles in fungi", Chapter III).

What to do after checking the presence of beetles

When find a log or stump or branche which are suspected there are hidden beetles, the first thing to do is to put the beating sheet under the trunk. Then start peeling and lifting the bark very carefully (preferably in pieces) to prevent that sudden movements or strokes could allow insects to escape. Also must look the inferior face of the bark.

Collection under bark and wood (Fig. 70)

A variety of beetle families live under dying bark of dead or decayed trees, as well as in fallen logs, stumps and even in branches on the ground. Also in living trees with thick and very cracked bark are enabling environments. Any time of year or place is a good to look for bark beetles (or their larvae).

Although it is very easy to catch beetles living in these environments, it is necessary to know the following items to optimize the collection outcome:

According to Speight (2005) small species of bark beetles can be found under thinner bark, large species need big space for their galleries that thick bark provides.

Trees with bark very attached to the trunk (or smooth) or with few crevices for insects to hide, are not very helpful for this activity.

Logs

Look for places where there are plantations, where the accumulated logs are exposed to sunlight but still retain moisture; these logs generally have a great variety of beetles although many species don't living in very sunny places but in quite dark. Put a beating sheet under the trunk and then with a machete break the trunk to look for the beetles (Fig. 71).

Collecting in Rotten wood

Many species of beetles are found in rotten wood, to collect these beetles must be broken into small pieces of wood on a beating sheet and review these carefully. Also the rotten pieces of wood can be put into a funnel Berlese. Larvae can be transported to the laboratory (to raise them) with the same pieces of rotten wood.

Beetles trapping in piles of trunks and branches

Countless species of Coleoptera are found in branches, trunks or tree stumps. To be able to collect these forest species, during the early spring cut some branches of different lengths and diameters, belonging to various native trees, leaving them on the forest edge or in a clearing (Fig. 72).

It is advisable to leave the branches piled in shady places, but also where sunlight comes for a few hours during the day.

Fig. 71 - Cutting a log with a machete to find beetles. *(Photo: Carlos Aguilar)*

Each pile of branches should have more than 2 sq. m and about 90 cm high, ensuring that each group belongs to the same species of tree, to know which insect species live in each tree (for this purpose we may put a sign stating the name of the tree to which the branches belong). If there is not interest to establish the relationship Species/ Tree, then the option of mixing several branches in each cluster is valid. Each pile of branches must be separated from the other at least 1 m away, to walk between them. It is important to locate them so that they can be examined both above and below without having to move the branches when searching Coleoptera.

To make this kind of "clusters of wood and branches" is necessary the approval of the site's owner and a special permit from the government if it is required by the law.

After 5 to 10 days, to check these branches, both day and night, trying not to move them suddenly to prevent beetles could escape.

During the night, most beetles are below every branch, whether perched or walking. At night use a flashlight. If this is collected by hand, imprison them carefully with fingers against the branch, without pressing too much, but quickly and forcefully, trying not to break them. After, reviewing the branches must be use the beating sheet during the day and if needed, the aerial net.

Fig. 72 - Collecting in clusters of wood and branches. *(Photo: Carlos Aguilar)*

The best locations for these "piles of trunks and branches" are the primary forests, but the secondary forests are also indicated.

This method does not harm the trees if it is responsibly done, trying to take only a few branches from each tree.

It is important to cut branches of various diameters to attract different species.

The accumulated branches can provide valuable material, sometimes more than 2 months since they were cut.

Several kinds of traps can be attached to the branch mounds according to the collector's taste and preferences (see below).

Fogging method

Those living in these vegetal organisms can be extracted by cutting the plants and immediately depositing them in plastic bags to prevent those insects could run away. Then spray inside the bags with an insecticide based on pyrethroid, so that most insects die or at least could not escape. When the effect of the insecticide is finished, take the insects which have left plants, and change them to bottles with 70% ethanol. But before removing the plant from the bag, it is best to look for those insects which have stayed among the leaves, many of them are very small (Stuntz et al. 2002, Yanoviak et al. 2003).

Spraying the canopy is not very effective for extracting the beetles associated with epiphytic plants, as a high proportion do not die by the action of the insecticide, and among those who die, many are left in the leaves of plants (Yanoviak et al., 2003; Luna 2005).

Also, to collect the beetles living in mosses, algae and small lichens those grow in vertical walls with water filtration, rocks and sticks that reach the surface of the water, those of visible size must be collected first, by hand; then the entire plant mass must be removed and put into a bowl or jar with ethanol for subsequent separation in the laboratory; or without ethanol but with a little moisture, if needed to keep alive beetles.

Fumigation of wood, trunks and stumps

Locate the stump or trunk which is going to be sprayed. Choose a fallen trunk but not lying on the ground; preferably with very rough and cracked bark, or nearly loose, or big fungi, molds, algae or lichens suspected to have beetle activity. This can be known by the tracks left by the beetles: sawdust or galleries in the wood. The logs, stumps, trunks or branches can be found in newly cleared areas, maintaining this condition for some months; and also places where there was a fire in or near a forest. Keep in mind that when the wood is too burnt (charred) it has no many insects. However, when starting a forest fire, the smoke and an odor of freshly burnt wood, attract many species of wood-eating beetles and their predators. But according to Tero (2007) the fire had a positive effect on beetle diversity only if an adequate number of retention trees were left on the logged site. It is interesting to know that approximately 40 species of insects in Northern Europe are considered to be dependent on forest fires and Majority of these species are beetles (Heliövaara and Väisänen 1984; Lundberg 1984; Wikars 1997).

After the wood is located, put a white solid sheet cloth in the ground, or simply a beating-sheet. Then, fumigate the stump with a spray. Turpentine(1) can be used which does not kill the beetles quickly, but makes them go out of their hiding places and fall into the cloth. This procedure is very easy and without any danger. This can also be done in fallen branches or semi-dry wood. Although in these cases it is advisable to take them to the laboratory for rearing the beetles which are inside the timber. The trunks, branches or stumps covered with soil are not appropriate, because almost no insects live there.

When located a good trunk under forests, in the recently fallen trees should be reviewed before the underside of the trunk for found the beetles. After doing that then you can wrap the trunk with a strong, white cloth or big plastic. The fabric is attached to the trunk by tips and in the Center (Fig. 73). When the trunk is covered with this fabric, then spray it with a pyretroid (Mapex, Baygon, etc.) by introducing the poison spray bottle between the trunk and fabric. Meanwhile other activity can be done, such as beating in the foliage or installing the traps. In approximately one hour we'll be back to the fumigated tree or trunk and proceed to uncover it carefully, spreading a sheet on the ground just below the trunk. It is better to do it with the help of another person, placing carefully

Fig. 73 - Spraying fungi on a trunk . *(Photo: Carlos Aguilar)*

1. Turpentine: This is a liquid used to dissolve paint, enamel or varnish. This product is dangerous if ingested and if it is in prolonged contact with skin. It is good product to get beetles to go outside of branches or stumps, applied with a sprayer. The insects will come soon out of their hiding places and then can be collected.

the contents of the sheet in a jar with ethanol. After saving the ethanol content, immediately spread the sheets on the floor just below the trunk. Then proceed to the observation of the bark, with the beating sheet underneath (so the dead insects fall in it), brush the trunk from top to bottom with a small brush, soft and dry of 5 cm. wide approximately. So insects that may be left on the wood or among fungi will fall to the beating sheet or on lower fabric. It is important to remember that some species hide in the bark cracks and are often undetectable without careful consideration.

With a soft forceps, a brush or by hand, remove the insects that can be seen on the beating sheet or on the fabric and save the beetles in a vial with ethanol. The sample must be labeled in the same place and separated later at the laboratory. The problem with this method is that many of the collected beetles live under the bark of the trunk (Not always in fungi) and most of the time they can not be associated with any specific fungus.

TRAPS FOR XILOPHAGOUS BEETLES

Xilophagous bottle trap

This trap (Fig. 74) is very effective to capture Curculionidae (Scolytinae, Platipodidae), Bostrichidae and Anobiinae. It consists of an inverted plastic bottle, two-liter capacity and 2 large windows open at the sides. To attract the beetles it has an uncovered inside container with ethanol mixed with sawdust of native trees. Specific pheromones can also be used as an attractant. This bottle is attached to a tree trunk or branch. Beetles attracted to the trap fall to the bottom drowning in soapy water or alcohol. This trap should be checked daily.

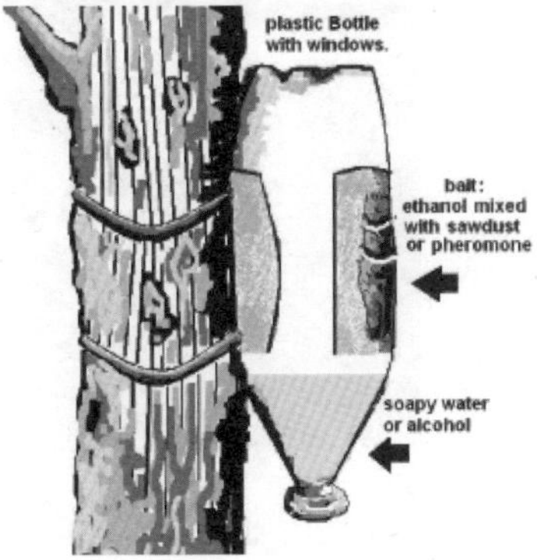

Fig. 74 - Xilophagous bottle trap. *(Drawing: Carlos Aguilar)*

Perforated bottle trap in trees

This trap is made with a plastic bottle (2 Lt.), painted with acrylic glue diluted in water; immerse it immediately in the native wood sawdust, waiting to dry during two to four hours. Then shake it and drill some holes beginning at 12 cm from the bottom up. The holes should have 10 to 12 mm in diameter each, separated by no less than 4 cm. (Fig. 75). Pierce the bottle bottom's center, pass a wire through the hole and bend it like a hook to hang it from the selected branch. In this way the bottle will remain inverted, with the cover down. Finally, through the lower hole (and the bottle already inverted) inject equal amounts of ethyl acetate and ethanol. This trap imitates the branches or trunks and the mixture of ethyl acetate with ethanol imitates the smell of sick or dying timber, which attracts various species of xilophagous beetles. The trap should be reviewed every two to three days.

Funnel-vane trap (Klein et al. 1973)

It consists of two or more vertical fabric sheets interlaced as a helix and the bottom is a funnel a container with ethanol 90 °. Among the sheets there are a curved plastic hose with ethyl acetate inside (Fig. 76). With this trap is possible to capture many species of wood-eating beetles.

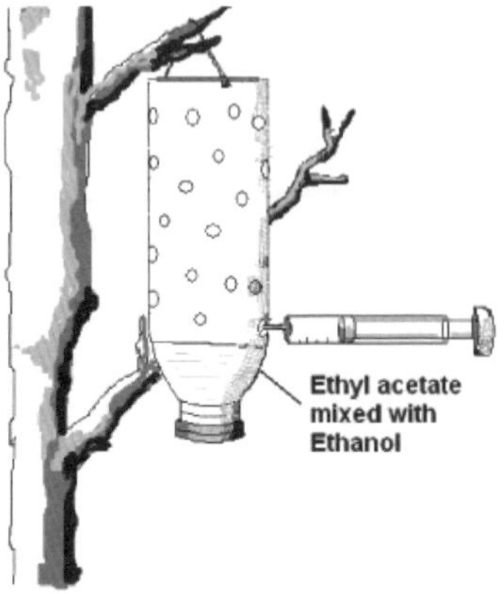

Fig. 75 - Perforated bottle in trees. *(Drawing: Carlos Aguilar)*

Small emergency traps

Pieces of cut branches can be covered with a fine mesh, which has a collecting bottle (Fig. 77). These traps are placed in the diseased branches hanging from trees or shrubs. But also serve those who are on the ground, but not very dry. The important matter is to find the branches containing larvae of beetles. This trap can be left on the tree or shrub live and review it a few weeks or months later, or cut the branch with a small saw and transport to the laboratory, then there put the trap and wait for the emergency beetles. This collection method allows a direct association between the collected specimen and their habitat. But, are difficult to standardize and many species are produced in quantities too low for statistical analysis (Oakland 1996).

Fig. 76 - Funnel-vane trap, with vertical fabric sheets interlaced as a helix. *(Photo: Carlos Aguilar)*

Big traps of emergency

Pieces of previously stacked trunks or branches can be covered with a tent (e.g. fully closed Malaise trap). Various designs of this type of traps from "fabric tents" in which wood is placed pre-cut into pieces (Owen, 1989, 1992; Grove 2000), or stores covering trunks (or part there of) fallen naturally (Yee et al. 2001). confinement, in a "store" or a container of pieces of dead wood or fungi, so the insects that emerge from these substrates are retained and may be collected and determined. Its advantage is that also they allow a direct association between the collected species and their habitat.

The trap shown here is fully closed Malaise trap (Fig. 78). Inside this trap put logs (or branches) recently cut, so when the xilophagous beetles (Buprestidae, Curculionidae, Cerambycidae, etc.) are born they will rise along the walls of the trap and coming in into the deadly bottle located at the top of the trap.

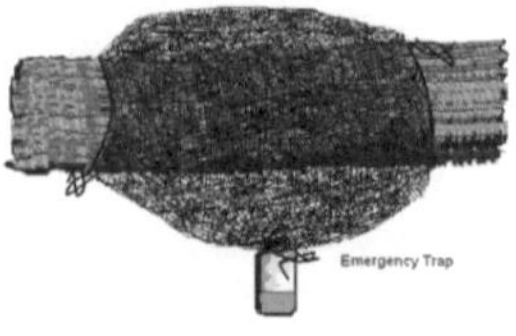

Fig. 77 - Traps of emergency. *(Drawing: Carlos Aguilar)*

Cardboard traps

This trap is to catch beetles that live within the wood or beneath the bark of trees. The trap consists of a corrugated card board wrapped in a spiral (see article "Sampling methods and traps"). The corrugations should be aligned and parallel. Roll the cardboard into a tight cylinder; be careful you don't crush the corrugations. This is hung from trees in-

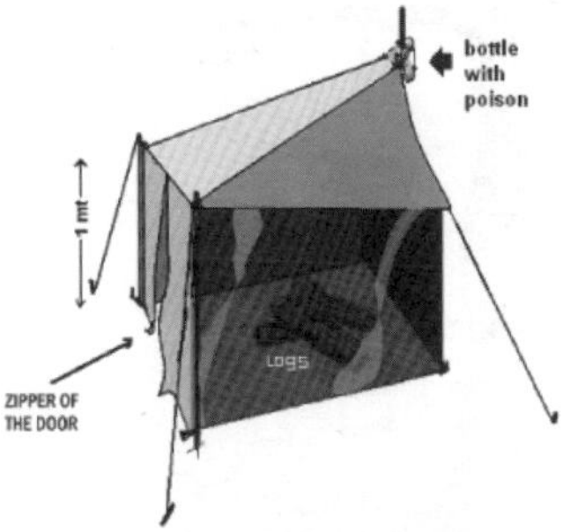

Fig. 78 - Fully closed Malaise trap with logs inside. *(Drawing: Carlos Aguilar)*

side and outside forests, also can be placed directly on the ground. Cardboard traps are good for monitoring organisms moving around in trees or shrubs especially those that are looking for places of refuge to pupate or overwinter. (Fig. 79)

Intercept flight trap in logs (modified from Cobos and Martin, 1987)

This trap consists of a quadrangular sheet of transparent acrylic (2 meters wide by 1 mt high). It has a wooden frame with two legs (1.80 mt long each) that keep it upright and attached to the ground (Fig. 80). The legs are buried at a depth of 40 cm, fixing under ground. At the bottom of the frame there are two containers suspended (one on each side of the acrylic) containing water with preservative liquid (see Article "Formulas", Chapter V). Under the trap, stacked on both sides, fresh logs are located, acting as attractants. The renewal of the logs is done about every two weeks. The beetles are attracted by the fragrant fumes of fresh wood hitting the transparent foil, making them fall into the water container. The traps must be checked at least once a week (Goldarazena 2004).

Fig. 79 - Corrugated cardboard trap. *(Photo: Carlos Aguilar)*

Lindgren funnel trap

The Lindgren trap consists of several plastic cones, arranged and connected on each other, so when insects collide with them, fall into a receptacle with alcohol placed under the cones, where they are collected. Traps should be suspended from a branch or between two trees.

This trap can be made with plastic funnels of approx.15 cm in diameter, each of them with a cut at their narrowest part (aproximately 4 cm diameter), so the hole is somewhat larger and insects can go through. Each funnel has three holes (see picture) through which the wires will pass and keep the structure upright (Fig. 81).

To assemble the trap, take the three wires and twist them in order to make a hook which will allow the set to remain suspended (about 40 cm); then move all three wires through the holes previously made in

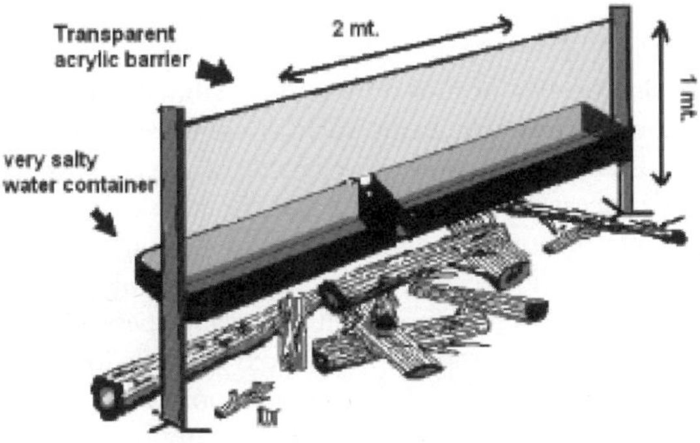

Fig. 80 - Intercept flight trap logs, to capture beetles xylophagous. *(Drawing: Carlos Aguilar)*

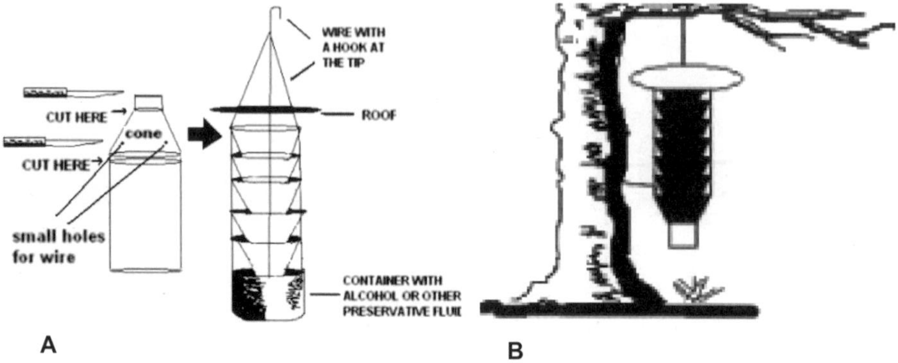

Fig.81 - Lindgren funnel trap (a) How to make plastic bottles (B) How to locate this trap. *(Drawing: Carlos Aguilar)*

the ceiling. In the same way, slide the cones, separated according to your choice. Finally the wires are passed through the receptacle, bending the tip of the wires up. Each funnel can be painted black or purple.

Some chemicals can be used as attractants such as alpha pinene and ethanol. In nature these chemicals are naturally volatilized from some trees when they are weakened, ill or dying. These traps simulate these trees through the artificial flavors of decaying trees, which are elected by a large number of Coleoptera which attack the wood. Other baits are pheromones and/or turpentine.

Also, only salty water can be used in the bottom, but in this case the cones should be painted in purple, blue o black, operating in a similar way to the colored dishes on the floor, but the Lindgren trap is located in one tree. Instead of water, a sticky tape can be placed on the inner edge of each colored funnel; this will make the insects get stuck on the paper and then removed with ethanol.

The problem with these trap is that they occupy a lot of space to be transported, but are effective to collect a variety of beetles such as Anobiinae, Buprestidae, Scarabaeidae, Bostrichidae, Chrysomelidae, Cerambycidae, Curculionidae, Anthribidae, Brentidae, Curculionidae (Scolytinae, Platypodinae), etc.

Alternatives

An alternative to collect xilophagous beetles occurs during dispersion flight phase. Traps are used with containers, containing attractant items such as aggregation pheromones and/or monoterpene compounds and/or alcohols of vegetable origin. This group of compounds is closely linked to the process of attraction to sick, weak or dying trees, such as the alpha-pinene compounds, beta-pinene, limonene or ethanol. The aggregation pheromones may be Pityol with which is possible catches Curculionidae (Scolytinae of *Pityophthorus* genus (Dallara et al. 2000)). Ipsdienol and Ipsenol attract many species of *Ips* and *Lineatina* genus (MacConnell et al. 1977). The compound 3-hydroxy-3-methylbutan-2-ona attracts *Trypodendron lineatum* (Francke & Heemann 1974). Sulcatol compound attracts species of *Gnathotrichus* genus (Byrne et al. 1974). And Chalcogran compound attracts *Pityogenes chalcographus* and *P. quadridens* (Francke et al. 1977).

Collecting beetles in epyphites, algae, lichens, mosses and fungi

Studying the chapter "Where do They Live...", you can see that large numbers of beetles are related to epiphytes, algae, lichens and mosses.

The first action to collect is the observation, e.g.: walking detecting bromeliads, orchids, algae, liverworts, lichens and mosses, which are abundant. The next step will be to find beetles above, below, inside and near these bodies.

Among the roots of many epiphytes, living many beetles and their larvae, so it's important to look carefully at them.

Note that many species of orchids and bromeliads are protected by laws and their collection is regulated. In these cases, they must resort to using methods that do not destroy or damage these plants, such as: brushing, pitfall trap, baited trap and / or shower the plant in situ.

Such activities are time consuming and need a hard work, but the results can be very rewarding for the collector interested in small families of Coleoptera because when looking at this kind of habitat, there are many possibilities of finding new species for science.

Some sampling methods are:

Pitfall and baited traps: To be able to collect vascular epiphytes insects without harming the plants, pitfall traps and baited traps can be placed in the vicinity of the epiphytes (Luna 2005).

Shower plant: Another noninvasive method is put a container under the plant, spill water over the plant, then the water of the container is strained to catch the beetles.

Another method can be spray with pyrethroid insecticide (which is based on water, to avoid damaging the plant) after grap with plastic bag the plant. A few minutes later remove the plastic bag and pour water on the plant to get the beetles in the same manner as described above.

Method of brushing: In the case of collect beetles living in bromeliads, mosses, algae and lichens growing on rocks, logs, stumps or branches, a jar or plastic container, wide mouth, can be placed underneath; then spray with a pyrethroid and then pass a broad brush by the foliage, so the beetles fall into the container and can be easily trapped. However, to catch these beetles living in mosses, algae and lichens the methods are the same than those described for the collection of beetles living in fungi.

REARING LARVAE LIVING IN WOOD

A good method of collecting certain beetles is to transfer the larvae to the laboratory in order to raise them. In this way many more species of beetle can be collected which cannot be found by traditional methods. There are many species which only live as adults for a short time and others which appear only in specific periods of the year. In both cases it is difficult to determine when the the adults of these elusive species will emerge.

Experienced people may be able to rear beetles from their immature stages in controlled environments, but when bark-dwelling insect pupae and larvae are removed from their natural habitats they are notoriously difficult to care for until adulthood (Speight 2005).

Detecting the larvae

When searching for larvae on branches or trunks it is advisable to look for dead or broken branches, preferably those hanging from trees, but not excluding those on the ground, which are not in direct contact with the soil. It is also important to check the sawdust waste in the vicinity of the branch or trunk. Branches covered in soil are unlikely to contain larvae.

To determine if these branches are home to any insects, the whole branch, including the tip and any damaged parts, should be examined using a knife.

Sometimes when a branch is opened lengthwise with a knife, it is possible to detect long galleries with what appear to be small packets of elongated dense wood; these are beetle feces and indicate their presence.

When larvae are found, do not distrurb the burrow, instead try to cut the branch far from the gallery where the larvae are. The branch should be cut across with a small hand saw and moved to the laboratory where the larvae will be reared. Take into account that many species of larvae beetles need different parts of the branch to complete their development.

Some larvae live under the bark and then migrate into the branch, excavating galleries to the tougher or softer places of the wood, depending on the need of the species. Branches or trunks with exposed holes, indicate that the adult beetles have emerged and flown away.

Burying branches and logs

The ends of small logs can be buried either horizontally or vertically to encourage larvae. Some logs may also be placed on the ground (baited or unbaited), and left for some time to allow the beetles to colonise them.

Removal of the bark from these logs some time later will reveal any species present.

According to Leather et al. (2005), to study the emergance of adult beetles, the cut ends of the logs should be covered with paraffin wax or plastic to prevent the logs from drying out. This material can be stored in controlled environmental conditions or simply be left in the forest until the birth of adults.

Arriving at the laboratory

Where possible, i.e. when the branches and trunks are not too large, they can be transported back to the laboratory in plastic bags. If they very small, they should be kept in plastic boxes with lids, such as tupperware or any other suitable container. Do not forget to label the branches or trunks carefully, with the name of the tree to which they belong.

On arrival at the lab, the samples must be stored and labeled correctly, then placed into the collapsible breeding boxes (Fig. 82 and 83) or Petri dishes covered with fine tulle (Fig. 84). The tulle must be secured properly to prevent the insects from escaping. Depending on the species, the adults may emerge within a few weeks but some may take a few months.

Fig. 82 and Fig.83. Collapsible boxes for rearing *(Picture courtesy of BioQuip Products. http://www.bioquip.com).*

Fig. 84 - Plastic container for the larvae's breeding. *(Photo: Jorge Barrett)*

From time to time spray the samples with water to keep them moist. It is important to try to reproduce the conditions and environment in which the larvae were found e.g. the light and moisture conditions.Try not to disturb the larvae in order to keep them in the same position as they were found in.

During winter or the driest season is the best time to search for larvae in branches and trunks. This is when most beetles will still be larvae or pupae. Adults usually emerge in spring or in the wet season; however, in most tropical countries adults and larvae are present in large numbers throughout the year.

Important notes

Andreas Herrmann (pers. comm. 2010) explains how to collect larvae in Germany:

"In the winter and early spring time we usually collect branches of trees and put them into (empty) aquariums or other kinds of glass containers in our home. In the warmth of our rooms the larvae of Cerambycidae, Buprestidae *and all other wood insects develop within a few weeks. You can also catch beetles, which normally live only in the unreachable areas at the very top of the trees. The larvae must be collected after the winter as they need to be frozen and defrosted at least once to develop."*

According to Jonsell and Hansson (2007), rearing bags have been shown to be one of the most efficient methods for collecting saproxylic species from dead wood.

Ulyshen and Hanula (2009) describe in detail the following method for collecting and rearing saproxylic beetles from dead wood: *"Branches are cut and tied together and suspended in large bags from wooden beams with synthetic rope and enclosed within large (170 l) extra-strength black plastic trash bags. In one bottom corner of each bag we attached a clear plastic collecting jar containing propylene glycol. To prevent mold problems, we continuously ventilated the bags using an electric blower (HADP9-1 Cast Aluminum Pressure Blower, of Americraft Manufacturing Co., Cincinnati, OH,*

USA). Air from the blower flows through a plastic PVC pipe (-10 cm in diameter) that runs and the length of the rearing facility near the ceiling. Each side of the pipe has rows of holes into which are inserted sections of clear vinyl tubing (0.95 cm o.d., 0.64 cm i.d.). Each section of tubing carries from the pipe to one of the rearing bags. The bags became inflated with air, thus forming effective funnels. Excess air escapes through a single small hole (-2 mm) drilled near the top of each collecting jar. Overhead fluorescent lights are left on at all times." It is probably not very important to control the temperature or humidity in the rearing facility, when environmental conditions are similar to the place of origin of the samples.

Ulyshen and Hanul (2009) also clarify: *"the screen window are opened along both sides of the facility to allow for air movement and to match ambient conditions as closely as possible."*

Cannibalism is a serious problem in the breeding of insects and requires the raising of individuals in individual containers. The diseases are also a problem: these can be caused by the introduction of an unhealthy specimen in a colony, unsanitary conditions or lack of food, and overcrowding. The boxes should be cleaned frequently and all dead specimens or unhealthy larvae should be removed. Take care not to injure specimens when transferring them to give them more food or when cleaning the cages. Small beetles can be transferred with a camel hair brush. Examine the breeding boxes carefully before placing the specimens inside to reduce the possibility of predators and parasites that can be introduced accidentally. The cages must also be safe from ants, rats and other predators.

Remember, when larvae become adults, they should be killed and stored permanently in the collection or on cotton-bed envelops (or any other preservation method), a small label must be enclosed, with the adult's date of birth and explain that they were raised under laboratory conditions. On another label indicate the country, department or province and specific location where the branch or trunk was obtained, the date, the collector's name, and the geographical coordinates of the collection place are recorded if possible. It is also very useful to Record the scientific name of tree from which they were born. Rearing larvae increases scientific knowledge on these species.

BEETLES IN FUNGI

Where are the fungi?

Fungi are present in an incredible variety of sizes, shapes and colors and are found in many different places. According to Sung-Oui et al. (2001) some groups of fungi beetle hosts are:

Agarics: Agaricaceae, Amanitaceae, Bolbitiaceae, Coprinaceae, Cortinariaceae, Fistulinaceae, Hygrophoraceae, Paxillaceae, Pluteaceae, Russulaceae, Schizophyllaceae, Strophariaceae, Tricholomataceae, etc.

Gasteromycetes: Lycoperdales, Nidulariales, Sclerodermatales, Tulostomatales, etc.

Boletes: Boletaceae, Aphyllophorales (mostly "polypores"): Bondarzewiaceae, Cantharellaceae, Coniophoraceae, Ganodermataceae, Gomphiaceae, Hericiaceae, Hymenochaetaceae, Polyporaceae, Sparrasidaceae, Steriaceae, etc.

Habitats of fungi

Forests, temperate forests, grasslands, mountains, deserts and any other environment available.

Most fungi in nature are located in moist and shady areas, until an average height of two meters, and at the bottom of trees or logs.

The woody ones grow on stems of old trees.

At sites with enough humus and leaf litter on the ground there are fungi.

Many species are found in the vertical walls of canyons.

There are many fungi beneath boards or logs.

Mushrooms grow also under the litter and animals' faeces.

There is also a wide variety of fungi in the inner roads of the forest.

Other fungi grow in very sunny places such as savannas or grassland and even in deserts, the latter almost always at the base of trees or shrubs, often under rocks or on the side where sunlight does not shine too much.

Generally (not always), in the habitats of many species of fungi it is very difficult to use the method of fogging trees. So, checking the fungi directly is the best methods to achieve a variety of beetles which are not normally get otherwise.

When to search fungicolous beetles?

The search of fungicolous beetles may be done at any time of the year, although the more humid and hot season is the most productive one. Remember that there are species of beetles passing the dry season hibernating in fungi.

Beetles associated with fungi

There are many families of beetles that spend their life cycle associated with various species of fungi, including Staphylinoidea, many Cucujoidea and Tenebrionoidea

Each type or species of fungus has a variety of specialized mycophagous beetle and / or other species that feed on micophagous beetles.

The fauna of beetles associated with fungi also varies according to the age of the fungi (pers. observ.). Fleshy fungi live a short period, while those of hard consistency, which live much longer. In hard fungi, the beetle populations become more stable as time passes. Some species of Coleoptera are attracted to very fresh fungi, and other species prefer older or rotting fungi. Generally, fungi are a valuable food resource for Coleoptera (especially in forests) and most species have specific beetles.

Previous steps before collect

When meeting with some fungi in the field, the first thing to do is visually inspect it without removing it. During the day, many fungi have no insects which could be visually located. However, if the same mushrooms are checked at night, many beetles will be found (pers. observ.). It is useful to put a white cloth under the fungus so that the beetles will fall on it trying to escape. The same fungus can be revisited on several occasions which will allow gathering a variety of beetles in the following days.

It is useful to put a white cloth under the fungi, so the beetles fall onto it when they try to escape.

The freshly collected mushrooms may be placed in paper bags, bottles or other suitable container. Samples must not be very wet and so collectors should avoid using plastic bags (Mueller, G.M., et al. 2007).

Before collecting the mushrooms, take pictures in situ is desirable, this will be helpful for subsequent identification.

Some species of fungi have a soft texture. If you are spending several days collecting in the same place, try not to remove the fungi or lichens or other organism, so you can return to the same place several times to review. For that reason you have to tilt the mushrooms (or lichens) a little, avoiding breaking and search in their basidiums or leaves.

It is best to check during the night with a flashlight. These are the most active hours of fungicolous Coleoptera. During the day most insects are hiding in the vicinity of fungus. Almost all beetles found in these habitats are perched under the fungi, or within them; in the case of mushrooms they are found in basidiums and sometimes on them.

Extracting beetles from the fungus

Due to the difficulty of extracting the beetles, plus the fact that apparently they bear no insects, and/or the short time to be working on the site, it is recommended to harvest the mushrooms and perform the extraction in the laboratory. But before the harvest it is very useful to take pictures in situ, and it is important to check carefully around, collecting a sample of the substrate where the fungus lives and, if it is fruiting on loose bark, it must be revised to look for beetles hiding there, using a knife or small axe (Fig.85). The bark must be carefully lifted, examined and removed the crust along with the fungus; all that goes into cotton or yarn bags preferably. Each sample should be kept in separate bags and labeled immediately.

If you want to extract the beetles in situ, fungi must be put inside an airtight box or bottles of suitable size and sprayed with an aerosol based on pyrethroid. Then the fungus is crumbled and dead insects are removed.

Fig.85 - Breaking a fungus with a knife to collect fungus beetles *(Photo courtesy Andreas Hermann).*

In fungi with very hard consistency, it is difficult to extract insects. In that case, the mushrooms must be kept in a large glass jar and sprayed with turpentine using a water spray, trying not to soak the fungus with turpentine, preventing drowning and allowing the beetles to go out.

Later, when no more bugs are going out, break the moss with a knife, slip the remains (use a colander or sieve of mesh 5 mm) on a white cloth or plastic and trap with a brush (fine tip and soft hairs), previously soaked in ethanol, the beetles falling on the fabric. It is also effective to put the mushrooms in a Berlese funnel or Winkler/Moczarski eclector modified with light (see Chapter III) locating below a jar with ethanol. Put a lamp (100 Watts) on the sample; the heat makes the insects go out and fall in the ethanol.

An incubator with good results is that of Oakland (1996). The mushrooms collected must be put in a cardboard or plastic box painted in matt black color. At the bottom of one side, place a transparent glass or plastic jar. Being more enlightened, it will attract insects by phototropism and when they leave the mushrooms will fall into the bottle (Fig. 86).

To capture the little beetles living in fungi it is not appropriate to use an aspirator, risk-

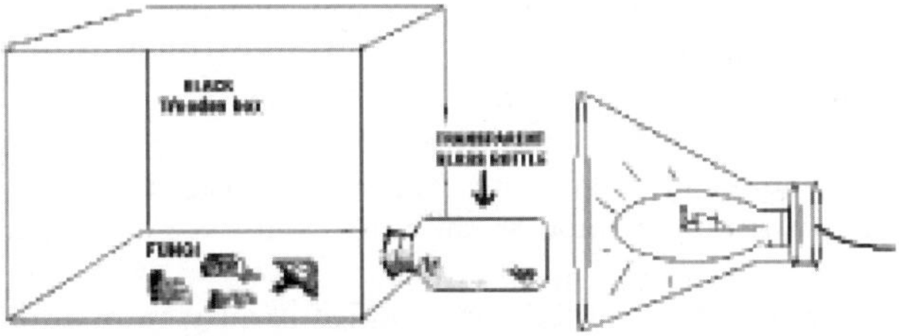

Fig. 86 Black box with transparent jar, used to collect the beetles that live in mushrooms (modified from Oakland 1996).

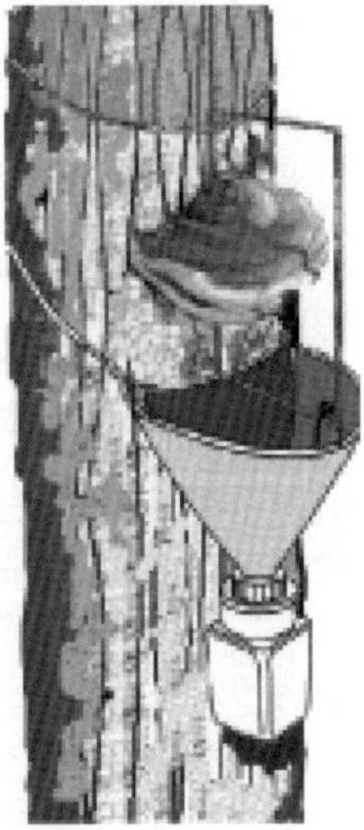

Fig.87 - Flight intercept barrier trap, for used on trunks with fungus (Modified from Kaila 1993)

ing inhaling toxic spores. A mechanical aspirator sampler can be used (see "Basic collection kit", Chapter V), but it is best to use a soft brush with the bristles previously soaked in ethanol, touch the beetle with the tip of the brush and so the insect will be attached to the brush and may be placed directly in a jar with ethanol or other poison -ethyl acetate, potassium cyanide, etc-.

When small beetles are collected it is helpful to use a hand lens (of 5X to 8X), in order to detect them among the remains of what is being extracted. This magnifier can be used either in the field or in the laboratory.

Other larger beetles such as Erotylidae, Endomichidae, Tenebrionidae, Melandryidae, Brentidae, etc. are nearly always found at night -sometimes during the day- walking among the fungi. They are collected with blunt soft tweezers or with bare hands.

If there are mushrooms on logs or stumps, it is convenient to put the beating sheet beneath before touching them, because the insects will drop rapidly when disturbed.

Another approach is to use flight intercept barrier. These traps can be placed on the trunks, near of the fungi or among them (Kaila. 1993). It is useful to put a few sheets as vertical barriers attached to fungi so that the insects go directly to the traps (Fig.87). These barriers can be done of cardboard, plastic or metal. With this type of traps many specimens can be caught, and with good luck interesting specimens are available, but these traps must work for several days and be reviewed at least one time every two days. Small jars with ethanol can be used as bait; this attracts the arriving beetles to rotten fungi (Jonsell et al. 2003). Also can be used antifreeze or simply water with soap can be placed in bottom of the trap

How to breed fungi in the laboratory

Is best to collect at least 3 or 4 mature, immature and complete specimens of mushrooms extracted from the substrate with a knife (Schigel (2008).

The fragile pieces are placed in cardboard boxes or wrapped in aluminum or placed in plastic containers. Woody or leathery specimens are placed in paper bags. Small containers of 1/4 lit - 1/2 lit or more can be used for small species, and can also be transported in washable cloth bags, which can even be suitable for breeding. Many fungi tolerate drying and can be grown for many years without problems, but others die within a few hours after drying (Mueller et al. 2007). Refrigeration is a good way to keep some material in the laboratory for a few days. Humidity in refrigerators is often very high and can lead to contamination by other fungi in the substratum. So refrigerators can become incubators because the substrates contain many other fungi. To freeze fresh samples can be useful, including with live insects; however, be careful if insects and fungi must be kept alive, because many organisms -including those in cold climates- appear to be relatively tolerant to cold. Refrigeration or freezing is a good method to kill beetles and make dissections (Mueller, et al. 2007).

When the journey is long, the mushrooms will be transported keeping them in the same position as they were found in their original location. Remember that they are very fragile, just as the larvae contained inside, This step requires great care.

Basidiocarps will be hand collected from appropriate sites where all basidiocarps will be examined for the presence of beetles (Sung-Oui Suh et al., 2001).

In the laboratory (and also in the field), keep them in collapsible cages lined with fabric. It is advisable to keep open the breeding box for a few days until the excess of moisture comes out and then keep it closed until the adults begin to emerge. In the breeding box agarics and watery polypores (Schigel 2008) must also be included to allow the larvae become into pupae and not to die earlier.

According to Schigel (2008) to prevent mold forming inside the moist boxes, sawdust can be used, which should be changed regularly. For pupae, it is also very good the wood sawdust, sand or other fine materials. Occasionally the mushrooms have to be sprayed with water, but not too much, to avoid the larvae death. Dr. Schigel (2008) recommends putting some dried Sphagnum in the brood chamber, which can balance the moisture after spraying with water. Avoid the green moss that spreads rapidly in the breeding box.

Mushrooms can be kept outdoors or in the refrigerator to 4 ° C, between 2 and 3 months, and later keep them between 2 to 3 months at room temperature for emerging adults. The Polypore intended to feed the larvae can be collected throughout the year, regardless of the area in which they grow (Schigel 2008). OBS: Remember that many mushrooms are poisonous or hallucinogenic, so you should wash your hands thoroughly after handling. While handling, avoid touching eyes, nose or mouth. So it is convenient to wear thin plastic or nylon gloves to perform these activities. However, even wearing gloves, wash your hands after handling fungi. It would be very useful to get an illustrated book or field manual to be able to recognize which species may be poisonous in the region to be visited. Moreover, there are certain countries where people gather mushrooms to eat; when those places are fumigated, the mushrooms have to be removed and destroyed after review, to prevent other people from becoming intoxicated. The place where mushrooms are grown must be totally separated from the rooms where the permanent collections of insects are, because many species of beetles living in fungi attack collections.

How to keep the specimens

To keep the newly collected fungi, Schigel (2008) recommends using Eppendorf tubes (1.5 or 0.5 ml), which are easier to use in the field and serve to keep them for a long time in a suitable room.

Each tube should be labeled preferably in the same location where the specimens were collected which will help to remember important information.

In the field, it is important to have a different method to that used at the laboratory to mark the tubes with specimens; permanent markers are suitable for that purpose but writing a label with India ink or a pencil in each sample is always better since other types of ink (including the permanent markers) disappear upon contact with water or ethanol.

It is advisable that each specimen has its own label, independent from the other specimens. But in the field all specimens trapped in the same fungus can be assigned to a single bottle. After reaching the laboratory, each specimen will be separated and labeled separately.

Some researchers write complicated codes in their collections, with dates, abbreviations and other data, but the simply numbering is easier and more reliable for a future database.

But at the collection site some numbers can be used which will temporarily replace certain data, for example: "Leg. John A. Smith" can simply be noted "001", but we must remember that number at the laboratory and change it to "Leg. John A. Smith" in the final label. For that purpose, you can have a notebook in which the numbers and their meaning were noted earlier. Many times in the collecting site there is not too much time to write long labels. Using a method similar to that of numbers, the description in a field notebook will be easier for each specimen.

Another method consists of a written label with lots of data to be filled:

- Specimen N°…
- Environmental conditions...
- Kind of soil …
- Height above sea level...
- Date...
- Specific location …
- Collector...
- Country...
- Fungal species or on what plant the fungus was found and/or beetle...
- Type of wood...

This type of list is useful but also risky as it can get wet with rain as you enter data, etc. Also a voice recorder or a computer (laptop) can facilitate the work, writing on a database all possible information, but you also risk of running out of batteries in the collection site, or the PC or the recorder fail just at that moment, or rain begins to fall, so it is better to write the data directly into a paper label, which will accompany the specimens collected, and these labels can be supplemented with some of the methods cited above.

The specimens obtained can be kept in vials with ethanol (90%), or cotton beds or other methods discussed in this book.

Special labeling fungus beetles

Each label (for each specimen) must include at least these data:

- Country
- Department, state or province
- Specific location
- Date (day, month and year)
- Collector's name.

And in another label the following data must be included:

- Type of environment where the specimens were collected (forest, prairie, desert, mountain, garden, etc).
- Where the fungus was collected (wood, soil, leaves, faeces, etc.).
- Form of fungus: size, color, smell.
- Form of lichen or other vegetation.

- Better if all this information is accompanied with a photograph taken in situ.
- Other information considered as relevant can be included: height above sea level, humidity, illuminated or dark place, etc.

Each method can be used according to the collector's choice, and they can even be combined. It is advisable to get many fungi belonging to many species, but do not abuse with the removal of these organisms in nature, since intact specimens should be left in the collection sites. So it will be possible to check them in the future and minimize the intrusion on balance and environmental preservation.

Materials

(These materials are optional, depending on the method of collection)

- Polyethylene or plastic bags -preferably dark- to store vegetal or fungi samples
- Sawdust
- Blunt soft tweezers
- Mechanical vacuum cleaner
- Knife, little machete or a small hand axe
- Flashlight
- Jacket or belt with many pockets
- Small vial containing ethanol 75%
- Toilet paper
- Aerosol poison (based on Pyrethroids)
- Small water spray (for hair, like those used in salons)
- Ethyl acetate
- Gloves
- Vial dropper for ethyl acetate
- White trays
- Buckets with water
- Deadly bottles of different sizes
- Berlese trap
- Interceptor bottle traps on trunks
- Antifreeze
- Ethanol (90%)
- Fine-tipped brush with soft hairs
- Extraction box with jar and whit external light
- Boxes or containers for storing mushrooms in the laboratory
- Turpentine
- Beating-sheet
- Berlese funnel
- Plastic bottles with water (2 lt)
- Little plastic container with pressure top
- Soft white cloth or plastic tray
- Large colander or sieve
- Soft brush (5 cm. approx. width)
- Hand lens (5x or more)
- Vials, cotton beds or paper envelopes
- Field notebook, pencils and paper labels

- Thin plastic gloves or surgical
- Canteen or bottle with clean water for washing hands
- Soap for washing hands
- Liquid preservative for pitfall traps

BEATING METHOD

One of the most widely used collecting methods for beetles is beating, combined with the use of a "beating beating-sheet" or "beating sheet" (Fig. 88). The method consists on locating the beating sheet under the branches of bushes or trees —taking care in order to avoid shaking the target plant— in which Coleoptera may exist. The branches are hit with a stick (approximately 80 cm. long and not too heavy). This way the insects get surprised and play dead, releasing themselves and falling off from the branches —in a reflex action that would normally make them reach safe shelter amongst the lower vegetation or the leaf litter— to finally falls on the beating sheet. There they can be picked up in two ways depending on the size of the insects: by hands (larger beetles) or with the aspirator (smaller specimens).

The hits should be quick and mild enough as to prevent both breaking off the branches and damaging the insects hidden amongst them. Many leaves, twigs and other plant debris usually fall onto the beating sheet together with the insects; it is necessary to carefully spread them over the surface of the beating sheet in order to get the beetles at sight. This can be partly done by gently shaking the beating sheet, so the insects can surface over the debris. The specimens should be picked up as quick as possible, most Buprestidae and Mordellidae, for instance, escape quickly after falling onto the beating sheet. Shadowing the beating sheet with the body usually helps in retarding at least for some seconds the reaction of these insects.

Fig. 88 - Practicing Beating sheet. *(Photo: Jorge Barrett)*

The collector must always take special care about snakes and wasp nests that might be eventually hidden amongst the vegetation. Some special attention must also be taken before beating a plant in bloom. It must be carefully checked for fast flying flower visiting beetles like many buprestids, cerambycids and members of other families that do not normaly play dead. They must be collected first either by hand or netting or previously sprayed with ethyl acetate or chlorophorm (even faster than ethyl acetate) before proceeding with the beating. Beating can be frustrating under windy weather for two main reasons: 1) many insects that fall on the beating sheet are difficult to collect since they are blown away by the wind before being picked up,

Fig. 89 - Tree in decayed process. *(Photo: Jorge Barrett, Uruguay)*

and 2) under strong wind many beetles (e.g. some lamiine cerambycids) are firmly seized and, when hitting the branches, they do not fall down easily.

It is convenient to keep the aspirator hanging from your neck by means of a long piece of string, thus allowing free hands while beating and avoiding loosing the aspirator itself. Remember that the aspirator must be periodicaly emptied in a larger vial containing toilet paper and some drops of ethyl acetate applied with a dropper. Always keep in mind avoiding putting many specimens at the same time in the same vial because they will bite and damage each other. It is recommended to have one vial for recently captured individuals and another one to periodicaly transfer already dead specimens. Care must be taken as to avoid mixing specimens from different places.

Although every kind of vegetation is potentially good to get different species, it is important to know how to choose the conditions that will yield a better collecting in terms of diversity and number of specimens. It is important to choose places where the vegetation is diverse: "Variety of plants is equal to variety of beetles." The isolated trees —either near the forest or in the middle of a savanna— are usually not good for beating, although some isolated species of Curculionidae and Chrysomelidae can also be found in these situations. A relatively undisturbed forest with a rich understory and spanning over at least a hectare in extension is a preferrable habitat. The quantity of species and individuals is directly related to the existence and density of the understory. A recently burnt forest is also very profitable for this activity since many species —mainly Cerambycidae— are attracted by the scent of burnt wood (they are also atracted by the scent of the wood being cut).

The best areas for beating are along the forest edge and along paths into the forest, especially cart paths, old paths, or recently open roads inside the forest of more than four meters (12 feet) wide, and with available understory along the roadside. The untouched, denser and poorly illuminated parts of the forest, away from the edge and paths are not good for collecting. Insects in these areas tend to spend more of their lives up in the canopy.

Trees and bushes having small leaves and with a lot of green foliage blended with sick or dry branches tend to be the most productive ones. It also gives good results to hit heaps of dry or burnt bamboo (not the big and thick branches) blended with other plants. Branches and leaves that form bridges among trees are very good places to beat. Branches that fall forward, forming cascades blended with creepers are also very profitable as also do very dense small branches with a mixture of many green and withered small leaves (there you can find Carabidae [Agra spp.], Cerambycidae, Elateridae, Brentidae, Curculionidae, Languriidae, Chelonariidae, Tenebrionidae, etc.).

Trees or bushes with very big leaves, don't generally throw good results, neither trees whose leaves are covered with earth and thus roadsides along heavily transited roads are

Fig. 90 - The ideal foliage for beating. *(Photo: Jorge Barrett, Uruguay)*

Fig. 91 - Way to grab the beating-sheet. *(Image given by BioQuip Products. (http://www.bioquip.com).*

Fig. 92 - Beating sheet. (Image given by BioQuip Products. (http://www.bioquip.com)

not appropriate. Beating method is neither very effective on trees with very big and isolated branches, nor those with a clean trunk and branching starting above 2 m. over the ground.

Lianas are also good for several species of beetles, especially Cerambycidae. It is convenient to make some wounds in these lianas and check them in the following days. It is suitable to visit the same plant in different dates since different species may appear every consecutive day. Beating can also be tried on fallen trees with part of the trunk and branches still away from the ground. The beating sheet can be placed under these sections and then they can be firmly but gently beated, so insects can fall on the beating sheet without suffering damage.

With some practice and experience we will know what places are more productive and what kind of vegetation is better for beating. The practice will also show us how to recognize and handle the insects that fall onto the beating sheet. Many species of cerambycids and curculionids, for instance, look like plant debris while playing dead.

The best hours for beating are early in the morning and at dusk. Many interesting beetles can also be found at night. Night beating would preferably be carried out by two people, one of them illuminating with a good lamp while the other one hits diverse plants. In this situation it is recommended to have ears protected by cotton plugs and mouth and nose covered with a mask, as many insects will come attracted by the light.

Materials

Beating sheet
Hard stick or branch (70 cm. or 1 mt. long)
Aspirator

Poison vials (several)
Soft Pincers
Belt or Vest for the vials
Dropper with ethyl acetate (like those for eye drops)

SWEEPING METHOD

A large number of beetles living in the foliage, flowers, grasses, etc. are often undetectable. In many grasses and small height plants it is difficult or impossible to use the beating sheet. In these cases, to collect the beetles that are in the grass or low vegetation (including shrubs) is very useful to use a sweep bag (Canaday 1987).

A variety of Coleoptera, mostly small families, can be collected with this instrument, such as Mordellidae, Elateridae, Curculionidae, Buprestidae, Chrysomelidae, Bruchidae, Anthicidae, Pedilinae, Cryptophagidae, Nitidulidae, Helodidae, Dascillidae, Cantharidae, Dermestidae, Cleridae, Melyridae, Dasytinae, Othniinae, Meloidae, Rhipiphoridae, Phalacridae, Coccinellidae, Lagriinae and several others. According to Gadagkar et al. (1990) and Young (2005) sweep bags can collect samples that are completely representative of the population or community of insects.

Fig. 93 - Sweeping vegetation. (Picture courtesy of Frank Rodríguez, Venezuela)

The sweep bag is like a butterfly net, but instead of tulle, it has a white, thick fabric resistant to abrasion, the bag is approximately 70 cm long x 45 cm diameter. The ring must be very hard and not flexible, with a handle about 70 cm long (Fig. 94).

This sweep bag is used to sweep all kind of vegetation, mainly those of low height as grass, bushes, etc. Although it can be passed through all plants, in the case of the tallest (trees or shrubs) it is preferable to use beating sheet. Be careful around thorns and burrs. Also avoid collecting from wet vegetation. The "sweeping-bag" should be plied in an energetic figure-of-eight motion, with sufficient forward movement to prevent overlap of sweeps. (Fig. 93). After 10 to 15 sweeping strokes, the bag should be examined and the specimens extracted. This can be done in a number of ways. Close the bag with your hands pressing the fabric below the ring, and then open it slowly and catch the beetles by hand or with an aspirator. It is necessary must be quick

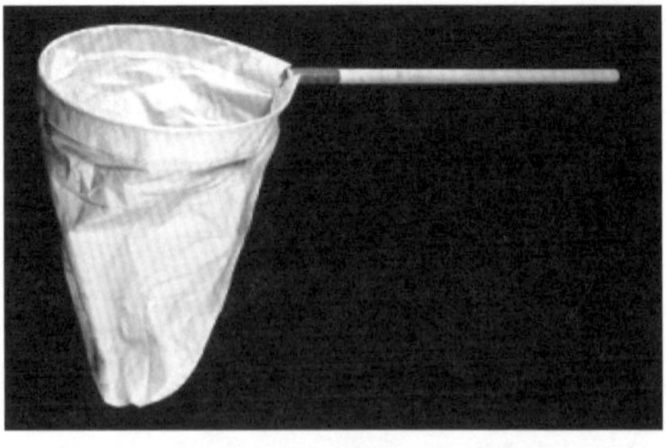

Fig. 94 - Sweep bag. (picture courtesy of BioQuip - www.bioquip.com).

when catching the trapped beetles since many are agile flyers. It is sometimes necessary to place the entire bag in a large plastic bag and kill all the captive insects with vapours of ethyl acetate.

According to Reling & Taylor (1984) and Young (2005) sweep bags may be towed by planes or trucks (see Article "Car-catcher", Chapter III).

Experience is the best way to determine the places where the sweep bag should be passed, which is achieved through practice. But to get a productive collection it is best to choose those open sites, inside or near forests (or savanna), where there are many small plants of various species, preferably those with flowers.

FOGGING TREES

Fogging trees is good method to get countless species of beetles. But, it is necesary the owner's consent and, in some countries, obtaining a special permit from the government.

According to Muir and Gambrill (1960) and Anderson et al. (1994) spraying tree are selective for larger insects that do not stick to the foliage, and among smaller beetles can be collected as much as 50% or less. This is because many small beetles live in the cracks, holes, curled leaves, epiphytes or under barks, and when one tree is sprayed these insects are not easily captured. In order to realize the number of species which can be found in the upper canopy of a forest, a study by Paul Araujo et al. (2001) is given as example: they sprayed several trees in the Ecuadorian Chocó, in a single habitat, and the results were published in 2005. Araujo and colleagues found 1,054 morphospecies of beetles. These researchers concluded that in the canopy the beetles are present in several microhabitats:

Dead branches with dry leaves suspended in the canopy: in those places the following families are present: Carabidae, Staphylinidae, Cerambycidae, Chrysomelidae, Endomychidae, Erotylidae, Tenebrionidae, Ptiliidae and Phalacridae, Cleridae and Scydmaenidae.

Edaphic microhabitats of the canopy: formed by the accumulation of decomposing leaves layers on trunks and large branches. There are Pselaphidae, Staphylinidae, Scarabaeidae and Curculionidae.

Sets of fungi: formed on decaying vegetable material suspended in the forest canopy may contain Leiodidae, Scaphidiidae, Eucinetidae, Lathridiidae, Phalacridae, Erotylidae and Endomychidae; also predators such as Carabidae and Cleridae.

Aquatic microhabitats of the canopy: inside bromeliads, branches' axils and hollows of the trees, there are Dytiscidae and Hydrophilidae.

Tree grasslands: they consist of herbs colonizing the canopy, with several species of herbivorous beetles such as Curculionidae, Chrysomelidae, etc.

Ant and termite nests: they may have Staphylinidae, Pselaphidae and Scydmaenidae.

Vertebrate nests: they are inhabited by Dermestidae, Cryptophagidae and Lathridiidae. This explanation shows that a variety of beetle species can be collected by fogging trees.

Steps to follow before fogging one tree (Fig. 95)

Should be considered the size of the tree, while smaller will be easier to fumigate. It also depends on the equipment used for fogging. There are simple equipments, such as backpacks used in the farms, there are spraying machines (Golden Dyna Fogger) and other more specialized equipments (gasoline or electric engine) sold in shops specialized in entomology equipment.

The steps:

1 - Choose a tree in bloom into the jungle or on the edges; may be also a tree in a savanna, where beetles flying can be observed.

2 - Remove the weeds under the tree.

3 - Extend a white cloth (sheets or plastic) on the ground at the foot of the tree. Or to avoid easily damaged for beetles and to prevent small insects from being left behind, can be used hanging funnels (Fig. 95), instead of sheets.

4 - Once cloths or plastics are put in due place, carry out the fumigation. Choose a windless, preferably during spring.

5- If many specimens are stored in a jar with alcohol, it is better to sacrifice them in ethanol 80%, mixed with ethyl acetate (equal parts) to prevent further hardening of insects. This is a semi-conserving solution and after a few days the specimens must be moved to alcohol at 75° where they can be kept forever.

Fig 95 – Collect with hanging funnels.

Quimical compound used as poisons

Depending on the tree to be fumigated, the adequate chemical mixture and methods are:

- 3% Permethrin in a diesel solution. This mixture is biodegradable and in just one minute it can be spray an entire tree. It has been used by Araujo et al. (2005).

Note: Prevent possible accidents using gas masks and avoiding the presence of children.

- **Aluminium phosphide**: it is a very effective quimical compound also, sold in tablets or capsules, in stores dedicated to agricultural chemicals. The Aluminium phosphide dissolves in water and in contact with the air releases a very toxic gas (Phosphine). This procedure must be done very carefully. Perhaps it is better to use other less dangerous contact quimical compound.

- Pyrethrin is a natural insecticide, derived from the Chrysanthemum flower (Pyrethrum). It has been used to control pests for over a century and is generally considered quite safe. Pyrethrin should not be confused with pyrethroids (such as Permethrin) which are synthetic insecticides. Pyrethrins are considered one of the least toxic insecticides to mammals. It is biodegradable and breaks down with exposure to light. According to Leather et al. (2005) fogging is typically used from 5 to 10 minutes of time in one location.

Materials:

- Gas mask (if the poison is dangerous).
- Plastic cloth or some soft but firm material enough to cover the ground below

the tree. These must be white.

- Bottles with ethanol.
- Backpack sprayer with a long hose.
- Sprayer bottle device. You can also use a sprayer with an electric motor.
- Field notebook.
- Pencils.
- Soft blunt tweezers.
- Shovel or machete to weed.
- Aluminum disulfide pills.
- Permethrin (or other quimical compound, see above).

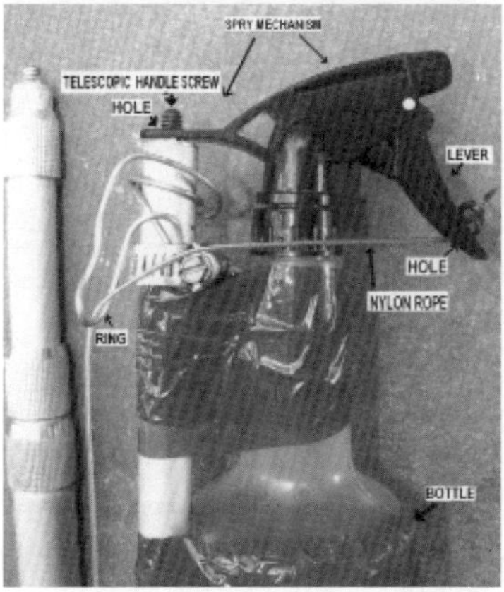

Fig. 96 - Aerial spray bottle (made by hand), detail. *(Picture: Carlos Aguilar, Paraguay)*

Sprayer bottle device for fogging trees

This spray bottle is useful for spraying small areas in the tops of the trees (Fig. 96 and 97) and manufacturing is very simple. The elements used are at low cost.

Materials:

1 Sprayer Bottle (1/2 lt),
1 Adhesive tape (scotch),
1 Telescopic handle of butterfly net (3 mt or more),
1 Metal clamp,
Nylon rope (several meters)
Piece of thick hard wire (20 cm).

How to make it

Fig. 97 - Spraying one tree with spray bottle device (made by hand). *(Picture: Carlos Aguilar)*

The wire must be tied in the telescopic handle and both ends are attached with metal clamp forming a loop. The lever of the bottle has a hole near the tip. The opposite end of the spray mechanism has a hole also where is inserted the tip (screw) of telescopic handle.

The bottle must be attached to the telescopic handle with “vinyl electrical tape”, but in a way that the bottle can be opened to introduce the quimical compound for fogging trees.

In the hole of the lever will be introduced one end of the “nylon rope” (which also goes through of the wire loop). The rope will run along the telescopic handle, so the operator can pull the other end of nylon rope, will trigger the mechanism that will produce the spray.

Join the different parts as shown in picture detail (Fig. 96).

VACUUM SAMPLING IN SHRUBS AND TREES

According to Leather et al. (2005), vacuum sampling can be used to collect insects from vegetation such as weeds, shrubs and small trees.

All machines work the same way, using a vacuum device to suck the insects into a collection device. Air is drawn through a suction tube and a collecting bag which captures the arthropods. The bag can be removed from the tube and its contents may be emptied on site.

According to Claire Ozanne (in Leather et al. 2005) there are two basic models of vacuum samplers:

Vacuum hose with a diameter greater than 20 cm (W- type) that include the Dietrick vacuum sampler (D-vac) (Dietrick 1961) and the vacuum sampler by Thornhill (Thornhill 1978). These vacuum samplers are effective for extracting insects that live in the lower parts of vegetation (Stewart and Wright 1995).

Vacuum hose with a diameter less than 15 cm (N-type) which are either purposefully-built or converted from garden leaf vacuums (Stewart & Wright 1995; Buffington & Redak 1998; Ozanne, in: Leather et al. 2005). According to Ozanne: the N-type models are recommended because they are quieter and the lighter weight makes the easier to manipulate and manage in the field.

Fig. 98a - Sampling pecan foliage (Picture courtesy of Russell F. Mizell, 2010).

The structure of a small household vacuum can be modified to collect insects, such as a model recently designed by Russell Mizell III (Fig. 98a and 98b) (pers. comm. 2010) that can be viewed online in: www.ufinsect.ifas.ufl.edu/hoover_sampling.htm.

Three main collecting methods can be employed: holding the nozzle over vegetation, searching an area of the canopy for a set length of time (Herms et al. 1990); or vacuuming the foliage along transect or across a 180° arc (Buffington & Redak 1998). The method will depend on the density and height of the vegetation, but in all cases the narrow-hosed models will be much easier to use than the heavy wide-hosed types (Leather et al. 2005).

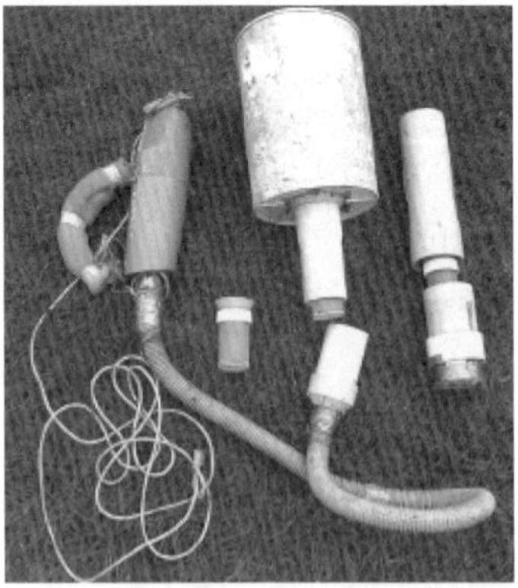

Fig. 98b - The picture shows the vacuum cleaner and the two types of collecting heads – large 5 gal bucket for pecan terminals and the 4" PVC for single stem plants. *(Picture courtesy of Russell Mizell, 2010).*

According to Henderson y Whittaker (1977) when vegetation is wet insects remain in the foliage, so the collection should be done with vacuum cleaners under dry ambient conditions. Aspiration can be performed on individual plants or directly onto such vegetation as in pastures or forest understories.

Samples should be cleaned by hand and this involves much work, because some insects are attached to the plant material and large plant fragments should be carefully examined.

The extraction of the samples using Berlese funnels or Winkler/Moczarski eclector or with a black box with jar and light is a good alternative (Sutherland 1996). However, according to Leather et al. (2005), it is better to keep the samples in ethanol before cleaning to prevent loss of specimens by predation. Later, the insects can be removed by immersing the samples in water (Dondale et al. 1971).

BEETLES ASSOCIATED WITH VERTEBRATES

In the Neotropical region some Scarabaeidae (*Uroxys*, *Trichillum*, etc), only on the body of sloths and other mammals; the fur must be carefully searched to collect the beetles, which is a challenging task. The beetles travel on the sloth until it descends from its tree to defecate, and the beetles then oviposit on the sloth dung. But most Coleoptera which live in association with vertebrates must be sought within and near the burrows of vertebrates which make nests underground, where it is most likely to find them using techniques detailed below.

Fig. 99 - Removing material from the bottom of a burrow, using a large spoon auger. (Photo: Carlos Aguilar)

It is not the intention of this book to teach how to recognize vertebrate burrows but we want to give some examples to provide the reader a foundation on the subject. On the other hand, to identify what type of burrow belongs to each species of mammal requires much experience in the field.

Different species of vertebrates (including some lizards and tortoises) dig different types of burrows and build different kinds of nests but often they do not differ much from one another. But with a little experience, they can be recognized thanks to the traces left by the animals living there. For example, the footprints on the floor at the entrance to the burrow, the arrangement soil expelled during digging, the marks of feet and dragging of their bodies (in the case of lizards), etc.

When the burrow has been abandoned for a long time, there are spider webs, the soil of the entrance is usually hard and the grass is tall.

Many paths constantly traveled by rodents can be identified by the tunnels left under the tufts of grass in open areas, or beneath the herbaceous layer in woods and closed forests. These paths are easily found in the fall because the vegetation does not quickly grow back (Villalba and Yanosky 2000).

In South America, some animals like *Myocastor coipus* Molina, build their burrows in cliffs, on the banks of rivers. Others like *Lagostomus maximus* Blainv, place different objects at the entrance to their burrows such as sticks and stones, and the surrounding area is completely clean of plants. The *Sylvilagus brasiliensis* (Linnaeus) build their nests in natural cavities (rocks, roots, tree trunks) and the entries are full of hair. *Euphractus sexcinctus* (Linnaeus) digs burrows with the entrance in shaped like a "U". Armadillos in the genus *Dasypus* and *Priodontes* have very characteristic burrows, as can be noted with a little practice. (Villalba and Yanosky 2000). Finally, all Pink-Fairy edentates (*Chlamyphorus*) are fossorial par excellence and dig large tunnels in dry and very sandy areas, where forest vegetation and Chaco forest prevailing. They rarely leave these tunnels, except at night. They move the earth to the sides of the entrance and when they come out of the den and leave marks that make a triple track (Cabrera and Yepes 1960).

Rodents, depending on the species, can be found in meadows, marshes, streams, salt marshes, forests, scrub, deserts, hill ridges, peaks of rigorous moor, near the sea and nearly every imaginable ecosystem.

Burrowing species have a marked tendency to return and reoccupy the same dens in a given season; sometimes these caves have many feces, whose state indicates whether these burrows have been temporarily abandoned (Cabrera and Yepes 1960) or are occupied.

The easiest way to collect Coleoptera associated with vertebrates is simply to examing the faeces (sometimes mixed with grass) which are near or inside their burrows. All this can be put in a Berlese funnel or Winkler/Moczarsky eclector.

It is also very interesting to dig open the burrow with a shovel and sift all the contents with a sieve of appropriate size. This is a hard work, but interesting species can be obtained. If there is a fecal pile or fecal chamber, most beetles might be here.

Another method consists of excavating the entire burrow. If it is large a bulldozer or mechanical excavator can be used and then its contents will be examined at the laboratory. The problem is that the procedure is very costly and, of course, demands a lot of work.

A simple method is to use the same feces of the vertebrates as bait and to bury the trap cup or jars (as a pit fall) with the bait at the entrances of the burrows. That will attract dung beetles that live in these places.

Those persons devoted to searching these dens, should do it in all types of environments, including high mountains which are the home to several species of rodents. There are species of beetles, such as some Scarabaeidae *(Onthophagus),* which are only known in a single site in America, and these have been collected in hills or mountains. Most species of Coleoptera in burrows (primarily in the Americas) are copro-necrophagous and only some species in this continent are adapted to living on certain types of feces or carrion. One example is a species of Scarabaeinae *(Tetraechma sanguineomaculata* Blanchard*)* that has only been found associated with the feces of "viscachas" *(Lagostomus maximus (*Desmarest*))* (Halffter and Martinez 1977), although it has recently been collected with white light in the Paraguayan Chaco (Aguilar 2003).

Other rodents *(Neotoma* spp*, Pappogeomys* spp*, Geomys, Thomomys, etc)* also have species of Scarabaeidae *(Ataenini and Eupariini and Onthophagus* spp*)* associated with their feces (Zunino and Halffter 1988) and Leiodidae (*Catops*, *Ptomaphagus*) (Peck and Skelley 2002). Two genera of Platypsyllinae live are in beaver lodges (Arnett et al. 1981). Some species of beetles live associated with bat feces. Moreover, other species live in bird nests (Histeridae, Dermestidae, Trogidae, etc.). A good way to collect they consists in beating the recently abandoned nests and take the beetles with an aspirator.

According to George Hangay (pers. comm. 2005), there are several species of beetles in the stork nests.

Hangay says that some representatives of other Coleoptera families (Leiodidae, Ptinidae, Staphylinidae, etc.) are also associated with burrows. He noted that he successfully used a wire tray shaped like a small flat shallow box, with a screen mesh 0.5 mm x 0.5 mm. To this is attached a thin and tough wire several meters long (to be able to subsequently remove the tray). Within the tray different types of baits are used (meat, cheese, feces, fish, decomposing fruit). This is pushed to the bottom of the hole with a long flexible rod (or plastic conduit), leaving the tray in the den for about 3 days. It is then removed with the cable. Then it is convenient to dig into the hole (where the faecal deposits are) if it is dislodged by the tenant vertebrates and capture the specimens with a sieve or colander.

According to Jason (pers. comm. 2005), who has collected in burrows of pocket gophers in the U.S.A., the older mounds should be located, and digging into them, proceed to take the lawn with an inch of superficial land and then place them in a Berlese funnel or Winkler/Mockzarsky eclector.

Albert Allen (pers. comm. 2005), also in USA, has obtained interesting results in ground squirrel burrows using a large spoon auger (Fig.99), digging deep into the trash at the entrance. According to Allen Scarabaeinae are present, as well as some Histeridae, Leiodidae, Tenebrionidae, and other families. He has collected many interesting species by pouring the contents of a burrow into thick polyethylene bags (80 pounds), taking them to the lab and extracting the fauna. To make this tool insert the handle of a spoon into the end of a PVC tube of 6 to 7 meters long and one inch wide (Warner 1997).

With regard to appropriate methods for the collection in burrows of pocket gophers, Richard Cunningham explained (pers. comm. 2010):

"For pocket gophers (mostly Thomomys), we find a lateral burrow by probing the ground between mounds with a thin tool such as a screwdriver, large forceps or steel rod. If we cannot find a lateral burrow we will dig into the mound itself to find the burrow. Once the burrow is located and excavated we place a gopher trap on both sides of the burrow to attempt to trap the animal. We bury a plastic container flush with the soil of the burrow between the traps. In the container we pour some ethylene glycol solution or anti-freeze. We bait the beetles by suspending dung over the container. I use a plastic spoon pushed into the soil at an angle or a small plastic cup filled with dung with a hole in the lip to put wire. The wire is then bent, pushed into the soil next to the container with the small cup suspended inside the container. We bait with human dung but others have used pig or other dung successfully. Bill Warner has used deer dung that he hydrates with water in the small cup. He said that it works very well, is easier to use (less messy) and carry on trips. It is of course less stinky. We then place a board over the exposed burrow and trap then cover it with soil. This keeps the burrow dark and the humidity correct. We check the burrow after a day or can leave it for weeks as the preservative will keep the specimens from rotting. Sometimes the gopher gets by the traps and buries the container. The soil in the container and around it should be strained with water through a screen as specimens can be in the soil. We have found that it is best to put in as many traps in a locality as possible as the inquiline populations seem to vary from burrow to burrow. You may get many specimens in on burrow and none in a burrow nearby."

Cunningham (pers. comm. 2010) said: If the gopher is captured and removed from its burrow beetles can be collected for a week or more. Sometimes pocket gophers bury the traps and no more beetles can be caught. If so, then try again.

According to Cunningham (pers. comm. 2010) Jim Saulnier has had great success rearing out specimens from the throw-out mounds of the pocket gophers. He locates a mound that has gopher dung pellets and plant refuse that has been pushed out as the gopher does its cleaning. He puts the mound material in a bucket, brings it home, puts it into 5-8 inch flat containers, covers them with cloth screen and waits for the specimens to emerge. He has found adults, larvae and pupae in the mound soil. He does keep the soil somewhat moist so the insects do not die.

Bill Warner (1997) explained that he used, various sizes of ladles attached to PVC pipes to scoop out soil from burrows of pocket gophers, ground squirrels and prairie dogs. And sifting the soil using a sifter with window screen bottom can produce many Aphodiinae (Cunningham pers. comm. 2010).

He even named the tool the "Plague Scooper" as one must be careful scooping out prairie dog and ground squirrel burrows due to fleas in the burrows which carry bubonic plague bacilli. Another way is to find the soil push-up or heap that the pocket gopher throws out of the burrow containing traces of excrement and vegetation that may be nesting material. Sometimes pocket gophers clean their burrows, and litter expelled contains eggs, larvae and pupae. Take this soil with manure, place it in drawers or containers and wait for the beetles to emerge. At the laboratory, the soil should be covered to prevent drying by spraying it with some water. The best season is after the winter rains, in early spring.

With regard to methods for the collection in nests of moles in Europe, Andreas Herrmann explained (pers. comm. 2010): *"At the end of each harvest season (autumn) the mole builds one single molehill which is much bigger than all its others; this big molehill lies over its nest. The nest itself consists of grass or leaves and is located almost 50 cm under the surface, so you have to dig in the earth. This means big efforts because at the end of the autumn the ground may be frozen and a normal showel is not hard and heavy enough for this purpose. Then you sieve the nest and select the beetles (which concentrate in the warmth of the nest for hibernating) at home in a warm room.*

There are several species of different families (e.g. Staphylinidae, Scydmaenidae, Histeridae, Leiodidae Platypsylinae, *etc.) which are dependent on moles and therefore can be caught only in this way".*

Cunningham (pers. comm. 2010), said that just after the winter rainy season (in North America) he has collected in woodrat *(Neotoma magister* Baird*)* middens, in the Allegheny Mountains, collecting the faeces and the soil beneath and those of the nest tunnels, resulting in many *Aphodius* spp. larvae (Scarabaeidae) after sifting the material. The beetles are concentrated because woodrats use a single spot a permanent toilet for their droppings.

In the burrows of Prairie dog *(Cynomys ludovicianus* (Ord)) and Richardson's ground squirrel *(Spermophilus richardsonii* (Sabine)), Cunningham has used a tool called the "pest scooper". It is a large spoon attached to a long stick or PVC pipe. The soil is taken out with a shovel and placed in a 5 gallon bucket. Aphodius can be caught, but be careful not to be bitten by fleas which transmit bubonic fever. It is advisable to dig in wet soil but not that which is dry and dusty, because in the burrows other dust borne diseases may be present.

A plastic box can also be used like those of video cassettes or CDs, making small holes on the sides. The bait is placed between two sheets of paper and then inside the box, sealing it with tape, without covering the holes in the sides, and tying it to a long rope. It is inserted into the nest by pushing it with a stick, so that it is lying down at the bottom. Next day or up to 3 days later it is retrieved. The problem with this box trap is that it produced dirty specimens because the feces or bait are in direct contact with the beetles, but several species of Aphodiinae and *Onthophagus* can be collected.

Collecting beetles living in burrows can be an extremely valuable and interesting activity to science, because in a very few countries has this kind of work been done, and certainly there are good possibilities to find new or very rare species.

According to Peck Stewart (pers. comm. 2010) recently, Dave Almquist has discovered that the beetles which are found in the burrows of gopher tortoises (*Gopherus polyphemus (*Daudin*)*) in the SE USA, do not locate the burrows by olfaction but by sight. They use the dark semicircle of the burrow entrance as a visual cue to locate their host's habitat. By making a trap which has a dark semicircle above a tray of glycol preservative, the burrow can be mimicked, and the beetles captured.

The species of mammal or reptile living in the burrow must be identified to establish which animal is related to the Coleoptera collected there. It is important to get a book showing the different nests and burrows of vertebrates to be able to record the full details of this type of collection.

Remember that it is essential write on a small label the basic data such as:

Country, Department, Province or State

Specific site or coordinates (if possible) showing the closest place and the distance of collection site (if the name of the place where the burrow is located is unknown)

Date and name of collector

In another label indicate some ecological data, for example:

Scientific name of the mammal occupying the burrow and the height above sea level. Type of environment: forest, savanna, paramo, etc.

The bait used, etc.

Any other interesting information

These data must be written with ink on a small label or labels made with a laser printer. Remember that the more information contained for the specimens will make them more valuable for science.

In the field, the specimens can be killed and shipped with traditional methods.

Finally, it should be taken into account that the transmitters of dangerous diseases such as Hantavirus are rodents. So it is helpful to find out if these and other similar diseases are present in the region where this type of beetle is going to be collected. We must also beware of Bubonic plague, contained in the burrows.

If the samples are going to be extracted with a strainer, it is better to spray before with poison and then proceed to strain the material. But do not spray when the beetles are collected with a Berlese funnel or other similar method, where it is required that the specimens do not die at the time of being taken from their hiding places. To see the fleas and prevent them from biting is convenient to use light-colored clothing, tucked into socks and pants and is better to work with white gloves (Warner 1997).

BEETLES ASSOCIATED WITH INVERTEBRATES

Peck Stewart

Many beetles live in direct or close association with various terrestrial invertebrates. There is a complex terminology categorizing these relationships, especially for the species living with social insects, such as inquilines, guests, predators, nest scavengers, and so on.

There are small coprine dung scarabs (*Zonocopris gibbicollis* (Harold) and *Z. machadoi* Vaz-de-Mello which live part-time in the mantle cavities of large *Bulimus* land snails in wet forests in South America. These scarabs emerge from the mantle cavity to oviposit on snail dung when the snail defecates. I have seen these in mountain forests in NW Argentina. This association parallels that of the Neotropical dung scarabs which live in the fur of tree sloths until the sloth descends from its tree when it defecates.

Another association is that of some scarab beetles which are specialist feeders on millipedes. These are *Sceliges* and *Onthophagus* in the Afrotropical region, and *Deltochilum valgum* Burmeister (and maybe related species) and undescribed *Canthidium* in the Neotropics. They locate their prey by using the quinine defensive compounds released by the millipedes (and may also use millipede dung for reproduction?). These can be caught in pitfall traps baited with large dead millipeds. Dung scarabs on West Indian islands which lack a terrestrial mammal fauna may (at least in part) use the dung of land snails or millipedes. This topic is largely not studied.

Scotocryptine beetles (Leiodidae Leiodinae) such as *Scotocryptodes*, *Scotocryptus* and *Parabystus* live in the nests of meliponine stingless bees in the Neotropics. The beetles probably act as scavengers in the nests and feed on fungi such as yeasts. Interestingly, adults of these beetles are all eyeless. They can be found by opening the bee nests, which usually occur in the ground or in tree trunks, or sometimes inside termite nests. To open the nests it is wise to wear a head covering, gloves, and taping shirt sleeves and pants cuffs closed to protect one from the irritating biting of the bees. Country people often make artificial nests for the bees to obtain their honey, and such nests can be made with an opening for easier access to search for the beetles.

Many beetles, especially Staphylinidae, live in termite nests throughout the world tropics. These nests must be dug open and the nest contents searched. The nests of Neotropical *Nasutitermes* termites can be sifted and the material put into a cloth bag for a day in the shade. Most of the termites then die or become moribund so that when the nest debris is spread on a cloth the next day, the beetles are more easily seen because of their more active movement. Additionally, I have found that by putting the termite nest siftings into a pile for a day or two, many Ceratocanthidae (Scarabaeiformia) can be extracted from the debris by hand searching or extraction in a Berlese funnel. Alternatively, the nest debris, if placed in front of a flight intercept trap, will also catch the Ceratocanthidae which fly to the broken nest debris.

There is a rich fauna of staphylinids, histerids and other beetles associated with many kinds of ants. It is often difficult to collect such beetles. Fairly easily collected are beetles associated with army ants (*Eciton* spp.) in the Neotropics. These are seen by merely lying on the ground along a moving raiding column of ants, (or especially a column which is shifting the position of the bivouac) and watching the column. Many of the beetles look remarkably like the ants so some practice is needed to form a "search image" to spot the beetles. I do not know if the columns of doryline driver ants in the Old World tropics of Africa have an associated beetle fauna.

Many Nemadini (Leiodidae, Cholevinae) occur in the mound nests of some ants such as *Camponotus*, *Lasius*, and *Formica*. These nests can be opened and searched. Alternatively, in the NE USA, it has proved successful to put boards on the top of the nest mounds in the spring season, and the beetles rise from the cool nest to the underside of the warmer boards where they may be found. *Ptomaphagus* (subgenus *Echinocholeus*) live in the nests of *Pogonomyrmex* harvester ants in the Southeastern and Southwestern USA. Since these ants have exceedingly painful stings it is risky to dig up their nests to find the beetles. Some inquiline beetles have been taken in pitfalls set at the side of harvester ant nests. This method might work also for beetles in the nests of New World attine leaf-cutter ants. If one can find the refuse piles of leaf-cutter ant nests, these often have many beetles in them, such as cremastochiline scarabs.

Stewart B. Peck, Professor of Biology
Department of Biology
Carleton University, 1125 Colonel By Drive
Ottawa, ON Canada K1S 5B6

COLLECTING IN INSECT NESTS

Hives of bees

To collect the beetles species that live in hives can be used small corrugated cardboard wrapped into a spiral or sticky boards with aggregation/attractant pheromones, both bait and traps can be located either in the apiary or inside individual hives. The pollen is collected in an area of the trap that is separated, therefore unprotected from beetle invasion from the bee colony (Stedman 2006). Also, can be used propolis into corrugated cardboard, but these traps must remain for a short time in or near the hives, this is to prevent pests in the hive, because of the beetles ability to successfully reproduce in pollen, especially pollen that it is unguarded (Douglas 2003).

Fig. 100 - Reviewing the straw for collect beetles. *(Picture courtesy Andreas Hermann, Germany).*

Fig. 101 - Hives of bees from Germany. *(Picture courtesy Andreas Hermann, Germany).*

Elzen et al. (1999) has tested beetle-trapping devices designed to work outside of colonies. They discovered that odors from hive products, plus adult bees, were attractive to flying adult beetles as this combination yielded the greatest number of beetles in the traps.

Also, larvae and adults beetles can be attracted to lights. Light can be mounted at ground level in a container (containing water with detergent) or buried in the ground. According to A. Herrmann (pers. comm. 2010) with respect to Dermestidae it is highly recommended also to sweep the plants (especially in their blossoms) in the surrounding of bee hives. Their larvae feed inside the hives on dead insects, while the adults after hatching visit the flowers to feed on pollen and meet the other sex for mating.

According to Andreas Herrmann from Germany (pers. comm. 2010) *"under the straw, that has been put into the hives by the apiculturist for protecting the bees against cold, the coleopterists in Germany have collected many interesting beetles which are associated with these bees. For this purpose the sieve is used together with a Winkler/Moczarski eclector* (Fig. 100). *This stuff can also be sieved with a metal grill on a white cloth. The beetles falling on the fabric can be caught with an aspirator or with a wet brush."*

Hermann also explain me: *"Most of the beetles which could be collected in this way are not bound to bees directly. They simply prefer the microclimate, the food (parts of dead insects). Also there are some predators and other arthropods which are concentrated there. Furthermore I could imagine that there is a special kind of mould existing, because some (rare) members of mould feeding families like e.g. Cryptophagidae and* Latridiidae *have been recorded especially from bee hives."* (Fig. 101).

Wasp nests

A method of collection for Buprestidae, Curculionidae, Phalacriidae, Nitidulidae, Chrysomelidae, etc. is to find the nests of wasps from the families Encyrtidae, Braconidae, Sphecidae, Cerceridae (the genus *Cerceris* spp.), because many of these wasps hunt Buprestidae, Curculionidae and others beetles which provide food for their larvae (Evans and Rubink 1978; Scullen and Wold 1969). For example: in Eastern North America is found *Cerceris fumipennis* Say (Hymenoptera: Crabronidae), this species is buprestid-hunting and is known now from more than twenty colonies of varying size in Canada. The wasps are most often nesting in open areas of hard-packed sandy soil surrounded by woody habitat suitable for their Jewel beetles prey. Also Many colonies are associated with somewhat disturbed sites compacted by human activity (baseball diamonds, parking areas, roadsides and the soil around campfire). Other beetles-hunting species are *Cerceris californica* Cresson and *C. conifrons* Mickel, *C. crotonella* Viereck and Cockerell (Evans and Rubink, 1978). When you find those nests containing buprestids or others beetles, please, you should be prudent in removing only a few specimens of these nests, because the abuse in the extraction could affect populations of these wasps. Androw, R. A. (1992) in a paper online (see references) relates his observations on Buprestids-hunting wasp.

ABOUT TRAPS

According to Leather and watt (2005), to sample an insect requires both a sampling technique and a sampling program.

Also, according to Leather et al. (2005) the sampling techniques are the methods used to gather information from a single sampling unit. The result of sampling depends on the equipment and techniques used in the collection. Leather et al. (2005) classified the traps in two types: passives and actives. Presented here is only a summary of this classification, for more details see the work of Leather and his colleagues (2005):

1.- Passive traps (for example: Malaise traps, intercept traps, barrier traps, etc.)

They depend entirely of chance. Passive traps allow unbiased estimates of insect populations because the insects are neither attracted nor repelled by the traps,

Leather et al. (2005) explain one example with aerial suction trap: the insects are driven by a motor of air in the trap and the collection tube is passive in the action, as the machine depend on insects flying in the area, and depend on the vacuum, because the insects are attracted for other reasons to the area where the trap is (for food, reproduction, humidity, etc) This method can be seen in the article "Vacuum sampling", chapter III. A major drawback of using passive traps is that they are not very useful at low densities. This is a particular problem when programs are designed to control the abundance of insect pest outbreaks, such as on the lists of quarantine (Leather and watt, 2005).

2.- Active traps (For example: baited pitfall traps, colored pan traps, light traps, etc)

Take advantage of the behavior of insects, luring them with chemical lures, baits, or even colors, these traps can be varied to give different trapping efficiencies and goals (Finch 1990). According to Leather and Watt (2005) the active traps give a partial estimate of the density per unit area and must then apply conversion factors. By using attractive traps, especially if they are being used to obtain an estimate of the population, it is vitally important to know if what range of traps are effective and whether there are directional as well as distance effects.

PITFALL TRAPS

Due to the varied habitats and feeding habits of many Coleoptera, it is necessary to use different types of traps.

It is important to know that necrophagous, coprophagous, and saprophagous beetles and their predators are often observed or used as model organisms in wildlife studies and environmental impact assessments worldwide. These beetles are easily surveyed with pitfall traps.

Different types of baits are used in these traps to attract beetles. These baits range from easily accessible materials like dung, meat, fruit, molasses, beer, rum, cider, and the like, and various chemicals such as acetic acid, phenol, and acetates. Depending on the region and the families of beetles targeted, the collector must select the methods and baits to be used.

The technique with pitfall traps was first developed by Hertz (1927) and shortly after by Barber (1931), who used open-top containers buried with the rim level to the substrate surface, so that anything falling into the container becomes trapped.

The pitfall trap is a container buried in the ground so that its mouth remains flush with surface. In the container various types of baits are introduced (meat, dung, fruits, etc.). When a beetle falls into the container (incidentally or attracted by the bait) can no longer escape. There are several models of pitfall traps. The most common consists simply in burying a plastic cup or something similar with bait inside. For example, for the collection of Carabidae (Cicindelinae among them), just put vinegar in a glass buried flush with ground and many species that are active at night will fall in it. Other models of pitfall traps are more sophisticated and each collector can choose the one that best suits his needs.

It is appropriate to check pitfall traps every day and if that is not possible, then some preservative fluid should be introduced into these traps to prevent the decomposition of captured specimens.

Pitfall traps made of small plastc containers

There are many models of these traps (Fig. 102). When using feces as bait, empty small plastic container will be useful enough. Cut plastic bottles or cans without lids can be useful for this purpose. These containers should be large and at least 15 cm deep. To prevent disturbance by vertebrates, put a piece of wire mesh with large holes over the trap or a roof. This will keep out scavengers while beetles pass between or below the mesh grid. Place a roof over pitfall, to protect the trap from rain or debris (Fig. 102-103).

Funnel pitfall traps

With a knife cut the bottles crosswise, a little above the centre, so that it is divided into two parts; one cup-shaped and the other funnel-shaped. The funnel thread section is cut, making a circular hole large enough to allow the largest beetles to go through. It is inserted inverted into the container, so that the narrowest part is facing the bottom and then join the two sides with wire (Fig. 102c and 103). With this device, vertebrates cannot enter their heads into the trap and eat the bait (especially when using carrion). Bait can be put directly at the bottom of the trap or inside a little bag which can hang on an interior wall of the trap preventing the beetles from contacting the bait.

Pitfall traps with window

To build these traps also plastic bottles can be used (1 L or more capacity) with their respective covers. With a knife, make a horizontal window of approximately 4 cm high by 5 to 10 cm wide, leaving an overhang which gets up to avoid water inside the trap when it rains. Four small holes should be made on the sides of the bottle, to be able to introduce two small sticks or twigs across the width of the trap; a twig will go up and the other lower, so they are crossed when seen from above. These branches will prevent scavenging vertebrates (like foxes, lizards and dogs) from reaching the bait or disturbing the traps (which is particularly problematic when using carrion baits). John Leavengood explains the following method: *"to take a square, thin piece of wood and hammer a nail through each corner. This is then placed over the trap and the nails separate the wooden panel from the soil and trap by about an inch. This also protects the trap from water accumulation."* (Fig. 102d). Rarely, animals other than dung beetles are attracted to the feces. When ready, the trap is buried so that the window remains flush with ground.

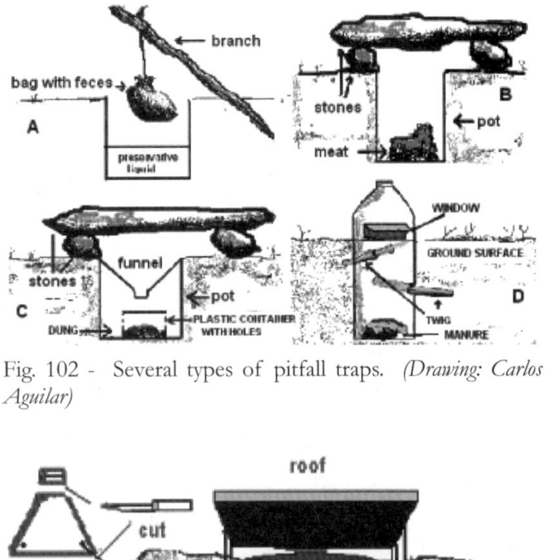

Fig. 102 - Several types of pitfall traps. *(Drawing: Carlos Aguilar)*

Pitfall trap with buried buskets

These traps are made by burying a bucket and placing a smaller box with the bait (inside the bucket), which can be covered with

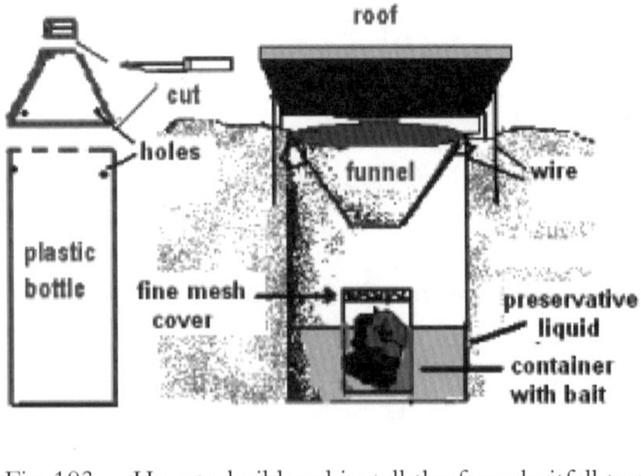

Fig. 103 – How to build and install the funnel pitfall traps. *(Drawing: Carlos Aguilar).*

Fig. 104 - Pitfall trap, with bait into a bag suspended above the trap. *(Photo: Bolívar Garcete B.)*

a grid when meat is used; the bucket can be covered with another grid with large holes to prevent other animals from eating the bait. The beetles will drop in the bucket and not in the bait receptacle (keeping specimens clean. An effective trap that requires some extra work consists in putting in forest barriers of cross-shaped black polyethylene (or tables) several feet long. At the ends of the barriers large containers with a capacity of 4 liters or more are buried (Fig. 105). These traps can be used with or without bait. This trap is effective to collect many species of flightless beetles of medium to large size, living on the forest floor (Carabidae, Tenebrionidae, Scarabaeidae, etc).

Baits for pitfall trap

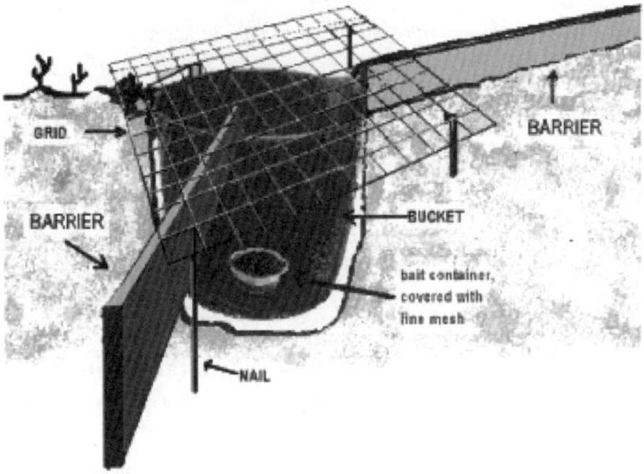

Fig. 105 - Pitfall traps with buried buckets and barriers. *(Drawing: Carlos Aguilar)*

Various, strongly aromatic animal products may be used as baits. These baits should be from native's animals of the sampling site. The efficiency of each type of bait depends on where the pitfall trap is located and the climatic conditions.

Although human feces are among the best baits to attract coprophagous dung beetles, the feces of pigs, monkeys, sheeps, cows, horses, some rodents and other omnivorous attract different species of beetles also. My friend Reinhard Foerster (from Argentina) sent me a specimen of *Eudynopus dytiscoides* (Schreibers) a very rare Scarabaeinae from Argentina that was collected in dung of sea lions. However, this species also may be collected with human feces. Also, Peck Stewart (2006) in Galapagos Island has collected beetles in droppings of giant tortoises and land iguanas, and guano and nest materials of sea-birds. And also, according to John Leavengood (and several colleagues) many dung beetles in the USA have been collected at deer feces only. Also, numerous clerids beetles of Korynetinae subfamily across the world are only found in the feces of certain ungulate species (Leavengood, pers. comm. 2010). On the other hand some feces mixed with other products, for instance pork feces mixed with molasses, are good baits.

Some Neotropical Scarabaeinae are not very choosy about their food[1]. Many are copro-necrophagous[2]; some are just necrophagous or only coprophagous the latter mainly in America (Paolo Gandini, pers. comm. 2010). In contrast to other continents like Africa, Asia, even Europe, where most beetles are strictly necrophagous or coprophagous, many species are adapted to reach only certain types of feces and not others (Hanski and Cambefort 1991). This is probably because in these continents the beetles have been able to evolve over time with the great herds of herbivores. For example, in Africa Gordon and Barbero (2008) observed that several species of Scarabaeinae prefer the feces of some herbivores: *Kheper aerates* and *Sisyphus ocellatus* Reiche (in impala dung); *Kheper platynotus* (Bates), *Aptychonitis anomalus* Gestro, *Onitis inversidens* Lansberge, *Onitis keniensis* Gillet, *Onthophagus nigricornis* Fairmaire and *Anachalcos convexus* Boheman (these prefer elephant dung). Many species of *Aphodius* (Scarabaeidae, Aphodiinae) are attracted by small droppings, but most species prefer to use large droppings, especially bovine dung (Cambefort and Hanski 1991). The Australian species of Scarabaeinae are mostly coprophagous (86%) and many of them are attracted to marsupial dung; only 8% are mycethopagous while less than 1% is necrophagous (Cambefort 1991). The Hybosorinae species that occur in the rain forests in South-East Asia arrive in great numbers to feed on carrion at night (Kuijten 1978, 1981, Hanski and Krikken 1991). In the Americas, cow dung is attractive for many species of *Ontherus, Isocopris, Copris, Ateuchus, Dichotomius, Onthophagus.*and even *Sulcophanaeus.*

According to Lumaret and Kirk 1991: *"Dung represents both a microhabitat with caracteristics changing gradually in time and resource which, although it may be abundant locally, is ephemeral in nature, only remaining suitable for dung beetles for a short period of time. The drying out of dung prevents any further utilization by most coprophagous insects."*

It's good to know that carnivore feces are not very attractive are not very attractive to many beetles, since they usually consist only of remains of bones, hair and/or feathers. However, in Italy Paolo Gandini (pers. comm. 2010) collected many Troginae under dry dog feces in a very sandy soil, Gandini also informed me that in South America (Uruguay) he has collected *Canthon ornatus* Redtenbacher in dog feces. Despite all this, in general in these feces do not reach many different species of dung beetles.

Small dead animals, such as birds, mice, frogs and reptiles are very attractive to Scarabaeinae sensus lato, Dermestidae, Cleridae (Korynetinae), Silphidae, Hydrophilidae, Tenebrionidae, Carabidae, Histeridae, Staphylinidae, etc.

The bovine large intestine is very attractive to many beetles while those of chickens and other animals are very attractive to Phanaeina and other species in the Neotropical region.

If using intestines as bait, we should note that this tissue really gives off a foul smell when decomposing, so it should be cut into pieces while still fresh, and when the journey to the collecting site is very long, put the tissue in an insulated container with ice. Once in

1. Nevertheless, the most of the Scarabaeinae prefer dung of omnivores over carrion, manure from cows, fruit or fungi decomposed. It is possible that these beetles detect certain compounds that cause human feces are more attractive to them as nitrogen, which require reproductively immature adults for feeding and maturation period, ending stage in the development of your muscular system and females culminate the maturation of their eggs (Hanski, 1991). An important part of the diet of dung beetles are the bacteria found in the different resources, omnivore dung has a lot of these, which can be essential in the metabolism of these insects, which is why they prefer this type of bait (Bustos-Gómez & Lopera, 2003).

2. Most of these insects are active at midday, but among necrophagous there are many of nocturnal habits.

place, these casings can be put directly at the bottom of the pitfall.

Dead fish attract other species than dead mammals. Baited traps attract the beetles either way. Also pitfall traps placed in trees at different heights within the forest, attract species that usually are not found in soil (Cambefort and Walter 1991) and there are certain species that are attracted under specific conditions. In this respect Andreas Herrmman (world expert in Dermestidae) told me in a private message:*"Some species of the genus Dermestes are not attracted by traps deposited on the ground but in the trees (for catching those species I usually wrap a little dead bird in a bunch of straw and tie this "construction" in the top of a tree). Kind of trap: when I fill the bottom of the trap with sand (so the bait is placed directly on sand instead of "normal" earth, straw, wood, plastic or metal) some scarabeids and histerids appear which obviously need the sand for digging; when the sand is missing they don't visit the bait at all."* Anyway, in the opinion of John Leavengood (and mine also), should be noted that in the traps, it is good to include sand/ soil to help prevent specimen desiccation and to allow the beetles to bury themselves for protection from other insects like ants. However, the absence of sand will not result in the absence of dung beetles if the trap is made properly. However, most likely some species of Dermestidae (and some other species of beetles) have specific ecological requirements.

In the Neotropical region there are other baits used to catch several species of Scarabaeidae *(Deltochilum* spp.*Canthon* spp.) and, according to Krell (2004), some African and tropical Asian *Onthophagus* are attracted to living Myriapods or millipedes belonging to the groups that roll (Diplopoda). Also in Paraguay, the author has collected several species of Scarabaeinae with traps baited with Diplopoda. These millipedes must be kept alive in a container; also pitfall traps with buckets described before can be used for this type of bait. Remember that after handling Diplopoda you must wash your hands thoroughly, because these animals exude cyanide acid which is poisonous, and sometimes even expel cyanide, but Diplopoda do not bite or sting. Dead Passalidae attract some species of "dung beetle". Interesting bait is dead Meloidae, which attracts *Notoxus* spp. (Anthicidae) (personal observ.). Dead snails are also good baits to put in the traps in some regions (John Leavengood, pers. comm. 2010). In Uruguay, Paolo Gandini (pers. comm. 2010) has collected many *Canthon bispinus* Germar, these beetles were pushing balls made with bodies of earthworms.

On the banks of rivers, lakes or seashore, it is interesting to use pieces of fish or dead shrimp as bait (Olivier Boilly, pers. comm. 2009). These baits can also be used in other environments such as savannas or forests. In Chile, for example, by using pieces of fish into the forest *Ceroglossus* (Carabidae) can be caught, as well as other species of small Carabidae and even some Scarabaeinae *(Dichotomius torulosus* (Escholtz)).

Some fruits are good baits, such as rotten bananas and other fruits with a bad smell as it decomposes. Fresh banana peels attract some U.S. dynastines and watermelon rinds attract some Rhyncopherine weevils (Leavengood, pers. comm. 2010). Also good to mention the use of wine or other fermenting fluids as baits. Thus, the more types of baits used, the more species of beetles captured. The author's brother collected many *Ceroglossus* in Chile using bananas as bait mixed with yeast.

According to Belcher (1989), novel bait for larval Elateridae consists of a mixture of wheat/corn: meat. This bait must be renewed at least once a week. But keep in mind that Trogidae beetles are attracted to the traps during the more advanced stages of meat putrefaction.

If feces are used as bait, it must be changed every 3 to 4 days; after that time feces dry and are significantly less attractive. The use of more traps in different places results in greater quantity and variety of the beetles obtained. Therefore it can be better to change the location of traps every week.

Where to locate and how to prepare the traps

The first thing to consider is the site where the pitfall traps will be placed and the climatic or environmental conditions of the chosen site. Some sites are not favorable. For example, in flood zones, traps are often lost with the flooding of rivers and streams and coprophagous / necrophagous beetles living in the soil are present only in periods without inundation.

The soil components may vary. It may be composed of stones, clay or sand. It may be argillaceous, or a mixture of various types. In very stony or very sandy soils very few species of silphid beetles (coprophagous or necrophagous) occur because in these places it is very difficult to build or maintain their burrows, which generally have a particular shape in each species.

Clay and black ground soils will house a larger number of species; the soil is sufficiently strong, not overly harsh, and penetrable to these insects.

To be able to recognize when soil is more or less suitable for these beetles, take some soil and wet it with some water so that it becomes a paste. This mass should take the form of a stick. If the stick disintegrates when it is lifted from one end, the soil is sandy. If it stays as a whole, the soil is argillaceous or black ground. To recognize if the soil is a mixture of sand with clay or mud, the soil must be bent until it takes a horseshoe shape. Finally, if the soil can take the form of a wheel, then it is composed primarily of clay or black ground, which is the best soil for the silphid beetles to make their burrows.

Problems often arise when the baits are eaten by animals (dogs, foxes, lizards and birds), when it rains heavily and traps are carried away by water, when disappearing under mud swept away by currents produced by rain or rising water channels, and when the traps are filled with sudden rainwater, losing an entire workday. However, if you want to get a variety of beetles, pitfall traps should be placed in all types of habitats, from the natural to the altered areas and also within and outside forests. More traps in more places means more species collected. It is not advisable to set traps in farmed land since few beetles can be found due to insecticides.

It is desirable to bury at least 5 to 10 traps or more in each habitat. They must be separated from each other between 10 to 20 m straight in a line in order avoids for getting the exact location. Alternatively, small flags or colored markers may be placed near each trap to facilitate finding them.

It is best to mix cow dung with some water to keep it moist (and releasing more odor), since many species of Scarabaeinae arrive quickly after the feces have been deposited by the cattle. Usually this happens at noon, in broad daylight. It is best to use these baits in open areas, either in a grassland or savanna. Results are not good using cow feces in forests, because the cows and beetles which feed on their feces live in open areas. Moreover, it is worth remembering that deer feces and generally the small fecal pellets are not very attractive, with rare exceptions[(3)] such as those of certain rodents and lagomorphs (hares and rabbits). But that depends on the site where the collection will take place. Experience will teach when to use these baits, but it is helpful to know beforehand which species of

3. In Eastern North America, deer faeces are attractive to these beetles, they also adapted to these droppings. (Leavengood, pers. comm. 2010).

Las Vegas, Mendoza (Argentina). Habitat for: Cerambycidae (*Acalodegma* sp.), Tenebrionidae, Carabidae, Buprestidae, Scarabaeidae, and other families. (Photo: Pablo Marano).

Avalanches and slope movements, such as landslides and mudslides can form crevices and cracks which serve as refuge for many beetles. General Roca, Río Negro, "Valle de la Luna", Argentina (Photo: Pablo Marano).

PLATE I

Some Tenebrionide, Chrysomelidae (*Sternoplax* and *Anatolica*), Carabidae (*Harpalus* spp.), Buprestidae (*Agrilus* ps.) occurr in the dune field from Ordos, 50km east of Baotou, Inner Mongolia, east limit of Gobian Desert.
(Photo courtesy of Oboilly Olivier).

Among the lichens and mosses that are on the branches and trunks of living trees, there are numerous species of beetle which can be searched with a flashlight, especially in the spring night. Misiones, Argentina. (Photo: Jorge Barrett).

PLATE **II**

General roca, Rio Negro, Argentina. This habitat is also known as "El Monte", here are several endemic species of beetles such as: Tenebrionidae (*Psectrascelis*), Carabidae (*Mimodromius*), some Cerambycidae, and the relictual Ommatidae and Cupedidae. (Photo: Pablo Marano).

Shangxiang-Sichuan Mountains, Chine. Habitat of: Lucanidae (*Platycerus*), Melolonthinae (*Parexolontha*), Cicindelinae (*Calochroa flavomaculata* (Hope), *Calomera angulata* (Fabricius), *Calomera funerea assimilis* (Hope) and many other species. (Photo: Olivier Oboilly).

Permanent pond in Capiatá, Dpto Central, Paraguay. When vegetation is submerged can be collected many species of aquatic and semi-aquatic beetles. On the ground near of this pond, There are Cicindelinae such as: *Brasiella, Odontocheila* and *Pentacomia.* (Photo: Jorge Barrett).

Stream in Caacupé, Dpto Central, Paraguay. Habitat of Dryopidae (1), Elmidae (1), Dytiscidae (2) and Gyrinidae (2). (Photo: Jorge Barrett).

PLATE **IV**

Stream in Naranjo, Dpto Cordillera, Paraguay. Habitat of 1) Hydrophilidae, 2) Dytiscidae and 3) Elmidae. (Photo: Jorge Barrett)

Hydrophilidae are swimming. Anthicidae and Cicindelinae (*Brasiella minarum* and *Pentacomia punctum*) are running on the ground. Capiatá, Dpto Central, Paraguay. (Photo: Carlos Aguilar).

PLATE **V**

Dunes near the sea in Fortín Santa Rosa, Canelones, Uruguay. Habitat of: 1: Dynastinae (*Tronistes rouxii* Burmeister), 2: Chrysomelidae, Curculionidae and Cantharidae. (Photo of Jorge Barrett).

Arroyos y Esteros, Dpto. Cordillera, Paraguay. Habitat of Carabidae: Cicindelinae: *Brasiella argentata* and *Brasiella* they are running on the flat rocks.
(Photo: Carlos Aguilar)

Santa Lucía Beach, Corrientes, Argentina. On this river beach can be found Carabidae (Cicindelinae such as *Tetracha* spp.) and among the vegetation there are some species of Chrysomelidae, Curculionidae and Tenebrionidae. (Photo Pablo Marano).

Piribeby, Dpto. Cordillera, Paraguay. This is a low forest, habitat of Cerambycidae, Carabidae, Cetoniinae, Rutelinae, Buprestidae, Lampyridae, Lycidae, many Dung beetles species such as: *Coprophaneis bonariensis* Gory, *Deltochilum* spp, *Dichotomius* spp, *Canthon* spp, etc. (Photo of Carlos Aguilar).

“Laguna Capitán” salt lake in Loma Plata, Chaco, Paraguay. Habitat of several species of Tenebrionidae, Carabidae, Scarabaeidae, Cicindelinae, Chrysomelidae, Cantharidae, Buprestidae and other families. (Photo: Jorge Barrett).

It's the Gaube lake in Pyrénées (France). According to Olivier Boilly, here can be found *Cetonia aurata* Linnaeus (Cetoniinae). (Photo of Olivier Boilly 2010).

Sandy area in Vallermí, north of Paraguay. Habitat of several species of Carabidae, Scarabaeidae (Melolonthinae, Dynastinae, Rutelinae, and Scarabaeinae), Tenebrionidae, Chrysomelidae, Curculionidae and others families (Photo: Carlos Aguilar).

Savanha near of "Loma Plata", Dry Chaco, Paraguay. In the leaf litter in the ground at the foot of the Carob trees (*Prosopis* spp.) usually there are interesting beetles fauna such as Anthicidae (*Notoxus* spp.), Staphylinidae, Scarabaeinae, Carabidae, Tenebrionidae, Pselaphidae and other families. (Photo of Jorge Barrett).

PLATE **IX**

Around Philadelphia City, Boreal Chaco, Western Paraguay. Habitat of Carabidae (*Brachygnathus* spp. and *Morion* spp.) and some Buprestidae species.
(Photo of Carlos Aguilar).

Savanna ("Cerrado" type), Northern Paraguay. In this habitat there are many species of Buprestidae in the flowers trees and on the ground occur Scarabaeinae such as *Phanaeus kirbyi*, *Ph. palaemon*, *Ph. thalassinus*, *Oxysternon palemo*, *Coprophanaeus jasius*, and several species of *Canthon* and *Dichotomius* (Photo Carlos Aguilar)

"Cerro Acahay", Department Paraguarí, Paraguay. This is an extinct volcano, with available habitats for beetles. Here can be collected Cetoninae *(Gymnetis* spp. *Holoserica* spp. *Inca bomplandi, Marmarina* spp.), Cerambycidae, Buprestidae, Carabidae, Tenebrionidae, Scarabaeidae, and many other beetles. (Photo of Carlos Aguilar).

The Mandraka Reserv on RN2 (60km east of Antananarivo), a good place with few primary forest (very rare in Madagascar because of the Eucalyptus), there are a lot of Cetoniinae: *Epixanthis novempunctata, Tetraodorhina ebenina, Coptomia pauliani, Liostraca bina, Pygora lenocinia*, and it's a place of lot of species of *Euchroea* ! and a lot of dung beetles but no *Carabus*, only a few tigers. (Photo and text, courtesy of Olivier Oboilly, 2010).

PLATE **XI**

Mbaracayú Forest, Northern Paraguay. Habitat of many different beetles such as *Macrodontia flavipennis* (Cerambycidae), Scarabaeinae (*Sulcophanaeus faunus*, *Coprophanaeus ensifer*, etc), Lucanidae (*Leptinopterus tibialis*), Dynastinae (*Megasoma janus*, *Heterogomphus* spp., *Strategus centaurus*, etc. *Cyclocephala* spp.). (Photo:Carlos Aguilar).

PLATE **XII**

Savanna (Cerrado type), Dpto Amambay, Paraguay. On termite mounds can be found several interesting beetles: Cicindelinae (*Cheilonycha auripennis* Lucas, *Pentacomia degandei* (Tatum), *Prepusa miranda* (Chaudoir), etc.), Aphodiinae, Lampyridae, and other beetles which are also within these nests. (Photo: Jorge Barrett).

Fallen tree in the jungle, Naranjo, Dept. Cordillera, Paraguay. Among the decaying wood there are many species of beetles: Passalidae, Histeridae, Brentidae, Silvanidae, Lucanidae, Tenebrionidae and several larvae of Scarabaeidae and Elateridae and other families. (Photo: Jorge Barrett).

PLATE **XIII**

Among the fungi (or nearby of them) there are many species of beetles, mainly at night (see "Beetles in fungi", Chapter III). Mbaracayú Forest Reserve - Dpto. Canindeyú, Paraguay. (Photo: Carlos Aguilar).

50km in the south of Chiang Mai (Thailand), picture taken from a train. Habitat of several Cetonidae species such as *Protaetia fusca* (Herbst), *Glycyphana chamnongi* Antoine, *Jumnos ruckeri* Saunders. (Photo courtesy of Olivier Boilly, 2010).

PLATE **XIV**

The mangroves are habitat for: Cerambycidae: *Aeolesthes holosericeus* (Fabr.) and many species of *Bledius* spp. (Staphylinidae), there are some Tiger beetles. Other families such as Chrysomelidae, Curculionidae are pests of tropical mangrove trees, feeding on flower buds, leaves, or bark. Northeastern Brazil. (Photo: Gonçalo Lopes, Portugal).

It's in Vatumandria (East Coast) near Indian Ocean (Madagascar). A place with a lot of heather plants. According to Olivier Oboilly, here there are *Polybothris sumptuosa* (Klug) (Buprestidae), *Hemilia striata* Gory & Percheron (Cetoniinae) and a lot of dung beetles. (Photo courtesy of Olivier Oboilly, 2010).

PLATE **XV**

Parque Municipal Monday, border between Paraguay and Brazil, Dpto Alto Paraná. During the day and night, On the underside of the trunk and branches of recently fallen trees, are several species of Cerambycidae, Curculionidae, Brentidae, Bostrichidae and other xilophagous beetles. (Photo: Jorge Barrett).

Giant ferns forest, in the Biological Reserve Mbaracayú, Dpto Canindeyú, Paraguay. Habitat of several Dung beetles species (*Dichotomius*, *Canthon*, *Deltochilum*, *Coprophanaeus*, *Phanaeus*, etc) and many different Cerambycidae species. (Photo Carlos Aguilar).

PLATE **XVI**

herbivores are present at the scene (this dung is easily available in local zoos). So it is useful to have a vertebrate identification guide to know what animals are living in each site.

The baits are placed directly at the bottom of the pitfall traps. But a cleaner method consists of placing the baits within a cloth of 15 cm X 15 cm or so. Take the four corners of the cloth standing them up, like a small bag which is tied to the end of a string. The remaining point is tied to the tip of a small stick about 20 cm long, which is buried in the ground at the edge of the pitfall trap, so that the bag is suspended in the air. It can also be placed above the bottle (Fig. 102a and 104), but it is better that the bag remains suspended inside the jar, so that the beetles are more likely to become and remain trapped.

To prevent the beetles coming in direct contact with the bait (so they do not get dirty), plastic containers that fit within the pitfall traps must be used, like the ones used to store photographic roll film (with tight caps). Small holes will be made on the lid using a hot wire. Put the bait (dung or meat) into these packages and place them into the pitfall traps. The inconvenience of this method is that later, leaving the collection site and removing the traps, you have to wash the bottles to prevent contamination of the visited place, but the big advantage is that the beetles will not come into contact with the bait and thus will remain clean. In these bottles one can also be put phenol, acetate, rum, etc. as bait. With these chemicals Geotrupinae are sometimes trapped, mostly in temperate countries. Also most European *Geotrupes* are coprophagous, but *Geotrupes stercorosus* (Scriba) often makes brood balls of moldy forest litter (Rembialkowska 1982) The Nort American species have significantly more varied food selection (Howden 1955a). Some species of Geotrupinae feed on fungi (Howden 1955a).

Many collectors use a variety of preservative liquids at the bottom of the trap, such as "antifreeze" or "glycol". This fluid is poisonous and slightly sweet and attracts dogs, pigs and other wildlife, which can be killed when ingested.

Furthermore, according to Sasakawa (2007) the samples preserved in ethylene glycol are not suitable for taxonomic studies, due to hardening of membranes of the beetle genital organs. But according to John Leavengood (pers. comm. 2010): *"specimens "might not" be suitable for taxonomic studies relying on internal morphology and certainly not for molecular studies. However, my colleagues and I have never had any problems collecting into this liquid and then using the specimens for research."*

It is not advisable to use formaldehyde or ethanol as preservative because the specimens become hard and brittle, but they can be softened by immersion in a potassium hydroxide solution (van den Berghe 1992). Ethylene glycol may be useful in taxonomic studies if it is mixed with other substances in the appropriate way, but there are very few formulas in the literature. For example, Schmidt et al. (2006) used a mixture of ethylene glycol diluted at 25% in salt water. They demonstrated that the evaporation rate was low; since over 75% of the sample volume was recovered and had strong effects of conservation (more than 85% of the samples were preserved without damage to appendages). Isono (2005) maintained traps for 11 weeks with dilute ethylene glycol and obtained similar results to those obtained by Schmidt and colleagues. According to Schmidt et al. (2006) and Gurdebeke & Maelfait (2002) in the long term and despite of evaporation, this fluid can damage the samples. Acetic acid at 10% mixed with brine and ethylene glycol can be used, but according to Sasakawa (2007) in the long term this will also damage the samples. However John Leavengood told me in a private message:*" My colleagues and I prefer ethanol as long as we do not need to dissect genitalia. The insects are flexible enough to pin and position and they are suitable for molecular research."* To avoid insect escape each trap is also enhanced with slippery stuff (applied to the interior rim of the container) a 1:1 mix of petroleum jelly and mineral oil.

It is advisable to check the traps every day. In the event that the samples will remain for days without being checked, it is necessary to use acetic acid at 10% mixed with brine and glycol (in very small quantities). Also, if environmental conditions are not extremely hot (so the liquid does not evaporate completely), use 50% of good quality vinegar, mixed with 40 parts of water, one tablespoon of salt and 10 parts of 75% alcohol. The beetles are preserved quite well in the liquid without becoming hard, as long as the traps are checked at least once a week (pers. obs.).

In countries with warmer climates, most of the year is appropriate to collect beetles with pitfall traps, however, is preferable to use these traps after a heavy rain. In temperate regions the best time to collect with pitfall traps is in the spring and early summer.

What is caught with pitfall traps?

In Neotropical wet forests, in wide roads but little traveled where the beetles have their flight paths, there are several large Phanaeina such as *Coprophanaeus. Phanaeus, Diabroctis, Oxysternon, etc.* Within the virgin or nearly virgin forest there are several species of *Deltochilum, Dichotomius* and Silphidae. Off the roads, in the understory (in the dark forest) rare species of *Deltochilum* and *Canthon* are found. Various rare species such as *Canthon* and *Deltochilum,* including several *Dichotomius* are adapted to undergrowth habitats, although places without undergrowth should not be ignored because some species of Scarabaeidae are also abundant there.

How to transport the specimens from the trap to the laboratory

When checking the pitfall traps, it is practical to carry a large jar with a tight cover, with a handle, wide mouth to be able to slip a hand inside, halfway filled with fresh sand or soil. A string or wire loop may be affixed to it such that one may carry the trap across one's back. All the beetles captured with the pitfall traps must go inside the bottle where they bury themselves in the sand or soil and clean themselves, and can be taken out of the bucket in the camp or laboratory.

Beetles should not be left together and/or alive for more than one day, especially because of scavenging beetles, which are cannibals and eat each other when they are confined. This cleaning method can be used for all those obtained in similar conditions and coming into contact with baits, provided that they are diggers.

The collector can set aside this bottle and sacrifice the beetles directly with a killing bottle, but doing so many specimens will be very dirty, depending on how the bait was placed in pitfall traps. If the bait was introduced directly into the trap and was in contact with the beetles, they should be washed by hand or with some liquid (like water or alcohol). It is very important to note that washing beetles with water is the best method if they were collected in the same day, to avoid rotting. After washing the beetles, they must be placed on a sheet of paper, so the water will evaporate and prevent the specimens from drying. After cleaning they can be immersed in flasks with the appropriate preservative fluid or in bed envelopes with all collection data.

Many species of different families (Leiodidae, Staphylinidae, Histeridae, etc.), including several scarabs of the genera *Pedaridium, Ateuchus, Canthidium, Urozys, Trichillum, Onthophagus,* Aphodiinae, etc., are very small beetles (less than 5 mm), and are also collected with feces or meat. When they are captured it is better to introduce them directly into a small vial with preservative liquid. Arriving at the camp (or laboratory) they should be changed to other bottles with new preservative liquid to keep them clean. They can be kept permanently or can be prepared dried in pins, but glued to small triangular-shaped points, or bottles with dry sawdust. Remember to include all data collected with each specimen.

To preserve the collected material

Very tiny beetles falling into the traps can be stored in a bottle with ethanol (75%) or another formula (see chapter V), also in paper envelopes, or cotton bed envelopes.

TRAPS WITH FRUITS

Use bottle traps to get many species of Coleoptera belonging to several families: Elateridae, Cerambycidae, Nitidulidae, Histeridae, Staphylinidae, Scarabaeidae as Rutelinae and above all Cetoninae (*Paragymnetis*, *Gymnetis*, *Corvicoana*, *Marmarina*, *Hoplopyga*, *Euphoria*, etc.) and Lucanidae, They can be made with thick plastic bags or polyethylene of suitable size, with the advantage of small size for transport. Pieces of wire which have a ring at each end are required (Fig. 106). These wires are placed in the plastic bags so the bags remain like closed bottles. A hole is made as a window on the upper side of each bag or bottle, through which Coleoptera will enter (fig. 107 and 108). The wire of the trap structure is prolonged and bent into a hook to be able to suspend the traps on the branches of trees.

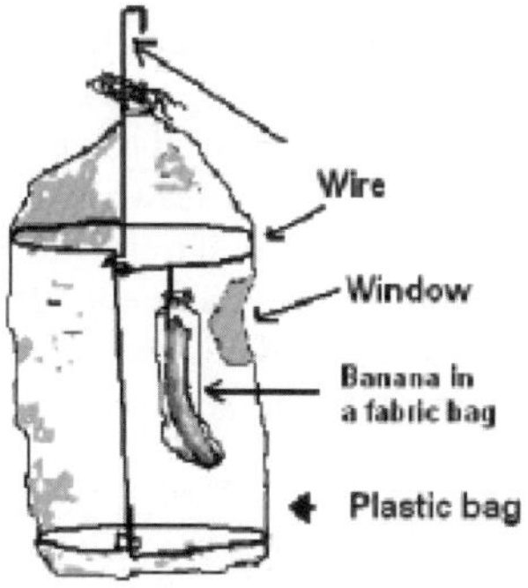

Fig. 106 - Fruit trap made with thick plastic bags or polyethylene. *(Drawing: Carlos Aguilar)*

In this type of traps half-decomposed fruits are used as baits, like bananas or pineapple chunks, both shelled, putting them into a thin sock or a plastic bag with a hole below to allow the putrid fluid of fruits drain, and to prevent them to dry very quickly. This bag is suspended inside the trap with a piece of string or wire.

Mix the fruit with malt or dark beer or even rum, all this accompanied with black sugar, not refined, with a very strong odor. Some yeast can be added, previously prepared. Crush all these ingredients, trying to obtain a mass rather thick. You can also use the liquid secreted by the wounds of trees, mixed with some rum, black sugar and other fruit outlined above, but in this case the mixture must ferment for about 3 days before using it in traps. You can also use pieces of watermelons, which attract Cetoninae of *Inca* genus. In these traps you can also find Nitidulidae, Histeridae, Cerambycidae, Elateridae, and some occasional Lucanidae.

The best places to hang the bottles or bags are trees along the roads within the forests. These roads must be at least 10 m wide and a great deal of low forest on the sides (Fig. 108). A good result can be obtained placing the bottles on the edge of a forest,

Fig. 107 - Bottle fruit trap. *(Picture: Carlos Aguilar)*

Fig. 108 - Placing a fruit bottle trap out on the edge jungle. *(Picture: Carlos Aguilar)*

right in the ecotone, for example, in the limit with a savanna. It is better to choose places slightly altered, though lower or secondary forests frequently lodge a good variety of species. The traps must not be too hidden among the foliage.

Hot days with slight wind are the best to use the bottle traps. We must seek the paths not heavily used by vehicles; there are not many beetles there because the leaves of trees and bushes are filled with earth. But we shall find insects in places little traveled by vehicles within the forests. Old and wide roads should be selected. The narrow and very new paths are not good for these traps because insects are just invading these areas and must wait the dying trees to emit an odor to attract them. Allemand & Aberlenc (1996) recommend that when using these traps in temperate forests, it is good to use as bait red wine mixed with beer (in equal parts) and 10% salt (NaCl). They gave more details on this type of traps, in an article available online (see bibliography).

Another type of trap that works is the aerial flight intercept trap described by Peck, S.B. and A.E. Davies 1980 (See "Flight intercept traps", Chapter III) this trap is hung in trees at different heights, starting from 3 meters (even in the canopy).

PURPLE TRAPS

This trap is very effective to collect Buprestidae, but also Carabidae, Eucnemidae, Mordellidae, Elateridae, Cleridae, Throscidae, Cerambycidae, Anobiinae, Trogossitidae, Nitidulidae, Curculionidae, Scolytinae and other families.

The trap consists of three purple plastic sheets, a little wavy, 24 inches long, linked together in a triangle. The outer walls of the trap have a non- toxic glue type (Fig 110), from which insects can not detach by themselves. It can be made with tanglefoot glue on purple PVC foam cut to size.

Fig. 109 - Purple trap, based in adhesive emulsion. *(Photo: Sergio García, Argentina)*

This trap uses no bait because its main attraction is the color; however some types of pheromones or sawdust are also used as attractants. The bait is suspended in a small bag inside the trap.

It hangs vertically from branches, near or attached to the trunks, preferably in open places within or along the forest, where it is suspected there is an abundance of beetles. It can be checked once a week.

Purple trap can be used throughout the spring until fall.

Be careful with the walls with adhesive; the dirt is easily transferred from the hands when you touch them.

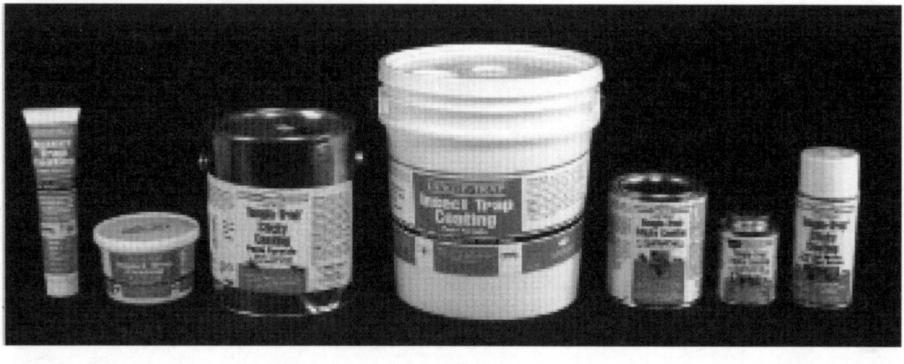

Fig. 110 – Tanglefoot glue. *(Picture courtesy of BioQuip, www.bioquip.com)*

The insects stuck in the traps are removed with liquid toluene, and they are soaked in water mixed with ethanol to be softened.

CONCAVE COLORED PAN TRAPS

Many flying insects settle onto surfaces whose tone and/or color contrast clearly with the background and different species and sexes of insects have different responses (Young 2005).

Knowing that flowers attract a large number of beetles, colored pans imitating flowers can be used to deceive and bring them into these traps.

The colored pan traps (fig. 111) must be: purple, blue, yellow and white and it is important to get them in bright colors. Many other authors have found that red traps are favored by some beetles (Dafni et al. 1990), The small dishes are the best since these is easily transported. These shallow dishes measure approximately 15 cm in diameter and 3 cm in height, or a little more and must be slightly concave, to be able to transport them comfortably one inside each other. To avoid creating a shadow inside, the inner walls of the pans must not be angled, because the shadows inside of pans are less attractive to beetles. The standard diameter of each pan trap must be about 20 cm, but the author has achieved good results with the small ones (only 6 cm diameter). The small size of these dishes allows placing them in groups of three or five, imitating a flower.

Fig. 111 - Colored pan traps on the ground (Chaco, Paraguay). *(Photo: Bolívar Garcete B., Brazil)*

The best place to put the pan traps is in the flowering season, when there are more insect pollinators, but this is relative because flowers are often seen out of the traditional spring season, anyway these traps attract insects at any time of year.

These traps are effective with an intense sunlight (pers. observ.). Is interesting to know, the beetles run away from the flowers when a cloud blocks the sunlight and, when the cloud passes, they return. This means that cloudy days are not suitable for using color pan traps.

The dishes must be placed in various sites, especially in open environments such as grasslands, scrubs or deserts, even in forest clearings. It is convenient to choose places with many flowering plants. For example, deserts are ideal to capture Buprestidae around the flowered cactus, although they also fall into these traps in different environments, like other small families as: Cantharidae, Melyridae, Cleridae, Dermestidae, Elateridae, Mordellidae, Chrysomelidae, etc.

The colored pan traps should be placed separated by a few meters or if there are many, then put together some of them in groups of 3 to 4, imitating the larger flowers. After being placed on the ground, then put crystalline water inside, mixed with some crystalline liquid soap to reduce surface tension. Pour the liquid without filling the pan traps, so the beetles can not grasp the edges and escape. The traps are left in place for a set time, of-

Fig. 112 - Filling water in a container. *(Photo: Carlos Aguilar)*

ten 2–4 days, after which their catches are filtered through a sieve and stored (Young 2005). It may be necessary to add a preservative such as sodium benzoate or antifreeze, if traps are emptied infrequently (Young 2005).

In many places the crystal clean water is scarce, especially in deserts and in the islands. The water should be completely transparent in order to maintain the color of the dish and do not lose its effectiveness. Plastic container (5 Lt.) can be used to transport the water (Fig. 112). The contents of the traps will be emptied with a fine mesh strainer and beetles will be saved in a jar with ethanol (Fig. 113).

Fig. 113 - Straining contents from the color pan trap. *(Photo: Carlos Aguilar)*

The traps also can be placed on the upper branches in the canopy (fig. 114). According to Usher (1990) and Bowie (1999) height of trap above or within vegetation has been found to influence the capture for insects in rainforest understory, heather and oil seed rape and therefore should be standardized. And according to Young (2005) this could be done in three ways, depending on the insect groups that want to collect:

Standardizing to just above the height of the understory so that visibility is consistent,

Standardizing to the mean height of the food resource (flowers or fruits), and

Standardizing to a defined height from the ground, perhaps determined as optimum in a pilot study.

Fig. 114 - These traps can also be placed among the trees at different heights. *(Photo: Sergio García, Argentina)*

FLIGHT INTERCEPT TRAPS

Malaise traps

One way to obtain many species of beetles is intercepting them in flight by putting obstacles in their usual flight paths. These are called flight-intercept traps. One of these is the Malaise trap (designed by Malaise in 1937), which consists of a kind of tent with a black colored vertical panel, provided with two side walls and covered by a roof.(Fig. 115-116).

The insects in flight hit against the panel and, deflected by the side walls, tend to fly upward moving towards the top as if through a funnel that directs them to a container with preservative fluid located in the more elevated corner of the roof.

The trouble is that the Malaise trap catches hundreds of butterflies, moths and other insects which will cover the beetles with their scales, forcing a tedious cleaning of the beetles trapped.

These traps are located in places where the beetles pass such as roads in forests and also in the clearings. Some investigators suggest a north–south orientation with the head of the trap facing the sun's zenith (Noyes 1989). According to Noyes (1989) traps set in exposed and sheltered situations in the same habitat type may have different efficiencies, due partly to differences in flight behavior and partly to microclimate variation. For this type of trap it is essential to choose a good place where it will be located.

Avoid places where people walk, because they almost always carry the traps away. Also avoid places where large animals pass, like cows or horses, etc., because these animals may collide with traps and destroy them. So it is best to choose private lands or places where you need permission to enter.

Fig.115-116 Malaise traps in the "Cerrado". *(Photo: Carlos Aguilar)*

They also can be placed in savannas, deserts or mountains. The best days are those with a slight wind, heat and humidity, day and night.

Malaise traps can also be combined with coloured pan traps with water and soap these can be placed on the ground under Malaise traps so that many of the insects attracted to the pan traps are collected into the Malaise trap. There are several versions of Malaise traps. Examples of Intercept traps:

Suspended malaise traps

Alternative designs of Malaise traps have been made to hang in low or high vegetation, these traps have four panels of fine mesh (Basset 1985; Springate & Basset 1996) (Fig. 117, 118 and 119). These can be suspended from branches with ropes within the canopy or across flight paths between trees.

Fig. 117-118. Suspended Malaise traps. (Image courtesy BioQuip Products. (http://www.bioquip.com)

Fig. 119 - Lucanidae was also captured placing traps in the treetops. (Image courtesy BioQuip Products. (http://www.bioquip.com)

One should keep Malaise traps for several days in the same place. If the place is a tropical habitat and environmental conditions are suitable, these traps will be full of insects after two or three days. With less ideal conditions, it may take up to a week to be filled.

Flight intercept traps

To catch many species of Scarabaeidae, Histeridae, Staphylinidae, Carabidae, Leiodidae and other families that fly at very low altitudes during the day and/or night, one can use a polyethylene film or fabric with fine mesh (green or black). It should be about 1-2 m wide x 1 m high. (Fig. 120)

The fabric is stretched between two trees in the forest or poles in the open field. Try to make the mesh panel vertical and as straight as possible, like a wall (so the longest side will remain parallel to the ground), and the lower edge of the fabric is separated from the ground about 10 cm, putting sets of shallow pans below. Stewart and Davies (1982) recommend filling the trays a 50/50 solution of water and ethylene glycol-based auto antifreeze, with a pinch of sodium benzoate. The insects hit the fabric and then fall into the pan/tray below and drown. Stewart and Davies (1982) recommend spraying the fabric with a knock-down insecticide. The beetles can be collected from the trays with a fine-mesh colander.

Fig. 120 - Flight intercept traps with trays with water in the bottom. *(Photo: Bolívar Garcete B., Brazil)*

This kind of trap is especially efficient in dense forests for sampling small, flying insects, such as small beetles that are otherwise difficult to collect (Peck and Davies 1980).

Note that this kind of trap does not need any type of bait. Install 5 to 10 of these traps, and place them where it is suspected that the insects fly, for example in a savanna, a grassland or a path or clearing within the forest, etc. Some people use baits with these traps, but it is better not to use them, so you can catch insects flying naturally through the site and are not attracted to known baits. In case of rain, a plastic roof can be put above. This trap can be checked every day or left running for a periods in excess of 2 weeks. These traps may collect many species of beetles, for example, in Paraguay, the author found several specimens of a new *Dichotomius* species, never collected otherwise. The author has used intercept flight traps near the ant nests (*Atta* and *Acromyrmex*) for collect *Dendropaemon* (Scarabaeinae), some Staphylinidae and some Histeridae.

Canopy window intercept trap

Intercept traps may also give very good results by placing them in between tree trunks at various heights (Fig 121). These models can be manufactured with plastic frames, thick nylon or even the clean sheets used for radiography (although, with radiographic sheets, the traps are much smaller). Build a small wooden frame (or other materials) to fit the film and place a tray under this frame with the formula recommended by Stewart and Davies (see above). One could also add two small containers with rotten bananas or some other bait (it is not always necessary to bait the trap). The trap also can be placed among the flowers in the canopy. The blade is attached to the bottom tray with wire and the trap is held in the branches with string or wire. The most difficult to do with this trap is attaching the tray in the branches, so that it won't move or spill the liquid preservative, but with patience the trap will be secured.

Fig. 121 - Canopy flight intercept trap hanging between trees. *(Drawing: Carlos Aguilar)*

Intercept bottle trap in logs or trunks

Intercept traps consisting of bottles attached to branches (Fig. 122-123) are useful to catch the beetles (such as Tenebrionidae, Alleculidae, Carabidae, Cerambycidae, Erotylidae, Languriidae, Anthribidae, Curculionidae, Trogossitidae, etc.) that walk during the day and night in the branches and trunks of trees.

These traps are made with two bottles. Add 90% ethanol to the collection bottle. The mouth of the funnel is affixed to the trunk of a tree. In order to fix the funnel on the surface is convenient to search for trees with a few irregularities of the bark. Besides, the traps have to be painted the same color as the bark of the tree (although this is not fundamental, but it is convenient). To do so, some soil can be glued on the plastic surface as shown in the Fig. 123. It is convenient to use 3 traps for each tree arranged on different sides of the trunk. It is also helpful to set small plastic strips or other material, or build some roads made with mud, wax or clay for insects to walk towards the mouth of the trap on an even surface. These devices should be well attached to the logs, using clay or mud at the edges.

Fig. 122 – 123 - How to put bottle-interceptor traps on the trunks of trees. *(Photo: Antonio Cantero, Paraguay)*

The problem with these traps is that they catch all kinds of small animals, sometimes even lizards and spiders. They must be reviewed at least every two or three days, but they can remain several weeks placed on the same tree. No need to use baits, because beetles used these traps as hiding places.

Note: the longer you run a trap in a given location, the more likely you are to eventually catch any particular rare species. According to Martikainen (2003) sometimes but not always, it is better to look by eye for most beetle families at their habitats without using traps. In any case, these recommendations are more valid for European boreal forests than for other biomes (GTLI 2007).

SOME DEVICES TO COLLECT BEETLES

Berlese funnel or Tullgren funnel

Berlese or Tullgren funnel, it is used to remove insects that live in organic matter, such as flood debris, fungi, bark and rot, small pieces of wood, etc. It consists of a funnel placed over a jar with ethanol 90º (Fig. 1). Inside the funnel has a grill and up the funnel is located a light bulb. On the grill must be download the materials collected, such as litter, fungi, grass, small pieces of bark, humus, and other plant waste. The insects are going to escape from the heat and then they fall down in the jar with ethanol. Many models of Berlese funnel can be easily made with disposable bottles. However, it is best to use large metal funnels, because with a large funnel can be processed more than 10 kg of organic material per day.

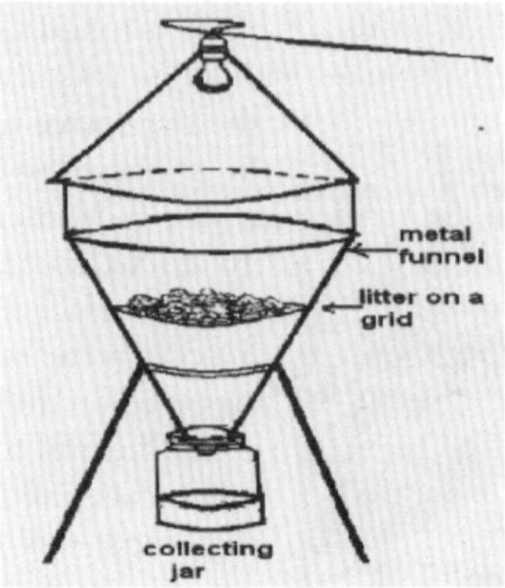

Fig. 124 - Berlese funnel used to remove beetles from litter and other organic or from inorganic materials.

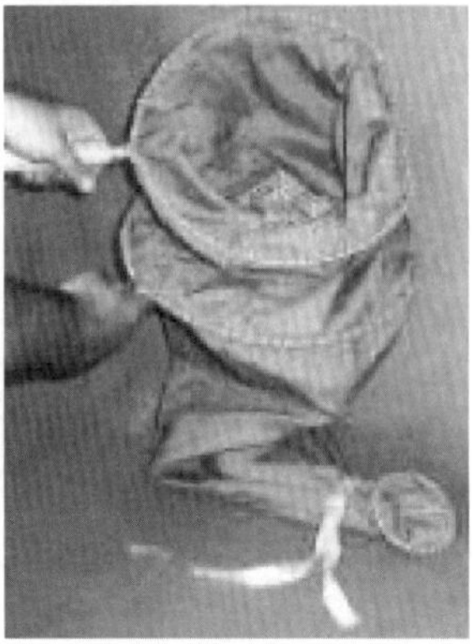

Fig. 125 - Inside of litter reducer *(Picture courtesy BioQuip Products - www.bioquip.com)*

Winkler/Moczarski eclector

This tool is made from a square or rectangle of canvas. Sorting is automatic, but it takes several days to sort all specimens.

It consists of two devices:

1. **The strainer (or litter** reducer) (Fig. 125- 126). This is a cloth bag with an open mouth at the top. Inside it has a grid like a sieve. Outside it has two short handles, one above and one below, in order to be able to handle it. At the bottom there is an opening that should be tied before using. Through the mouth, put the material that the collector wants to strain and shake with both hands. The smaller material will pass through the grid falling to the bottom. The largest materials remaining in the bag and on the grid are thrown away. The materials strained at the bottom should be placed in a cloth bag, to be later transferred to the following device:

2. **The separator:** this is a bag similar to that described previously. But inside there is a small flat bag

Fig. 126 - Sifting litter with the strainer. *(Photo: Carlos Aguilar)*

with holes. The size of the selector can vary to suit individual needs. But a good size according to Hangay and Dingley (1985) is: 300 mm x 300 mm x 700 mm (Fig. 127). A 6 mm x 6mm mesh is held inside the bag (Fig. 128). Put in the mesh bag the litter and other materials previously strained. At the bottom, the external bag has a ring which attaches to a bottle with preservative liquid. This instrument is hung in a place enlightened by the sun. Insects descend running away from the heat and fall to the lower bottle. Wheeler and McHugh (1987) modified this tool to collect in fungi and litter. They changed the inner bag by a grid (with a light bulb above the outer bag), resembling a Berlese funnel. (Fig.129).

The Winkler/Moczarski eclector may be limited to situations of low relative humidity, probably because it is based on drying to remove the insects. However, Besuchet et al. (1987) dispute this idea, suggesting it is the normal activity of insects in response to disturbance that is responsible and have found success in both dry and humid environments. Strain the leaves of the forest or other organic material with the litter reducer and put the sifted material in a Winkler/Moczarski eclector (Fig. 126 and 127) or a Berlese funnel (Fig. 124).

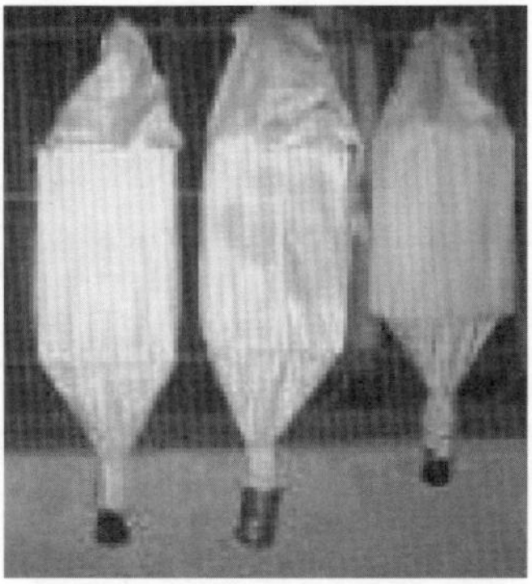

Fig. 127 - Winkler/Moczarski eclector. (*Photo: Carlos Aguilar*)

Fig. 128 - Inner view of Winkler/ Moczarski eclector *(Picture courtesy BioQuip Products. http://www.bioquip.com)*

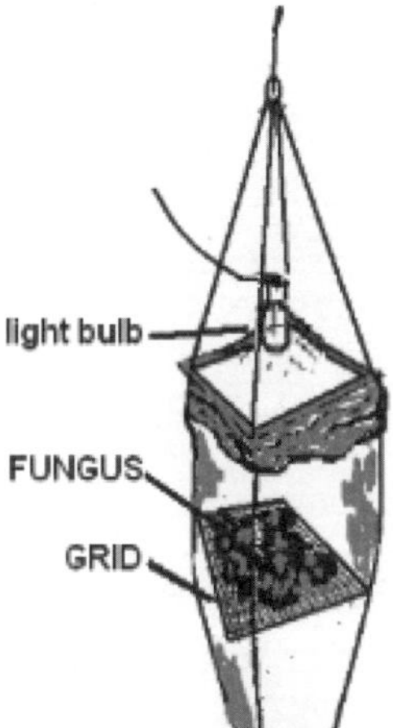

Fig. 129 - Winkler/Moczarski eclector with light (Draw modified of Wheeler and McHugh 1987). *(Drawing: Carlos Aguilar)*

Fig. 130 - Sieving leaves litter *(Picture courtesy Andreas Hermann, 2010)*

CHAPTER IV
WHERE DO THEY LIVE...

TAXONOMIC LIST OF THE BEETLES FAMILIES

This classification is based on Hava 2007 and Lawrence et al. (1999a, 2002 and 2010). This list does not include fossil families.

Suborder Archostemmata
Ommatidae
Crowsoniellidae
Jurodidae
Cupedidae
Micromalthidae

Suborder Myxophaga
Lepiceridae
Torridincolidae
Hydroscaphidae
Sphaeriusidae

Suborder Adephaga
Amphizoidae
Aspidytidae
Carabidae
Gyrinidae
Haliplidae
Trachypachidae
Meruidae
Noteridae
Pelobiidae
Dytiscidae

Suborder Polyphaga
Series Staphyliniformia
Superfamily Hydrophiloidea (incl. Histeroidea)
Hydrophilidae
Histeridae
Synteliidae
Sphaeritidae

Superfamily Staphylinoidea
Hydraenidae
Ptiliidae
Limulodidae
Leiodidae
Scydmaenidae
Silphidae
Agyrtidae
Staphylinidae

Series Scarabaeiformia
Superfamily Scarabaeoidea
Lucanidae
Passalidae
Glaresidae
Pleocomidae
Diphyllostomatidae
Geotrupidae
Ochodaeidae
Scarabaeidae
Belohinidae
Glaphyridae
Trogidae
Hybosoridae

Ceratocanthidae

Series Elateriformia
Superfamily Scirtoidea
Eucinetidae
Clambidae
Scirtidae
Declinidae

Superfamily Dascilloidea
Dascillidae
Rhipiceridae

Superfamily Buprestoidea
Buprestidae

Superfamily Byrrhoidea (incl. Dryopoidea, Psephanoidea)
Byrrhidae
Elmidae
Dryopidae
Lutrochidae
Limnichidae
Heteroceridae
Psephenidae
Cneoglossidae
Ptilodactylidae
Chelonariidae
Eulichadidae
Callirhipidae

Superfamily Elateroidea (incl. Cantharoidea, Artematopodoidea)
Rhinorhipidae
Artematopodidae
Brachypsectridae
Cerophytidae
Eucnemidae
Throscidae
Elateridae
Plastoceridae
Drilidae
Omalisidae
Lycidae
Telegeusidae
Phengodida
Rhagophthalmidae
Lampyridae
Omethidae
Cantharidae
Podabrocephalidae
Ototretinae
Ototretadrilinae
Pterotinae
Cydistinae
Neocrowsonia
viatorius
Stenocladius
Harmatelia

Series Derodontiformia
Superfamily Derodontoidea
Derodontidae
Nosodendridae
Jacobsoniidae

Series Bostrichioformia
Superfamily Bostrichoidea (incl. Dermestoidea)
Dermestidae
Endecatomidae
Bostrichidae
Ptinidae

Series Cucujiformia
Superfamily Lymexyloniodea
Lymexylidae
Limexilonidae

Superfamily Cleroidea
Phloiophilidae
Trogossitidae
Chaetosomatidae
Metaxinidae
Cleridae
Acanthocnemidae
Mauroniscidae
Phycosecidae
Prionoceridae
Melyridae

Superfamily Cucujoidea
Boganiidae
Byturidae

Helotidae
Protocucujidae
Sphindidae
Biphyllidae
Erotylidae
Monotomidae
Hobartiidae
Cryptophagidae
Agapythidae
Priasilphidae
Phloeostichidae
Silvanidae
Cucujidae
Myraboliidae
Cavognathidae
Lamingtoniidae
Passandridae
Phalacridae
Propalticidae
Laemophloeidae
Tasmosalpingidae
Cyclaxyridae
Kateretidae
Nitidulidae
Cybocephalinae
Smicripidae
Bothrideridae
Cerylonidae
Alexiidae
Discolomatidae
Endomychidae
Coccinellidae
Corylophidae
Latridiidae

Superfamily Tenebrionoidea
Mycetophagidae
Archeocrypticidae
Pterogeniidae
Ciidae
Tetratomidae
Melandryidae
Mordellidae
Rhipiphoridae
Zopheridae
Ulodidae
Promecheilidae
Chalcodryidae
Trachelostenidae
Tenebrionidae
Prostomidae
Synchroidae
Stenotrachelidae
Oedemeridae
Meloidae
Mycteridae
Boridae
Trictenotomidae
Pythidae
Pyrochroidae
Salpingidae
Anthicidae
Aderidae
Scraptiidae
Ischaliidae
Rhizonium
Aprostomis
Lagrioidinae
Afreminae

Superfamily Chrysomeloidea
Cerambycidae
Chrysomelidae
Orsodacnidae
bruchidae
Megalopodidae

Superfamily Curculionoidea
Nemonychidae
Anthribidae
Belidae
Attelabidae
Caridae
brentidae
Curculionidae

Order: Stresiptera
Stylopidae

WHERE DO THEY LIVE? WHAT DO THEY EAT? HOW TO COLLECT THE DIFFERENT BEETLES FAMILIES?

According to their eating habits Coleoptera are divided into 5 trophic groups: The **herbivorous** and **algivorous** (primary consumers), **fungivorous**, **detritivorous** and **carnivorous** (secondary consumers or higher level) (Marinoni 2001).
To indicate the number of known species in each family worldwide, has been used the reference of Department of Entomology of Texas A&M University (http://insects.tamu.edu/research/collection/hallan /test/Arthropoda/Insects/Coleoptera/Family/Coleoptera1.htm) because in this web site is constantly updating the list of known species of beetles worldwide.
Nevertheless, most of the time was included the bibliographical references, which gives the number of species known until the date mentioned in the reference.
Extra information has been included for each subfamily, giving relevant information on some species and general methods of collection for each of them. In some families are not given extra information to each of its subfamilies, because in those cases, the sampling methods are the same for almost all species of this group. Some subfamilies are explained in more detail, for example Cicindelinae subfamily, because them requires several specific sampling methods.
Remember that this book is not intended as a book of taxonomy. However I think this chapter will help you acquire the basic knowledge on the number of known species, their distribution, biology and general collection methods for each of the families of beetles that are known throughout the world.
In a book of this size is impossible to include all information currently exists on every family, so to obtain extra information about some families or specific groups, you can consult the bibliography at the end of this book.
This classification is based on Hava 2007 and Lawrence et al. (1999a, 2002 and 2010). This list does not include fossil families.

Suborder ARCHOSTEMMATA

OMMATIDAE: Only 2 genera, with 2 species of *Tetraphalerus* in Brazil and Argentina and 6 species of *Omma* from Australia (Neboiss 1960; Vulcano and Pereira 1975; Lawrence 1999). They are nocturnal and live in the wood of dead trees and / or dying. Some adults are specialized to eat pollen and are often found in flowers (Grimaldi and Engel 2005). Larvae are suspected to develop in plant roots (Sarmiento 1969; Hörnschemeyer et al. 2006). They can be found in rainforests as well as in arid grazing or open woodland. Adults are clumsy fliers (Atkins 1958). According to Hangay and Zborowski (2010) in Australia occur in Western New South Wales, Victoria and South Australia in dense forests, but they also live in pastureland and in drier areas in Australia, they probably develop in subterranean, decaying, fungus-ridden wood, such as roots or fallen trees. They can be collected with flight intercept traps or purely by chance (Hangay and Zborowski 2010).

CROWSONIELLIDAE: *Crowsoniella relicta* Pace, this species is only known from Central Italy (Pace 1976; Lawrence and Newton 1995; Grimaldi and Engel 2005). They have endogenous habits, because the beetles were sieved in calcareous soil from old chestnut trees (Holger 2003).

JURODIDAE: There is only one present species *(Sikhotealinia zhiltzovae* Lafer) known by a single specimen found in Russian Siberia and considered a "living fossil" (Лафер 1996) plus another fossil species (*Jurodes Ignoramus* (Ponomarenko 1985)). *Sikhotealinia zhiltzovae* Lafer is the only known beetle with three ocelli (Lesch & Beutel 2004). At first, scientists included this family among the Adephaga, classified as carnivorous. However, now they are included as phytophagous in Archostemmata. Indeed, their feeding is unknown, but their anatomy shows that they are phytophagous.

Comparing Jurodidae with other Archostemmata, Kirejtshuk (1999-2000) said that these Coleoptera seem to be beetles of the wood or they have borers larvae, like other Archostemmata. However, their long and simple tarsus with strong claw indicates that they were and are associated with lifestyles outside on the trunks and branches of trees in the freshwater basins, rather than other habitats.

CUPEDIDAE: 30 species worldwide (Grebennikov 2004). This is a regarded as the most primitive of beetle families (White 1983). With two subfamilies, one is monogeneric: **Priacminae** erected for *Priacma serrata* (Leconte) from North America and **Cupedinae**, with eight genera known from all zoogeographical regions of the World, with the exception of New Zealand and Europe (Neboiss 1984; Grebennikov 2004). Feed on dead fungus infested wood, and adults probably live in dead subterranean wood such as tree roots, buried structural timber, rotten cores of living trees or stumps (Crowson 1962; Ren and Jingjing 2006). Larvae feed in moist, rotting soft woods and harwoods, and sometimes in moist timbers of buildings. They are no injurious, because eat decaying wood (White 1983). Adults and larvae occur together boring into rotten wood the bark of hard and soft wood logs or branches prefering firm and moist wood previously attacked by fungi (Evans and Hogues 2006).

Rarely captured, adults can be exposed by chopping infested logs and stumps. Sometimes found on vegetation or flying in sunlight near infested wood. *Priacma serrata* (LeConte) from U.S. probably feed on fungus in rotten wood. Live under tree bark in coniferous forests and during daylight hours, sometimes males are attracted to odor of laundry bleach (White 1983). They are collected by beating from dead branch and sweeping low vegetation, also other *Priacma* species have been caught with flight intercept traps (Grebennikov 2004). According to Hangay and Zborowski (2010) In Australia they also live in pastureland and in drier country.

MICROMALTHIDAE: This family is represented by a single species (*Micromalthus debilis* LeConte) which is native to the Eastern United States and Belize (Downie & Arnett 1996; Philips 2001), and accidentally introduced into other countries through trade wood (Lawrence & Newton 1995), but according to Keith Philips (2001) it can sometimes be difficult to ascertain whether a species is native or introduced, as Whitehead and Wheeler (1990) pointed out. It has been reported from British Columbia, and also has been reported from New Mexico, Brazil, Cuba, Hawaii, Hong Kong and South Africa and has eventually been trapped in water reservoirs (and flood debris) with Berlese funnel.

These beetles are very strange because the larvae are able to reproduce (they are paedogenic). Its development is extremely complex, because the larvae may procreate live larvae or lay eggs from which larvae are born that may become adult females (Barber 1913; Pringle 1938; Scott 1938; Pollock & Normark 2002).

The species is associated with structural rotten timbers. The larvae usually eat fungus of infected wood and sometimes are found in building wood, being wood borers in hardwood. They are found under bark and dead wood and the larvae sometimes in decaying wood of *Quercus*, in decayed oak, chestnut and pine logs or and in the wood of old house's basements, but are not considered as serious pests. Many of species are endemic, known only in limited areas (Neboiss 1960, 1984). Some of them eat wood fungi (Lawrence 1991). They are attracted to light and attracted to laundrys bleach containing sodium hypochlorite (Evans and Hogue 2006). Also they can be collected with flight intercept traps (Philip 2001).

Suborder MYXOPHAGA

LEPICERIDAE: Only 2 species are known: *Lepicerus bufo* Hinton from Mexico and *Lepicerus inaequalis* Motschulsky from Central America and Venezuela (Shepard et al. 2005). They live in sandy areas at the edges of rivers and streams, the larvae are aquatic. One of the species is found above 1,500 meters above sea level. They live in moist sand or gravel deposited by streams or rivers, feeding on algae. They are collected in stream banks, in places slightly sandy. *Lepicerus bufo* Hinton lives between 900 to 1,320 meters above sea level (Shepard 2005).

To collect they the garbage of plants can be sift on the edges of rivers and plant debris accumulated by seasonal flooding in the hills creeks. Search among the emergent vegetation. Look carefully at the sand on the banks of streams, pouring water on sand or pouring sand into a large container of water (like a bucket) and wait they climb to collect manually . Pitfall traps can also be used. *Lepicerus inaequalis* Motschulsky has been collected relatively high on the bank of a stream (about 0.3 m above the water line) and near where the marginal vegetation meets the bare sand surface (Shepard et al. 2005). Also this species has been collected in "damp piles of trash left by alternate rising and receding of the river" (Hinton 1933: Cited in: Shepard et al. 2005) and Chandler (1973) have collected they under stones in a dry and somewhat sandy place. *Lepicerus bufo* Hinton has been collected from the air-water interface among water plants (Arce and Gutierrez 1988). Sometimes the two species have been found coexisting in a single stream (Hinton 1933). Generally, *Lepicerus inaequalis* Motschulsky has been collected in dry sand deposits along the coast. According to Shepard et al. (2005) it is not easy to find them. A long observation of the surface of the sand is required. Shepard and his coleagues say that eventually some specimens will move a few steps and they will reveal their presence. The aforementioned authors also recommend extending a white cloth on the soil, on the banks of streams and place a white light on it. The beetles will approach walking on the cloth.

TORRIDINCOLIDAE: Comprising about 60 species in seven genera (Beutel & Vanin 2005). They are distributed in Africa, Madagascar, Asia and South America (Jiří & Fikáček 2008). They feed on algae. The larvae and adults are found on wet rocks washed by shallow and fast waterways covered with algae with rich submerged vegetation (Endrödy-Younga 1997). According to Léonard & Dessart (1994) they found African species pollinating the flowers of riverweeds (Podostemaceae). This plants grow on rocks in rapids and waterfalls of rivers (Jiří & Fikáček 2008). According to Reichardt (1973) species of *Ytu* live in hygropetric habitats, mainly on the seepages along roads which are usually exposed, with rather thin fi lm of water and often fluctuating water regime. Larvae of *Ytu* differ from the remaining Neotropical torridincolids, which are usually found at sides of streams and rivers (Reichardt 1973; Vanin & Costa 2001). Similar habitats at sides of streams and rivers are known for *Torridincola* (Steffan 1964, 1973) and *Delevea* (Endrödy-Younga 1997).

Satonius kurosawai (Satō 1982) widely distributed in the Japanese islands of Honshū (Prefectures Aichi, Fukui, Fukushima, Hyōgo, Miyagi, Naga, Niigata, Shimane, Tottori and Yamagata) and Shikoku (Ehime Prefecture), this species has been collected in hygropetric habitat, on wet surfaces of rocks in streams and seepage habitats (Satō 1982; Hayashi 2007; Hayashi & Kadowaki 2007).

On Paraguay, one specimen of torridicolid also was found on the wall of a small bank where the water ran (Ulf Drechsel, pers. comm. 2010).

Torridincolidae are most frequently found in aquatic and hygropetric habitats on rocks in stream rapids, alongside streams and waterfalls or below, often on sites covered with algae (Reichardt 1973; Beutel & Vanin 2005), these environments are easily found, but as these beetles are very small, they are rarely collected.

HYDROSCAPHIDAE: Distributed in Europe, South Asia, Philippines, Cebu Island, Africa, Madagascar, USA, Mexico, Panama and Brazil. Hydroscapha recently has been collected between Venezuela and Colombia (Hall & Short 2010). According to Fikáček and Šípková (2009) three genera are known (*Scaphydra*, *Yara* and *Hydroscapha*) with 22 species. Some of them live in the sandy banks of the rivers covered by aquatic vegetation. Others live in flowing water. Species of *Hydroscapha* occur as larvae and adults in streams, hot springs and pools, feeding upon algae (Vanin et al. 2005). The larvae are entirely aquatic and like the adults, eat the algae that grow on rocks in shallow and fast water (Hall & Short 2010). They have also been found among filamentous algae from pools and hot springs, where they can tolerate temperatures up to 45° C. They also live on algae attached to rocks in rapid water course and sandy margins of streams. Spangler (1981) has found them in hygropetricous hábitats in land springs of the road embankments. They also live in mud and can be collected by sweeping water in their habitats or can be collected with a colander.

SPHAERIUSIDAE: 21 known species are distributed in America, Africa, South Asia, Australia and Europe (Arce-Pérez and Moron 2008). Only one genus is known *Sphaerius* (= *Microsporus* Kolenati) (according to Reichardt 1973). They are hygropetric, found in a large variety of moist environments, such as sand, mud, under stones, among plant roots and leaf litter, in moss of swamps and litter on the sides of water bodies (Arce-Pérez and Moron 2008) they also are found along the banks of streams and rivers and they eat algae. In Mexico one species has been collected in the side of a stream with water lilies, submerged grass, with a few mosses and a few algae (Arce-Pérez and Morón 2008). Also they can be collected with pitfall trap (Telnov et al. 2005). The species *Omicrus ingens* Hansen and Richardson (from Puerto Rico) has been collected on bromeliads and is one of the few larvae known in this subfamily (Hansen and Richardson 1998).

Suborder ADEPHAGA

AMPHIZOIDAE: There are only 5 known species in this family in the single genus *Amphizoa*, with one species in China, another from Korea and 3 in Western North America (Rivera et al. 2002). Adults and larvae are predators but they will also scavenge dead insects. They live near fast flowing rivers and on wet soil and are associated with mountain streams. Adults are usually found on rotting wood piles and other decaying plant material found in eddies and backwaters of streams. But, they are unable to dive for a long time (Arnett et al. 1980). Adults do not swim (Ribera et al. 2002). The larvae cling to floating debris with head down, ready to dive to capture prey. After molting three times, the larvae transform into pupae in the mud or sand, some distance from the shoreline. During the day, adults and larvae are slow, but become more active at night. When caught, the adults excrete a fluid from the anus, which smells of rotten wood. They are found most often at the edges of rapid cold currents. There they live under rocks or coarse gravel from the banks or cling to the exposed roots below the walls of banks (Evans and Hogue 2006). To collect them the foam with floating debris which accumulates in eddies and backwaters may also be strained. Adults are found in masses of pine needles and branches swept onto shore. Adults are sometimes found in large numbers at the base of falls, and occasionally found in ponds and lakes, but those specimens were probably carried downstream (Evans and Hogue 2006).

ASPIDYTIDAE: It is a family of recently discovered aquatic beetles, with patchy distribution. Two species have been described**,** *Aspidytes niobe* Ribera from Western Cape, Province of South Africa and *Aspidytes wrasei* Balke from Shaanxi, Province of China (Beutel & Vogler 2002; Ribera et al. 2002.). Adults and larvae are found exclusively in the thin film of water that sometimes seeps through the walls of rocks. They feed on small aquatic invertebrates. Although they are true aquatic beetles, they are not adapted for swimming or diving (Alarie and Bilton 2005).

CARABIDAE: All over the world 1,840 genera are known with more than 40,000 described species (Thiele 1977; Eitnar et al. 1989). Carabidae**,** popularly called ground beetles, are abundant insects in wetland ecosystems, with less diversity in arids temperate environments (Erwin 1985). They belong mainly to three ecological groups: Hygrophilous, living on the edges of streams or ponds; arboricolous, living in logs or leaves, and geophilous, which live in the soil without being associated with water. The latter may constitute elements of the soil macrofauna, and be either endogenous or cavernicolous.

Most species of Carabidae are terrestrial. While some feed on plant material, most are predators. Many live among the rocks along river banks, some live in swamps or marshes and on the beaches of the ocean. Adults and larvae are found under rocks, sometimes in galleries dug in the ground or in natural holes, in leaves in the forest. Most of the larvae and adults are strictly carnivorous; however, they may eat fruits and spores of macrofungi. Many live in the cracks formed in dried mud. There are diurnal and nocturnal species. Many species occupy specific places either on beaches or in forests, places that are not occupied by other species of Carabidae (this to avoid competition between them). Some Carabidae prefer to live on river banks and others in not flooded wetlands within the forest. There are other species on the edge of slow moving waters. Others live among leaf litter on muddy lands. There are also species living in the mountains 2,500 meters above sea level, such as some Scaritini, Psydrini, Pterostichini, etc.

Many species are collected manually, but also with various traps: Barber, Berlese funnel and Winkler/Moczarski eclector, with pitfall traps, with sweeping, beating and pitfall traps with baits of fermented substance or meat (Spence and Niemela 1994).

Other species may be collected "treading" loose vegetation and leaf litter in swamps; in moist areas of these sites, ground beetles try to escape when flooded and then they can be collected manually.

Among the litter and humus of forests many species of Carabidae are hidden and a good way to catch them is to put plant material, even with some soil, in a large bag; then empty it over a bowl or bucket, and sometimes it can be emptied in a hole made for this purpose on the floor or in a pool of clear water near the place where the humus was obtained. When it is empty the trash in the water Carabidae try to escape by swimming toward the edges so it's easy to catch them manually. The same procedure can continue until the bag is empty.

Also many Carabidae are hiding in the plant material that floats on the water. A similar method consists in putting in large plastic bags the plant material floating in the water, for example the one in lagoons, backwater of rivers, streams, beaches, etc. This material is emptied in a large container (bucket or basin) with water and pushied it to the bottom. The Carabidae will try to escape and can be easily collected. There are also species of Carabidae that approach to feed on the slightly rotten fruits on the forest floor, especially those of trees belonging to the genus *Ficus*.

Citrus trees such as oranges, lemons, etc. are not very attractive to the majority of Carabidae. According to Kipling (pers. comm. 2005) *"places having clusters of flowers dropped on the ground are excellent for collecting various carabids, especially among the piles of petals along the trails in wet forests and mainly at night. Some species are found in piles of leaves and plant litter accumulated between rocks or in logs and wood blocked in rivers and streams; many rare species are found under rocks of medium to large size, located along the shores of big rivers. Many rare and unusual species of* Carabidae *are found in large numbers in drier forests and in open grassland but near water (artificial lagoons or where the water is accumulated), in ravines in the same places, and also in ravines on the banks of rivers and streams, both at day and at night. Especially look in the cracked land in places where the mud is almost dry. There we can see many small species of* Carabidae *often accompanied by other families, such as Anthicidae. Interesting species are also available by examining the rotting sticks or logs that are on the mud or on wet sand. A few species are close to, or fly close to burned wood or are attracted by forest fires. At night you must walk with a flashlight along the paths and near wet areas to be able to collect the* Carabidae *walking on the ground."*

Many species are strictly arboricolous (genus *Agra*, many of the Bembidiini tribe, etc.) and live mainly in the foliage, and some species live mainly in the upper canopy. Thus, the best method to collect these species is by beating the foliage, using an beating sheet under the clusters of branches passing from a tree or shrub to another, forming bridges between them, where the carabids prefer to walk. Also, is helpful to use the sampling methods by fogging the canopy or shrubs.

Several species are found under bark of trees and fallen logs. Some species can be found walking on saline, both day and night. Others live in sand dunes or in caves. In arid habitats and during the day, some species that coexist in the same place are rarely found together. For example, certain species of *Brachygnathus* spp. and *Morion* spp. live only under the fallen cactus, others only under sticks and still others are under fallen termite mounds (Garcete 2001); other species live only in rotten logs (Catapiesini, Pterostichini).

Some species are attracted by ultraviolet light and others by white light (Brachinini, Pterostichini, Galeritini, Apotomini, etc.). At night, traps should be placed near wet areas.

The pitfall traps with buckets buried are very productive. A variety of species are collected with pitfall traps and little pieces of meat can be used as bait (in forests) and small pieces of fish on the shores of rivers or streams. Is also useful the following mixture: Water 60%, ammonia 40%, or just vinegar, but these traps should be checked daily, however, according Ulyshen et al. (2005): *"Pitfall trap data give an incomplete picture of the carabid community and should be interpreted carefully. Additional methods, such as use of Berlese funnels and litter washing (Spence and Niemela 1994), collection from lights (Usis and MacLean 1998), and deployment of flight intercept devices (Liebherr and Mahar 1979; Paarmann and Stork 1987), should be incorporated in surveys to better ascertain the species composition and relative numbers of ground beetles. Flight intercept devices, like pitfall traps, have the advantage of being easy to use and replicate, but their value to carabid surveys is largely unknown."*

Carabids can be sacrificed with potassium cyanide, ethyl acetate or putting them directly in a mixture (equal parts) of alcohol and white vinegar.

There are two subfamilies of carabids that deserve to be treated as a separate item, only to facilitate the study of his capture:

---Subfamily Paussinae: The species are a Pantropical group, with 22 genera and about 565 described species (Lorenz 2005). About 10 species are found in the Mediterranean Sea area, including 2 in the European area. Both adults and larvae are myrmecophiles (obligate guests of ants) and they have morphological adaptations to live in ant nests, at least during part of their life cycle (Nagel 1979, 1997; Di Giulio & Moore 2004). They have glands, which may be in the head, thorax or on the tip of his abdomen, which secrete an aromatic substance that ants live. The larvae and adults of Paussinae feed on ants and termites (Deuve 2001).

---Subfamily Rhysodinae: They are found on every continent in wooded areas. This subfamily, known worldwide, includes 170 species in 20 genera. There are more species in New Guinea, Philippines, Indonesia and Northern South America. In Europe there are 2 species and they are very rare. Adults and larvae live under bark and in moist dead wood. These beetles are found in rotten wood and presumably feeding on slime mold plasmodia (Grimaldi and Engel 2005). According to Hangay and Zborowski (2010) in Australia they live under the bark of hardwood trees (ash, beech, or elm and pine). Specialized structures in their jaws allow them to eat mushrooms. Adults do not make galleries; they just stay among the layers of decaying wood, without leaving visible traces of their passage, but the larvae dig short tunnels. Adults must be sought under the bark of dead conifers and in logs and stumps. They prefer moist, partially decayed wood, either as logs or in parts of the roots of dead stumps. Search for larvae under slime mold fruiting bodies which grow in soft and spongy wood. The places with yellow slime mold plasmodia are excelent to look for Rhyzodinae (Evans and Hogues 2006). They may be collected manually or by taking pieces of rotten wood where the larvae live, to rear them in the laboratory.

---Subfamily Cicindelinae: Presently, there are around 2,600 known species of Cicindelinae worldwide (Pearson et al. 2007) with 331 being Neotropical (Reichardt 1977).

Here, the reference is made to the American species of Cicindelinae, because the author has more experience with the sampling methods of Neotropical species, but surely, with the methods explained here, you can also collect a majority of the species living on other continents. Below there are some aspects of biology and techniques for collecting adults of this fascinating group of beetles.

Places where adult Cicindelinae live:

The habitats where they live are varied but almost always wet, for example:

Sand or beaches on the banks of rivers, lagoons or permanent and semipermanant ponds and lakes.

The sand dunes near the seashore.

In sandy eruptions far from water.

Perched or walking on the beach or big rocks on the banks of streams, canals and waterways in general.

Roads within savannas or grasslands.

Walking on earth in wet paths in and around forests. They may be seen in the path, on its shores or in the middle.

On forest trails, perched on plants at low altitudes, often between the grass and even among bamboos.

On several highland sites, such as hillsides.

In areas with clay or mud near lakes or ponds.

On the trunks of some trees on the sides of watercourses.

On large and damp flat stones, either on hills or near water courses, even those that emerge from streams or rivers.

In the mud formed near streams or moist ravines.

On termite mounds.

Walking besides of columns of ants (usually those belonging to the *Esciton* spp.).

Most species are active on damp soil, clear of vegetation.

Perched on plants within the forest.

To collect a variety of Cicindelinae it is important to visit different kinds of environments, inside and outside forests, including deserts and prairies. Sometimes, in an area of a few square meters, for example, in the Neotropical region, different species of *Brasiella, Cylindera, Cicindela, Pentacomia,* Odontocheila, etc. can be found. Many nocturnal or crepuscular species hide during the day under rocks or sticks on the ground, even between the cracks formed in the dried mud. It is best that the places to be visited are little traveled by people or vehicles. When human activity is low, they are more likely to find Cicindelinae. But if the place is visited by persons or vehicles, is best wait until they are gone, as the Cicindelinae usually return to their places of activity when everything returns to calm.

Cicindelinae in the some genus exhibit strong niche partitioning (Wyatt Hoback, pers. comm. 2010).

Generally, each species occupies a particular territory which is not invaded by other species (Pearson et al. 2007). For example, if there is a road of red land (loam), certain species of Cicindelinae will be found there; farther on, on the same road, the color of the land could change (due to different chemical composition of the soil) and there another species could be found. This sometimes happens a few meters away. Some species of tiger Cicindelinae prefer shady places and others prefer sunny places. Some species only become active at certain times of the day. That's why some of them can only be collected in the morning when the sun is just coming out (many *Pentacomia, Cicindela, Brasiella*), others only during the peak of solar light (*Pentacomia, Cheylonicha, Cicindela, Brasiella*), others at twilight (*Tetracha, Iresia*). Many species of *Odontochyla* are found in the dark roads of the forest, others are found almost only at night (*Tetracha, Iresia, Amblycheila, Bucephala, Omus, Oxycheila*) and others are inside or outside the forest at anytime of the day.

There are species of *Cheylonicha, Brasiella* and *Prepusa* that spend most of their life on the termite mounds. Their flights tend to be circular, returning to the same or a nearby termite mound.

When frightened, they hide in holes in the termite mounds and go out only when they think the danger has passed. They are rarely sitting on the ground, except on the surface of the termite mound.

Most species living in trees (*Iresia* spp.) only run on the trunk and are crepuscular or nocturnal, and very fast (pers. observ.). Other nocturnal species live on sandy beaches or open areas and (*Tetracha* spp.) run generally on it.

Many species of *Odontochyla* live in the clearings and roads within the forests, and make very short flights of only a few meters. They are very fast, but stop every few moments (as if watching their territory). When you want to catch these species, they run away in any direction, but stop again, seeming to calculate their next move.

There are some species that are very difficult to find: *Ctenostoma schmalzi paraguayensis* Nidek and *Ctenostoma ichneumoneum ellipticum* Naviaux (both Cicindelinae from Paraguay, they are running between the columns of ants of the genus *Esciton* spp and apparently feed on the insects that these ants capture during their excursions.

To collect *Oxycheila* should look on the banks of streams, rivers or at the bottom of humid ravines, where the soil or sand is black and muddy, or on the rocks.

Best time to collect Cicindelinae:

In the tropics, spring is the best time to find them, although some species can still be found in late summer. They are present throughout the year within the forest in tropical and equatorial regions. In temperate countries they are found only in the warmest time of year. There are some species that live in the mountains at high altitudes and can withstand temperatures lower than usual, but cannot be found during the winter. In many places there are several species that live together but some of them live temporarily apart and have different time of appearance.

In temperate zones two life forms are known, Spring-Fall or summer-active species. For Spring-Fall, adults emerge and feed then aestivate before reemerging and mating (Wyat Hoback pers. comm. 2010).

Larvae from usually permanent burrows and in temperate regions require between one and three years to complete development (Wyatt Hoback, pers. comm. 2010).

Adults of Cicindelinae are not active on cold, rainy or cloudy days. Their activity depends on temperature, humidity and light intensity (pers. observ.). However, it is not easy to collect them when the weather is warmer, since they are more active and swift when the temperature rises.

The species of genera *Cicindela, Cylindera, Odontochyla* and *Brasiella* are easier to find on sunny days when the temperature exceeds 70° F. They are found on beaches, trails and roads, dunes, grasslands, marshes, alkali flats, mud flats, road cuts and banks, forest clearings and roads within the forest. When walking in places where they live, suddenly they can be seen when flying a few meters. These beetles are active fliers and runners, making it essential to use a net to catch them.

How to catch the Cicindelinae with an aerial net:

Almost all Cicindelinae are very agile fliers and run very fast, but there are those that cannot fly (*Tetracha, Omus, Amblycheila* and other). However, these are also very fast. When either perched on the ground, a branch, a termite mound or anywhere else, they can see with their large eyes everything that moves around. They even recognize the various objects around them, so the collector must walk very slowly without making sudden movements to prevent them escaping. If possible, a single person should make this work because when there are two people the chances of escaping increase. Many species, due to their small size, can be easily confused with flies when viewed from afar. Looking at the way they fly they seem to be Diptera, family Asilidae, but very soon the collector learns to identify them.

To collect Cicindelidae it is important to have good vision and to be very agile. The method with nets used to collect the diurnal species is the same used to collect butterflies. It is preferable with a light handle and thin white tull. Other collectors prefer a green or black tull, but from personal experience, the author believes that white is the best, since dark colors hinder the detection of the insect when trapped against the ground. To find them you should walk softly and slowly in their habitats, hiding the net behind the head. When you locate a tiger beetle try to keep some distance from him, because when they discover their pursuer they turn to his direction to watch and wait for the right moment to escape. When you can get close enough to the insect, the net must be lifted very slowly behind the head (no sudden movements) and with one hand (usually right) take the tip of the handle, and with the other hand pinch out the fabric of the back of the net. Slowly bring the net as close as possible to the insect, above it, and deal the final blow against the ground, capturing the beetle.

When the blow is given, from top to bottom, you must immediately release the back of the net, pinning the animal between the ground and the net. Then the collector may approach to the net as quickly as possible and lift the back fabric of the net, so that the beetle could walk up. When the beetle goes up, the net must be raised quickly and, with a fast move, shake it in the air (like catching something on the fly) so that the beetle will go to the bottom, driven by inertia. With your free hand close immediately the mouth of the net and the beetle will be caught. Very carefully the insect is removed from the bag and passed to the killing jar. You must throw it quickly into the jar and close it immediately because these beetles are extremely fast to escape. You can also put an open killing jar into the net and catch it there. The presence of Cicindelinae can also be investigated from a few meters away in places where there are large rocks with flat surfaces. These rocks are located on hills or in valleys, near or in streams (where these rocks crop out on the water surface) or in damp places. The same method with the net described above can be applied here.

The Cicindelinae carry out short and fast flights at low altitude, 1 to 2 meters of the ground, and sometimes a little more. Many species (mainly nocturnal) only run on soil or sand. The flying species rarely keep a flight that lasts more than 2 to 3 seconds. They always go in a straight line, keeping away from the collector (*Brasiella, Odontochyla, Cicindela, Pentacomia* and other). Sometimes they hide among the plants at the roadsides. In that case, the collector must walk on the plants or move them with the net to scare the beetles and force them to return to the clear floor, where you can catch them.

Other sampling methods:

Some species of *Cicindela* can be captured with pitfall traps, but as these beetles quickly escape out of the buried traps, then it is better to use the pitfall traps with funnel which can also be used to trap other species of tiger beetle which run on the ground at night.

Preservative such as soupy water or antifreeze car can also be used in traps (Wyatt Hoback, pers. comm. 2010).

Cicindelinae are also found near the water springs where there are interesting species of *Oxycheila*, but these populations are often not abundant. Some other species roam at night on roads within or outside forests, jungles or deserts. These are relatively easy to see from afar. To catch them must walk slowly, illuminate them with a flashlight and then take them with your hands. Other species live in trees (*Iresia* spp.) and occasionally are found at night on the trunks of living trees that grow near streams or rivers.

In temperate zones, summer forms often come to lights and can be captured frequently with light traps (Wyatt Hoback, pers. comm. 2010).

In areas where there are sets of termite mounds, the collector must approach very slowly and see if they have any *Cheylonicha, Prepusa, Brasiella*, etc. *Cheylonicha* are manually caught with one swift motion, putting the palms on the insect without destroying it. Then, very carefully, take it with the fingers. If Cicindelinae discover the collector they will escape. To avoid this, the collector must lie on the termite mound (lying face down on it). If you do not make sudden movements, the Cicindelinae will approach again or leave the recesses of the termite mounds. However some species like (*Prepusa* spp.) and the smaller beetles (*Brasiella* spp.) are difficult to collect because they quickly run away, making it essential to use nets to catch them.

Cicindelinae apparently do not distinguish colors very well; they probably see only in black and white. It seems that this happens with *Iresia lacordairei* Dejean and also *Oxycheila pochoni* Mandl found on very dark nights in black mud or on rocks by streams near of the jungle (pers. observ.). The author has collected these two species in Paraguay and the method employed is simple: Use a flashlight with a sheet of red cellophane paper glued in front of the bulb, illuminating the insect that is perched on a tree or on the ground and proceed to capture it by hand. However, other species such as *Oxycheila labiata* Wiesner, *Tetracha martii* (Perty), *Tetracha cruciata* Brullé, *Phaeoxantha testudinea* (Klug) including *O. pochoni* Mandl and other nocturnal species can be collect with a flashlight without using cellophane paper. Probably only *Iresia lacordairei* Dejean does not distinguish colors very well (?). To determine this, it is necessary to conduct more experiments. However, *Iresia* is very rare in nature, which hampers this demonstration. Many species of Cicindelinae arrive to ultraviolet light or mercury white light.

The nocturnal species as *Megacephala* spp., *Oxycheila* spp., etc. do not fly even if it is very hot, being easier to catch them manually with a head lamp or flashlight. But Cicindelinae usually prefer running and flying when frightened, so it is better to use aerial nets and pitfall traps to collect them.

The small size species (*Brasiella, Cicindela, Odontocheila, Pentacomia, Cylindera*, etc.) are found in diverse habitats, and only practice will allow a collector to recognize them in the field.

To collect various species of *Tetracha* at night, also, must visit the white sand beaches or lands along rivers, streams, marshes or bare soils. They are also on the ground near illuminated advertising signs in areas with street lighting. They can be collected with pitfall traps, using pieces of meat or chicken bones as bait. During the day they may be found under boards, logs and stones. Many species are sometimes seen walking on cloudy days, even when the soil is wet after rain.

Can be useful also bury pitfall traps with baits of diced meat, fish, or just vinegar. To find out where to bury these traps first of all must locate the sites where Cicindelinae abound.

Some species of *Cicindela* may be attracted to lights at night. Also, at night, many species of *Cicindela* can be manually caught while on the ground in the area illuminated by the light trap. We must also examine other bright lights such as street lighting and commercial signs.

North Amazon species have larvae that usually occur in bare soil with access to moisture, these areas often flood and the larvae are adapted to survive immersion (Wyatt Hoback, pers. comm. 2010).

The *Amblycheila* spp are nocturnal and can be searched for with a flashlight and walking the arid areas and grasslands. However, it is better to collected them with funnel pitfall traps. Often they hide under the land of vertebrate burrows. Be cautious when approaching the termite mound or the vertebrates burrows because many times there are snakes in these nests such as Rattlesnake (*Crotalus* spp.) or Jararaca (*Bothrops* spp.) that take refuge in those places, especially in savannas and grasslands (Paul et al. 2002; Canevari et al. 2007).

How to kill the Cicindelinae:

Two bottles with any of the usual poisons in killing jars can be used to sacrifice the tiger beetles

(Ethyl Acetate or Potasium Cyanide). They can also be introduced directly into Ethanol. The best bottles are those with a pressure lid.

Do not put many specimens simultaneously in the same bottle, as they bite each other. They should be introduced one at a time and wait until the beetle is stunned by the poison. Only then put another beetle in the same bottle. It is essential that the killing jar contains enough toilet paper so that the insects can walk on it and do not bite or dirty each other. After entering a few specimens in the first killing jar, remove them and put them in a second killing jars (preferably with Ethyl Acetate) using the first bottle for the next collected insects. In addition, from time to time we must ensure that the Ethyl Acetate did not evaporate. To verify this must open the bottle, bringing it near the face, shake the air with your hands and smell it; if it does not give off any odor, then must be put a few drops of this liquid with a dropper.

How to collect the larvae of Cicindelinae:

Tiger beetle larvae may spend between 1 to 3 years before transforming into adults, and as adults they live only a few months; males usually live only a few days or weeks. In contrast, some females have lived, in laboratory conditions, up to 5 months (Paul J. Johnson 1996/1997). Some larvae make galleries on the ground near water (*Tetracha, Odontocheila*, etc.). The larvae of *Omus* spp. live in round holes, on hard ground, usually on the slopes and foothills of mountain ranges (White, R. 1983) and feed on prey that walk around their burrows. However, in most species larvae make their burrows in the ground, in areas near the adults. Burrows of Cicindelinae vary considerably in size, sometimes measuring more than 1 / 4 inch in diameter. The burrows are distinguished by having a smooth edge, with sand or soil away from the entrance, at a distance of 1 / 2 inch or more. According to Paul Johnson (1996 / 1997) the larvae of some *Amblycheila* spp. are quite large and elongated, spend all their development in the depths of their tubular burrows, and can range between 2 to 3 feet below ground. These galleries are difficult to locate among the many other holes of 6-8 mm in diameter in the ground along the edges low cliffs (Paul J. Johnson 1996/1997). The larvae of some Cicindelinae build nests in the form of towers which are visible from afar. The larvae are very shy and hide quickly at the slightest sign of danger. It's always better to look for them at night lighting the holes with a flashlight, so they can be discovered, looking out their hiding places, brandishing their large jaws, waiting for an unprepared prey, such as an ant, a beetle or even a cricket. Often they are very abundant in areas where they live.

To extract the larvae a little piece of gum can be used (previously immersed in hot water to melt). The gum sticks to the tip of a thread (or straw), and moves very slowly near the den of the tiger beetle, so that it looks like prey. If the larva attacks it with its jaws, it will stick to the gum and can be quickly put in a jar. You can also enter a straw into the burrow and use it as a guide for digging a hole with a small shovel or a sharp knife. Many holes are quite deep, even several feet deep, so be patient to remove the larvae. When the collection is finished you have to leave the place as it was at first, spreading the land over the holes to avoid too much damage to the habitat.

OBS: It is important to note that some species of Cicindelinae have small populations, mainly the species living in trees, black mud, termite mound or dunes, so we should not take out many specimens of these species. In some countries, some species are listed as endangered.

GYRINIDAE: A family of worldwide distribution. There are 13 genera and about 995 species. The adults are predators and breathe the open air, in part by the antennas which are modified in respiratory organs. They are found on the surface of clean water, in ponds, streams, rivers and lakes, where several species mating, swimming in circles in dark places. Others are commonly found along the banks of these aquatic environments or among the aquatic vegetation of the vertical cracks of walls in some streams. Some species of *Gyretes* live in mountain streams. Pupae are found in emergent plants along the coast. Adults feed on insects trapped in the water. Larvae are immersed in water and they are predators. Adults are very good swimmers, when they are frightened they dive quickly and can remain hidden under the water for a considerable time. They hide at the bottom on stones or plants. Larvae live underwater. They must be collected with an aquatic net, both day and night and go to the lights in large numbers. When hooked, they expel a foul smelling milky fluid (like Dytiscidae) secreted by prothoraxics glands.

HALIPLIDAE: The crawling beetles are found worldwide, with 206 species in 5 genera (Vondel 2005 and 2007). Adults are aquatic and can be found in lentic habitats. Some species are developed on the margins of slow streams. Others prefer calm waters on the banks of rivers and lagoons, or wander on aquatic plants found in these locations. There are species living in brackish water (Evans and Hogues, 2006). The *Haliplus* spp. lives in dense vegetation in habitats with ephemeral or semi-permanent water like road ditches or swamps. Many species live attached to algae such as *Spirogyra* spp, *Chara* spp. and *Nitella* spp. (Spangler 1981). Both adults and larvae are predators and some of them feed on filamentous algae found in shallow water. Many of them may be facultative scavengers. Species of *Brychius* genus are found in the eroded parts of the streams, and others live in permanent, clear, nutrient-rich ponds and lakes or in slow flowing backwaters of rivers (Evans and Hogues 2006). In North America *Haliplus* and *Peltodytes* occur in weedy ditches, ponds and lakes and *Apteraliplus* found only in vernal pools. Those of *Brychius* genus are found along the banks of streams, rivers or lakes (Evans and Hogues 2006). They can be collected with lights near their aquatic habitats, or by sweeping with an aquatic net, the collector must come quietly to a watercourse where these beetles are being active. Places with shallow water, such as permanent ponds, lakes or streams with plenty of aquatic vegetation or algae, will be very productive to get many species of crawling beetles.

TRACHYPACHIDAE: 2 genera and 6 species known: *Trachypachus* genus with 3 species in Western North America and 1 species in Eurasia, and the *Systolosoma* genus with 2 species from Chile and Argentina (*Systolosoma breve* Solier and *Systolosoma lateritium* Negre). These beetles are reminiscent of Carabidae. They are diurnal, predators or scavengers. Larvae are campodeiform type and as adults, live in dry soil in shady places, especially among the litter. Two Canadian species (British Columbia): *Trachypachus holmbergi* Mann occurs from Alaska to Saskatchewan and has been collected in synanthropic localities, such as urban flower gardens at the University of Alberta and on loose loam or conifer duff. Another species is *Trachypachus gibbsii* LeConte which is found only in British Columbia. *Trachypachus slevini* Van Dyke lives in Moolack beach along the Oregon coast and is found on the cliffs of the sandy beaches of Pacific Ocean. The beetles run in the talus slope, consisting of small particles, at the base of the cliff of exfoliating clay. The beetles run on the road for short periods under the sun and then hide in piles of garbage plant. They can live also higher, in the hard clay of the cliff. The other species are *Trachypachus zetterstedti* (Gyllenhal) from Northern Europe, *Systolosoma breve* Solier and *Systolosoma lateritium* Négre, live among the leaf litter of conifers in Chile. *Systolosoma* has also been collected in Río Negro, Argentina, during the summer, with pitfall traps (with vinegar) and Malaise traps (Sackmann 2006).

MERUIDAE: Found only in one place in Venezuela, this omnivorous species (*Meru phyllisae* Spangler & Steiner, 2005) live in low water flows over rocks and gravel at the site where it is very calm and illuminated by sunlight. The larvae are unknown (Spangler and Steiner 2005). With only 0.8 mm long and they are considered the smallest of Adephaga species.

NOTERIDAE: A family of worldwide distribution, with 242 species in 14 genera. The subfamily **Phreatodytinae** is considered as distinct family by certain authors, are restricted to Japanese groundwater habitats. Adults and larvae feed on roots in submerged aquatic plants living in water pools and running water. Archangelsky (2001) suggests that adults are predators. They are found in lowland areas; several biotic and abiotic factors determine the distribution of family members; they have been collected in lentic habitats essentially permanent, fresh and turbid water, located in sunny places with abundant aquatic vegetation. However, in the fast flowing lotic environments noterids populations are poorly represented. They are found in caves and also been collected in brackish water holes made by crabs (Spangler 1981). To collect them an aquatic net should be used and look for them at different times during the day, covering all physiognomically and ecologically different microhabitats that can be found. They are also collected with the water trap bottle. Lights attract many different species.

PELOBIIDAE: Only a single genus is known, *Hygrobia* with 7 species: One European and North African (*Hygrobia hermanni* (Fabricius)), one Chinese (*Hygrobia davidi* Bedel) and five Australian: *Hygrobia australasiae* (Clark), *Hygrobia maculata* Britton, *Hygrobia niger* (Clark), *Hy-*

grobia wattsi Hendrich (according to Esat & Balık 2008); and *Hygrobia cretzschmari* (Heyden & Heyden). They are predators; larvae and adults live mainly in stagnant water and use to walk on it. In Europe they have been found in a small pond situated 610 m above sea level. The bottom of the pond was muddy with rich vegetation. They are caught with aquatic nets with very fine tull. *Hygrobia hermanni* Britton were found on dog excrement (Castro & Cárdenas 2003).

DYTISCIDAE: The Predaceous Diving Beetles have a worldwide distribution. 180 genera are known, including 4040 species in 11 subfamilies (Nilsson 2001, 2004 and 2006; Jach and Balke 2008). Adults and larvae are predators, live together mainly in lentic aquatic ecosystems such as shallow lagoons with emergent vegetation, ponds, temporary pools, ditches, streams and dams. Others live in crab caves, hollow trees, backwaters of rivers, in the water accumulated in the axils of some plants, hot springs, salt and alkaline marshes, fens, grooves filled with water, wetlands, artesian wells, underground streams, river and streams banks and hygropetricous habitats.

Dytiscids have not been found in open seas or excessively alkaline waters, from sea level to 5000 meters above sea level; they live in fresh and clean water, but some of them can tolerate polluted waters; most live in shallow water; they have to come to the surface to breathe, but some have been found 3.5 meters below a stream (Spangler 1981; Balfour-Browne 1940). Some may live in very turbid water saturated with organic material. Almost all the species are skillful swimmers (Machado 1989) which reduces the chances of capture. Mercury lamps and UV-light attract different species of Predaceous Diving beetles. They can be collected with an aquatic bottle trap and with "bright foil traps" (see "Aquatics and semi-aquatics environments", Chapter II). Many species are hidden in submerged objects and between the open spaces between the aquatic vegetation, there can be captured with the aquatic net and by sweeping as deep as possible. Among the dense aquatic vegetation there are many different species. So walking on water, in their habitats, is possible to collect a great variety of beetles. Don't abuse this technique, because many aquatic plants may be destroyed. A collection method is to pour a little oil or kerosene in small wells or pools, this stifles the beetles when they come to the surfaces and are easily collected. According to Andreas Herrmann (pers. comm. 2010), in Germany many members of the subgenus Dytiscinae could be lured with pork liver. For this can be usedmodified plastic buckets and submerge them in ponds.

Suborder POLYPHAGA

Series STAPHYLINIFORMIA

Superfamily HYDROPHILOIDEA (incl. Histeroidea)

HYDROPHILIDAE: A large family with about 3,150 species worldwide (Short & Hebauer, 2006). According to Archangelsky et al. (2005) there are 9 subfamilies: **Horelophinae, Horelophopsinae, Hydrophilinae, Sphaeridiinae, Hydrochinae, Helophorinae, Epimetopinae, Spercheinae** and **Georissinae.**

Adults are omnivorous, eating spores, algae, rotting leaves and other vegetation. Others live in manure, or *in* decomposed organic matter or decaying fruit. Some of them are predators (for example, snails) and eat carrion (Evans and Hogue, 2006). These families have 16 genera categorized as "incertae sedis" and at present 8 subfamilies are accepted.

---Subfamily Horelophinae: Subfamily from New Zealand, with only *Horelophus* genus, is extremely rare, possibly hygropetric (Hansen 1991). They are rarely collected and only adults are known.

---Subfamily Horelophopsinae: With the only genus *Horelophopsis,* from Yapen Island in New Guinea, they are probably terrestrial (Hansen 1997).

---Subfamily Hydrophilinae: With about 50 genera, adults and larvae mostly living in stagnant water, running water or in hygropetric habitats (Jäch 1998). This subfamily includes the *Tropisternus* which are very common in the humid months. Prefer shallow aquatic habitats with plenty of organic material or algae. In some species females stridulate to attract males (Arnett, et al. 2002). *Tropisternus* can be collected with light traps.

---Subfamily Sphaeridiinae: With worldwide distribution and 106 genera, it is the largest subfamily in species of all Hydrophilidae, most of which are terrestrial; rather few species are aquatic / riparian but in general they live under stones near water, among plant roots, in damp places with much humus, litter or moss in marshy sites. They can frequently be encountered in all kinds of rotting organic matter: Fruit, fungi, dung, carrion etc. (Gentili et al. 1995). Some tribal representatives like Omicrini and Coelostomatini live under bark, in moss covering dead trees and vegetable waste. The species *Omicrus ingens* Hansen & Richardson, from Puerto Rico has been collected in bromeliads and is one of the few larvae known in this subfamily.

---Subfamily Hydrochinae: Represented in all the major zoogeographical regions (Hansen, 1987, 1991).With the only genus *Hydrochus* with a worldwide distribution. About 200 species have been described (Makhan 1995 and 2001; Hansen 1999; Short & Hebauer 2006). These are mainly aquatic beetles: *Hydrochus spangleri* Hellman in Steiner was collected in a vernal pool in USA (Steiner et al. 2003). In Paleartic region *Hydrochus ignicollis* (Motschulsky) were collected from slowly running water, especially very cold even in summer, eutrophic and well vegetated (Incekara et al. 2004). It is reported that *Hydrochus ignicollis* (Motschulsky) avoids shaded areas; it is rarely found in deciduous woodlands and is generally found together with *Hydrochus brevis* Herbst (according Hansen 1987).

A European species (*Hydrochus nitidicollis* Mulsant) is known as the Beetle of river stones because they are found in association with pebbles and other sediments exposed in river banks; they are also found in wet substrates above and below the water mark. In Europe they are in danger of extinction, like other European species, due to destruction of aquatic habitats (UK BAP, 2003; Goddard 1999; Incekara and Bouzid 2007). Most *Hydrochus* are easy to collect since they cling to plants or debris on the banks of ponds and lakes. Some species are immersed in the grass in open places with shallow water. They can be collected with light traps, by means of a colander or with aquatic net.

---Subfamily Helophorinae: Consisting only of one genus *(Helophorus)* this genus comprises more than 190 species (Incekara et al. 2004a), represented in the major zoogeographical regions: Palearctic, Nearctic, Neotropical and Ethiopian (Balfour Browne 1958; Chiesa, 1959). *Helophorus abeillei* Guillebeau is an alpine and subalpine species (1200 - 2800 MASL), usually found along with *Helophorus ponticus* Angus, *Helophorus daedalus* d'Orchymont and sometimes *Helophorus faustianus* Sharp, of Turkey, and they mostly prefer small springs with muddy or sandy bottoms. *Helophorus arvernicus* Mulsant, from Austria, Denmark, Finland, France, Italy, Norway, Russia, Spain, Scotland, Sweden and Turkey (Balfour-Browne 1958; Chiesa 1959) are found at the edges of running fresh water. Adults are found mainly in the spring but also in summer and autumn. The *Helophorus* are found in shallow waters among the rubble. They penetrate into the moist soil in the sea coast. Helophorinae larvae have been rarely captures; adults are most frequently found. Larvae and adults are herbivorous and live in silt water. They cling at underwater vegetation and are not active swimmers.

---Subfamily Spercheinae: With 60 known species (Hebauer 1997). Larvae and adults generally live in stagnant water, in shallow streams and grassy margins of small lakes and permanent ponds, swamps and marshes (Spangler, J.P. 1981). All known species belong to the *Spercheus* genus widely distributed in the world; in the Neotropical region only two species are known. These beetles live in shallow, still waters such as ponds, dams, in bore drains, stock water troughs or slow moving creeks and streams, staying close to the underside of the waters' surface. Adults are highly specialized as scavengers and eating plants and decaying vegetal material. Larvae are known worldwide as filtering of debris with a specific mechanism for the carrion. But in Australia, larvae are predators, attacking a variety of snails, worms, small crustaceans and insect larvae. Adults and larvae are clumsy swimmers that walk along the underside of the surface film.

---Subfamily Epimetopinae: Consisting of 3 genera and 27 known species from South America, tropical Africa and tropical Asia (Ji & Jäch 1998b). Adults are aquatic. Larvae seem to have riparian habits because they have no respiratory specializations for living in the water (Archangelsky 2007).

---Subfamily Georissinae: With a single genus (*Georissus*) they have been found in all major biogeographic regions. 77 species have been described so far (Hansen 1997: N°48; Bernhard et al. 2006; Fikáček & Trávníček 2009). The representatives of the genus occur in all zoogeographical regions, with the highest diversity in the Old World tropics (Hansen 1999). Only 2 species are known in South America: *Georissus humeralis* (Pic*)* from Argentina, Bolivia and Brazil, and *Georissus amrishi* Makhan, from Suriname they has been collected at light.

The species are generally riparian in all stages. They are supposed to have "reverted" to terrestrial habits from aquatic ancestors. Adults and larvae inhabit wet muddy or sandy banks of streams, rivers and standing waters, sometimes also on wet habitats far from permanent water bodies (Fikáček & Trávníček 2009). Adults are saprophagous feeding most probably on decaying organic particles found in the mud (Hebauer 1998) and are also common on the shores of the sea. Larvae are predaceous, preying on various invertebrates, larval Diptera and Nematoda (Hansen 2000). *Georissus* species can be collected by placing the sand where they live in a container with water and catch them as they float. The South American species reach the lights and some of them have been collected in very disturbed secondary forests.

Most hydrophilids are collected sweeping with an aquatic net submerged in the vegetation in the shallow margins of ponds, streams and lakes. The habitats with vegetal submerged organic material and algae are very productive for collecting beetles, must be walk in these habitats at night with a flashlight and daytime search also among the moss and other dense plant material.

In the cold forest, can be collected by hand in clusters of wet moss covering tree branches, those in wet rocky slopes or soaking the moss in a bucket with water and collecting beetles that float in the surface.

Hydrophylidae living on the rocks can also be collected spraying their habitats with insecticide (pyrethroid). A knife or screwdriver is useful to scrap the surface of rocks, soil of ravines or near waterfalls.

Also, it is good to use the aquatic bottle trap in several places (streams, ponds, lagoons, etc). Another sampling technique is to immerse the aquatic substrates and vegetation, in order to obtain the species that cannot swim and stay trapped in the water surface.

It is also useful raking the wastes which are floating on the water or on the edge and sometimes thrown by water on the beaches. This waste must be moved to a white sandy beach or light-colored earth and shed inside a container with water; or alternatively into a hole previously digged in the sand and full of water. The beetles attempting to escape can be extracted with a strainer or the fingers. Another option is to put small portions of organic materials on a wire or plastic mesh, that sits over a bowl or white tray and the artificial wastes are irrigated and the artificial shower. The beetles will fall to the bottom of the container and can be caught with a fine sieve (such as those used to brew tea). The artificial shower is made with a plastic bottle (1 lt. or more), to which many small holes have been practiced at the bottom. The bottle is filled with water and serves as a shower to wash the organic materials gathered on the banks of watercourses. Sometimes it is inconvenient to carry a large container to the collection site. However, it is advisable to have on hand the shower, container, strainers and grid, to be able to use all techniques of collecting and obtain a greater representation of species present at each site.

On the banks of any still water habitat, with an unobstructed beach, pour water out with a bucket or with your own hands (including gullies). The water that runs off can be collected with a strainer; so several species of Hydrophylidae and other beetles that live underground in these habitats are trapped. But also Hydrophyilidae swimming in spilled water obtained from a lake, pond, river, stream, etc. will fall into the earth where they can be captured with a strainer and washed with the shower. The smallest may be removed by a hand aspirator.

The semiaquatic species living in the beaches, on the banks of watercourses, are also collected by immersing the land or sand where they live, in a basin with water and when the beetles are trying to escape, they can be caught by hand or with a fine sieve. The very fast swimming species such as

Berosus, Tropisternus, Hydrophilus and other genera, usually escape when the water is thrown out of their habitats, so it is better to use an aquatic net and catch them directly into the water.

There are many and varied *Berosus* species; some prefer shaded habitats but others tolerate open shallow water; in some groups there are a choice of substrate and may even change of habitat (Oliva 2007).

Derallus species occur in very shaded habitats, often with floating debris (Oliva 1981). Most Hydrophilidae are collected sweeping with an aquatic net submerged in vegetation in the shallow margins of ponds, streams and lakes. The habitats with vegetal submerged organic material and algae are very productive for collecting beetles. Walking along these habitats at night with a flashlight is a good method. Search within the moss and other plant material dense. In the cold forest, can be collected by hand in clusters of wet moss covering tree branches, those in wet rocky slopes or soaking the moss in a bucket with water and collecting beetles that float in the surface. Hydrophylidae living on the rocks can also be collected spraying their habitats with insecticide (pyrethroid). A knife or screwdriver is useful to scrap the surface of rocks, soil of ravines or near waterfalls.

Also, can be used the aquatic bottle trap. Another technique is immersing aquatic substrates and vegetation, in order to obtain the species that cannot swim and stay trapped in the water surface.

It is also useful raking the wastes which are floating on the water or on the edge and sometimes thrown by water on the beaches. This waste must be moved to a white sandy beach or light-colored earth and shed inside a container with water; or alternatively into a hole previously digged in the sand and full of water. The beetles attempting to escape can be extracted with a strainer or the fingers. Another option is to put small portions of organic materials on a wire or plastic mesh, that sits over a bowl or white tray and the artificial wastes are irrigated and the artificial shower. The beetles will fall to the bottom of the container and can be caught with a fine sieve (such as those used to brew tea). The artificial shower is made with a plastic bottle (1 lt. or more), to which many small holes have been practiced at the bottom. The bottle is filled with water and serves as a shower to wash the organic materials gathered on the banks of watercourses. Sometimes it is inconvenient to carry a large container to the collection site. However, it is advisable to have on hand the shower, container, strainers and grid, to be able to use all techniques of collecting and obtain a greater representation of species present at each site.

Pour water on the beach of still water habitat (including gullies). The water that runs off can be collected with a strainer; so several species of Hydrophylidae and other beetles that live underground in these habitats are trapped. But also Hydrophyilidae swimming in spilled water obtained from a lake, pond, river, stream, etc. will fall into the earth where they can be captured with a strainer and washed with the shower. The smallest may be removed by a hand aspirator.

The semiaquatic species living in the beaches, on the banks of watercourses, are also collected by immersing the land or sand where they live, in a basin with water and when the beetles are trying to escape, they can be caught by hand or with a fine sieve. The very fast swimming species such as *Berosus, Tropisternus, Hydrophilus* and other genera, usually escape when the water is thrown out of their habitats, so it is better to use an aquatic net and catch them directly into the water.

There are many and varied *Berosus* species; some prefer shaded habitats but others tolerate open shallow water; in some groups there are a choice of substrate and may even change of habitat (Oliva 2007). *Derallus* species occur in very shaded habitats, often with floating debris (Oliva 1981).

One of the strainers should be large, light, with plastic bottom (or flexible metal mesh), with very small holes, preferably with a metal rim and handle or hard plastic. For terrestrial species of Hydrophilidae, must be put on a white dish small amount of organic waste (wet cow dung, wet leaf litter, rotten carcasses, etc.) and with the shower spray a little water on them. That assists the separation of the smaller species. A Berlese funnel is very useful to obtain specimens from the litter.

Other traps for collect hydrophilids are pitfall traps or Winkler/Moczarski eclector. Many species that not available with other methods can be trapped with lights located near the aquatic habitats. Review the street lights and shop windows, the windows of houses, hotels, etc.

HISTERIDAE: According to Penati (2009) there are approximately 350 genera and over 3,800 species are known worldwide (Majka 2008), but are at their most diverse in the tropics (Michael Caterino, pers. comm. 2010). Many species live in dung, between decomposing organic materials, under the bark of dying trees, corpses, bones, fruits and where resin flows. Adults and larvae of most species are predators of soft-bodied insect larvae and eggs, mainly of cyclorrhaphan Diptera found in dung (Kovarik & Caterino 2005), some histerids are also attracted to dung; prey on Scarabaeinae larvae (Reichard 1941; Bickardt 1916a).

Several species of Histerinae (*Margarinotus)*, Saprininae (*Euspilotus* and *Saprinus)* apparently prefer carrion (Kovarik and Caterino 2005). Many histerids occur in decaying materials of all kinds, or under the bark of trees. Species found in decaying organic matter are predaceous, feeding on fly larvae, while the sub-cortical species are predaceous on various wood and bark-boring insects (Majka 2008). Histerids are also located in corpses, old fungi, under bark and decaying logs in wet places, in burrows of mammals, and also in bird nests, particularly those that are constructed within tree cavities or nest boxes and caves (Kovarik and Caterino 2005). Other species are in the colonies of social insects (Kovarik & Caterino 2000; Majka 2008). In America, several histerids species can be found in rotting Agavaceae plants, in yucca, in rotting cactus, in succulents and *Euphorbia* in xeric areas (Yélamos 1993; Kovarick and Caterino 2005).

Almost any decaying organic matter attracts histerids, for example, in Paraguay the author found some species (unidentified) on rotten bananas and grapefruits that probably prey on the Nitidulidae (or their larvae or Drosophila flies), who come to those rotten fruit. Also most beach histerids are associated with decaying algae (Caterino, pers. comm. 2005). In the Southern Paleartic, rotting roots of *Ferula* (Apiaceae) and *Astragalus* (Fabaceae) attract several histerids (Kovarik and Caterino 2005).

Some Paleartic and Neartic species regularly inhabit damp organic material within hollow living tree (Kovarik and Caterino 2005).

The following Histeridae subfamilies are known:

---Subfamily Abraeinae: 429 described species (Kovarik & Caterino 2005), all species are very very small and all are predaceous. According to Caterino (2009) most species of the group are associated with rotting vegetation and many species are found under the bark of dead and dying trees (especially Plegaderini and Acritini), and in leaf litter. *Plegaderus*, *Eubrachium*, *Teretrius* and *Teretriosoma* are associated with the galleries of various woodboring beetles (Kovareki and Caterino 2005). A few species are known to be associated with ants' nests. *Abraeus, Chaetabraeus*, *Aeletes*, and *Acritus* are attracted to dung and probably prey on mites (Kovarik & Caterino 2005). In addition, *Aeletes* and *Acritus* are associated with rotting tree trunks. Some species *Chaetabraeus* are found in stored products (Kovarik and Caterino 2005).

---Subfamily Chlamydopsinae: According to Caterino & Dégallier (2007) comprises 174 species and they have been described from Australia, India, Japan, Taiwan, Fiji, New Caledonia, Indonesia, and Malaysia. All species are obligate associates of ten ant genera from four subfamilies (Myrmicinae, Dolichoderinae, Ponerinae, Formicinae) and a single termite (*Eutermes*) are reported as hosts of Chlamydopsinae (Caterino & Dégallier 2007; Tishechkin & Caterino 2007; Penati 2009). To collect them can be used flight intercept traps; with pyrethrum fogging of canopy and understory, and also sifting forest litter (Tishechkin 2009).

---Subfamily Dendrophilinae: 416 species worldwide (Kovarik & Caterino 2005). Nearly cosmopolitan with respect to its zoogeographical distribution (Mazur 1997); the species live in habitats of decomposing organic materials and in bark burrows where they prey upon xylophagous beetles and other wood-boring insects (Ye-Jun & Hong 2007). Some species have been also found in nests of birds, mammals and rodents (Bickhardt 1917; Yélamos 2002). According to Caterino (2009), *Dendrophilus pygmaeus* (Linnaeus) appears to be an obligate myrmecophile associated with *Formica rufa* (Linnaeus). *Platylomalus gardineri* (Scott) has been collected (in Spain) in a fallen coconut trunk, and under the bark of a stump (*Populus nigra* Linnaeus) (Juan de Ferrer 2005). In New World some *Carcinops* were collected from the flowers of palms of genus *Attalea* (Kovarik & Caterino 2005). Several species of *Carcinops* and *Dendrophilus* are found in stored products (Kovarik and Caterino 2005).

---Subfamily Histerinae: Species worldwide (Kovarik & Caterino 2005). This subfamily is very diverse in form and numbers of species, containing probably around one-half of all histerid species (Caterino 2009). They are found in dung, also associated with dead wood and other rotting vegetation, fungus, vertebrate nests, and the colonies of social insects (Caterino & Vogler, 2002). Those of the genus *Platylomalus* live under bark of fallen trees or stumps, among decaying vegetation, in the wounds of trees and under the bark of recently dead trees, especially in *Pinus* spp. The flat species of *Hololepta* are found under the bark of trees and in logs which are dead between 6 months to 1 year or longer, they move slowly and can easily be caught with forceps. *Hololepta reichii* Marseul has been collected in nests of bees Meliponinae Brazil (Coletto-Silva et al. 2008). In New World *Oxysternus* has been found in association with palms of the genus *Astrocaryum* and *Platysoma conditum* Marseul, was discovered to feed on larvae of weevils and Eulophidae (Hymenoptera) in mature cones of *Cycas* (Kovarik & Caterino 2005). Some species of *Phelister* and *Hololepta* are found under bark and decaying logs in wet places, the author of this book found species of this genus in rotten stump palm (*Copernicia alba* Morong) in the Paraguayan Chaco.

Some species (*Margarinotus striola* (Sahlberg), *Hister unicolor* Linnæus, *Hololepta lucida* LeConte) are attracted to sap fluxes of living trees (Reichardt 1941; Kovarik & Caterino 2005). Several species of *Atholus* and *Hister* are found in stored products (Kovarik and Caterino 2005)

---Subfamily Niponiinae: With only the genus *Niponius*, with 22 known species (Mazur, 1997), species distributed in Japan. However, some species are also known from scattered locations throughout southeastern Asia, from India to tropical Australia (Caterino 2009). All are cylindrical in shape and are apparently specialized bark beetle predators and according to Lewis asserts that they do not enter transverse galleries, but rather tunnel under bark (Gardner, 1930; Lewis, 1885).

---Subfamily Onthophilinae: With approximately 10 known genera Caterino (2009) and 83 species (Kovarik & Caterino 2005) with worldwide distribution. According to Caterino (2009) also they are largely restricted to the Old World tropics and Africa through Southeast Asia to Australia. Some species generally attracted to decaying materials from dung to compost, while others are found only associated with fungus or with the nests of specific vertebrates and other are mymecophilous.

---Subfamily Saprininae: contains over 600 species in about 40 genera (Caterino 2009), worldwide distribution. The majority of the species are moderately xerophilous, connected with open land. Species inhabiting forested areas or deserts (Slawomir and Masahiro 2003). They occur predominantly on carrion and dung, some in rodent burrows and bird nests (Slawomir and Masahiro 2003). Among genera found in rodent burrows are *Euspilotus, Eremosaprinus, Geomysaprinus and Aphelosternus* (Evans and Hogues, 2006). Many Saprininae are also found in nest of the ants (Majka 2008). Some genera are restricted to inland dunes, while others are found in sandy riparian and beach habitats (Caterino 2009).

Euspilotus bisignatus Erichson, from Chile conceal themselves in crevasses in partially dried dung and thus are able to capture adult flies that come to the dung to feed and ovoposit (Carlton et al. 1996; Kovarik and Caterino 2005). *Gnathoncus* sp. is found in birds' nests. Adults and larvae attack immature stages of insects living in these habitats (Costa 1952; Penati 2009). Several Asian *Saprinus* prey adult flies on dung (Reichardt 1941). Certain plants such as *Dracunculus vulgaris* Schott and *Amorphophallus* spp. from Old World, and other species, have flowers that emit a fetid fragrance, which attracts carrion insects including histerids (Thérond 1931; Reichardt 1941). According to Kovarik and Caterino (2005) several *Saprinus* species are attracted to these flowers. It is unlikely that these beetles act as pollinators as they sometimes destroy the flowers (Reichardt 1941). Several species of *Gnathoncus* and *Hypocacculus* are found in stored products (Kovarik and Caterino 2005).

---Subfamily Tribalinae: According to Caterino (2009) they are about 200 described species. Most of the Neotropical species (*Plagiogramma, Epierus* and *Idolia*) are associated with dead wood. Species *Tribalus* (of the Old World) are more varied, including some dung associates. Some are found beneath the bark of older dead trees, in rotting wood, in piles ow sawdust and very few adults of *Epierus* and *Tribalus* feed on fungal spores (Kovarik & Caterino 2005).

---Subfamily Hetaeriinae: They are 341 species from Holartic and Neotropical Regions (Kovarik & Caterino 2005). They live with ants and termites often attacking the larvae of these insects (Kapler 1999; Mazur, 1997; Fabio Penati 2009).

Most species are found in the Neotropical region, primarily with army ants (Formicidae: Ecitoninae) (Helava et al. 1985).

---Subfamily Trypanaeinae: Neotropical, Afrotropical, Australian and Oriental Region 190 species (Kovarik & Caterino 2005). *Pygocoelis* with approximately 10 known species including *Pygocoelis africanus* (Lewis) found under bark of fallen trees, this species is known from central and Westem Africa (Yélamos, T. 1992); *Trypeticus* with 100 known species, distributed in: Indian subcontinent, Nepal, Sri Lanka, Indo-China, Malayan Peninsula, the entire Indo-Malayan Archipelago including the Philippines, New Guinea, Northern Queensland, Taiwan, Southeast China and Japan (Kanaar, 2003), are probably oligophagous, if not monophagous (Lewis, 1884) Their distribution depends on the presence of their xylophagous prey species (Kanaar, 2003) live in the burrows of bark beetles and other wood-boring insects, where they prey upon the larvae and sometimes also upon the adults (Bickhardt, 1913, 1916). The flight intercept traps is a very useful to collect *Trypeticus* (Kanaar, 2003).

Most species can be manually collect in their habitats, or with Berlese funnel, Winkler/Moczarski eclector.

They are also collected with pitfall traps located inside and outside the forests, baited with meat, decomposed fruit or dung, or straining the sand of sea beaches and sand from the roots of desert plants (Evans and Hogues 2006); putting intercept flight traps near nests of ants or termites; several species can be collect with lights traps.

SYNTELIIDAE: Known from China, India, Japan and Mexico; the only genus is *Syntelia,* with 9 species (Hansen1997; Zhou and Yu 2003; Newton 2005). Most *Syntelia* species are found at moderate to high elevations (1,500 - 3,900 m) of mountains in subtropical or tropical zones (Newton 2005). *Syntelia histeroides* Lewis also live at lower elevations; both adults and larvae have been collected under the moist bark of decaying logs of various trees, also in summer at sap flows of young *Quercus* spp.and *Ulmus* sp. (Lewis 1882; Mamaev 1974; Öhara 1994). Larvae and probably adults of this species prey on larvae of xylophagous beetles (Newton 2005). *Syntelia westwoodi* Salle, live in high desert areas of the south-central Mexican plateau at 1,700 - 3,000 m. Adults have been found several times in the moist fermenting interior of large decaying columnar cacti, prey on larvae of Diptera. (Newton 2005). Adults and larvae are predators of other insect larvae living under bark (Zhou and Yu 2003).

SPHAERITIDAE: 5 species are known in one genus (Newton 2005). Adults and larvae probably feed on larvae of Diptera (Newton 2005). *Sphaerites politus* Mannerheim (from USA, Canada, Alaska and Japan) can be locally common at carrion (Keen 1895) this species has been found also in bear dung and in or near compost (Hatch 1961) and can be collected also with unbaited pitfall traps and flight traps in old-growth conifer forest (Newton 2005). *Sphaerites glabratus* Lobl (from Holarctic region and China) live in Northern conifer forest, sometime at high altitudes, adults are attracted to sap flows on birch stumps, and have been observed to feed on the sap and mate there (Newton 2005). *Sphaerites dimidiatus* Jurecek (Holarctic region and China) has been collected with a colander in an alpine meadow (Newton 2005). They can be are collected in liquid resin of birch trees in mixed forest. The remaining species (*Sphaerites nitidus* Löbl and *Sphaerites opacus* Löbl & Háva) are known from only a few localities in Central China (Löbl 1996; Löbl & Háva 2002; Newton 2005). Adults of *Sphaerites* feign death when disturbed (Kovarik cited in: Newton 2005).

Superfamily STAPHYLINOIDEA

HYDRAENIDAE: Worldwide distribution. Family with more than 20 genera and more than 1000 described species (Jach 1995c). Adults of most species are aquatic (stagnant water, running water, hygropetric) or riparian. Some appear to be amphibious (facultative aquatic) or riparian. A few species are known to live exclusively in hyper saline marine rock pools and a few species are strictly terrestrial. Larvae are usually not aquatic. They live in aquatic and terrestrial environments, on rocks or in humus of hygropetric habitats, near waterfalls in streams. Also in the rocky beaches of the sea and sandy margins of streams, in small water pools, perched on branches, moss, herbs, leaves, rocks or logs stacked in aquatic environments, and even in seabirds' nests. Also found in water tanks located in places with abundant vegetation (Lawrence & Newton1995). Some species are found in the sand of rivers, ponds and wells.

Although most live in fresh water, they are also found in brackish inland or coastal waters (Spangler 1981). Adults can be phytophagous, predatory or saprophagous. Larvae are terrestrial, phytophagous or predators and live in organic material near the water margins. Adults of Limnebiinae subfamily are found in moist vegetation along the edge of streams and in decaying Sphagnum moss in bogs and fens. Both larvae and adults are aquatic and phytophagous. According to Spangler (1981) to collect Hydraenidae the following methods should be used: "*In areas with stagnant water they are captured by pushing the water against the edges of the habitat and washing the substrate. The specimens floating on the water surface reach the beach floating or pulled under* water *film. They can be caught with the fingertips, with an aspirator, a brush or a thin sieve*". They can be collected passing fine water net by aquatic vegetation and also with light traps.

PTILIIDAE: A family with generally terrestrial world-wide distribution. About 560 species are described so far (Jach 1998). Some species live in mammal's nests (Cephaloplectinae) and birds' nests. They eat microorganisms, mold, mildew, spores and mushroom hyphae growing in these habitats (Lawrence & Newton 1995). Others are in the litter on the forest floor, in decaying logs, piles of decaying vegetation, decaying fungi, under the bark of trees (Genaro y Gutiérrez 2001). They can also be found in the ground (Dybas 1980). A number of species it is reported to be riparian/littoral, living in seaweed on the shores of the sea (Jach & Ji 1998). In other moist organic material such as bat guano in caves (Hall 2001) in U.S. many of these beetles were found in nests of an alligator, a natural compost heap that has probably been the habitat used by these beetles for several million years (Dybas 1980).

Members of *Nanosella* genus are the world's smallest beetles they live in spore tubes of on the underside of Polyporaceae fungi (Dybas 1980). According to Dybas (1980): *"The smallest member of this family is an undescribed species from South America that reaches just 0.30 mm long, and not much bigger than a single celled Paramecium; however, they have composite eyes, antenna, complex mouthparts, wings and all other essential structures of their larger relatives; a complex organization packaged in a very small space."* Ptiliids are collected with Berlese funnel trap (Dybas1980) and Winkler/ Moczarski eclector. Species of the genus *Acrotrichis* live among leaf litter in coastal south-east Australia. They can fly but rarely are attracted by lights (Hangay 2010). In New Zealand *Acrotrichis* has been collected from compost, carrion, and leaf litter in gardens and paddocks (Colin 1982).

The author of this book has captured ptiliides with pitfall traps, using only 80 ° ethanol (mixed with a little glycerine). Another method is by sifting bark and other litter plants which are under the logs. This can be done with a hand grid (fine mesh 0,5 cm) on a white cloth.

According to Colin (1982) the use of vapour (ethyl acetate, etc.) to kill ptiliids should be avoided, otherwise pubescens becomes disturbed, sculpture is obscured and the beetles frequently become greasy and damaged. Specimens should always be introduced alive into 70% ethanol in which they can be stored indefinitely. For more information on mounting specimens of Ptiliidae for collections, consult the work of Colin (1982) (see bibliographical references).

The following subfamily is explained in more detail:

Limulodinae: now placed within the Ptiliidae: About 30 species are known distributed in North and South America and Australia (Campbell 2009). The members of this group live obligatory in anthills, where they feed on exudations from the body of ant adults, larvae and pupae In only one ant nest of several species Limuloidae can be found, even different genera (Arnett and Thomas 2001) and are neither carnivorous nor a scavenger of the ant nest (Arnett1968). The literature confirms an association with *Pheidole* ants (Charles 1980). According to Wilson et al. (1954) *"Paralimulodes present a very bizarre appearance in life. When resting them superficially resemble mites. They are usually to be seen riding on the heads and abdomens of the workers and only rarely run along the ground in a foraging column of the ants".* In Argentina only *Paralimulodes* has been found (Wilson et al. 1954), and the genus *Cephaloplectus* has been found in Panama (Seevers and Dybas 1943). To collect them, the ant nest must be carefully examined and watch the ants when they move out carrying these tiny beetles on their bodies. When it found an ant with its host, both insects must be put in a single vial with ethanol and label it immediately.

LEIODIDAE: A generally terrestrial family, of worldwide distribution. It includes 6 subfamilies (Lawrence & Newton 1995) in 342 genera and about 3750 species. Some are found under bark or dead wood. Many have saprophagous habits and are scavengers in moist habitats in forests, feeding on mold and mildew. Some of them are carrion-eaters; others live associated with ants or the nests of bees of the subfamily Meliponiinae (Aponte 1996; Yanega & Lescmer 1994). Others live in vertebrate burrows on their hosts as ectoparasites of small rodents or insectivores mammals or beavers (*Platypsyllus castoris* Ritsema). Some feed on the body of globular fungi (puffballs). In Anistomini the larvae and adults are mycophagous restricted exclusively to slime molds. A lot of them eat rotting fungi, but only a few feed on polypore fungi. Others live in dung and carrion and probably feed on the microorganisms of decay. Many Cholevinae are scavengers (Leptodirini) and inhabit boreal, temperate and tropical forests. Most Catopina (Cholevini tribe) are saprophagous, eating organic material, vegetal or animal; a few species eat mushrooms; some live in ant nests of *Camponotus*, *Formica*, and *Lasius* ants. Many are found in carrion along with Silphidae and Staphylinidae, in decomposed material like humus, dung and fungi, among leaf litter on the ground and below ground at a shallow depth, between the bark of logs, and in nests of small mammals. Many species are found in caves (Hatch 1929; Portevin 1929).

Leiodinae (Sogdini, Leiodini), Catopocerinae and Coloninae eat hypogean fungi. Platypsyllinae are parasites on small mammals including mice, moles, shrews and beavers. They have been collected on the body of these mammals and/or in their burrows. Occasionally they are found in some Hymenoptera nests.

Dissochaetus sp. has been collected in caves in Central America and *Nemadus horni* Hatch is frequent in caves in the USA. This subfamily has some "aquatic" species: *Platypsyllus castoris* Ritsema must be regarded as an obligate ectoparasite because it is normally found on the body of beavers (*Castor fiber* Linnaeus and *Castor Canadensis* Kuhl) and therefore it is often submerged in water inside the fur of the host beaver. Another species, *Silphopsyllus desmanae* Olsufiew, has a similar behavior because it lives on the Eurasian water mole (*Desmana moschata* Linn.). The members of Camiarinae subfamily are distributed in south temperate regions; some of the genera such as *Zearagytodes* from New Zealand are associated with polypore fungi (Leschen 2000).

A good technique for collecting Leiodidae is the use of pitfall traps in either closed or open forest (Peck and Davies 1980) with dung, carrion, fungi, smelly cheese, etc., as bait. To place also strategically these traps in the burrows of vertebrates (see article "Coleoptera associated with vertebrates", Chapter III). Also, fogging with pyrethrum insecticide rotting logs or stumps with bark is a productive method or to collect mushrooms, humus or vegetable waste and put it in a Winkler/ Moczarski eclector or Berlese funnel trap. A few come to lights. Several species are transported on the body of ants and to collect them is necessary to see if the ants have deformed head or body, in these deformities are usually the beetles that are above them. So, when is discover an ant transporting a commensal, both must be captured and put into a vial with 75% ethanol and all possible data collection or carry the live ants and their parasites to the laboratory. The use of large area flight intercept traps has captured many previously rare or new species.

Leiodids can also be found together with stingless bees of the Meliponinae Subfamily.

SCYDMAENIDAE: All over the world 84 genera are known and about 4900 species (Newton & Thayer 2004-2005). They are found on rocks, under bark and mainly in wet moss; at the bottom of the litter layer in forests; in decaying logs and tree stumps; in hollow trees and piles of sawdust. Some species live in caves. *Cephennium* usually live in leaf litter, rotten wood and humus at the base of trees in mountain areas and deciduous forests. *Cephennium* it is specialized in preying on oribatids mites. Scydmaenidae tend to be locally common, possibly following the distribution of mites. Some species live in ant nests and this is not so strange because normally many ants keep "ranches" with a variety of domestic animals, parasites and commensals. Scydmaenids are generally nocturnal and found in organic litter (Klimaszewski & Watt 1997). Utilise appropriate collection methods, such as litter samples with the invertebrates being extracted via a Berlese funnel.

SILPHIDAE

M. D. Ulyshen

This small group of relatively large-bodied beetles consists of two subfamilies, 15 genera and approximately 175 species worldwide (Peck, 2001 and references therein). While most species are restricted to the Northern hemisphere, especially Eurasia, the family can also be found in South America, Africa, Southeast Asia and the Australasian region. The 74 species of burying beetles belonging to the genus *Nicrophorus* are the most familiar members of the family, famous for burying small vertebrate carcasses and for their complex behaviors, including parental care. In temperate regions of the Northern hemisphere, the frequent coexistence of ten or more silphid species at a given location is the result of complex resource partitioning, with important implications for their collection. Although few species specialize on particular carrion types (e.g., fish vs. birds, Shubeck 1976) or decay conditions, members of the Silphinae generally utilize larger carcasses than the Nicrophorinae (Anderson 1982). Many silphid species show strong habitat associations (e.g., forests vs. open areas) or are only active during certain periods of the year (Anderson 1982) or at certain times of day (Shubeck 1971). While most species are concentrated near the ground, some can be captured more readily in suspended traps (Shubeck 1970). Of particular interest is *Nicrophorus pustulatus* Herschel, a North American species found to be much more common high in the forest canopy than near the ground using carrion-baited flight intercept traps (Ulyshen et al. 2007). The authors suggest *N. pustulatus* utilizes carrion in tree cavities to avoid competitive interactions with more aggressive species on the forest floor, but this has not been investigated. It should be noted that, despite the common name for the family, not all silphid species utilize carrion, however. Of the subfamily Silphinae, for example, species of *Aclypea* are phytophagous and species of *Dendroxena* are predators of caterpillars or snails (Peck, 2001). Certain Silphinae species, in addition to being attracted to carrion, can sometimes be collected at Phallales fungi, apparently drawn to the strong odors they produce. Similarly, although *N. pustulatus* is attracted to suspended carrion-baited traps (Ulyshen et al. 2007), the species was recently found attacking black rat snake eggs (Blouin Demers & Weatherhead 2000, Keller & Heske 2001). Consequently, although carrion-baited pitfall trapping is highly effective for most species, this method alone is not always adequate to fully or efficiently sample the fauna. Collecting at lights and carefully searching appropriate habitats can yield those species rarely or never associated with carrion. Finally, it is important to mention that extreme care should be taken when operating carrion traps in areas known to be inhabited by rare or threatened species. For example, *Nicrophorus americanus* Olivier of North America has declined over 90% of its former range and was placed on the U. S.

Endangered species list in 1989 (Sikes and Raithel 2002). Live trapping techniques have been developed to protect this species (Bedick et al. 2004) and should be used in lieu of traditional traps in areas where it may occur.

M. D. Ulyshen
USDA Forest Service,
320 Green Street, Athens,
GA 30602-2044, USA

AGYRTIDAE: Distributed in the following regions: Holartic, Paleartic, Neartic, Mexico and New Zealand *(Zeanecrophilus). There are* 3 subfamilies: **Pterolomatinae**, **Necrophilinae** and **Agyrti-**

nae. They are 62 species known (Ruzicka & Schneider 1995; Peck 2001: Prieto & Pérez 2002).

They are scavengers feeding on decaying organic material, found in wet and cold habitats, in beach and river drift, in the margins of mountain streams and fields of snow at high altitudes (Evans and Hogue 2006). Snowfields are inhospitable places but they are very appropriate for species that tolerate low temperatures because on the snow there are many immobilized or dead insects. *Apteroloma caraboides* Fall, found in the Sierra Nevada (California), and *Pteroloma nebrioides* Brown were collected from shaded leaf litter beside a mountain stream. They are active during the fall, winter and early spring. Most species are found in the mountains but some of them can live on the coast (Evans and Hogue 2006). They are rare in collections and are captured in carrion during the cold seasons of the year. According to Evans and Hogue (2006) they can be also collected by sifting straining debris on the banks of rivers, sifting fungi or from material under rotting bark where *Ipelates* or *Agyrtes* specimens can be found. *Apteroloma* are found on lake beaches, in the moss and gravel on the banks of mountain streams or under rocks located on the edges of the mountain snowfields. *Lyrosoma* is found in beach wrack and bird burrows in the Aleutian Islands.

STAPHYLINIDAE: One of the largest families worldwide, with over 51000 species in over 3,393 genera (Newton & Thayer 2004-2005), in 32 subfamilies worldwide (Newton et al. 2001; Yunyun & Dong 2009). Each year nearly 400 species are discovered (Herman 2001) and probably over 75% of tropical species have not yet been described. They live in a great variety of habitats: Banks of rivers, streams, ponds and lakes. They occupy almost all moist environments in the world, many are located in the subtidal interstitial zone (Moore and Legner 1976) but none of them is truly aquatic because they do not live in open waters, although some winged adults have been seen in sea surface, far from land, probably because they are blown away with the wind. Moreover, this is a demonstration of their high dispersal capacity (Howard and Thomas 1999).

Others are found under bark on decaying organic material (fruit, wood, etc.). The vast majority live in humus in forests; others live in the foliage, including some species only found in the canopy. Some feed on mold, others (Neophonus spp.) eat fungi (Thayer 1987; Newton.1984) or find prey larvae of flies and other organisms in fungi. Many live in caves, underground tunnels of vertebrates, in very small cavities in the soil or dig small burrows. Other species live in nests of vertebrates (turtles, birds and rodents) and there their preys are mainly larvae of fleas and flies. Some Amblyopinina live on the skin of rodents in Central and South America (Ashe and Timm 1987). For example, some *Amblyopinus* species are ectocomensals of rodents (*Oryzomys albigularis* (Tomes), *Peromyscus nudipes* (Allen) and *Reithrodontomys creper* Bangs) and for years it was thought that they were parasites of these rodents and feeding on their blood. Now, however it is believed that the rodents are used only as a transport for moving from one nest to another. Very probably they lay their eggs in the nests of these mice and the larvae of these beetles prey on arthropods present in these nests. According to Nalini and Nathaniel (2000): *"This beetles leave the host during the day when the host become inactive and roam throughout the nest they climb back onto the mouse and take up theirs characteristic position behind or betwen the ears when the mouse becomes active at nightfall".* Nalini and Nathaniel (2000) found that the beetles eating fleas and other true ectoparasites in the nest during the day. Many adults and larvae of Staphylinidae attack almost any prey. They are frequently found feeding on pollen inside the corolla of flowers. Some attack land snails and slugs (Tod 1973; Fisher et al. 1976), but they are not very specialized in this type of prey. An almost commensal relationship with snails has evolved in some Asian Aleocharinae such as *Zyras sagax* Cameron, in which the adult beetles enter the mantle cavity of *Ryssota* (Pulmonata: Helicarionidae) and perhaps feed on mucus or feces of the snail (Auffenberg 1986). Its distribution in arid environments is restricted to humid microhabitats. Some species are fungivorous and others have phytophagous or saprophagous habits. One of the most interesting groups, from a biological point of view they are those who live as commensals (mutualistic or symbiotic) with ants and termites (Hölldobler 1971).

Larvae are predators and can be found in piles of decaying plant material, also on carrion or dung. There are several methods to collect them: Berlese, Winkler/Moczarski eclector, pitfall traps, lights, malaise trap, fumigation, flight intercept traps, etc.

Subfamilies of Staphylinidae

---Subfamily Tachyporinae: Adults and larvae are found in various wetland habitats such as litter, decaying vegetation, fruits, and beneath the bark of fallen trees and mushrooms. They are also very common beneath dead wood. Adults and larvae move rapidly and most are generalist predators, although some of them are mycophagous (Howard. and Thomas 2008).

---Subfamily Oxytelinae: They dig in wet sand along the seashore and in mud at the edges of lakes, rivers and streams; they are also found in salty soils temporarily wet. They are primarily saprophagous, feeding on algae and diatoms. Each gallery in which they live is marked by a small mound of sand or mud (White 1983). To collect these beetles put the soil with the galleries' into a bowl with water; when the Oxytelinae come to the surface they may be collected. Other species are subsocial: *Platystethus* adults care for their immature stages, and adults of the *Bledius* genus maintain tunnels to lay their eggs (Herman 1986). Both larvae and adults are saprophagous, fungivorous or algivorous. Some *Apocellus* spp. has been seen eating flowers (Chittenden 1915). Adults and larvae live among decaying plants, in dung of vertebrates, in tunnels which they construct in diatom-laden sandy or muddy shores (Herman 1986) or in nests of vertebrates (Hu and Frank 1995b).

---Subfamily Habrocerinae: Includes only two genera, one *Nomimocerus* (Coiffait & Sáiz) with six species restricted to Southern of South America. Adults are saprophagous and mycophagous on molds, feed spores, conidia, hyphae, arthropod cuticle and detritus (Newton 1984). And *Habrocerus* distributed worldwide (Newton 2006), adults ad larvae are saprophagous and mycophagous on molds (Newton 1984), has been collected in old straw and wood chip piles and on angiosperm *Conopholis Americana* (Linnaeus), a root parasite of oaks (Newton 1984).

---Subfamily Pseudopsinae: With worldwide distribution, 4 genera are known: *Zalobius:* Some larvae and adults has been collected (in California) from flood debris and along stream edge *(Grebennikov 2005), Nanobius: Both adults and larvae of some species has been collected with Berlese in leaf and log litter in California (Grebennikov 2005)* (Newton 1982; Herman 1983) and some *Asemobius* and *Pseudopsis* has been collected in fungi (Newton 1984).

---Subfamily Staphylininae: Adults and larvae are found in many habitats including forest leaf litter, decaying plant materials, fruit, fallen fungi, carrion, dung, moss, piles of grass, under bark of dead trees, in washed-up brown algae and on seashores (Howard et al. 1986). Some live in mammals' nests, so is useful to put pitfall traps in these nests also; in moist areas, such as marshes, swamps and wet clay (White 1983). Some are collected with pitfall traps.

---Subfamily Euaesthetinae: Adults and larvae live among decomposed plant material on banks of aquatic habitats. Adults do not climb vegetation. The mouthparts suggest that adults and larvae are predatory (Howard and Thomas 2008).

---Subfamily Empelinae: Contains only *Empelus brunnipennis* (Mannerheim), from North America. This specie has been collected in forest leaf litter, in fungi and in moss at stream edge (Newton et al. 2001).

---Subfamily Glypholomatinae: With 2 genera: *Glypholoma* of Chile (Norte Chico and Magellanes) and Argentina. They were found between 220 and I70 meters, nearly all by sifting ground litter in forests of *Nothofagus, Libocedrus* and other trees. Genera *Proglypholoma:* Also from Chile (Newton 1975).

---Subfamily Apateticinae: With 24 species in genera *Apatetica* and *Nodynus.* They live worldwide, but mainly distributed in South East Asia (Proshchalykin 2008). Some species are saprophagous (Ananthakrishnan 2004).

---Subfamily Olisthaerinae: Worldwide distribution, only 2 species known: *Olisthaerus substriatus* (Paykull) and *Olisthaerus megacephalus* (Zett.) both species have been collected under bark of dead conifers (LeConte 1862; Arnett and Thomas 2001).

---Subfamily Solieriinae: Family with a single genus (*Solierum*) distributed in Chile. Feeding habits and larvae are still unknown (Beutel 2005).

---Subfamily Proteininae: Adults and larvae are saprophagous or mycophagous. They live in fungi and decaying vegetation, and sometimes are found in carrion and in caves. The European

Megarthrus genus has been collected in cow dung and mushrooms (Stan 2006).

---Subfamily Phloeocharinae: With 7 genera, including *Charhyphus* known in Europe. They have underground habits (endogenous). In Spain they have been collected under rocks buried in the soil, and in soil samples washed and placed in a Berlese funnel (Hernando 2002).

---Subfamily Trichophyinae: With only one genus, *Trichophya*, that includes currently 15 species from paleartic, Nearctic and Eastern regions (Assing 2003). They are rare in collections. *Trichopyia pilicornis* (Gyllenhal) is associated with wood chippings and sawdust in woodland of pines. This species had been collected with car-catcher and with flight intercept traps (Cooter and Barclay 2006).

---Subfamily Oxyporinae: Adults and larvae feed only on fleshy mushrooms, and they digest their food preorally as in Steninae, Euaesthetinae, Paederinae and Staphylininae, unlike the other subfamilies (Howard and Thomas 2008; Leschen and Allen 1988). They have been collected in mushrooms and rotten log near streams.

---Subfamily Trigonurinae: With 5 genera widely distributed. They can be collected on carrion and fungi.

---Subfamily Omaliinae: Adults and larvae occur in leaf litter, rotten fruit, moss, and under bark of dead trees. Adults of several species and larvae of a few occur in flowers. Many genera are predators and feed on freshly killed small insects; others are phytophagous feeding on flowers, or saprophagous feeding on decaying fruits.

---Subfamily Protopselaphinae: With a single genus *Protopselaphus,* from Malaysia. They can be collected with Berlese funnel or with Winkler/Moczarski eclector from forest litter (Sakchoo-wong et al. 2008).

---Subfamily Megalopsidiinae: *Megalopinus* is a specialized genus found in rotting logs, where fungi are growing. Adults preorally digest their food, using a rotary-mill, a method for liquefying tis-sues and chewing their prey (Lesch and Newton 2003).

---Subfamily Osoriinae: Adults and larvae move very slowly. They are found under bark of dead trees, fallen timber (Bohác 1978), in roots and humus. Many are found in sandy banks of streams. Others are *saprophagous* but can also be *mycophagous*, and some eat the roots of living plants. They can be collected with pitfall traps and sifting litter.

---Subfamily Piestinae: Adults and larvae live under bark of conifers. They are slow-moving beetles living together under the bark of dead trees and cut banana stalks. They are *saprophagous* or *mycophagous.*

---Subfamily Neophoninae: With 1 single genus (*Neophonus*) which has been collected in Ar-gentina and many localities in south-central Chile. Adults live in forest foliage and are mycophagous Staphylinidae. They eat primarily foliicolous fungi (Thayer 2009).

---Subfamily Micropeplinae: Adults and larvae occur in leaf litter and are associated with fungus infecting dead trees.Their eating habits are not known at present.

---Subfamily Megalopsidiinae: Adults are predators and are associated with fruiting bodies of fungi growing on decaying wood, but are not abundant (Howard and Thomas 2008).

---Subfamily Steninae: Adults are active during the day in sunny places on muddy or sandy shores of lakes and streams. Some live along the banks of lakes, rivers, ditches. They run on the ground and climb low vegetation. Others are not restricted to aquatic habitats and they climb trees. When dipped into water the glands in the abdominal tip produce a chemical substance that breaks the surface tension of water (as a detergent would do) and so the beetle can escape from their predators. Adults and larvae prey on small soft-bodied insects such as Collembola (Weinreich 1968). Several species live in cow dung and others can be captured with pitfall traps.

---Subfamily Paederinae: Lives in decomposed plant material and are often found under objects on the ground. Adults (with normal anterior tarsi) and *Palaminus* (which have expanded anterior tarsi) rise on vegetation. Adults and larvae are predators. One species of *Myrmecosaurus* is associated with *Solenopsis* ants (Tschinkel 1992; Wojcik 1980). Larvae of only one genus (*Astenus*) are known to spin silken cocoons (Howard and Thomas 2008). They run quickly when disturbed.

---Subfamily Leptotyphlinae: Due to their small size and cryptic habits nothing is known about their behavior and foods. Leptotyphlinae members are fairly uniform in shape and appearance. Adults are 1 to 2 mm long, blind and the lack of hind wing (Howard and Thomas 2008). 525 species have been described in 5 tribes and 44 genera. Larvae can be distinguished by the combination of their small size (0.128 mm maximum documented in head width), along with the complete absence of eyes (Grebennikov & Newton 2008).

---Subfamily Aleocharinae: They are among the smallest Staphylinidae, with about 1200 known species (White 1983). They are found in almost all moist habitats: Under stones, dung, mushrooms, rotting vegetables, under bark, in oozing sap or in moss, in rotting fruit, living plants leaves, flowers. *Aleochara* spp., are found in fungi, carrion and dung. Other genera inhabit the shores of lakes, rivers and seas, caves with guano, or dung of epigeic vertebrates. Some live in termite and ants nests. *Tachiona* are found inside of webs that cover some species of Hepialidae (Lepidotera) burrow of living *Trema* (Ulmacea) trees (Nalini et al. 2000). Some resemble their hosts and have a behavior that allows them to interact with social insects. Others live in nests of birds and small mammals. Larvae are parasitoids of diptera pupae. But most are generalist predators and some of them specializing in well-defined prey. *Gyrophaena* spp live in fungi and are gregarious (White 1983) and some have a subsocial behavior (Ashe 1987). Some can be collected with pitfall traps.

---Subfamily Dasycerinae: 17 species are known to in the Northern hemisphere, Southeast Asia and Europe. According to Wheeler (1984) and Tomáš Kopecký (2006) they are found mainly among the fallen material, logs, leaves and twigs with fungi at the foot of trees in deciduous forests. They probably feed on soil fungi and fungal mycelium of wood. The best way to collect them is looking under the rotten wood between the "moldy fungi" of various colors (white, green, blue and yellow).

---Subfamily Scaphidiinae: Adults and larvae are found among dead leaves and rotting wood and are associated with slime molds (Myxomycetes) and other fungi. They are strictly mycophagous (Lesch 1988), feeding especially on Polyporus woody fungi. They are also found on and under the bark of trees. In New Zealand, the greatest diversity is in the Northern South Island where many allopatric species of the genus *Baeocera* feed on fungi. In that country, these beetles are found in Polyporacaea, Corticiacaea and Myxomycetes (Loebl and Leschen 2003). The same authors explain that the species of *Brachynopus* spp.

are reared on corticioid and polyporoid Basidiomycetes which are invested upwards. Some species of *Scaphisoma* adults are commonly found on mushroom (Loebl and Leschen 2003), for example, Hawkeswood (2003) has found several adults of *Scaphidium exornatum* (Oberthur) feeding on the fresh basidiocarp of a pale brown *Poria* sp. (Polyporaceae) growing on a moist rainforest log, at Mt. Glorious, about 30 km west of Brisbane, Queensland (Australia), also Hawkeswood has observed *Scaphidium punctipenne* Macleay feeding on fungal filaments and small fruiting bodies of a white *Poria* species (Polyporaceae) growing on a rotten *Eucalyptus* log in a residential garden in Brisbane, Queensland. Scaphidiines species can be collected with Malaise trap, Winkler/Moczarski eclector, with Berlese funnel, by spraying trees, with beating sheet and with light traps. They have also been collected with pitfall traps using a simple mixture of water, detergent and ethylene glycol (Sackmann 2006).

---Subfamily Pselaphinae: Including over 8400 species, they live all around the world. They are found in tree trunks, under stones, in stumps, under bark, in humus, walking on the ground, in the forest litter, in organic material and moss. The supertribu Clavigeritae are found in ant nests (Costa 1952). Larvae and adults of are predators feeding on mites and small insects (Collembola, etc). Most species are found in leaf and wood litter of the forest floor (Newton et al. 1999) usually in a small number of individuals. Many species are obligate associates of ants or termites (Kistner 1982). They can be collected in flood refuse or by putting humus into a Berlese or Berlese funnel funnel, Winkler/Moczarski eclector, sifting litter on a white canvas or by sweeping and beating vegetation. Also is helpful to look under the bark of trees, under rocks and fallen logs. They can be collected with pitfall traps, white light and blacklight and flight intercept traps and fumigation.

---Subfamily Brathininae: They are now included within Omaliinae (Thayer 1985). There is only one genus, *Brathinus,* with one Japanese species and three in the United States and Canada. They live in grass roots growing near water and also in caves (Peck, 1975). Very rare in collections, perhaps the best way to collect them is taking samples of grasses growing in damp places and put them in a Berlese funnel or Winkler/Moczarski eclector. Adults have been collected during the summer months. As adults, these beetles may pass the winter in caves in Southern Appalachians.

To collect Staphylinidae Thayer and Newton (2004) used the following formula which can be used in intercept traps and pitfall traps: "1 gal. (3.9 L) trap fluid. Usually will be used about 49% propylene glycol, 49% water, 1% propylene phenoxytol, and 1% liquid dish detergent (or similar, unscented if possible). The exact proportions are not critical, and some people report satisfactory results with as little as 10- 11% propylene glycol. Add the detergent after mixing all other ingredients to reduce foaming. Those same entomologists used the following methods to collect Staphylinidae: "Flight intercept trap, pitffal traps, UV-light traps, substrate collection, sifting, Berlese/Berlese funnel processing and by hand-collecting ("niche collecting") pyrethrin spraying, fogging, blacklighting, beating vegetation". All these methods have been widely discussed in this book. Many microhabitats have Staphylinidae in which these beetles can be collected manually: Fungi, dead wood, under bark of standing trees and dead trees, all vegetation, flowers, trash on the banks of streams, among the rocks and gravel on the edge of the courses water, debris, on algae and bits of wood thrown into the ocean beaches, also in fresh dung and carrion.

Series SCARABAEIFORMIA
Superfamily SCARABAEOIDEA

LUCANIDAE: The family, represented by 8 subfamilies (Benesh 1960), consists of about 800 species of worldwide distribution except Antarctica (Mizunuma and Nagai 1994).

They eat decaying organic material and many live in tree stumps. Some feed on the sap that flows from certain old or sick trees. Most of them are *saprophagous.* Some rare species (*Brasilucanus* spp.) from Brasil and Paraguay live in termite nests (personal observation; Ratcliffe 1984; Vulcano and Pereira 1961). Adults of some small-sized species feed on flowers (Ratcliffe in: Arnett et al. 2002). Larvae of all species live in wooden stumps or decaying logs, so they are not considered as pests. They lay their eggs in cracks in the bark, especially near the roots and they are more common in wooded areas. Several species are found inside trunks and dying trees. *Sphaenognathus* is distributed in South America and all the known species live in the Andes, between 2,000 and 3,500 m. (Bartolozzi & Onore 2006). According to Bartolozzi & Onore (2006) *Sphaenognathus xerophilus* Bartolozzi & Onore has been collected in a very dry area, on a mountain with scrub: Small shrubs, ferns and grasses. Larvae of this species and other *Sphaenognathus* species are found in the soil (Bartolozzi & Onore 1993). Adults, larvae and pupae of *Ceratognathus argentinensis* Martínez, had been collected in decaying basal branches about 2 meters tall, in *Nothofagus dombeyi* (Mirbel) in Southern Chile and Southern Argentina. Such species have also been collected spraying the tops of the trees (Mondaca and Paulsen 2004). According to Geoff (2008), *Lamprima latreillei* Macleay (from Australia) is active by day and comes to flowering of *Pittosporum* and similar plants. *Rhyssonotus nebulosus (*Kirby*)* (also from Australia) can be found under rotten logs.

Several methods have been used to collect the species of this family. For example, in countries outside Europe, window traps have been used (Hammond 1997; Grove 2000); the same traps have been effective in Europe collecting *Sinodendron* spp. (Barbalat 1995) and also for *Ceruchus* spp. (Simila et al. 2003). *Platycerus caprea* (De Geer) (according to Barbalat 1995) have been collected with small Malaise traps suspended in tree canopies. *Platycerus caraboides* (Linnaeus) (according to Barbalat 1995) have been collected with colored pan trapping. Many species of Lucanidae can get caught in traps with decayed fruit or black sugar and molasses, hung in very old trees. In South America, to collect some Lucanidae such as *Leptinopterus tibialis* (Eschscholtz), Mr. Reinhard Foerster explain me: *"We must review the host plant, known in Argentina by the vernacular name of "Cambá-Acá" (Guazuma ulmifolia*

Lamarck*) (pers. comm. 2009). This is a small tree 5 meters tall, with a trunk diameter of 30 cm. It should not be confused with "Camba-rá" (*Lantana camara* Linn.*). By reviewing their trunks we will find some holes with sawdust. In early November, the holes are covered with small tubes of mosquito netting cloth nailed with tacks. This is the method to capture individuals that will emerge as adults from mid-November to mid-December. They are also attracted by fermented fruit, such as rotten bananas, adding beer or old fermented cider."*

The author of this book has collected several specimens of *Leptinopterus tibialis* Eschscholtz in the resin flowing from the wounds of the shrub known as "Kokú" (*Allophylus edulis* (A. St.-Hil., A. Juss. & Cambess.)) in the early morning before dawn.

Dorcus spp. and *Lucanus* spp. have been collected in Europe in traps baited with rotting fruit and sap fermented with honey (Allemand & Aberlenc 1991; Paolo Gandini, pers. comm. 2008). Light traps are good for catching *Pseudolucanus* spp. but its effectiveness in the collection of *Lucanus* is subject to debate because it is important the kind of lamp used (UV light is ineffective; mercury lights are better). However, several species of Lucanidae come to lights in South America, including Chyasognathini and *Sphaenognathus* spp. in Peru.

Grove et al. (2006) have collected in logs - rolling the following species: *Lissotes cancroides* (Fabricius), *Lissotes curvicornis* (Boisduval), *Lissotes obtusatus* (Westwood), and *Syndesus cornutus* (Fabricius). *Moreover, Lissotes latidens* Westwood is thought to live at the soil-log interface (Meggs and Munks 2003) and Grove et al. (2006) have collected live species, with pitfall traps around large logs. They did not use bait, but they has put plastic barriers inserted between the log margin and the trap, for that the beetles coming into to the trap. This beetle is listed as endangered under the Tasmanian Threatened Species Protection Act 1995 and under the Commonwealth Environment Protection and Biodiversity Conservation Act 1999.

The large species of stag beetles should be killed in a cyanide bottle, wrapped in cotton envelops and carefully dried. The smaller species may be preserved in ethanol solution.

PASSALIDAE: 63 genera with 791 species are known, nearly all of which are tropical (Asia, Africa, Australia, and especially in the Neotropical region, except Chile), while in Canada there are only a few species (Schuster in: Arnett et al. 2002). The family does not occur in Europe. Adults and larvae live inside rotting wood, and they feed on this wood and the feces of adults, which are predigested by microflora, essentially an external rumen (Reyes-Castillo 1970; Reyes-Castillo and Halffter 1984). There is a species that lives in the detritus chambers of leaf-cutter ant nests (*Atta* sp.). This species is found on the Mexican side of the Rio Grande. The bess beetles are captured by hand, breaking rotting logs found in forests, savannas or grasslands. Sometimes they come to lights. A method to kill Passalidae is to put all of the specimens collected in the same trunk in zip bags and add few drops of acetone. These beetles, in general, are more or less well known taxonomically and are present throughout the year anywhere within tropical forests, so they are very useful for a rapid ecological assessment. They are collected very easily and in a short time (from 3 hrs per day during 15 days) it is sufficient time to have an inventory of a community of dung beetles in a small area (Cano 1993; Schuster et al. 2000).

GLARESIDAE: Represented by only one genus, *Glaresis*, with 59 species which are distributed in all continents except Australia and Antarctica (Scholtz 1982).

Adults are found in sandy places, and it is suspected that they feed on underground fungi. They are commonly collected in large quantities using ultraviolet light. Adults stridulate when held by hand (Scholtz et al. 1987).

PLEOCOMIDAE: The family includes 2 genera, *Acloma* (19 spp.) and *Pleocoma* (30 known species), distributed from Southern Washington to Northern Baja California in Mexico. Males are winged and females are wingless. Adults do not eat and are active after the rains of autumn, so their common name is the "rain beetles". Larvae have a slow development spanning between 8 to 12 years, moulting several times. They feed on roots of live grasses and trees and are well adapted for digging galleries (Ellertson & Richter 1959). Larvae eat the outer parts of roots and are deep under ground in the same host plants.

When males emerge, they start looking females that are flying near ground level (tracking the smell of their pheromones), and when encountered, many males surround a single female. No one knows the duration of the larval stages of several species, but some species may have up to 9 larval molts and only after 8 to 13 years do they reach the adult stage. Pupation occurs in late summer in a simple, elongated cell. After pupation, adults of both sexes emerge from the soil more or less simultaneously. In other species adults are born at the beginning of autumn or winter rains, while others are active in mid-winter or early spring. The adult activity corresponds closely with rain or melted snow, altitude and some specific climatic conditions, depending on the species. Some species require a certain amount of rain to develop. Although some adults fly at dusk, others fly on rainy nights (Fellin 1981). Adults live for a short time, depending mainly on environmental conditions. Males arrive in large numbers to lights (Hovore 1972, 1979). Many species develop colonies, which can provide many specimens when discovered. Each place has its own colonies mainly in mountainous sites, where new species can be discovered. Female pheromones can be used to attract males (Evans and Hogues 2006).

DIPHYLLOSTOMATIDAE: This family includes only the genus *Diphyllostoma* with 3 species, which are endemic to the west coast of California in the United States (Fall 1901, 1932). Females differ from males by having very small eyes and only vestigial of wings. Almost nothing is known about their behavior except that adults are diurnal (Jameson and Ratcliffe in: Arnett et al. 2002) and can be collected by aerial net and sweep netting.

GEOTRUPIDAE: With global distribution, this family includes 68 genera with about 620 species ((Scholtz and Browne 1996; Jameson in: Arnett et al. 2002). **Geoptrupinae** (including Geotrupini, Lethrini and Taurocerastini) are found in the Holartic region, and in America they are distributed from Canada to El Salvador. There are also Taurocerastini (*Taurocerastes patagonicus* Philippi) in Southern Chile and Argentina. **Bolboceratinae** (including Athyreini and Bolboceratini) are distributed in Australia, Africa, South America and Central America to Canada. They may have saprophagous, coprophagous or fungicolous habits. Adults of most species provision larvae in earthen burrows with dead leaves, cow dung, horse dung, or humus. Some feed on hypogeous fungi and some Athyreini live in galleries in very hard soil (*Neoathyreus, Parathyreus* and some *Athyreus*). Bolboceratinae also live in the deep galleries on the ground, which may be near trees. Some species of this subfamily can be detected by the presence of a mound of fresh earth, with an open circular aperture (Saiz et al. 1989). Adults of *Bolborhinum* (from Argentina and Chile) occur mainly in early winter (in July) and are active until the autumn in the month of May. The species of *Bolborhinum* live near forests and in clearings surrounded by low shrubs (Mondaca & Smith 2008). In South America the adults of the genus Bolborhinum are largely nocturnal or crepuscular, appear mainly in July and are active May. They are found in the proximity of forests and in clearings surrounded by low shrubs (Mondaca & Smith 2008). The adults spend most of the time underground sheltered inside their burrows and can construct galleries in sandy soil or sometimes can also be found in argillaceous, volcanic, or rocky soil and it is possible to find more than one species of *Bolborhinum* sharing the same habitats (Mondaca & Smith 2008). Most species of *Bolborhinum* are moderately attracted to artificial light, especially to low intensity ultraviolet light (Mondaca & Smith 2008). Apparently many Bolboceratinae (*Athyreus* spp., *Neoathyreus* spp., *Parathyreus* spp.) and many *Bolboceras* species prefer to fly during hot, humid, rainy days (pers. observ.).

In Australia according to Matthews (1972) the best sampling method to collect geotrupides is marsupial entrails. In North America, Woodruff (1973) has collected geotrupides with pitfall traps baited with malt and propionic acid, putting water into the pitfall traps and a small amount of that solution. Baits consisting of one teaspoon of powdered yeast in one pint of water are effective for several species of Geotrupinae, and when iare used malt and molasses, the bait appeared to be more effective after fermentation had begun, usually within two days after mixing (Woodruff 1973). According to Woodruff, the malt is especially attractive for several burrowing beetles (e.g. *Mycotrupoes* and *Peltotrupes*) in North America. According to Woodruff (1973) the burrows of Peltotrupes or Mycotrupes are usually verticals and the most are marked by a large "push up" of soil at the entrance. When a burrow is detected, must be followed by inserting a straw into the hole, after should be digging along side to extract the beetles are inside.

The author of this book has collected also several species of geotrupids on the roads in and out of forests, in Paraguay, with an aerial net, by hand-collecting and with light traps, particularly on rainy days and/or immediately after completion of the rain, but always in extreme humidity and heat, both day and night.

OCHODAEIDAE: With global distribution, 104 species are known in 13 genera (Arrow 1912; Scholtz and Evans 1987; Scholtz et al. 1988; Paulsen 2007). Little is known about the adults and larval biology. Most species prefer sandy areas (dunes in sandy beaches, deserts, etc.). Many can stridulute. Adults of one species have been associated with the deposits of detritus of harvester ants. Basidiomycete's spores were found in the stomach contents of an adult specimen (Carlson 1975; Carlson and Ritcher 1974; Deloya 1988). Most species are nocturnal and, according to Woodruff (1973), adults spend the day in galleries dug in the ground and eating mushrooms. They are attracted by street lights, sometimes in large quantities. However, some adults are active during the day and on rare occasions they are collected with sweep nets, Malaise traps or seining flumes. In North America, Woodruff (1973) has collected Ochodaeus frontalis Leconte, in pitfall traps baited with amyl acetate.

SCARABAEIDAE: With approximately 25,000 species in 1,927 genera, they are found worldwide, in almost all available habitats. Here we are including several subfamilies which other authors separate as other families of Scarabaeoidea. In adults, food varies depending on the subfamilies to which they belong: Some are phytophagous (Melolonthinae, Rutelinae, Dynastinae, Cetoniinae, Trichiinae); saprophagous (Aphodiinae); coprophagous (Scarabaeinae, Aphodiinae); necrophagous (Scarabaeinae). *Zonocopris gibbicollis* (Harold) and *Zonocopris machadoi* Vaz-de-Mello live on the body of living giant land snails (*Megalobulimos* sp., and many *Uroxys, Trichillum* and even some *Canthon* live on the bodies of certain mammals such as sloths or monkeys. *Bradypodidium bradyporum* (Boucomont) was taken from the anus of *Bradypus infuscatus* Wagler (Vaz-De-Mello 2008), *Bradypodidium adisi* Ratcliffe has been collected from the fur of the pale-throated sloth (*Bradypus tridactylus* Linnaeus). Another dung beetle species, *Uroxys gorgon* Arrow, is also phoretic on sloths. Phoretic on sloths - it's unknown if this species specializes on sloth dung or feeds on other material in the fur (Ratcliffe 1980; Vaz-De-Mello 2008). Most eat fermented sweet substances and others eat flowers. They can be found in bromeliads (some *Canthidium, Aphengium*, etc.); in vertebrate burrows (Aphodiinae and *Onthophagus*) and many species live in ant nests (*Holocephalus* and *Lomanoxia*). Some species of *Deltochilum* and *Canthon* are predators of miriapods. *Canthon dives* Harold and *Canthon virens* Mannerheim, both use ants as food and oviposition sites. Wich feed on male and non-gravid female keaf-cutter ants *(Atta* sp.) but deposit their eggs in the abdomen of gravid female ants soon after their emergence from the nest (Borgmeier 1937; Navajas 1950; Costa 1953; Halffter and Mathews 1966; Mariconi 1970; Hertel and Colli 1998). *Canthon virens* Mannerheim and *Deltochilum kolbei* Paulian prey on Diplopoda (pers. obs.) also *Deltochilum valgum* Burmeister (Canio 1998). Larvae of Scarabaeidae may be saprophagous, phytophagous, coprophagous or necrophagous. Although some species of Cetoniinae are predators of ant eggs and larvae, bees and termites (Howden 2008), although most of them prefer to feed on pollen, rotting fruit and other fermented plant substances. Cremastocheilini (*Genuchinus ineptus* (Horn), *Cremastocheilus* spp.) are found in nests of ants (Cazier 1961). The larvae of *Cremastocheilus* and some *Euphoria* feed the vegetable debris of the Ants colony of the genus *Formica.* As adults, all species are presumed predaceous on ant larvae and in some cases pupae (Ritcher 1958; Mynhardt and Wenzel 2010).

Aphodiinae is a subfamily with world wide distribution, and many generra has been described, for example, some interesting species, among others, cited by Peck (2006) of some Caribbean islands are: *Rhyparus spilmani* (Cartwright and Chalumeau) from Dominica, Guadeloupe islands, has been collected at light trap and by sifting plant debris; adults and immatures of *Saprosites wirthi* (Cartwright and Chalumeau) this species is endemic to Dominica island, has been collected from decaying heart of Euterpe palm.

According to Woodruff (1973), in North America, many different species of Aphodiinae were collected from leaf mold or pine needles and were collected by scarping away the surface litter and inspecting the soil surface. Many other also were collected with Berlese funnel, and *Psammodius* species were collected by pulling up plants near beaches, they were found under card boards also and the windrows of debris behind the high tide marks.

Orphninae is other subfamily, includes about 80 species worldwide (Arrow 1912; Paulian 1984). Little is know of the ecology and biology of this subfamily. In the New World, adults are collected at low to mid-elevations. They have been observed on banana and cacao plantations on banana stems and have been collected in flight intercept traps. The adults, pupae and larvae of *Aegidium cribratum* Bates was found under rotting logs (Morón 1991). *Aegidium dominicensis* Cartwright and Chalumeau, endemic to Dominica Island, the adults found in banana trash and rotting banana trunks (Peck 2006).

Dynastinae usually are collected with light traps. Some species are found in specific habitats, for example, females of *Thronistes rouxii* (Dejean) live under the sand on some beaches of Uruguay. Many *Dyscinetus* and *Diloboderus abderus* Sturm live in the grass. As a curious case, some years ago, millions of them appear in South America.

The species of *Cyclocephala* are found on flowers (on the day and at night) and at night can be attracted to lights. Most dynastine larvae feed on roots and many others on decaying wood or the fungi within. In other species, larvae live under the soil near the roots of some trees and feed on fallen fruit (*Megasoma joergenseni* Bruch). Other species like *Heterogomphus*, *Strategus*, etc., feed on the roots of certain plants. Many dynastine adult males, especially the large ones (*Megasoma, Dynastes*, etc), do not eat as they spend all their time searching for females to breed (and die shortly thereafter). However, females live much longer than males and this is probably due to the fact that they fly less than males, expending less energy, although egg production consumes large amounts of sugars and fats. Since females are not engaged in searching for males, they have more time to eat and regain energy lost during reproduction. According to Peck (2006) *Hemiphileurus laeviceps* Arrow from Dominica, Guadeloupe, adults are attracted to lights at several localities and larvae has been collected from rotting logs and trees (mango, Inga, etc.). The representatives of Rutelinae are usually collected in flowers from which they feed (*Rutela, Macraspis, Chasmodia,* etc.), or can be collected in traps with fermented products like fruits, molasses, etc. Many species come to the lights (*Pelidnota, Strigidia*, etc.). *Leucothyreus guadulpiensis* Burmeister from Dominica, Guadeloupe and Marie Galante islands, adults feed on Citrus leaves at night (Peck 2006). *Macraspis tristis* Castelnau (from Dominica, Guadeloupe) adults attracted to flowers of *Poinciana* (flametree); larvae in decaying trees (Peck 2006). Members of Melolonthinae live in a variety of habitats such as pastures (*Liogenys* spp., *Serica* spp., etc.), on flowers (*Macrodactylus, Phyllophaga, Serica*, etc.), on branches, walking on the ground (*Apterodermidea*, etc.), in fermented vegetable material, etc. They can be collected on a variety of shrubs and herbaceous plants using light traps, sweeping and beating. The Anaidinae are found in carrion or feces and feed on fungi. Many come to lights.

In North America, Woodruff (1973) have collected Scarabaeinae with liquid baits (containing a small amount of the attarctant in water) inside pitfall traps, include proprionic acid, butyric acid, asafoetida, amyl acetate, and fermenting solutions of malt, molasses, and yeast.

BELOHINIDAE: This family is known by a single species *Belohina inexpectata* Paulian. The six specimens known are from the original type series. This species is found only from Southern Madagascar "the side of the road between Beloha and Tsihombe". The specimens were collected in March 1957 after "very strong rains" (Paulian 1959). Nothing is known about their biology, except that they cannot fly.

GLAPHYRIDAE: The family includes 5 genera and 200 species worldwide. They are widely distributed in the Holartic region. One of the genera is found in America (Chapin 1938) and even Chile, but apparently absent in Australia. They fly both day and night, and during the day they are often confused with some Hymenoptera (metallic bees) for their brilliant colors. They often have colored bands on the abdomen, which is very setose, resembling that of a bumble bee. They often fly between flowers and foliage in sandy areas. Larvae live free on riparian and coastal sand dunes, where they feed on decaying leaf litter and detritus that is layered in the sand. Larvae of *Lichnanthe vulpina* (Hentz) are often a plague of cranberry fields in the northeastern United States (Carlson 1977 1980; Richter 1966; Westcott 1976). Members of this family can be collected with an aerial net, Malaise trap or light trap on some hot nights. **TROGIDAE:** There are about 300 species in 3 genera, on all continents (Jameson 2002, 2009; Sholtz 1990). The genus *Trox* is widespread in the Holarctic and Ethiopian regions; the genus *Omorgus* occurs primarily in arid regions in the Southern continents, and the genus *Polynoncus* occurs throughout South America. Adults and larvae are saprophagous or necrophagous. They can be found in carrion and also in mammals and birds nests, where they feed on feathers, dung and bits of skin. To collect them the dried remains of carrion should be watched. Or to put directly on the ground a piece of leather with dried remnants of skin, which must be kept on smooth ground, inside and outside the forests. Keep them in these places for several days and check them frequently, looking at the ground under these pieces of skin or animal carrion. Sometimes they are attracted to the mercury lights during warm nights. Also check recently abandoned nests of birds or mammals. These beetles have been observed on some beaches in Mexico in dead marine turtle hatchlings or the rotten eggs of these reptiles (Rosado-Hernández and Deloya 2002). Paolo Gandini (pers. comm. 2010) has collected many *Trox* sp. under very dry dog feces in a sandy place in Northern Italy. **HYBOSORIDAE:** Cited sometimes as Hybosorinae (subfamily of Scarabaeidae). More than 570 species have been described in 71 genera (Paulian 1982; Howden and Gill 1988a; Howden and Gill 1988b; Howden and Gill 1995; Howden and Gill 2000; Nel and Scholtz 1990). They are distributed almost worldwide (Allsopp 1984; Martínez 1994; Ocampo 2006) but in many collections there are diverse species that have not yet been described. *Liparochrus* occurs in Australia, New Guinea, New Caledonia, the Loyalty Islands, Lord Howe Island and New Caledonia (Paulian 1980; Allsopp 1984). *Phaeochrous emarginatus* Laporte occurs in Africa, Australia and India. They feed on fungi and are also found in feces, carrion, rotting wood or under bark, and some are found in nests of ants or termites. Some species are predators (Ocampo 2006). Larvae are collected from decaying plant material and under bark.

Adults can be collected in a great variety of habitats and using many collecting methods. When disturbed, they bend their head and pronotum thus concealing the entire ventral side. When contracted in this manner, they resemble spherical seeds. This behavior probably helps them to evade or defend against predators. Hybosoridae, in general, can be collected in fungi, under bark and branches of dead trees, and with pitfall traps using carrion as bait. Occasionally they are attracted to light traps. Larvae have been collected under bark and others in the galleries of Passalidae (Ocampo 2002; Ocampo 2006).

CERATOCANTHIDAE: the species are widely distributed in the tropics, but European species are not known. In Australia only three species have been cited. They feed on hipogeous fungi (Nel and Scholtz 1990), or on rotting wood. Larvae feed on secretions of wounded trees. They are generally found under bark of fallen logs. Several species live associated with ants. According to Mynhardt and Wenzel (2010) the genus *Scarabaeinus* have been found in abandoned and decaying termite mounds. *Nasutitermes* nests are especially attractive to several species of *Astaenomoechus*. To collect Ceratocanthinae from these termite mounds, an active nest must be broken and left on the ground in the open air and after several days it must be carefully checked (Angel Solís, without date). *Germarostes allorgei* (Paulian) adults has been taken at many sites by beating dry vegetation and sifting leaf litter in forest in Dominica and Guadeloupe Islands (Peck 2006). Adults can be collected in a great variety of habitats and using many collecting methods. When disturbed, they bend head and pronotum thus concealing the entire ventral side. When contracted in this manner, they resemble spherical seeds. This behavior probably helps them to evade predators.

Series ELATERIFORMIA

Superfamily SCIRTOIDEA

EUCINETIDAE: Ten genera are known and according to Bullians & Leschen (2004) there are, approximately 50 species in World-wide: *Bisaya nossidiiformis* Reitter from Central Asia, *Eucilodes* (Caucasus, Japan and Russia), *Eucinetus* (widely distributed), *Eucinetella* (China), *Euscaphurus* (Western North America), *Nycteus* (widely distributed), *Subulistomella* from Japan, and *Tohlezkus* from Malaysia, Taiwan, Cameron Highlands region, Turkey, Finland, and Eastern North America *Jentoz-*

kus plaumanni Vèt distributed in the neotropics (Young in: Arnett et al. 2002), *Noteucinetus* from Chile and New Zealand and *Proeuzkus* from Thailand and Nepal (Bullians & Leschen 2004).

Eucinetidae live under bark and dead wood, humus and litter. They eat Basidiomycets fungi. Larvae and adults are Mycophagous restricted exclusively to slime molds and fruiting bodies of Basidiomycetes, Agaricaceae, Boletaceae, Coniophoraceae (Bruns 1984; Weiss and West 1921, Wheeler and Hoebeke 1984). Adults and larvae of *Eucinetus* species have been recorded feeding on Myxomycetes or basidiomycete fungi (Hoebeke et al. 1987; Lawrence & Newton 1980; Lawrence 1991; Weiss & West 1921; Wheeler & Hoebeke 1984). Habitats of Eucinetidae include beech and broadleafforests and coastal scrubland (Bullians & Leschen 2004).

They can be collected in rotten wood, under the bark of decaying logs, by sifting litter or with flight intercept traps. According to Dugdale (cited by Bullians & Leschen 2004) in New Zealand, *Noteucinetus nunni* Bullians &. Leschen was taken from fungi on wet log in two collections, with one from a Mucor-like fungus and the other possibly from a polypore and by pyrethrin-fogging fungus on *Nothofagus* logs in Southern portion of temperate Chile or mixed forests of *Nothofagus* and *Araucaria*. Most specimens of *Noteucinetus chilensis* Bullians & Leschen were captured using flight intercept traps. *Noteucinetus latipennis* Bullians & Leschen has been collected in leaf and log litter forest in *Nothofagus* spp. and shrubland in the Northern portion of temperate Chile (Quillota and Talca Provinces) Bullians & Leschen (2004).

Eucinetidae can be collected with Malaise traps, flight intercept traps, pitfall traps, Berlese funnel, Winkler/Moczarski eclector, sifting litter or by fogging trees and spraying dead logs.

CLAMBIDAE: Is a small family of 6 genera and 175 species worldwide which has been revised by Endrödy-Younga (1959, 1960a, 1960b, 1961a, 1961b, 1961c, 1965, 1974, 1978, 1981, 1990). According to Lawrence and Newton (1999), Lawrence et al. (1995) and Albert Allen (pers. comm. 2010), there are three subfamilies:

---Subfamily Acalyptomerinae: With single genus *Acalpytomerus* from Central America, Eastern & Southern Africa, Mauritius, Japan, India, Sri Lanka, Southeast Asia and Australia. The larvae and adults have been collected in leaf litter, haystacks and other accumulations of organic debris (Lawrence et al. 2000).

---Subfamily Calyptomerinae: Temperate parts of the Holarctic region (introduced in South Africa and Tasmania) with single genus and *Calyptomerus (*5 species). They are found in haystacks and feeds on molds.

---Subfamily Clambinae: With four genera, *Clambus* (cosmopolitan): Hoebeke (1984) have found *Clambus* spp. on slime molds (*Arcyria stipata* (Schwein)), *Stemonitis fusca* Roth and *Stemonitis axifera* (Bull)) and on molds such as Zygomycetes (*Mucor* sp.), other specie *Clambus bornemisszai* Endrody-Younga has been collected in log-decay (*Eucalyptus* sp.) in Tasmania (Grove and Bashford 2003); *Sphaerothorax* (Chile, New Zealand, Australia and Tasmania): *Sphaerothorax tasmani* (Blackburn) has been collected in log-decay (*Eucalyptus* sp.) in Tasmania (Grove and Bashford 2003); *Clambidus atomus* Fauvel (New Caledonia) and *Loricaster (*from North America and Europe).

Clambids are rarely collected, adults are usually caught when they fly in the afternoon and generally are found in the dark side of the forest canopy, under bark and dead wood feed on spores of slime molds and decaying organic material mainly on the spores of fructiferous bodies of Myxomycetes and Ascomycetes fungi (Young 2002). They can be collected by sifting leaf litter, and also reviewing piles of decaying vegetation and using a light traps, Berlese funnel or Winkler/Moczarski eclector, pitfall traps, flight intercept traps or fumigating the treetops. In Atlantic Canada many specimens has been collected in coniferous forests; many were collected in old-growth stands, perhaps indicating that these beetles favour such forest conditions (Majka and Langor 2009). In Asia the majority of specimens of *Clambus* were extracted from forest litter and sifted from mushrooms, probably from the surrounding compost, It is remarkable that in Vietnam few specimens has been collected at light whereas in Europe and in tropical Africa were hardly ever collected at light (Endrödy-Younga 1978).

SCIRTIDAE: Family with 3 subfamilies, over 30 genera and 1,000 species described, with worldwide distribution (Yoshitomi 2005; Lawrence and Yoshitomi 2007), but numerous new genera and species are still remained (particularly Oriental and Neotropical Regions). Adults are terrestrial and larvae are mainly aquatic (some terrestrial larvae were discovered). Pupae are usually terrestrial, but pupae of *Hydrocyphon* spp. (old world genus) are commonly collected under water. Adults are phytophagous, live in vegetation of stagnant water or flows. Many come to light and others are collected by beating or sweeping vegetation and other techniques. The genus *Scirtes* and some other small genera have enlarged hind femur (capable for jumping), but majority (e.g. *Cyphon*, *Elodes*, *Microcara*) have normal hind femur.

Larvae feed by filtering particulate detritus from submerged surface (Grimaldi and Engel 2005), and live in various wetland environments, e.g. phytotelmata, river, lakes, streams, marsh, and groundwater.

---Subfamily Scirtinae: with over 30 genera and 1,000 species (Yoshitomi, pers. comm., 2010)

---Subfamily Nipponocyphoninae: with single Japanese genus *Nipponocyphon.*

---Subfamily Stenocyphoninae: with single Chilean genus *Stenocyphon.*

DECLINIDAE: 1 genus with 2 known species in Russia, Korea and Japan. This family is closely related to the family Scirtidae (Lawrence & Yoshitomi 2007). They feed on pollen; immature stages are unknown. Adults of *Declinia relicta* Nikitsky, known in Russia and Korea (Lawrence et al. 1993) have been collected with pitfall traps. Some adults of the Japanese species *Declinia versicolor* Sakai & Sato were collected in dwarf bamboo leaves, together with species of *Cyphon sannoides* Yoshitomi (Scirtidae) (Sakai & Sato 1996, Yoshitomi unpub.).

Superfamily DASCILLOIDEA

DASCILLIDAE: According to Lawrence and Newton (1999), Lawrence et al. (1995) and Albert Allen (pers. comm. 2010) there are two Subfamilies,

---Subfamily Dascillinae: Occur throughout the North Hemisphere, Australia, Greater Antilles, with one species of Coptocera recorded from Old Calabar, Nigeria. The greatest diversity is in Asia. Larvae are scarab-like found in soil, and apparently feed on roots. Unlike the females of this group are generally flightless, with reduced elytra.

---Subfamily Karumiinae: Occur in more arid parts of Western North America, Mexico, temperate South America, North Africa, and Eurasia. Some are found in termite nests, larvae occur in moist soil, associated with subterranean termites.

Dacillids adults can be collected with sweeping bag, passing it by vegetation near the water and also on trees that are not near water.

RHIPICERIDAE: This family is composed of 5 genera and little more than 50 species (López-Colón 2001): *Arrhaphipterus* (found in Paleartic region), *Chameorhipis* (tropical, subtropical and Southern Africa), *Polymerius* (Chile), *Rhipicera* (Australia, New Caledonia and South America) and *Sandalus* (North America, South America, Asia, Eastern and Southern Africa). Adults are found on tree trunks, usually gathered around a female. They are also found in the soil between humus and litter. In North America, some species are gregarious and are found in late summer on the trunks of American elm (*Ulmus americanus* Linnaeus), shingle oak (*Quercus imbricaria* Michx.) and other trees. Males and females copulate on the bark, resting on nearby grass or flying nearby. Females lay many eggs in the cracks and crevices of the bark preferably in areas where there are a lot of cicadas (Evans and Hogues 2006). Its ovipositor is about as long as its body. Larvae attack the underground cicadas and eat them. The triungulin first instars apparently attach themselves to cicada nymphs before the nymphs enter the soil (Elzinga 1977; Lawrence & Britton 1994). The members of this family are rare enough and to find them it is necessary to check the trunks of trees and the base of shrubs in spring and autumn. In California they fly in late summer (Evans and Hogues 2006). They can be collected by sweeping foliage of trees. Some come to white lights.

Superfamily BUPRESTOIDEA

BUPRESTIDAE: There are slightly more than 14,600 valid species, known from the mainly of the world's temperate and tropical biogeographical zones (Bellamy and Nelson in: Arnett et al. 2002).

Only few develop in living trees while the majority attack weakened, stressed and recently felled hosts (Gardner 1957a). Some buprestids in the temperate zones are fairly or highly host-specific while a number of tropical species are polyphagous (Schabel 2006). In some species the life cycles are said to take up to

35 years, it is longer than for any other insect (Picker et al. 2002).

Adults are perched and ovipositing among the branches and trunks of dead or alive trees and shrubs. Most species feed on twigs, pollen and nectar. Larvae drill galleries in the wood, feed on the inner bark and outer wood in addition to the roots of healthy or diseased trees. Several species produce galls on the branches. A few live in pine cones or herbaceous plants (Bellamy in: Arnett et al. 2002). Little is known about the biology of African buprestids. For example in Zambia, they have been caught throughout the year, but some species were limited to or concentrated in the dry hot season, when trees are stressed, as well as during the early rainy seasons (Löyttyniemi and Löyttyniemi 1987a).

Some *Chrysobothris* and *Agrilus* feed on fungus (Bellamy in: Arnett et al. 2002). They are collected with an aerial net, by beating and sweeping vegetation or with Malaise trap. They are very active at the peak of sunlight. When beating in the vegetation, the collector quickly should make a shadow with his own body on the beetle, so the insect don't have chance to escape, because when the insect feels the shadow, it relaxes. Many species are attracted to freshly cut wood. The species attracted by flowers are often attracted to yellow blooms of composites flowers and mustards flowers. Some of these species are also attracted by colored pan traps (Evans and Hogue 2006.). To collect many species one of the best methods is by sweeping and beating vegetation early in the morning. Many species can be collected carrying branches with larvae to the lab and waiting that the adults emerge. Very little is known about the galleries of the Neotropical species. According to Rojas and Gallardo (2004) the tunnels made by Buprestidae larvae tend to have same form as the thorax of the same, almost three times wider than high, this allows to distinguish the galleries made by the Cerambycidae. Wölfgang Barries, from Austria (pers. comm. 2009) explains that the *Agrilus* genus often makes superficial and zigzag shaped galleries in the branches. In contrast, the *Agriloxia* genus makes more thick galleries and almost with the form of a very long letter "S"; the larva is at the bottom of the gallery. Sub-family **Schizopodinae** (False Jewel Beetles): 7 species are known which can be found in the deserts of California, Southern Nevada, Southwestern Arizona, and Baja California. Larvae may feed on the outer parts of roots (Nelson & Bellamy 1991). Adults of *Schizopus laetus* LeConte feed on flowers of desert plants; *Schizopus sallei* Horn is found in the leaves of herbs; adults *Dystaxia* have been collected on the foliage of *Quercus* spp., or occasionally on *Juniperus*, with *Dystaxia murrayi* LeConte sometimes found on *Chrysothamnus*. Adults of *Glyptoscelimorpha* occur on the foliage of *Juniperus* spp. (Bellamy in: Arnett et al. 2002). Adults are collected on the flowers during the day with a beating sheet under oaks or by hand, and swept from herbs and dry grasses in spring and summer (Evans and Hogue 2006).

Members of this family can be captured with an aerial net, Malaise trap and light trap in some hot nights.

Wolfgang Barries (pers. comm. 2010) also explain me that most of Buprestidae larvae leave a trail of very fine dust at the entrance to the galleries, which facilitates the identification of such galleries in the wood. Wolfgang says that to catch the species that roost in the very thorny branches (or trunks) should be use a great bunch of cotton which is placed in the palm of his hand. With this cotton bunch quickly press the specimens that are walking among the thorny branches or cactus. However, the best traps to catch these beetles are the colored colored pan traps, purple traps and Lindgren Funnel Traps.

Superfamily BYRRHOIDEA (incl. Dryopoidea, Psephenoidea)

BYRRHIDAE: According to Johnson (2002) about 290 species are known worldwide. Adults and larvae feed on moss and lichens, and some larger plants. Larval and adult *Amphicyrta* feed on the roots of a variety of herbaceous plants. Most Byrrhinae feed on the leaves and rhizoids of mosses and liverworts (Bryophyta) and a few have been associated with lichens or higher plants. All species are bryophagous and very specific with their microhabitats. Many species are associated with *Ceratodon purpureus* (Hedw.) moss in coarse sandy soils. *Amphicyrta,* adults and larvae, feed on succulent leaves and stems of forest and meadow herbs, and deciduous shrubs (Johnson 2002). Some *Cytilus* and *Byrrhus* species were reported as incidentally feeding on conifer seedlings, grasses, and clover grown in moist areas with abundant mosses. Other species are strongly associated with certain forest communities and they are more polyphagous on moss hosts (Johnson 2002). All *Curimopsis* spp. are generally restricted to cool temperate and boreal environments, are bryophagous and are usually found in sandy soil which is covered with fine mats composed of mosses. Adults of most species are active on their hosts, day and night; and under bark, in wood chips, leaves, edges of rocks, in mud areas, etc. while the sun is shining (Johnson 2002). They fly in spring and early summer. Often they can be found on beaches (Johnson 2002).

Adults of many species can be found next to their host plants when there is low light or at night or during very sunny day on logs, leaves, stones, etc.

ELMIDAE: There are about 120 genera and more than 1,200 species (Jäch & Koda 1995). Adults and larvae are usually considered to be aquatic; however, adults of many Larainae species are often encountered a little above the water line or in spray zones of water falls and cascades. Members of this family generally live in lotic habitats: Only a few species are encountered in lakes or ponds. They live in shallow fresh water courses (*Hexanchorus* spp.) and are found in a variety of natural submerged substrates, such as dead leaves and emerging twigs (*Phanocerus* spp.), roots of aquatic plants (*Hintonelmis* spp.), riparian vegetation, aquatic moss (*Xenelmis spp.*), decaying trunks and branches (*Heterelmis* spp.), stone surfaces, sandy depth (*Austrolimnius*), stony depth (*Cylloepus* spp. *Heterelmis* spp.) among others. In Lake Titicaca they live among the cattails. They feed on decaying plant material and can not live in acidic or contaminataed water.

Spangler (1981) says: "Larinae live out of water, near the air-water limit, in white water rapids of large or small streams and waterfalls. In these habitats the beetles are hanging on the rocks next to the current, in aggregates of several hundreds, forming compact beetle plaques with several centimeters long and wide. Other species live among branches and grass, piles of sticks rolled with the flow, inside or outside the aquatic habitat. Some females, maybe ovipositors, can be located in the water. They are found by spotting the silvery air spaces developed when water slips over their bodies... ". Most live where the water flows very quickly and take refuge among the stones and sticks of those places. Some species (*Hexanchorus, Heterelmis*) can live in the mountains over 3500 meters above sea level.

Many species live attached to rocks or branches. When frightened, they quickly fly from the water surface. Then you must be fast enough to catch them with the net. The vast majority can be caught with ultraviolet lights, although white mercury lights also attract many species. However, many species cannot fly so they must be collected directly in their habitats. William Shepard (Essig Museum) suggests using an aquatic net to collect many species and visiting different streams is recommended. According to Shepard 2006, the Elmidae like other Byrrhoidea can not live in highly acidic or polluted waters.

DRYOPIDAE: Their distribution includes Palearctic, Neotropical, Afrotropical, Oriental and Nearctic regions. About 24 genera and 240 species have been described so far (Koda & Jach 1995). Larvae are generally terrestrial in this family, and in about 80% of the species adults are known to be aquatic (lotic and lentic habitats). Some species are possibly riparian. They live in decaying plants in very humid places. They can not swim. One species (*Guaranius carlosi* Spangler) is terrestrial living on leaves or shrubs in Paraguay. Many *Helichus* spp. live under logs and partially submerged rocks in streams and creeks, and on twigs and gravel in streams of lower regions. *Elmoparnus* spp. are found in waterfalls and near water, between branches and grass hanging across streams, and have been collected in lowlands between clusters of leaves and twigs floating in slow waters and sandy bottom. *Dryops* spp. may be trapped in wet grass, leaves and twigs protruding from the water in streams and rivers (Shepard and Aguilar 2010). They feed on decaying aquatic plants. Larvae eat roots, very small plants and dead animals in the water; many are terrestrial and stay near water. They live in lakes, streams and rivers, often in the aquatic vegetation on the banks of these watercourses. They can not live in acidic or contaminated water.

LUTROCHIDAE: 15 species are known (http://insects.tamu.edu). They exist only in the New World and are located from North America to Paraguay and Brazil. They are aquatic and are found in small streams and fast-moving water, especially waters rich in calcium. Larvae crawl at the bottom of shallow streams and make galleries in calcareous algae. They feed on diatoms and decaying wood. They can not live in acidic or contaminated water (William Shepard, pers. comm. 2005). To catch them move strongly an aquatic net (with very fine but firm pull) through emergent vegetation in streams. Some species are collected with light traps. Adults are quick to fly when captured (Shepard, pers. comm. 2005).

LIMNICHIDAE: 5 subfamilies with more than 35 genera and more than 225 species of Limnichidae are known world-wide (Wooldridge 1986; Shepard 2002). Larvae are found in moist soil and debris, near bodies of water and wet sand (ponds, lagoons and streams). The ecology of many of the members of family Limnichidae is still unknown, but most seem to be riparian or semiaquatic (Hernando & Ribera 2003). A few species within riparian genera are secondarily terrestrial, mostly inhabiting tropical forests (Ribera & Hernando 1999; Hernando & Ribera 2001), with only one genus known to be entirely terrestrial, *Limnichomorphus* Pic (Satô 1966). Adults live at the water's edge. Larvae and adults are saprophagous and feed on deteriorated plant material. They can be found in driftwood, logs dumped on the beaches, on rocks, accumulations of sticks in the area where the water hits the falls, waterfalls, walls of dams, rapids and branches and herbs hanging on the banks of waterways. *Martinius* spp. and *Mexico* spp. live in the wet edges of muddy banks, coastal lagoons and mangroves. Larvae of these 2 genera are terrestrial (Spangler 1981). To collect limnichids on muddy banks and mangroves is helpful to run rapidly passing the net near the ground (Spangler 1981).

Try pouring water on sand or mud and when the beetles go swimming they can be caught by hand or with a sieve. Also make beating the vegetation with the net as if it were a beating sheet. We must also remove the vegetation and plant litter present on the banks of streams and rapids, and examine it carefully. Most species come close to the ultraviolet or mercury lamps where thousands of specimens can be trapped in a single night.

HETEROCERIDAE: This family is divided in 2 subfamilies: Elythomerinae Pacheco, 1964 (exclusively from Australia) and Heterocerinae MacLeay, 1825, with an almost cosmopolitan distribution (Lawrence & Newton 1995). With 5 genera know and about 200 species (Mascagani 1995). Adults and larvae live in small holes in the mud, clay, sand banks, banks of streams, ponds, lagoons and fresh and/or salty water locations, in the leaf litter on the ground, even in decaying wood and moss (Silvey 1935). According to Katovich (2002): The ecology of heterocerides is apparently closely tied to the rapidly changing riparian habitat, and according to Kaufmann and Stansly (1979): Frequent fluctuations in water level may cause heavy mortality in eggs, pupae, and larvae, but adults are not usually affected. Larvae feed on algae (diatoms) and decomposing organic material, in some species larvae and adults are necrophagous. They have a long adult life stage, and overlapping generations. (Katovich 2002).

Adults are most commonly associated with riparian habitats, although they are strong flyers and can be found some distance from water (Katovich 2002). They are not adapted to live underwater, then pour water in places where they live, which will make these beetles, flee to the highest places and then may be taken with fingers or a fine strainer. On warm nights thousands can reach the white or ultraviolet lights.

PSEPHENIDAE: An almost cosmopolitan family (not recorded from New Zealand) with

4 subfamilies, 28 genera and about 150 described species (Jäch 1998). Since they live in the water line and into the rapids, they are considered aquatic beetles. Adults have a very short life, a very fast flight and seldom are abundant. All species are probably algivorous and/or detritivorous (Shepard 2002), and can not live in acidic or contaminated water (Shepard, pers. comm. 2005). They are rare in collections (Archangelsky 2001). Adults (in North America) occur only in the summer months and are generally riparian, except females may lay eggs on rocks below the water (Shepard 2002). According to Spangler *"Larvae and adults of* Psephenidae *live on rocks in rapids and creeks. However, streams washed by the action of periodic flooding are generally uninhabited by* Psephenidae. *By contrast, streams that have a regular flow and are rarely washed by the floods may have a high population of larvae of* Psephenidae*".* Larvae feed on plants and microscopic animals, and it is believed that adults feed on plants (White 1983). Some species are found on driftwood.

Adults can be taken with an aquatic net in vegetation of moving or standing water or taking them directly from the surface of submerged rocks. In contrast larvae are most abundant and easy to collect by hand turning the stones; also passing the aquatic net under stones that have been turned, and into the rapids gravel. In California some adults are found on rocks and plants on the banks of streams in late spring and summer (Evans and Hogue 2006).

CNEOGLOSSIDAE: Distributed in Mexico, Costa Rica, Nicaragua, Panama, Colombia, Peru and Brazil. A single genus (*Cneoglossa*) is known, with 10 species (Crowson 1972). Aquatic habits, they prey snails. Larvae of *Cneoglossa edsoni* Costa, Vanin & Ide were found in decaying brambles wood fallen in streams of shallow running water (Costa et al. 1999). Larvae, pupae and adults were reared in the laboratory (Lawrence et al. 2000).

PTILODACTYLIDAE: A moderately large and diverse family with 5 subfamilies (Lawrence & Newton 1995), approximately 25 genera, nearly 500 species (Lawrence & Stribling 1992), and with almost world-wide distribution. Terrestrial and aquatic; many live in riparian zone; saprophagous habits, feeding on organic and plant decaying material, leaves and rotten wood. The aquatic larva of *Anchytarsus bicolor* (Melsheimer) has xilophagous habits (Le Sage and Harper 1976). The detritivore species *Anchytarsis palpalis* (Champion) eat decomposed leaves in lotic situations (Spangler, 1983). Other species (*Anchycteis velutina* Horn and *Araeopidius monachus (LeConte))* have been found in rock galleries eating the roots of emergent plants (Leech and Chandler 1956). They have also been found drilling softwoods (Dudley and Anderson 1982). Some terrestrial species with larvae have been collected (with Berlese funnel) in litter and decaying wet wood (Ivie, M. cited in: Stribling and Seymour 1988), so it is assumed that they feed on fungi (Stribling and Seymour 1988). Many are attracted at lights. Others are collected by beating vegetation or with an aquatic net.

CHELONARIIDAE: They are distributed throughout the warmer regions of America, Eurasia, Africa, East Indies, New Zealand, New Guinea and Northeastern Australia (Dalla Torre 1991). There are over 220 described species in 3 genera. Undoubtedly many species of tropical regions have not yet been described (Leschen and Early 2004). *Brounia* is endemic to New Zealand (Klimaszewski & Watt 1997). *Pseudochelonarium* (6 spp.) is endemic to India, Sumatra, Borneo and New Guinea (Paulus 1969) and *Chelonarium* is mainly found in the neotropics, but also in Queensland, Australia (Lawrence & Britton 1991, Lawrence and Newton 1995). They live in leaf litter and debris at the bases of orchids, Epiphytes plants, in nests of mound-building ants, discarded mounds of ants (*Atta* spp), occasionally under bark, walking on tree trunks and fallen branches,and even in pastures. Larvae feed on detritus in moist or dry litter, especially at the base of trees, orchids and other plants; in refuse heaps of ants or under bark. Some entomologists say that larvae of North American *Chelonarium puncatatum* Fabricius live with ants, but this hypothesis is being withheld for now, since some specimens have been seen under bark, but away from the ants, although some South American species apparently are associated with ants and termites (M. A. Ivie in: Arnett et al. 2002; Spangler 1980 and 1991). However, adults of many species of Chelonariidae feed on pollen and nectar from flowers. In New Zealand a rare species (*Brounia thoracica* Sharp) has been collected on flowers of *Melicytus ramiflorus* Forst & Forst (Leschen and Early 2004). According to photographs, the method of collecting used by these authors consists in beating the foliage with an aerial net (not using a beating sheet) with the mouth upward. This beetle from New Zealand is very fast to get away, but in South America various species of *Chelonarium* spp. collected by the author of this book on the edge of the jungle are not as scary and he has used only a beating sheet. Some species can be collected with a Berlese funnel, with Malaise trap, with sweeping bag and by beating vegetation, especially at night. Several species come close to the lights.

EULICHADIDAE: A small family with 2 genera and 21 described species (Jach 1995d). Adults are strictly terrestrial, feeding on decayed plant material. They have a short lifetime and are found only a few days in summer (Shepard 2002); they have riparic habits and live in the same areas as larvae, which live under rocks in streams of moderate to large size, especially when they have detritus and leaves (Shepard, pers. comm. 2005). There are species of *Eulichas* in China and Southeast Asia.

Tenocolus scutellaris LeConte is restricted to California, USA. Larvae of Eulichadidae are aquatic and are found in streams among dead leaves.

CALLIRHIPIDAE: Distributed in the Afrotropical, Neotropical, Indo-Australian, Asian region and North America. 9 genera are known and about 151 species (http://insects.tamu.edu). In North America, larvae of *Zenoa picea* (Beauvois) live in rotten wood, especially white rotten wood and usually with white rot fungi. Adults have a very short life (Young 2002) and are active at night.

Callirhipis lherminieri Laporte, from Dominica, Guadeloupe, St.Vincent islands, has been collected at lights and in malaise traps; larvae found in rotten wood (Peck 2006).

Superfamily ELATEROIDEA (incl. Cantharoidea, Artematopodoidea)

RHINORHIPIDAE: With only one known species *Rhinorhipus tamborinensis* Lawrence (from Southern Queensland, Australia (Lawrence 1988). Adults have been collected on the sides of the jungle near a small stream (Lawrence 1988). These beetles were also collected on the leaves of an introduced noxious weed, crofton weed (*Eupatorium adenophorum* Spreng) (Hawkeswood). The beetle has also been found near O'Reillys Guest House in the Lamington National Park and part of this vegetation is *Eucalyptus* forest woodland (Trevor 2006).

ARTEMATOPODIDAE: With distribution in Neotropical, Neartic and Paleartic regions. 8 genera with 60 species are known (Majka 2010). According to Lawrence and Newton (1999); Lawrence et al. (2010) and Albert Allen (pers. comm. 2010) three Subfamilies: **Allopogoniinae** Crowson 1973b (as tribe; as subfamily according to Lawrence 2005a), 1 genus *Allopogonia villosa* (Horn) *(in* Lawrence 2005) **Artematopodinae**, 7 genera North and South America, Italy, Eastern Siberia, China, and Asia. *Artematopus* larvae found in soil, and observed to feed on dead insects in the laboratory. **Electribiinae:** With single genus *Electribius* (nothing known about biology) from Southern Mexico, and Guatemala, also known from Oligocene amber in Northern Europe.

Adults of Soft-bodied Plant Beetles feed on moss found on granite rocks along streams in open woods. Adults of *Macropogon rufipes* Horn and *Eurypogon harrisii* (Westwood) have associated with moss (*Paraleucobryum longifolium* (Hedw.)) where it grows on non-calcareous boulders and cliffs; sometimes on tree trunks, stumps, and rotten logs in Maritime Provinces from Canda, except for the Eastern shore of New Brunswick (Ireland 1982, 2007; Majka 2010). *Macropogon sequoiae* Hopping has been found in several genera of moss (Cooper 1991; Young in: Arnett et al. 2002; Lawrence 1982). *Macropogon piceus* LeConte have been found in 75 year old red spruce forest (in New Brunswick and Nova Scotia, Canada) with flight intercept trap (Majka 2010).

Several larvae species have been associated with mattresses of moss growing on granite rocks, where they tunnel into the rhizoids. Adults are collected by sweeping and beating forest foliage (Young 2002).

BRACHYPSECTRIDAE: 4 species are known in one genus (*Brachypsectra*), distributed in North America, Asia, Northern Australia and Southern India. In North America larvae can be found beneath bark, under pieces of wood or in rock crevices, and they are slow movement's predators in ambushing (Young 2002). *Brachypsectra fulva* LeConte, from Texas, secretes a liquid that attracts certain spiders (White 1983). When a spider or other small arthropod is walking nearby, the *Brachypsectra* larvae remains motionless and flattens its body against the resilient surface anterior end and spiniform ninth tergum are abruptly raised, pinning the prey between the spine and the perforate, falciform mandibles (Young 2002). Adults have been found under bark, are usually very rare and probably with a short life (Fleenor and Taber 1999, Lawrence 1982, 1991; Young 2002).

CEROPHYTIDAE: In the *Cerophytum* genus 10 species of the Rare Click Beetles have been described in the Paleartic, Neartic and Neotropical regions and, there are still some new species known from the latter region which have not been described (Johnson in: Arnett et al. 2002). Albert Allen (pers. comm. 2010) he told me that has in his personal holdings a cerophytid from Cameroon (Africa) was missed in the world revision.

The rare click beetles are among the litter and under ground debris (Steiner 2000). Larvae are found in various types of angiosperm trees decaying wood, where they feed on fallen xylem plates (Buysson 1910; Mamaev 1978). *Cerophytum* larvae were collected from decaying Elm wood (*Ulmus*) and Alder (*Alnus*). Cerophytids can jump as Elateridae. Adults are usually collected with flight intercep traps when they fly in the afternoons in the lowlands and mid-elevation forests (Johnson 2002). They can be caught passing the sweeping bag by vegetation over fallen logs. Several species are attracted at lights.

EUCNEMIDAE: 191 genera have been describes and according to Robert Otto (pers. comm. 2010) the approximate number of described species should be around 1600, worldwide. According to Lawrence and Newton (1999); Lawrence et al. (1995), Albert Allen and Robert Otto (pers. comm. 2010) there are 9 Subfamilies:

---Subfamily Perothopinae: California (2 spp.) and Pennsylvania to Florida (1 sp.), the biology of immature stages are unknown.

---Subfamily Phyllocerinae: North America, Southeastern Europe, Caucasus, Central Asia to Southern Africa, and Madagascar, the larvae of Phyllocerus species occur in the soil and feed on plant roots;

---Subfamily Palaeoxeninae: With single genus *Palaeoxenus* California. *Palaeoxenus dohrni* (Horn), endemic to California, according to Evans and Hogues (2006) adult and larvae are found in the lower portions of incense-cedar (*Calocedrus decurrens* (Torrey)) stumps and under the bark of old trunks of bigcone Douglas-fir (*Pseudotsuga macrocarpa* (Vasey)) during spring and summer.

---Subfamily Phlegoninae: A single genus from Central and South America. Sometimes they have been collected with Winkler/Moczarski eclector.

---Subfamily Melasinae: Occur in all major regions of the world, adults are ususally collected on plants surfaces, at lights, or in flight-intercept, larvae of most taxa occur in dead wood and are capable of moving reasonably solid wood, by forcing their bodies between wood fibers, they are liquid feeders, so actual diet difficult to determine. It is possible they feed on the plasmodium of Myxomycetes, or digest plant tissue extra-orally.

---Subfamily Eucneminae and **Macraulacinae:** Are worldwide in distribution with similar biology as Melasinae.

---Subfamily Anischiinae: Is composed of one genus, *Anischia* and *Perothops* known from the Neotropics, New Guinea, New Caledonia, Solomon Is., Australia and central Africa, adults were collected in Panama from a fungus on the surface of log, adults and associated larva were sifting rotten wood and litter and associated with bracket fungi in New Caledonia. One larva of both *Anischia* and *Perothops* has been described already (Lawrence et al. 2007).

---Subfamily Pseudomeninae: 2 genera: *Schizophilus* (from North America) species very rare, adults have been observed on the bark of *Fagus*. Larvae were collected from decayed logs in Southern Wisconsin (Otto and Young 1998) other genus is *Pseudomenes* (from Australia).

Eucnemids adults are among the vegetation, under bark or under stumps. Larvae are found in the interior of very hard wood, damaged, where they feed on fungi. In general, larvae develop in the white part of the rotten and very wet and soft wood, and pupate in the most hard and dry parts.

Some *Melasis* spp., *Hylis* spp. (Melasinae) prefers living in conifers (Muona 2002). The paleartic species of Phyllocerus develop in the soil (Muona 2002). Some species jump like Elateridae and in general they are very good fliers. Evans and Hogues (2006) argue that these beetles are important indicators of forest biodiversity in North America because they seem to play an important role in the interaction between trees, fungi and forest regeneration. Jyrki Muona (2000) has also made the argument on the importance of eucnemids in the forest ecosystem. He touched on the matter in the revision of the Nearctic Eucnemidae.

According to Robert Otto (pers. comm. 2010) the vast majority of species live in forests and jungles, particularly in forests with varied structure and where there are clusters of dead wood that serve them for breeding. Adults are found in the canopy and on or under the bark of trees. The larvae are present in dead wood wet or develop in the soil near the roots of dead and dying trees.

These Coleoptera are rarely collected; they are usually collected by beating or sweept vegetation, by hand-collecting searching in the habitats where they live, and are also caught with Malaise trap, Berlese funnel and lights traps.

THROSCIDAE: Widely distributed in the world. 6 genera with 155 species are known (www.insects.tamu.edu, 2010). Adults live among the foliage and feed on pollen and mold. They have special organs on its body with three types of symbiotic bacteria (Johnson in: Arnett et al. 2002). Found largely in woodlands and forests, especially with a diverse forest structure and plenty of dead wood for breeding. Adults are found in the tree canopies and on/under tree bark. Larvae are either in moist dead wood, hard seasoned wood or thrive in the soil near the roots of dead or dying trees.

They are caught with pitfall traps (Malaise, flight intercept, etc.) or simply by aerial nets; this is done at the time of eruption of flying which is usually in the evenings. They can also be collected using a beating sheet or sweeping bag. Others come to the lights. Flowers should also be examined, samples of humus, near mammal burrows, rotting wood, in storage facilities of various grains and different foliages. Larvae have been found in blocky red-rotted portions of oak logs, in fungus and soil samples, and in grass tussocks. Larvae of European species *Trixagus dermestoides* (Linnaeus) were found in a variety of soils, always near tree stumps and under the litter. They are liquid feeders on outher portions of ectotrophic mycorrhyzal roots (Burakowski 1975).

The very unusual *Neocrowsonius* is found in termite colonies of endemic *Hodotermes viator* (Latreille), Cape region of South Africa (Albert Allen, pers. comm. 2010).

ELATERIDAE: Family with worldwide distribution. According to Johnson (2002) were 10,000 species known, but now, according to the same Johnson (ww.insects.tamu.edu) there are 600 genera with 12,271 species known.

The biology of this family varies considerably. Adults eat plant tissue, moss and decaying substances and feed on foliage, flower petals or pollen (Schabel 2006). Larvae are found in vegetable garbage, litter, humus and decaying plant or animal tissues and usually larvae are found in the soil. Most feed on roots, seeds or plants inside; others are predators and live in grooves on the surface of termite nests located in the Brazilian savanna and use bioluminescence to attract prey. In the soil they prey upon other insects larvae (Schabel 2006), or partially rotten wood, feeding upon cerambycid Coleopteran larvae. However, the natural habitat of the larval stage of numerous elaterid species is either grass land or pasture (Caldèr 1996). Accoding to Lawrence et al. (2010) the next subfamilies are known:

---Subfamily Agrypninae: Are cosmopolitan, usually found on bark, leaf surfaces, on leaf litter, but little is known of their feeding habits. Larvae occur under bark or stones, in rotten wood, leaf litter of soil. Some agrypnine larvae, particularly the larger, wood-inhabiting forms, are predaceous, but others occur in the soil, feed on roots of various plants, and may damage crops. Tetralobini (sensu Costa et al. 1992, 1994; Lawrence et al. 2010) occur throughout the warmer parts of the Old World from central and Southeastern Africa and Madagascar to Southeast Asia, the East Indies, New Guinea and Australia. The larvae are associated with the nests of termites. They have been recorded from the nests of Termitidae (*Nasutitermes* and *Macrotermes*) and Rhinotermitidae (*Coptotermes*) in Africa and Australia.

---Subfamily Cardiophorinae: With world wide distribution, the larvae are predaceous occur in soil and they are very distinctive, with a very slender, worm-like body and peculiar, externally toothed mandibles.

---Subfamily Denticollinae: Cosmopolitan subfamily found on foliage or bark surfaces and may feed on plant shoots or tips, nectar or possibly epiphytic microflora, larvae are liquid feeders utilizing extra-oral digestion, but they may be either predaceous or phytophagous; they live under bark or rocks and in rotten wood, leaf litter or soil. Larvae are serious pest of agricultural crops.

---Subfamily Elaterinae: With world wide distribution, but they are most abundant and diverse in Oriental and Holarctic regions, they are relatively uncommon in South America, Africa, and Australia, with none in New Zealand. Adults are found as most typical elaterids, larvae of many have been recorded in soil feeding on roots causing damage to cereal and forage crops and sugar cane.

---Subfamily Eudicronychinae: *Eudicronyhcus* occurs in India, Sri Lanka and central to Southern Africa, Anisomerus and Tarsalgus restricted to Africa.

---Subfamily Hemiopinae: With *Hemiops*, *Parhemiops*, and *Plectrosternus* occur in India, China, Southeast Asia, and East Indies. Four specimens of the genus *Hemiops* were collected by the Malaise trap set in a natural forest of Northern part of Okinawa Island in Japan (Ôhira and Makihara 2007).

---Subfamily Lissominae: Occur mainly tropical and subtropical, but some Drapetes occur in cooler parts of Holarctic region. Lissomine larvae feed in decaying wood;

---Subfamily Negastriinae: Occur in most regions of the world, but they are absent from New Zealand. Adults are often found on vegetation, and larvae liven in leaf litter or soil often in riparian habitats;

---Subfamily Physodactylinae: From South America, Africa, India, Southeast Asia and Australia (A. Allen, pers. comm., 2010).

---Subfamily Pityobiinae: *Pityobius* restricted to North America where the larvae and adults live under bark and in the wood of rotten trees, they appear to be facultative predators, and larvae will feed on mealworms but can be reared on moist wood frass found in tunnerl of termites and wood-boring beetles in Australia species of Parablax and *Parasaphes*, and *Wynarka sylvestre* Calder have been reared from larvae collected in rotting tree trunks in wet sclerophyll forest areas.

---Subfamily Semiotinae: From Western Hemisphere, where larvae are usually found in moist, sappy areas under bark and they have been observed feeding on fly larvae. Most species can be collected with light traps and with bottle fruit traps in trees (pers. observ.).

---Subfamily Oxynopterinae: The genus *Oxynopterus* hose larvae have been reported as predaceous on termites (Kalshoven 1955).

---Subfamily Subprotelaterinae: Are known from Southeast Asia, New Guinea and Australia. Nothing is known on the biology of *Subprotelater* genus (A. Allen, pers. comm., 2010).

---Subfamily Thylacosterninae: Are more or less tropicopolitan in distribution, occur in the tropics of America, Africa, Asia and Australia, and larvae feed in dead wood. In Australia *Cussolenis mutabilis* Bonvouloir from Bengal was found in decaying logs that were being attacked by eucnemids (Gardner 1936).

---Subfamily Subprotelaterinae: With single genus *Subprotelater* from south East Asia, New Guinea and Australia. In Japan they have been collected with light trap (Makihara and Ôhira 2006).

---Subfamily Campyloxeninae (Costa 1975) and **Morostominae** (Dolin 2000) were added as subfamilies by Lawrence et al. (2010).

---Subfamily Cebrioninae: 281 species are known in 20 genera. Occur in most parts of the world, but they are absent from Australia and New Zealand, more abundant in arid parts of Holarctic Region (Albert Allen pers. comm. 2010). The Cebrionid have lifestyles that make them appear evasive. Their larvae as "wireworms" live in the soil. Adults remain buried and come out of the ground to mate only during heavy rains. Apparently cebrionids emerge in large quantities at the same time. Larvae live in the soil and it is ignored what they eat. Females are rarely collected. Although females have well developed wings, they have never been seen flying. Males are attracted to lights, especially on rainy nights.

Peter Kate (pers. comm. 2009) has collected many Click beetles of small size sheding water in the gravel that is on the shores of rivers and streams, and even in the sand. When the beetles come up they can be caught with a net or by hand. Many species are under bark of dead trees, most are nocturnal and many come to the lights; others are collected by beating and sweeping vegetation or with Malaise trap; some species are attracted to decaying fruit.

---Subfamily Prosterninae: (This subfamily is not included in Lawrence et al. (2010)), world wide distribution. Most species can be collected with light traps.

PLASTOCERIDAE: A single genus *Plastocerus* with 6 species from Asia Minor and Southeast Asia of unknown biology (Albert Allen, pers. comm. 2010). They resemble the Elateridae family, but it is doubtful that relate to them. Males of this genus are not uncommonly taken in Southern California, either in flight at dusk or by beating or sweeping (Fall 1909).

DRILIDAE: 6 genera and 121 species are known worldwide, mostly in the tropics (Ladislav & Milada 2006). The *Drilus* genus (33 spp.) is distributed in Africa, Europe, Minor Asia, Central Asia and Sri Lanka. The 12 species of *Malacogaster* are spread through Southern Europe, the Canary Islands Africa Asia Minor. *Euanoma* (7 spp.), ocurring from Greece to Southern Russia, Georgia. The *Sellasia,* with 65 species distributed in Africa and with a few species in India. *Pseudeuanoma* (3 spp) is located in Eastern Europe and Asia Minor and *Paradrilus opacus* Kiesenwetter from Spain (Albert Allen pers. comm. 2010).

Drilidae show a marked sexual dimorphism. Males have wings and females are completely wingless and with a larviforme look. Most are nocturnal or crepuscular. Some species of this family are important predators of snails and they are useful to keep the snail populations potentially harmful. At least females of *Drilius, Malacogaster* and *Sellasia* genera are predators of land snails. Larvae and adults of *Malacogaster passerinii* Bassi (from Italy, Spain and Portugal) feed on gastropods (Mollusca) of *Helix* genus (Bahillo & López 2005). They kill the snail with a poisonous bite; they can even drill into the shell, secreting digestive enzymes and absorbing the snail partially dissolved. In general, females live in the soil, with limited movements and can be found under rocks or empty snail shells. Adult males can often be found in the flowers or vegetation. Females probably use pheromones to attract active males (Hansen and Larsson1973). In Northern Europe these beetles are very scarce.

OMALISIDAE: The *Omalisus* genus with 10 species (Ladislav & Milada 2006) is distributed in Southern Russia, Poland, Ukraine, the Balkans, Belgium, Italy, France, Germany, Croatia, Bosnia and Greece, and *Thilmanus* with 2 species is distributed in France, Sardinia and Corsica. Almost nothing is known about their biology. Probably most species are predators. Larvae emit light but not adults (Lawrence 1991b). According to Day (2009) the other families of luminescent Elateroidea derive from Omalisidae. *Omalisus fontisbellaquaei* Geoffroy lives among leaf litter in forests and are predators of millipedes of the *Glomeris* spp. genus (Bocak and Milan 2008).

LYCIDAE: Over 4,600 species in 6 subfamilies have been described (Ladislav & Bocáková 2006). They are most abundant in the tropics. Just in Northern Mexico 76 species have been described, but several species remain undescribed (Miller S. in: Arnett et al. 2002). From Paleartic region 400 species have been described, most of China and Japan. But the greatest diversity is found in the tropics. Adults are found under bark, in leaves and flowers, feeding on nectar, pollen and possibly honeydew; many larvae feed on fungi and many are predators and live under bark of fallen logs and leaf litter. They feed on rotting wood also but also eat liquid material (Evans and Hogues 2006). Sometimes adults are found congregated on the plants.

According to A. Allen (pers. comm. 2010) and Lawrence et al. (1999a) the next subfamilies are known:

---Subfamily Duliticolinae species of *Duliticola* and *Platerodrilus* they are known from India, Southeast Asia and East Indies. Larvae and larviform females of *Duliticola* species are well known inhabitants of tropical forests, where they usually occur on the surfaces of logs. Their peculidar form has earned them the name "trilobite larvae". Their feeding habits, like most Lycidae, they are uncertain, however they appear to suck juices from rotting plant material (A. Allen, pers. comm. 2010).

---Subfamily Calochrominae: *Calochromus* is an Oriental genus known from forest or shrub habitats. Some others genera are Lygistopterus and Macrolygistopterus from Neotropial Region (A. Allen, pers. comm. 2010).

---Subfamily Erotinae: Worldwide distribution, except the Neotropical Region, penetrating into Papua part of the Australian region as well (Kazantsev 2004). Some Neartic genera are *Lucaina, Caloptognatha, Lygistopterus and Dictyoptera,* the last is a Holarctic genus with 4 species occurring in North America (North of Mexico) in coniferous and/or mixed coniferous forests (Miller in: Arnett et al. 2002).

---Subfamily Leptolycinae: Neotropical, Etiopical and Oriental regions. These beetles are rarely collected, they have larviform females, and the species tend to be localized (Miller 1991).

---Subfamily Ateliinae: According to Vladimír & Ladislav (2009) is known only from males, and probably that females are neotenic, as was found in the Lyropaeinae. Larvae are unknown and, if similar to other net-winged beetles, develop in rotten wood and soil with a high content of organic matter.

---Subfamily Lycinae: Worldwide, except Australia Metriorrhynchinae Paleartica, Oriental, Etiopica and Australian Region (A. Allen, pers. comm. 2010).

Lycides are found under bark, in spring and early summer they can be collected with Malaise trap, anyway, they are easy collected with an aerial net or by beating and sweeping vegetation, because they fly slowly and clumsily for short distances, among the foliage, on the roads within the forests and among the low forest.

TELEGEUSIDAE: This family is restricted to the new world. 10 species in 3 genera are known (http://insects.tamu.edu). According to Lawrence et al. (2010) some molecular evidence (Bocakova et al. 2007) suggests that this group does belong within Phengodidae, as suggested by Crowson (1972). The winged males, which at night are usually attracted to lights (Zaragoza-Caballero 2008), are short-lived and probably do not eat. Females are much larger than males and are larviform. They are found from near sea level to 2,100 meters above sea level. Larvae may live in soil and are predators (Lawrence et al. 2010). Some species have been collected with Malaise trap and others with pitfall traps (pers. observ.).

PHENGODIDAE: With 31 genera and 243 species described (http://insects.tamu.edu), and according to Lawrence et al. (2010) there are 3 subfamilies: **Phengodinae**, **Mastinocerinae** and **Penicillophorinae**. They are distributed from Southern North America to Southern South America.

Larvae are predators, feeding mainly on millipedes, insects and other arthropods living in soil and litter (Evans and Hogues 2006) and they can be found under logs, beneath the bark, or other objects on the ground (O'Keefe 2002). Larvae of phengodides are nocturnal and more active on humid nights (Viviani and Bechara 1997). Males are winged, have a very short life and may not eat. Females are much larger than males and are completely larviform, feeding on millipedes and other insects (Tiemann 1967, 1970; Miller 1997; Eisner et al. 1998.

To collect the adults is good sweeping the wet soil beneath the plants, where the millipedes are active, because these invertebrates sometimes indicate the presence of phengodids larvae (Evans and Hogues 2006). The females can be collected under bark of dead logs and in fallen leaf litter. The author of this book has collected some luminescent species with light traps in the Paraguayan Chaco and others with beating and sweeping vegetation. Males are easily attracted by the lights. Larvae are attracted to card boards and other flat objects placed on the ground. Pitfall traps with cross-shaped barriers are also effective for collect these beetles. In North America *Zarhipis* occur in desert areas where the males are often attracted to blacklights (O'Keefe 2002). Adult females have been collected under loose bark also (Miller 1997).

RHAGOPHTHALMIDAE: 9 genera with 35 species (http://insects.tamu.edu). Occur throughout the Eastern Palaeartic and Oriental regions from Japan to India, Sri Lanka, and East Indies. Members of this beetle family have bioluminescent organs (also larvae) and are closely related to the Phengodidae and Lampyridae (Branham and Wenzel 2001) and were often included in one of these families as a subfamily (O'Keefe 2002). Albert Allen (pers. comm. 2010) informed me that the females are apterous, occuring in soil and litter and are predaceous. Males may be attracted to lights at night.

LAMPYRIDAE: A family with world-wide distribution, but absent from Australia and New Zealand. About 2,000 species have been described in 92 genera (Mcdermott 1966; Lloyd 2002). But now are known as 109 genera (http://insects.tamu.edu). About one half of these species occur in Tropical America and approximately two to three times as many fireflies are yet to be described (Lloyd 2002).

They are known by the popular name for fireflies. Most live hidden in the bark, foliage or litter during the day. Many fireflies do not produce light. Usually these species are diurnal or day-flying, such as those in the genus *Ellychnia.* A few diurnal fireflies that primarily inhabit shadowy places, such as beneath tall plants or trees, are luminescent. They are usually found on flowers or vegetation on the stream's banks (Evans and Hogues 2006). Larvae of some species can be found at the base of some grass or on leaves of shrubs (like adults) and some *Prionocyphon* can live at elevations above 3,500 meters above sea level. Most Lampyridae are terrestrial, but a few species, in their larval stage, have colonized freshwater aquatic environments (Stehr 1991). Larvae of species such as *Luciola cruciata* Motschulsky in Japan and *Luciola ficta* (Olivier) in China are strictly aquatic and provided with gills. Other larvae have been reported as facultatively aquatic (e.g. *Pristolycus kanoi* Nakane, from Taiwan) and others are known to be riparian/litoral (Jäch 1998). Adults and larvae of *Luciola afrogeneia* Ballantyne & Buck were reported from the spray zone of coastal coral rocks in New Guinea (Lloyd 1973; Jäch 1998), larvae were observed underwater in rock pools as well as above water (Jäch 1998). The aquatic / riparian larvae are reported from a variety of habitats: Running water (streams and rivers), paddy fields, small ditches, seepages, water falls, etc. The females and larvae of some lampyrids species are presumed to live in burrows or be inquilines of ants (Lloyd 2002). Light production in fireflies is due to a chemical reaction that occurs in some species (*Luciola* spp., Lampyris spp. and *Phausis* spp.) prey Gastropoda mollusk (Helicidae). *Lampyris* larvae attack the tentacles and other soft tissues of the snails and inject a poison that paralyzes them and then kill the snail (Carrero 2008.). Both *Lampyris* as *Phausis* before attacking the soft tissues of snails eat the mucus secreted by the gastropod. Other predatory of helicids are *Lamprophorus tenebrosus* (Walker), *Photuris pennsylvanicus* DeGeer, *Luciola cruciata* Motschulsky and *Luciola vittata* Castelnau. So these lampyrids species are useful for gardens as biological control. But in many species adults apparently do not eat (Evans and Hogues 2006). Fireflies are sometimes poisonous to vertebrate predators. This is due to a group of steroid pyrones (lucibufagins or LBGs), which are similar to cardiotonic bufadienolides found in some poisonous toads (Eisner et al. 1978).

They have specialized light-emitting organs, usually on the lower abdomen. The enzyme luciferase acts on luciferin in this organ to stimulate light emission.

Lawrence et al. (2010) include the following subfamilies:

---Subfamily Psilocladinae: *Cyphonocerus* from China, Japon, Vietnam, India, Nepal, (Ming-Luen et al. 2006). *Cyphonocerus* can display weak light signals, while *Pollaclasis* have pheromone signals like their relatives (Stanger-Hall et al. 2007).

---Subfamily Lampyrinae: Larvae are predaceous, feeding on a variety of soft-bodied invertebrates, such as earthworms, snails, and slugs, etc. It is likely that most species do not feed as adults. Some females can be anelytrous or brachyelytrous. *Pyractomena* larvae sometimes glow several feet above ground on woody vegetation.

---Subfamily Luciolinae: Are restricted to the warmer parts of the Old World, occuring from Southern Europe and Asia Minor to Southern Africa, Madagascar, Japan, Southeast Asia, East Indies, New Guinea, Australia, and some Pacific islands. In almost all Luciolinae both sexes are winged and luminescent and synchronous flashing has been observed in some groups. They seem to be all "flashing" (as opposed to "continuous-glow") fireflies. Feeding of adults is unknown, but larvae are predators of small land snails (Lawrence et al. 2000). They are not thought to be luminous although *Brachylampis* California has the last 3 ventrites pale in color.

---Subfamily Photurinae: Several *Photuris* species which acquire nutritional substances as well as defensive compounds for their eggs from their predation on other fireflies (Eisner et al. 1997), *Photuris* larvae are the emitters of the most commonly seen glows and on damp and rainy nights especially (Lloyd in: Arnett et al. 2002). *Bicellonycha* and female Photuris are known for mimicking the mating flashes of other fireflies for the sole purpose of predation.

Bicellonycha, previously known from the American tropics, occurs along certain waterways in South Central and South Eastern Arizona at elevations of 4,000'-6,000' meters (Cicero 1982). According to Vadim (1996) *Bicellonycha ornaticollis* (Blanchard) commonly occur in open marshy areas with *Bicellonycha lividipennis* (Motschulski); occasionally can be found in adjacent forests. Adults are active from October to February in Campinas (Brasil); males emit single yellowgreen flashes. Females are found on vegetation and occasionally are seen flying and flickering.

---Subfamily Amydetinae: With 12 genera. *Cladodes* from Chile, emits light only in the larval stage and adult females have some form of larvae that are walking on the ground or in rotting logs.

Many Lampyridae are collected by beating and sweeping vegetation during the day. Adult males were collected with ultraviolet light traps and pitfall traps "baited" with light emitting diodes set up to mimic the females advertising signal (Branham 2003). To attract *Microphotus* and *Pleotomus* males, Evans and Hogues (2006) described the next method: Green, red and yellow lights must be combined (emitting diodes, LEDs) on a piece of plywood. Arrange the LEDs so that they are on one line and with 6 inches apart with one another. Wire the LEDs in series and connect to an on/off switch.Two AA batteries are used as power source. Lights will turn on and off at random periods from 4 to 60 seconds. Evans and Hogues continued to explain: Males can be active for a relatively short time during the mating season (late spring through midsummer), so be sure to try this method at different times during spring and summer and throughout the night. Some species of *Photynus*, *Microphotus* are attracted to lights also.

OMETHIDAE: With global distribution. 8 genera with 32 species are known (Ramsdale in: Arnett et al. 2002). Lawrence et al. (2010) include the following subfamilies:

---Subfamily Driloniinae: With single genus *Drilonius* occur in the Eastern Palaearctic and Oriental regions from Japan to Java and Sumatra. Nothing is known about the biology of immature stages (Albert Allen, pers. comm. 2010).

---Subfamily Matheteinae: With two genera *Matheteus* occurs from central California to southWestern Oregon, and two species of *Ginglymocladus* occur in montane and coastal California. *Matheteus* adults have been collected on foliage, but nothing is known about their biology or immature stages. In Oregon, usually in late April, adults of *Matheteus theveneti* begin their activity which lasts between 10 to 15 days (Fender 1969).

---Subfamily Omethinae: Are restricted to the Nearctic and Eastern Palaearctic (Japan) regions. Nothing is known about their biology or immature stages. The females of several genera are also unknown

(Lawrence et al. 2002). *Blatchleya* and *Troglomethes* from North America are collected in the foliage during the day. *Symphyomethes* adults have been extracted from forest floor debris in forest soil. Adults emerge in the spring and summer and are probably all short-lived (Ramsdale in: Arnett et al. 2002).

All species of Omethidae are presumably univoltine (Ramsdale in: Arnett et al. 2002). Adults are active in spring and summer and appear to be short-lived. Nothing is known about their feeding habits (Albert Allen, pers. comm. 2010).

CANTHARIDAE: About 135 genera and 5,000 species are known world-wide (Lawrence 1982). According to Lawrence (2010) includes the following subfamilies:

---Subfamily Cantharinae occur primarily in the Holarctic and regions, extending to Southern Mexico, North Africa, India, and Sunda Archipelago in the Old World. Larvae are abundant in soil and litter, and are also either predaceous or phytophagous (A. Allen, pers. comm. 2010.

---Subfamily Silinae occur mainly in tropical regions, with most species in the Northern Hemisphere, with diminished numbers in South America, South India, Africa, and Northern Australia. Like most cantharids, adults occur on foliage or flowers and larvae are predators in leaf litter (A. Allen, pers. comm. 2010).

---Subfamily Dysmorphocerinae occur primarily in Southern Hemisphere, In the New World extend from Arizona to Chile, and in the Old World known from Southern Africa, Australia, New Guinea, and New Zealand. Like most cantharids, dysmorphocerines are diurnal and often swept foliage, while larvae occur in litter (A. Allen, pers. comm. 2010).

---Subfamily Malthininae occur from Canada to Brazil, throughout the Palaearctic and Oriental regions, and south to central and Eastern Africa, India, East Indies, with Caccodes to Hawaii, possible introduced *Malthodes* into New Zealand. Malthinini larvae are likely predaceous, and adults are often found on foliage in shaded situations (A. Allen, pers. comm. 2010).

---Subfamily Chauliognathinae: Occur throughout the New World, Australia and New Guinea. Adults are found during the day and are often gathered in large numbers on foliage and flowers, shrubs or trees (*Chauliognathus* spp), in the trunks of trees in forests, often on any prominent object. Breeding is the reason for this congregation (Webb 1986). Larvae are usually found in leaf litter or other accumulations of vegetable debris (A. Allen, pers. comm. 2010).

Soldier Beetles also can feed on nectar, pollen, and even fresh foliage (Lawrence & Britton 1994). Larvae are predators (Borror and Delong 1969) found in litter or fungi. Larvae pupate in earthen cells in the moist soil below the surface and emerge between spring and summer (LeSage 1991). Moreover, larvae can be active in autumn and winter in the field (Andrew 1996). Soldier beetles are usually brightly colored, and have a well-developed system of defense glands (Albert Allen, pers. comm. 2010).

Adults can be collected manually in the flowers, leaves and branches of shrubs, trees and plants in general. Many are caught by beating and sweeping vegetation, pitfall traps or with Malaise trap. Some species come to the lights.

INCERTAE SEDIS

The following families, subfamilies and genera are not classified in any particular family (sensu Lawrence et al. 2010):

PODABROCEPHALIDAE: Single genus *Podrabocephalus* from Southern India with biology unknown (Allen, pers.comm. 2010).

OTOTRETINAE: (Formerly in Lampyridae) (Branham and Wenzel 2001; Bocakova & Janisova 2010), molecular analyses of some elateroid taxa (Bocakova et al. 2007; Stanger-Hall et al. 2007) supported this view as well as the inclusion of the Ototretinae in the Lampyridae (Bocakova & Janisova 2010), but in this book are included as incertae sedis according to classification of Lawrence et al. (2010). They are known from Western North America, Eastern Palaearctic, Japan, and China to India, Southeast Asia, Philippines and East Indies. Little is known on the biology of ototretines. They are not thought to be luminous (Lawrence et al. 2000; Bocakova & Janisova 2010). Many are collected with by beating and sweeping vegetation during the day.

OTOTRETADRILINAE: Species of *Ototretadrilus* are known from India, Southeast Asia, Philippines, and East Indies. *Ototretadrilus* are known only from males and nothing is known about immature stages or biology (Albert Allen, pers. comm. 2010).

PTEROTINAE: (Formerly in Lampyridae) (Branham and Wenzel 2001) with Nearctic distribution, *Pterotus obscuripennis* LeConte occurs from Western Washington and Oregon to Northern Lower California; *Pterotus curticornis* Chemsak, is known from Riverise Co. in Southern California. Larval *Pterotus* feed primarily on slugs, while adults may not feed (Lawrence et al. 2000). *Pterotus* do not produce any light but are nocturnal (Evans and Hogues 2006). Larvae are found in moist areas such as humus and under rocks and they are predators. They can be collected by beating and sweeping vegetation during the day. *Pterotus* are attracted to lights.

CYDISTINAE: (Formerly in Rhagophthalmidae). *Cydistus* are known from Turkey, Syria and Iran. Only male are known and it is likely that females are larviform (Paulus 1972; Lawrence et al. 2000, 2010).

***Neocrowsonia viatoricus* Kistner & Abdel-Galil:** Also was placed by Lawrence et al. (2010) as incertae sedis (formerly a Throscidae genera) Distributed in Tropical Africa; adults have been collected in the nests of termite (*Microhodotermes viator* (Latreille)) in Cape Province, South Africa; immature stages are unknown (Kistner & Abdel-Galil 1986).

Stenocladius (15 species) formerly in Ototretinae (Lampyridae), genus from China, Japon, Vietnam. India, Nepal, (Ming-Luen et al. 2006) they have only plesiomorphic pheromone signals (Stanger-Hall et al. 2007). They are apterous and larviform (green 1912)

Harmatelia: Genus formally included in the subfamily Ototretlnae (Lampyridae), occurs only in Sri Lanka (Lawrence et al. 2000). Males have small, paired light organs on ventrites 1 to 8 and on the mesopleuron (Green1912), Females are apterous and larviform (Green 1912). Larvae are unknown (Lawrence et al. 2000).

Series DERODONTIFORMIA
Superfamily DERODONTOIDEA

DERODONTIDAE: 4 genera and 29 species are known (Háva 2006).

Larvae and adults usually live together, sometimes in large quantities, under bark and dead wood, feeding on the soft tissues of fresh or rotting mushroom bodies, and the products of their metabolism.

Lawrence et al. (2010) include the following subfamilies:

---Subfamily Peltasticinae: *Peltastica* spp. from Canada, Japan and USA, are specialized to eat the fermented sap under the bark. This fermentation is produced by bacteria and/or fungi which have been collected there (Háva 2006).

---Subfamily Derodontinae: *Derodontus spp.* from Paleartic and Neartic regions. They feed on fresh fruiting bodies of many species of Basidiomycetes fungi (Leschen 1994), also found in decomposed moss, "muddy" fungus and "watery" fungi. They are abundant during the colder periods of the year. Adults and larvae are commonly collected in the fall and spring and often they can live in frozen sporocarps.

---Subfamily Laricobiinae: *Laricobius* spp. from Paleartic and Neartic regions are predators attacking aphids of Adelgidae family found on logs, so they can control those plagues (Lawrence and Hlavac 1979). In *Nothoderodontus* are know two species from New Zealand, one species from Australia and Tasmania and three species from Chile (Lawrence 1985). *Nothoderodontus* are associated with sooty-mould fungi (Ascomycota), which form thick black "wefts" of mycelia of up to six different species (Hughes 1972: 226), generally on the surfaces of foliage, branches, tree trunks, and often surrounding rocks and soil (Gimmel et al. 2009). These growths are associated with sugary exudates of leaves or with the "honeydew" produced by sternorrhynchous Hemiptera, primarily in association with *Nothofagus*, but other plants as well (Klimaszewski & Watt 1997; Carlton & Leschen 2007; Gimmel et al. 2009) for example, *Nothoderodontus gourlayi* Crowson, from New Zealand, has been associated with sooty molds (Ascomycetes: Capnodiaceae) (Hava 2006).

Derodontidae are rare in collections and to catch them fungi and under the bark must be carefully examined. They can also be collected with a Winkler/Moczarski eclector and fumigating logs containing Ascomycetes fungi.

NOSODENDRIDAE: Family distributed worldwide, with a single genus (*Nosodendron*) and 68 known species (Háva 2009 in: *www.insects.tamu.edu*). They are placed in two genera. *Nosodendron* has 66 species (worldwide distribution) and *Nosotetocus* with 3 species from Colorado. According to Cannings and Scudder (2006) the genus *Orphimus* is placed in the Dermestidae by some entomologists, this family has 4 species with 3 in North America, including a Miocene fossil species. They live in rainforest areas, among mudflows and stale wounds of trees attacked by yeast. Probably they feed on bacteria, fungi and fermented products. The larvae of *Orphilus* burrow in dead wood (Cannings and Scudder 2006).

They can be collected with beating sheet placed under the tree trunks, spraying the trunks and with pitfall traps.

JACOBSONIIDAE: with 3 genera (*Sarothrias, Saphophagus* and *Derolathrus*) and 21 species is distributed in the Australo-Asian region, up to the Malagasy Archipelago and Madeira Islands,

Solomon Island, Fiji Islands, New Zealand, USA (Hawai, and Florida) and, Guadaloupe Barbados, Montserrat and Virgin islands in the Neotropical Region (Háva & Löbl 2005; Peck 2010). The small size of these beetles suggests that they might be capable of long distance dispersal, aided by air currents, and that the species might be widely distributed (Peck 2010). Adults and larvae are found in leaf litter, rotten wood, in fungus, in bat guano and under bark (Lawrence 1991; Klimaszewski and Watt 1997; Lawrence et al. 1999a). *Sarothrias* seems to be adapted to myrmecophily (Arnett et al. 2009). *Saphophagus* was collected (adults and larvae) in New Zealand, under fairly tight and rather sappy bark of dead trees, usually where the cambial region was dark stained, all probably feeding on fungi of the sphaerialean type (Wheeler and Blackwell1984). Jacobsoniidae species are probably more common and more widely distributed than is currently known, and it seems possible to discover new species (Peck 2010). According to Peck (2010) some *Derolathrus* species are found in guano in caves, others are collected in leaf litter (with Berlese extractions), and can also be collected with flight intercept traps (Philips et al. in: Arnett et al. 2002), sticky board traps (blue or yellow) in Subtropical and warm temperate forests and some *Derolathrus* are found in bracket fungus on trunk of large living *Sapindus* tree (Peck 2010).

Series BOSTRICHIOFORMIA
Superfamily BOSTRICHOIDEA (incl. Dermestoidea)

DERMESTIDAE

Andreas Herrmann

This family is distributed worldwide and somewhat 50 genera and 1,300 species are known so far (see the regarding catalogue of Jiri Háva from Czech Republic (www.insects.tamu.edu). They are devided into 6 subfamilies: **Dermestinae, Trinodinae, Megatominae, Orphilinae, Marioutinae and Thylodriinae** (Lawrence et al. 2010).

Some *Trogoderma* species are predators of wasp and bee larvae and spider eggs (Cannings and Scudder 2006). Very few species, mainly of the genus *Dermestes*, live in carrion in the last stages of decomposition, for example *Dermestes ater* De Geer, *Dermestes frischi* Kugelann and *Dermestes maculatus* DeGeer has been found in carrion in the Hawaiian Islands (Goff et al. 1986). A few species of *Dermestidae* attack collections of stuffed animals (e.g. *Anthrenus* and *Trogoderma,* seldom to *Attagenus*). *Dermestes maculatus* DeGeer is used in museums to clean dried tissue from bones. Some larvae are found in nests, leaf litter, branches and dead wood but nearly all larvae are bound to parts of animal origin. So if you find a larva e.g. of the genus *Globicornis* under bark, then it is searching for dead insects or spiders to feed on, or a place for pupation. Most genera are scavengers and feed on animal or dried plant material, such as skin and animal hair, feathers, dead insects and natural fibers. The adults of many species could be found in flowers feeding on pollen. Members of *Dermestes* spp. are found in animal carcasses or in the surrounding of insect assemblages. They do not occur in single dead insects directly but (a very few species) in insect nests where they most likely feed on died insects. Other Dermestidae occur in mammal burrows, nests of birds, bees or wasps. *Thaumaglossa* spp. breed in egg cocoons of the genus *Mantis* spp. Some species of Dermestidae can be found in debris on the floorboard or pollen stores of dead hives (Stedman 2006). Although feeding on vegetable matter is atypical for the family, some species, especially of the genera *Trogoderma* and *Anthrenus*, can complete their entire developmental cycle in stored grain and cereal (Hinton 1945). Tolerance to very low humidity enables many dermestids to infest various stored foods and raw materials, such as dried fish, meat, hides, etc. (Kiselyova and Mchugha 2006). Elsewhere they are valuable recyclers.

Many species of the genus *Thorictus* (most likely all of them) are myrmecophilous. Moreover, Hava and Lenoir (2008) collected *Thorictus martinezi* Hava & Lenoir over the heads of worker ants of the genus *Cataglyphis* and studying their behavior in the laboratory. They observed that the laboratory the beetles were maintained for several months at the head of a worker ant. Háva and Lenoir noted that these beetles clinging (with jaws) to an antenna of the ants and remain in the same for long periods. In nature, generally, these workers ants are in the nest (with the beetle on their heads), but can go to feeding sites with their load. Occasionally the ants tried to expel their cargo with the front legs. When beetles are free, they move at the site of foraging ants or in the same nest, or disappear for several days, probably hiding among debris from the nest (Háva and Lenoir 2008).

Some Dermestidae can be collected with pitfall traps baited with foul meat or dead animals (little mammals, birds, fishes respectively parts of them). It is also helpful to look inside the windows of houses or buildings in the spring. Other species are among the carpets inside the house. The species living among flowers can be collected with a sweeping net, flight intercept traps or by hand-collecting. They can be also collected by sweeping the flowers near beehives. Many adults come to the lights. Also, must be reviewed the carrion and the various habitats where they live. After collecting these beetles it is important to keep their scales and hair free of dirt that is conducted by immersing these beetles in a solution of detergent and warm water. When abdominal body fluids are present, they should be washed with 70% alcohol and dried with absorbent paper towel or tissue paper, carefully trying not to scrub off the hairs or scales, which are necessary for the correct identification of many species.

The cuticle of dermestids mostly is simply dark or brown, in some cases showing red or bright spots. The surface usually is covered with hairs or (rarely) is completely naked. In the genus *Anthrenus* the specimens have scales instead of hairs. The hairs respectively scales often form red, yellow or whitish spots or fasciae. Especially in males the antenna of nearly all species builds a club of enlarged segments, most genera have an ocellus between their eyes.

It is not easy to recognize all members of this family since the several genera look extremely different from each other, they don't have a "typical habit" like e.g. Elateridae, Carabidae or most other families have. An overview upon the world species is provided by my internet site http://www.dermestidae.com/Abbildungen.html, which shows colored photos of somewhat 700 world species of Dermestidae.

Andreas Herrmann
Bremervörder Strasse 123 - 21682 Stade-Germany (Deutschland)-Email: herrmann@coleopterologie.de
Web: http://www.dermestidae.com

ENDECATOMIDAE: Distributed in North America, with 4 known species in only 1 genus (http://insects.tamu.edu). Adults and larvae are found feeding on fungi growing on dead trunks of birchs (*Betula* sp.) (Crowson 1961; Zahradník 2006), Petr Zahradník 2006 cites the following species and their hosts:

- *Endecatomus lanatus* Lesne, from Siberia, Japan, Jilin (Lesne 1934). Host plant is *Phellinus* sp. and *Polyporus* sp., *Xantochrous radiatus* (Sow.).
- *Endecatomus rugosus* (Randall 1838:26) United States, Canada. Host is *Polyporus gilvus* Schwein.
- *Endecatomus reticulatus* Herbst (Europe) found in Birch bracket (*Polyporus*).
- *Endecatomus dorsalis* Mellié, from United States; Endecatomidae can be collected checking the mushrooms, loose bark, or fumigating logs.

BOSTRICHIDAE: Adults and larvae feed on Bracket fungi. But adults and larvae of bostrichid are principally hardwood borers (Beiriger and Robert 1996); they may also attack roots and seeds of various plants. To study the Bostrichidae must consult the work of Pierre Lesne (see bibliographic references), he and he alone, can be considered the father in the study of this family. The subfamilies are:
---Subfamily Dysidinae: One species of Southern Asia and Australasia (*Apoleon edax* Gorham) and a Neotropical genus with two species (*Dysides obscurus* Perty and *Dysides platensis* Fairmaire). *Dysides* sp. has been seen attacking wood dry and even furniture in Nicaragua (Maes & Ivie, un-

dated).

---Subfamily Polycaoninae: 2 genera and 28 species distributed primarily in the Americas, and some Asian species. *Melalgus confertus* (LeConte) attack many plants (branch and twig borer) weakened or injured stone fruit tree, especially prunes, are sometimes attacked. They bore into the dead wood of trees and shrubs outsides the orchard or brush piles (Strand 1999). According to Fall (2009) *Polycaon stoutii* (LeConte) is commun throughout maritime Southern California, once taken in numbres from beneath the loose bark of Eucalyptus in early summer. *Polycaon confertus* leConte is branch borer and, has been collected in grape vines (Fall 2009). Some species can be collected with light traps.

---Subfamily Bostrichinae: 51 genera with 376 species distributed throughout the world. They are xylophagous, attack dead wood, dry, furniture, buildings, etc. (Maes & Ivie, undated). Some host plants mentioned by Maes & Ivie (no date) are: Cotton remnants, Anacardiaceae (*Lanea, Mangifera, Parishia*), Bombacaceae (*Bombacopsis, Bombax*), Euphorbiaceae (*Manihot, Hura, Endospermum*), Lythraceae (*Lagerstroemia*), Malvaceae (*Kydia, Gossypium*), Meliaceae (*Azadirachta*), Mimosaceae (*Pithecellobium*), Myrtaceae (*Psidium*), Poaceae (*Zea*), Polygonaceae (*Coccoloba*), Sapotaceae (*Achras*), Burseraceae (*Boswellia, Canarium, Garuga*), Caesalpinaceae (*Cassia, Koompassia, Poinciana*), Combretaceae (*Terminalia, Anogeissus*), Dipterocarpaceae (*Anisoptera, Dipterocarpus, Parashorea, Shorea*), Fabaceae (*Dalbergia, Pterocarpus*), Fagaceae (*Quercus*), Rutaceae (*Citrus*) and several other plants.

---Subfamily Psoinae: 6 genera and 13 species found worldwide. Larvae of *Polycaon stoutii* (LeConte) pierce several hardwoods and adults can emerge from the polished table tops, showing their way by means of the damaged wood beneath the plate (White 1983).-

---Subfamily Dinoderinae: 47 species in 6 genera distributed worldwide. *Rhyzopertha*, *Dinoderus*, *Prostephanus* and *Apate* are found in grain and other stored products (like flour). Some species also attack wood (such as *Prosopis*, Anacardiaceae: *Spondias*, *Schinus),* Burseraceae (*Bursera),* Caesalpinaceae: *Delonix, etc.* (Maes & Ivie, undated). Other example is *Dinoderus minutus* (Fabricius), from Cuba, Dominica, Grenada, Guadeloupe, Hispaniola, Puerto Rico, St. Vincent. Old and New World tropics, adults and larvae bore into dry (not living) bamboo and canes, and the adults also bore into many kinds of timber, plants, and vegetable products (Peck 2006).

---Subfamily Lyctinae: 70 species in 11 genera with worldwide distribution. Each region has its own indigenous wildlife, but in almost every country there are many introduced species. Adults and larvae eat dry plant material or stored plant products; they make galleries in dead wood, rotten wood or dried plant material like roots of dry grass and dead wood. They are also associated with felled oaks. Adult cause more damage to the wood through the production of exit holes when they emerge. According to Rojas and Gallardo (2004) the wood damage appears as small holes in wood surfaces attacked, giving the appearance of having been riddled (shot of ammunition).

Lyctinae are rarely found in nature, but they can be collected by examining dead branches of live trees or recently dead. They are more frequently found in recent dry wood than on the old wood, so it is best to check the stored wood that has not been painted or varnished. *Lyctus caribeanus* Lesne, from Dominica, Guadeloupe, Hispaniola, Puerto Rico, Central America and Mexico, adults can be collected at lights and adults and larvae boring in wood of various trees (Peck 2006).

---Subfamily Euderiinae: With *Euderia squamosa* Broun from New Zealand. According to E. S. Gourlay (cited by Crowson 1961): They are found by beating *Nothofagus* branches (and under bark in some trees), but not on flowers, generally in the early part of the summer (chiefly November and early December). The larvae has been collected in indistinct galleries in moist or sappy bark of dead, or injured living *Nothofagus* trees one time in a daamged living trunk of Rimu (*Dacrydium cupressinum* Lamb.).

Theirs life cycle takes one or several years (Crowson 1961).

Bostrichidae are generally collected by beating vegetation. At night the fallen logs must be checked with a flashlight, since many nocturnal species use to walk among them. They can be collected with native trees sawdust trap, or pieces of wood. Usually are found boring in the twigs and smaller branches of Leguminosae. Some adults can also be captured with an aerial net when flying during the spring. Other species, which live within fungi or inside the timber, are so well hidden that it is better to transport these infected organic materials, raise the larvae at the laboratory and wait until adults emerge. Several species are caught with lights.

PTINIDAE: There are approximately 230 genera and more than 2,200 species known worldwide (Lawrence 1991, Lawrence & Viedma 1991), and the recently incorporated Anobiinae subfamily (sensu Lawrence et al. 2010, see below) together account for 3,700 species in 422 genera. With ten subfamilies: **Eucradinae**, **Ptininae**, **Dryophilinae**, **Ernobiinae**, **Anobiinae**, **Ptilininae**, **Alvarenganiellinae**, **Dorcatominae** and **Mesocoelopodinae**. Undoubtedly many species remain to be described (Philips and Bell 2009).

Many species are found in mills, warehouses, barns, shops, museums and hospitals where they feed, in the adult stages and in the larval (scavengers), on a wide variety of products, including grains, seeds and medicines. They are usually minor pests because their populations rarely grow too much. Others feed on manure accumulated in vertebrate nests and some live in mammal, bird and bees nests. Some taxa seem to be specialized on rodent or hyrax dung while others, such as Mezium, utilize bat guano in dry caves (Philips and Bell 2009). They are probably scavengers feeding on dry animal and vegetable material.

Larvae bore in bark, dry wood, twigs, vines, pods, seeds, woody fruits, galls, fungus, dead flower stalks. Sometimes in the young stems or shoots of growing trees (White 1962; Lawrence et al. 1999). According to Philips and Bell (2009) *"Several species of Ernobius and Dryophilus attack conifers; the former can also be found in cones. A large number of species in the genera Caenocara* Thomson *and Dorcatoma* Herbst *can be found in fungal fruiting bodies, including puff balls and woody shelf fungi."*

One species of Ptinus (from Australia) is reported to feed on arthropod remains and spider eggs in nests of a salticid spider (Hickman 1974).

Gnostus genus has three species, distributed from Bolivia and Brazil to central Florida (USA) and the Bahamas, and they are all associated with ants of the *Crematogaster* genus[(1)] (Lawrence and Reichardt 1966). Some species of Fabrasia from South America have been found together with ants of the *Camponotus* genus (Lawrence and Reichardt 1966). The *Gibbium psyllioides* (Czempinski, 1778) species live in many places where there are remains of plant or animal organic material, either inside the houses, barns, museums etc. (Coutin R. 1966).

The members of this family can generally be collected with Berlese funnel, Winkler/Moczarski eclector, Malaise trap and some species come to the lights. Philips and Bell (2009) explain the following methods to collect the Ptinidae: "Often, individuals are found by beating dead branches and twigs of various trees, particularly hardwoods, and sometimes flowering shrubs. Species of Dorcatominae subfamily can be found associated with fungi. The Ptininae and Gibbiinae also occasionally have been collected in relatively large numbers by baiting pitfalls with dung, carrion, or other substances. In rare instances, species can be numerous and will build up huge populations in areas of dense resources, as stored food products in a neglected household cupboard and in areas of accumulated dung, such as in caves or rock shelters".

Specific sampling methods for Anobiinae subfamily

Usually regarded as a family, but was recently included in Ptinidae by Lawrence et al. (2010). Worldwide with distribution and about 1,500 species are known (Schabel 2006) in 192 genera. Most larvae feed on dead or dying wood, eating the cones, seeds, galls or fungi (Booth et al. 1990). Larvae of most species bore into dead hardwoods and softwoods. Some species feed on the young stems or shoots of growing trees. Others are serious pests of stored products, including drugs, tobacco, seeds,

1. According to Philips and Bell (2009) "Within the Ptinidae. Most of these, known as the ectrephines, are found in Australia (Lawrence & Reichardt 1969) and include the genera *Diplocotes*, *Ectrephes*, *Enasiba*, and *Polyplocotes*. New World symphiles include *Coleoaethes*, *Fabrasia*, and *Gnostus* and several are associated with the ant genera *Camponotus* and *Crematogaster* (Lawrence & Reichardt 1966; Philips 1998 b).

spices, cereal products, leather and museum specimens (Arnet et al. 2003). Depending on species, the about 100 eggs are laid on wood or seed, from which the larvae bore round tunnels into the substrate (e.g. *Priobium*). The slightly curved, white grubs have short thoracic legs. There are species that attack the bark, and other species attack the wood, the latter left the wood turned into fine powder, and when the adults emerge they produce a large number of small holes, like those produced by the attack of Lyctinae (Bostrichidae) larvae, although general sawdust produced by the Anobiinae is of greater granulation and sandstone shaped (Rojas y Gallardo 2004). Species that attack the bark, they fill their galleries with small pellets, which can have different colors because they feed on bark and of some outer layers of xylem also (Rojas y Gallardo 2004). Pupae are found just under the surface of whatever substrate the larvae develop in (Schabel 2006).

They can be collected by sweeping and beating, with Berlese funnel, Winkler/Moczarski eclector, traps baited with pheromones or sawdust, and another quimical compoust, with cardboard traps and hand-collectig. In Paraguay, the author of this book has collected many species reviewing flowers and others were collected with UV light and mercury light, mainly on moonless night. He author has collected many different species also by beating and sweeping in all kinds of vegetation and with Malaise traps located in different habitats, both outside and inside the forest, even in very small and upset villages near secondary forests. These specimens were studied by Toskina (1993, 2000) but many others have not yet been studied and probably among them there are many new species for the science.

Series CUCUJIFORMIA

Superfamily LYMEXYLONIODEA

LYMEXYLIDAE (= Lymexilonidae):10 genera with 66 species are known (http://insects.tamu.edu), with 4 subfamilies: **Hylecoetinae**, **Lymexylinae**, **Melittommatinae** and **Atractocerinae** (Lawrence et al. 2010), distributed throughout the world. They live in decaying wood, on tree trunks or under bark (Young 2002). Some are pests to ship timber (Wheeler 1991). Adults of *Melittomma sericeum* (Harris), from Eastern United States, the chestnut timberworm, can be found in *Quercus* logs (Young 2002). According to Young (2002) adult females carry the fungal spores in a mycangia near the end of the ovipositor and deposit them in a sticky matrix with their eggs and when the larvae emerge, the first instar carry the fungal spores of the wood into on their bodies. Larvae of all the species are xylophagous, living in deciduous as well as coniferous trees and live in symbiotic association with fungi and microbes (Farkač et al. 2005).

They are probably short-lived and thus are not commonly encountered (Young 2002) and they can be collected with Malaise trap, by hand-collecting, The author of this book has collected a few specimens with beating vegetation and fumigating logs. Some come to the lights.

Superfamily CLEROIDEA

John M. Leavengood, Jr.

PHLOIOPHILIDAE: A single species is known from England, France and elsewhere in Europe (Hervé and Roge 2006). The species is widespread in Europe but it is rare. Adults and larvae of *Phloiophilus edwardsi* Stephens have been found associated with the *Phlebia* fungus, and are recorded from various localities in Britain (Crowson 1964). *Phloiophilus edwardsi* is found in winter, from October through early spring. They develop in orange comb fungus (*Phlebia merismoides* Fr.) in dry hardwoods and in the symptomatic red areas caused by the fungus (*Stereum rugosum* (Pers.)) in confiers and *Corticium quercicola* Jülich. It can also be collected in the spring on sunny surfaces of freshly cut branches. With the resurgence of fungi in April, larvae are more common than adults.

TROGOSSITIDAE: This family is represented by 53 genera and 374 known species of worldwide distribution. There are two subfamilies: Peltinae and Trogossitinae. They are predators of wood-boring insects of all life stages including the eggs. Some eat vegetable waste and fungi. The larvae of several species are found in conifer roots. Their larvae are predators of wood-boring beetles and live in galleries within the wood or can be micofagous. *Temnochila virescens* (Fabricius), a bark beetle predator from southeastern North America, was imported into Australia in 1981 as part of a biological control program for *Ips grandicollis* (Eichhoff), the five-spined engraver beetle (Lawson and Morgan 1993). In Europe *Seidlitzella procera* (Kraatz) can be found under the bark of black pine and *Temnochila caerulea* (Olivier) has been found in the early hours of the morning on logs and dry stumps of Spanish fir. *Nemosoma attenuatum* Van Dyke, from North America, has been previously described as an associate of twig beetles in the genus *Pityophthorus* (Curculionidae: Scolytinae) as well as other bark beetles (Nikitsky 1974). In India *Melambia crenicollis* (Guerin) has been observed preying upon the *Sinoxylon* (Bostrichidae) in galleries in dying sissu-trees. It was not found to be abundant (Stebbing 1914). *Calitys scabra* (Thunberg) is found under the bark of decaying logs in Canada (Eaton and Kaufman 2007). *Tenebroides mauritanicus* (Linnaeus) is often encountered in barns, warehouses and mills in cereal products, cocoa, nuts, seeds, and dried fruit, but does not cause significant damage. The adults and larvae are carnivorous, but the larvae may also feed on damaged grains (Bentancourt et al. 1999). *Lophocateres pusillus* (Klug) is a cosmopolitan species which breeds on amilaceous substances, cereals and dried fruits. *Tenebroides* are readily collected under bark, at lights and in light traps.

To collect trogossitid beetles, fronalin, Ipsdienol, Ipsenol, turpentine and/or Pinene can be used as attractants, which are available at stores where products against agricultural pests are sold, or in some laboratories where such chemicals are produced (Sánchez et al. 2008).

CHAETOSOMATIDAE: Three genera and 10 species are known. The two species of *Chaetosomodes* occur in New Zealand (Broun 1921). Among them is *Chaetosomodes halli* Broun, which is considered a pest of conifers (Commonwealth of Australia 2004).The genus *Malgassochaetus* is from Madagascar, with 8 known species (Menier 1991). Larvae are predators of beetles living in galleries within the wood. Chaetosomatid adults may be attracted with frontalin, Ipsenol, Ipsdienol or turpentine of pine.

METAXINIDAE: This New Zealand family is represented by only one known species: *Metaxina ornata* Broun (Kolibac 2004; Johnson et al. 2008). They are predators of margarodid scale insects *(Ultracoelostoma assimile* (Maskell)) and sooty mold on Southern beech (*Nothofagus solandri* (Hook.)) (Morales et al. 1988; Johnson et al. 2008; Gimmel et al. 2009).

CLERIDAE: Members of all subfamilies for which there are natural history accounts are known to be entomophagous as immatures and as adults, at least in part. Exceptions are adults which may use pollen as an alternative source of protein. Most entomophagous taxa are predacious on bark beetles and other wood-boring beetles.The larvae of many species from different subfamilies have been recorded as predators of Hymenoptera. Links to plant hosts are not strongly species-specific. Many species have been reared from numerous plant hosts.

Collecting clerids effectively generally requires finding trees infested by their prey species or careful rearing. Historically, clerids have been collected more by luck than by calculated decisions in locality or host. Malaise traps are particularly effective for capturing the locally common species of Clerinae. Hydnocerinae is perhaps the next most abundant subfamily in such traps, but much less so than Clerinae. With the exception of few tilline genera (e.g., *Cladiscus*), other clerid subfamilies are not consistently rendered from Malaise traps. However, Epiphloeinae would readily come to such traps if accompanied by or atop a pile of cut mahogany or oak limbs. Similar tactics utilizing volatilizing plant material are known to attract Cerambycidae, Scolytinae and Buprestidae. Likewise, it could be expected to attract their clerid predators. Other baited traps have not proven effective unless they employ frontalin, ipsenol or turpentine (or a combination of them), which attract some clerid species reliably.

Cleridae are most diverse in the tropics and the known biodiversity "hot spots". They demonstrate high endemism in southeastern Asia in the Wallacean islands and the surrounding regions. In North America the species with the most restricted distributions are typically known only from mountain chains in Arizona or the Florida keys. The same is expected for the less-studied Caribbean species. A significant number of species are known from Costa Rica, but limited study in adjacent countries makes comparisons of biodiversity difficult. Much is left to be learned regarding the Cleridae.

In order to identify clerids to genus or species one should seek literature by Barr (*Araeodontia, Neorthopleura, Aulicus*), Bartlett (*Apteropilo, Eleale*), Chapin (Philippines), Corporaal (*Stenocallimerus, Cladiscus*), Foster (*Trichodes*), Gerstmeier (*Trichodes, Olesterus, Diplocladus, Strotocera, Xenorthrius, Tillicera, Omadius, Tillus, Tilloidea, Falsotillus, Flabellotilloidea* and genera related to *Clerus*), Horn (*Cymatodera*), Kim (Korea), Knull (northeastern United States), Kolibáč (Thaneroclerinae, *Thanasimus, Korynetes,* Australia), Lea (*Lemidia*), Leavengood (southeastern United States, Hydnocerinae*), Mawdsley (*Omadius, Trogodendron, Aphelochroa, Orthrius, Clerus, Gastrocentrum*), Menier (*Evenoclerus*), Opitz (= Ekis) (*Colyphus, Perilypus, Dologenitus,* Korynetinae*, Epiphloeinae, Enopliinae*, North American genera), Rifkind (*Cymatoderella, Enoclerus*), Solervicens (*Eurymetopum, Silviella, Natalis, Epiclines, Eurymetomorphon, Notocymatodera,* Chile), Winkler (Afghanistan, Saudi Arabia, Cleropiestinae, Dieropsinae), and Wolcott (*Chariessa,* Costa Rica, Hydnocerine genera). The asterisked (*) taxa imply that ongoing comprehensive work is being conducted. Diagnostic literature of the aforementioned authors is not included in the already exhaustive bibliography.

Cleridae includes the following Subfamilies

---Subfamily Anthicoclerinae, (6 species): The two genera, *Allonyx* and *Anthicoclerus*, of this subfamily are distributed from Europe to southeastern Asia.

---Subfamily Clerinae, (~1600 species, including Cleropiestinae and Dieropsinae): Clerinae occur on all continents except for Antarctica. In the New World, over 200 described species of *Enoclerus* are predacious on bark beetles, weevils and other wood-boring beetles. They have been associated with pine, spruce, elm, cedar, white oak, hickory, sweet gum, maple, peach and other hardwoods infested with various genera of Scolytinae. The larvae typically hunt larval wood-borers inside galleries while adult *Enoclerus* remain on the bark surface feeding on adult scolytids. These beetles may be observed moving along branches of vegetation with movement patterns similar to ants, particularly those of the genus *Camponotus* (Hymenoptera: Formicidae) (Rifkind 1997: 298). In fact, most Clerinae exhibit aposematic coloration and mimicry of velvet ants, *Myrmex* weevils, various ants, Aradidae and fireflies, to name a few. While clerines are most commonly associated with wood-boring beetles, some species rely on termites or wood-boring moth larvae for sustenance. *Trichodes* and *Eleale* are omnivorous genera frequently encountered feeding on pollen (Foster 1976: 65). *Trichodes* and *Scrobiger* have demonstrated apivory (Knull 1951: 300; Foster 1976: 66; Bartlett 2009: 79). *Thanasimus dubius* (Fabricius) has been collected in great numbers by using frontalin, the sex pheromone of the Southern pine beetle, *Dendroctonus frontalis* Zimmermann. *Placopterus thoracicus* (Olivier) has been reared from various nest cells of various Sphecidae and from the galls of Cynipidae (Foster & Barr 1972: 123; Eliason & Potter 2000: 555). The immatures of some *Trichodes* complete their development on the egg pods of grasshoppers or within the cells of bees and aculeate wasps (Opitz 2002: 268; Knull 1951: 271).

---Subfamily Enopliinae, (~550 species): Enopliinae occur on all continents except for Antarctica. The North American firefly-mimicking *Chariessa pilosa* Forster has been recorded from a wide range of trees infested with Cerambycidae, Buprestidae, Curculionidae and Scolytinae (Hoffman, 1940: 56; Knull, 1951: 307). After infiltrating the galleries of its prey species the larvae exploit the prey species' pupal cell, as it does not construct one of its own (Bøving and Champlain 1920: 638). *Neorthopleura* and *Cregya* species have similar biology and have been collected en masse in South America. Many members of this subfamily mimic bark (e.g., *Cregya quadrinotata* (Spinola) and *Pelonium leucophaeum* (Klug)). *Pelonium leucophaeum* (Klug) has been reared from cypress and juniper trees infested with Cerambycidae (Bøving and Champlain 1920: 638; Opitz 2002: 279). Beetles of this subfamily are often collected at lights.

---Subfamily Epiphloeinae, (~300 species): This subfamily includes mimics of bark and fireflies restricted to New World distribution with very few temperate species. In North America, *Madoniella dislocatus* (Say) is commonly found in the galleries of many wood-boring insects. Their most common prey are bark beetles associated with various hardwoods (Opitz 2002: 278). Adults are diurnal and can be found on dead branches of trees, especially on oak. They have been recorded from trees infested with Cerambycidae, Zopheridae (Colydiinae), Buprestidae, and, most of all, Scolytinae (Chapin 1917: 29; Knull 1951: 302; Mawdsley 1999: 44). *Madoniella* is a speciose, cryptically-colored genus distributed throughout the Western Hemisphere. *Pyticeroides*, *Acanthocollis* and *Ichnea,* to name a few, are largely represented by mimics of Lycidae and Lampyridae.

---Subfamily Hydnocerinae, (~500 species): This group is composed of generalized ant mimics which occur throughout the world except for Europe and Antarctica. Testaceous members of *Isohydnocera* have been compared to wasp ants (*Pseudomyrmex*). *Isohydnocera* and *Phyllobaenus* span the Western Hemisphere. They have been reared from the galls of Tenthredinidae, Cynipidae and Gelechiidae (Sabrosky 1934: 65; Knull 1951: 284; Eliason and Potter 2000: 555), and from plants infested with Cephidae, Curculionidae, Noctuidae, Scolytinae, Cerambycidae, Buprestidae and Tortricidae (Eikenbary and Fox, 1968: 1219; Yearian and Warren, 1964: 263; Bøving and Champlain, 1920: 636; Pechuman, 1937: 9; Knull 1928: 314, 1951: 280). There are even records of adults emerging from nest cells of Megachilidae. A particularly interesting observation was made regarding two currently undescibed Costa Rican species. The larvae occupy *Piper cenocladum* (Piperaceae), in which they feed on the food bodies that the plant produces. The larvae then feed on adult and larval *Pheidole bicornis* Forel (Hymenoptera: Formicidae). It is believed that they chemically mimic their ant prey (Letourneau 1990: 215; Letourneau & Dyer 1998: 162; Dyer and Letourneau 1999: 5072, 2003: 62; Letourneau et al. 2004: 2145). The Andes restrict *Eurymetopum* and *Silviella* to Chile and Argentina. The Australian genus *Lemidia* has been collected from *Eucalyptus* blossoms and *L. obliquefasciata* (Gorham) was reared from *Brachyscelis strombylosa* Tepper (Hemiptera: Eriococcidae) (Lea 1901: 484). Adults have also been observed feeding on Psyllidae on *Acacia* and larval Sesiidae (Bartlett 2009: 226). The African and Asian genera are very poorly understood regarding biology. About as many species are known from Madagascar (e.g., *Evenoclerus*) as from mainland Africa (primarily *Cephaloclerus* spp.).

---Subfamily Korynetinae, (78 species): Perhaps the best known species, *Necrobia rufipes* DeGeer is associated with rotting meats, fish and cheese, cured ham and bacon, mammal tissue, guano, bone meal, coconut, garlic, grains, silk, cotton, rattan and salt. However, it does not feed on many of these materials. It instead feeds on other pests that infest products or decaying animal matter (Mallis 1997: 680; Peck and Thomas 1998: 86). It is a known predator of the cheese skipper, *Piophila casei* (Linnaeus) (Diptera: Piophilidae), and blowfly and dermestid larvae (Simmons and Ellington 1925: 845). Most members of Korynetinae are associated with decaying animal corpses. *Corynetinus* and *Prosymnus* are associated with cow dung. While the species of *Necrobia* are generally cosmopolitan, other groups are distributed as the ungulate mammals to which they are associated.

---Subfamily Tarsosteninae, (70 species): Distributed throughout the world, little is known of the species of this group beyond the cosmopolitan *Tarsostenus univittatus* (Rossi). This species has been recorded from a wide range of plants and furniture infested with Bostrichidae, particularly species of *Lyctus*.

---Subfamily Thaneroclerinae, (31 species, Thanerocleridae *sensu* Kolibáč): Members of this subfamily are often associated with stored products and their pests, on which they are predacious. The cosmopolitan species *Thaneroclerus buquet* Lefebvre has been encountered in rice, tea, coffee, tobacco, spices and drugs infested with *Araecerus fasciculatus* DeGeer (Anthribidae), *Lasioderma serricorne* Fabricius (Anobiinae), *Tribolium ferrugineum* Fabricius (Tenebrionidae), and *Sitophilus oryzae* (Linnaeus) (Curculionidae) (Corporaal 1939: 350; Knull, 1951: 286). *Ababa tantilla* (LeConte) has been encountered in tobacco bales from the Philippines and other goods in Brazil, Mexico and the United States (Corporaal 1939: 358). Adults of *A. tantilla* have been reared from touchwood (Polyporaceae: *Fomes igniarius* (Linnaeus)) in Texas (Knull 1951: 287). In Florida *A. tantilla* has been collected on *Ilex vomitoria* Aiton (Aquifoliaceae) and reared from the shelf fungus *Phellinus* sp. (Polyporaceae) inhabited by Ciidae. *Allothaneroclerus* and *Neoclerus* have also been associated with tree fungi. Members of *Zenodosus* and presumably other genera as well, overwinter as adults under the bark of scolytid-infested spruce and oak trees.

---Subfamily Tillinae, (~500 species): The Tillinae includes 72 genera which are strongly biased to Africa and poorly represented in South America and Australia. *Callotillus eburneocinctus* Wolcott, a species precinctive to the Florida Keys, has been reared from buttonwood and poisonwood (Foster and Barr 1972: 123). In India *Tillus succinctus* Spinola has been reared from bamboo infested with Bostrichidae (Gardner, 1937: 34). In North America *Cymatodera bicolor* (Say) has been reared from Florida dogwood, among other tree species, infested with Cerambycidae, Anobiinae and Scolytinae (Bøving and Champlain 1920: 626; Knull 1934: 207, 1951: 277). *Cymatodera undulata* (Say) exhibits similar behavior, but has also been observed feeding on larval Chalcidoidea in the galls of Cynipidae (Eliason and Potter 2000: 555; Balduf 1926: 29). *Lecontella brunnea* (Spinola) has been reared from nest cells of Vespidae, Sphecidae, Eumenidae and Megachilidae (Foster and Barr 1972: 123).

ACANTHOCNEMIDAE: This family is represented by only one species, *Acanthocnemus nigricans* (Hope 1845), which occurs in Australia (including Tasmania) and New Caledonia, and has been introduced to Burma, Thailand, India, Madagascar, New Guinea, Zambia, Algeria, Cyprus, Corsica, Sardinia, Sicily, Portugal and Spain (Alonso-Zarazaga et al. 2003; Lawrence and Britton 1994). This species has a special organ in the first thoracic segment which apparently is used to detect forest fires (Schmitz et al. 2002). Adults and larvae may feed on some ascomycete fungi (growing in the burned wood and soil) which appear after a fire (Vrålstad et al. 1998; Wikars 1992; Muona and Rutanen 1994). Burn sites probably serve as breeding grounds and, presumably after copulation, females lay their eggs in freshly burnt wood or soil (Schmitz et al. 2002).

MAURONISCIDAE: Represented by five genera and 29 species distributed from North America to Chile and Argentina (Majer 1994th meetings 1995; Lawrence et al. 1999). The known genera are *Amecomycter* (from Argentina, Chile and Perú), *Mauroniscus* (from Argentina, Boliva, Peru and Ecuador), *Mecomycter* (from California), *Mectemyctor* (from California) and *Scuromanius* (from California, Arizona, Texas, Mexico and Colombia). In North America they are found in late summer or early fall on desert flowers. *Mecomycter majeri* Howell has been found in flowers of *Encelia.*

PHYCOSECIDAE: Represented by a single genus *(Phycosecis)* with seven species: *Phycosecis litoralis*Pascoe, 1875, *Phycosecis algarum* Pascoe, 1875, *Phycosecis ammophilus* Lea, 1899, *Phycosecis atomaria* Pascoe, 1875, *Phycosecis discoidea* Pascoe, 1875, *Phycosecis hilli* Lea, 1921 and *Phycosecis limbata* Fabricius, 1781. The wingless adults and larvae are found in sand dunes along the coast of Australia, New Caledonia, New Zealand, and Vanuatu. According to Hangay and Zborowski (2010) they can be found on carcasses (particularly of birds) along the shores of the sea. Presumably they are predaceous and probably do not feed on carrion but on the other small organisms visiting the carcass (Jäch 1998; Hangay and Zborowski 2010). For example, *Phycosecis atomaria* Pascoe run over the surface of the sand during the day feeding on carrion and invertebrate corpses (Harris 1970a, 1970b). Adults and larvae dig galleries near their food source. Pitfall traps and sand sifting are suggested means to collect these beetles. Alternatively, shoveled sand may be submerged in water and manipulated until the beetles float to the surface.

PRIONOCERIDAE: Distributed mainly throughout the warmer regions of the world: Africa (except Madagascar), Saudi Arabia, Israel, India, Laos, China, Sumatra, Java, Micronesia, Japan, Philippines, and Australia. This family has 158 known species in the genera *Prionocerus, Idgie and Lobonyx.*

Two species of *Prionocerus* are known: one from India and southeastern Asia and the other extending to tropical Africa and Asia. The genus *Idgie* occurs throughout Africa and from Asia Minor, and across Southern Asia to Japan, the Philippines, and the Sunda Archipelago.

The genus *Lobonyx* l (16 species) occurs from Southern Europe and Northern Africa to Asia. Adults of *Lobonyx gracilis* Reitter (Algeria, Morocco and Spain) are diurnal pollen-feeders (Crowson 1964; Bahillo & Colón 2003). They are active from early April until late June (Bahillo & Colón 2003). This species has been collected on flowers of various *Cistus* species, and larvae develop in *Quercus* (Crowson 1964). The species *Lobonyx aeneus* (Fabricius) from Algeria, Morocco, Spain and Portugal, has been collected in May on *Asphodelus* and flowers of *Cistus* (Graells 1842; Bahillo de la Puebla & Colón 2003). *Lobonyx ruficollis* Raffray (in Spain) were collected in May on *Asphodelus* (Ignacio and Sánchez-Ruiz 2002). Larvae of many species are found in soil, leaf litter and under bark. The gut contents of a larva from Sri Lanka had numerous insect fragments. In Spain, a larva of *Lobonyx aeneus* (Fabricius) was found in "a pupal cell at the base of an oak tree (Lawrence et al. 2008). These authors also indicate that in India a larva of *Idgie* was eating a dead wasp and refer that these larvae have been seen attacking dead pupae and live larvae of Pyralidae.

Therefore, presumably adult Prionoceridae, along with some Melyridae, are pollenophagous and larvae are predaceous or saprophagous (Gardner 1929; Crowson 1964; Constantin 1990). Members of the family may be collected by sweeping and beating vegetation or using Malaise traps.

MELYRIDAE: Melyridae is a large family of worldwide distribution which comprises five subfamilies (Lawrence & Newton 1995), 289 genera and over 5,800 species, which are most abundant in dry, temperate regions. According to Lawrence et al. (2010) this family include four subfamilies: **Rhadalinae**, **Melyrinae**, **Dasytinae** and **Malachiinae**. One of these subfamilies (Malachiinae) contains several species which are found along sea coasts: the North American *Endeodes* spp. (wingless species, with very short elytra) and some members of the widely distributed genus *Laius* Guérin-Méneville. *Endeodes* are reported to occur "in rock crevices and along sandy beaches of the Pacific coast" (Jach 1998). Jach (1998) collected the type-series of *Laius ambonensis* Wittmer in large numbers on a stranded tree on a sandy beach of Pombo Island (Moluccas, Indonesia). Adults are mostly polyphagous, are found in large numbers on cone-bearing and flowering plants or foliage and feed preferentially on pollen. Many larvae are scavengers and live in sandy beaches while others live under bark or in litter and soil. Larval Malachiinae can be omnivores, scavengers or predators (Arnett et al. 2002). *Collops* and *Malachius* can feed on pollen, nectar and small arthropods (Evans and Hogues 2006). Many species of *Collops*, a large-bodied and attractive genus with about 70 New World species, have commonly been associated with agricultural pests, particularly in fields of alfalfa, sorghum and cotton. *Malachius aeneus* (Linnaeus) is a pest of economic importance. Several species are found near 4,800 meters above sea level in the Andes. They may be collected manually, with pitfall traps, Malaise traps, Berlese funnels, Winkler traps, beating and sweeping vegetation. A few species come close to lights.

The species of *Laius* known in Micronesia live on the shores of the sea, at or below the high tide, and can be seen running on sunny days (Bryant, In Champion 1921). The same author cites the comments made by Mr. Semper in Mindanao: "Lives in cracks and holes of sandstone that reach pretty far out into the sea, quite covered during the flood-tide but dry during the ebb." Bryant (IN: Champion 1921) and Wittmer (1958) noticed the same behavior in *Laius palitus*, from Madagascar, Indo-Australian region, Japan and Zanzibar.

John M. Leavengood, Jr.
University of Kentucky
Department of Entomology
S-225 Agricultural Science Center Building – North
Lexington, KY 40546-0091
Email: tokay@ufl.edu, tokaytokay@uky.edu
This contribution represents the Kentucky Agricultural Experiment Station,
College of Agriculture,
Publication no. 10-08-099.

Superfamily CUCUJOIDEA

BOGANIIDAE: Includes 5 genera with 11 species (www.insects.tamu.edu). According to Lawrence et al. (2010) there are known two subfamilies:

---Subfamily Paracucujinae: *Metacucujus goodei* Endrody-Younga, from South Africa, they are associated with cycads (*Encephalartos natalensis* Dyer & Verd) and *Metacucujus transvenosi* Endrody-Younga, are associated with cycads too (*Encephalartos transvenosus* Stapf & Burtt); another species such as *Metacucujus encephalarti* Endrody Younga (from Australia) are found in flowers of *Encephalartos lanatus* Stapf & Burtt (Jolivet and Verma 2005). *Paracucujus rostratus* Sen Gupta & Crowson from Australia are associated with angiosperms (Sen Gupta & Crowson 1966).

---Subfamily Boganiinae: Two genera from South Africa and Namibia and three genera of Australia and Tasmania (Endrody-Young and Crowson 1986). They feed on cycads flowers and *Eucalyptus* spp (Crowson 1990). *Boganium armstrongi* Sen Gupta & Crowson (from Tasmania) and *Afroboganium* (5 spp. from Namibia and South Africa) are associated with angiosperms (Sen Gupta & Crowson 1966; Endrödy-Younga & Crowson 1986) To collect them a sweeping bag can be passed on vegetation where they live.

BYTURIDAE: With 7 genera and about 25 species (www.insects.tamu.edu). In two subfamilies: **Platydascillinae** and **Byturinae** (Lawrence et al. 2010). They are known from the Palearctic Region, North and Central America. Adults and larvae feed on pollen, fruits and leaves, plants of genus *Rubus* and Rosaceae, including apples, blackberry and loganberry. *Byturus tomentosus* (De Geer), in Europe is a serious raspberry plague (Gordon et al. 1997). Other species attack graoevines and a wide range of other plants with serious damage to the fruit; more rarely on hawthorn (*Crataegus*), apple, pear, strawberry. For example, adults of *Byturus urbanus* (Lindeman) feed on raspberry and blackberry (*Rubus* sp.) and larvae gnaw the base of the receptacle, dig sinuous galleries in the young fruit and eat the drupes. These beetles are one of the most important pests of raspberry in shady locations in Europe. Larvae of *Byturus unicolor* Say are known to develop in the fruit heads of species of *Rubus* and *Geum*, but adults are collected on a wide variety of outher flowers (Goodrich 2002). *Xenasia grisescens* (Jayne) can be collected by beating oak branches and reared from oak galls and from Montery cypress (Cupressus macrocarpa hartw) (Goodrich in: Arnett and Thomas 2002). Byturidae generally can be collected by beating host plants or manually, removing them from the fruit.

HELOTIDAE: Distributed in Africa, South Asia, Sumatra, Java, Borneo and Japan. There are 102 known species (Wegrzynowicz 2000; Lee 2007) in 5 genera: *Helotay, Neohelota, Afrohelotina, Metahelotella* and *Strophohelota* (Kirejtshuk 2000). Some species are very rare in collections. There is very little information on the biology of these family members (Lee and Satō 2006). According to Oliff (1883) and Fukuda (1943) adults and larvae of *Helota gemmata* Gorham feed on sap exuding from trees. However, in Taywan, Lee and Satō (2006) report that when they used traps baited with decomposing fruit they could only collected *Helota thoracica* Ritsema. *Neohelota* are associated with ants (Lee 2007) however a good method to collect them is by sweeping flowers (Lee and Satō 2006).

PROTOCUCUJIDAE: Distributed in Australia (3 spp.) and Chile (4 spp.) (Slipinski 1998). It has a unique genus: *Ericmodes (= Protocucujus)* (Slipinski 1998)). Phytophagous habits, adults and larvae take place in the mould fungi under the bark or in the wood of decayed trees. In Chile they have been found attacking the shrub *Ulex europaeus* Linnaeus (Osorio and Cerda 1984).

SPHINDIDAE: Worldwide distribution. 9 genera and 59 species are known (www.insects.tamu.edu).

According to Lawrence et al. (2010), have four subfamilies **Protosphindinae**, **Odontosphindinae**, **Sphindiphorinae** and **Sphindinae**. Larvae and adults are mycophagous, restricted exclusively to slime molds and fungi found on tree trunks. They develop in the sporophyte of slime molds (myxomycetes). They are usually included in lists of saproxylic fauna. Among some species (in Valdivia, Chilean forest) *Protosphindus bellus* Burakowski & Slipinski (Protosphindinae) adults are found under bark of *Nothofagus*, decaying wood and litter (Grez and Elgueta 2003). In Europe, *Aspidophporus* and *Sphindus* are related to lower fungi (also with mold and dust). Sphindidae are generally collected with pitfall traps, Malaise, Berlese funnel, Winkler/Moczarski eclector and lights. Other collection methods are explained in the article "collecting in fungi", Chapter III.

BIPHYLLIDAE: Worldwide distribution, 7 genera and 197 species are known (www.insects.tamu.edu). Larvae and adults are found under bark and feed on organic materials resulting from the decomposition of wood, or they feed ascomycetous fungal spores living on the surface or under the cortex (Lohse & Lucht 1992). In Europe, adults and larvae of *Daldinia* genus are found in branches of deciduous trees.

They can be collected by sifted from dead leaves at base of trees or from organic material, or fumigating logs, looking under dead bark, on fungi, also using a Berlese funnel, a Winkler/Moczarski eclector or by hand-collecting with a brush dipped in alcohol. Other collection methods are explained in the article "Beetles in fungi", Chapter III.

EROTYLIDAE: Worldwide distribution. They are 3,200 species known and more than 280 genera (Wêgrzynowicz 2002). Leschen (2003) proposes six subfamilies: **Erotylinae**, **Cryptophilinae**, **Languriinae**, **Paraxonothinae**, **Loberinae** and **Xenoscelinae**.

The lifes of almost all species, both adults and larvae, is linked to large fructifications of the basidiomycetous fungi, which usually bear fruit on diseased trees or dead wood (Wegrzynowicz 2002). Some species feed on live plants, others feed on live or dead plant tissue, and others eat decaying plant debris. They can be found among the leaf litter or in decaying wood under the bark. They can be polinivorous, or exceptionally, eating animal products. Many are pests of stored food products and a few species *(Loberopsyllus* sp.) live associated with other animals such as rodents. *Macrophagus* spp. has been found in bee nests. *Lepidotoramus* sp. develops on butterfly cocoons (Leschen 2003).

About some subfamilies:

---Subfamily Erotilinae: included the largest amount of species and they all seem to be exclusively associated with basidiomycete fungi that grow on dead or diseased tissues of plants.

---Subfamily Languriinae: It is included as a subfamily of Erotylidae (Crotch 1873; Chapuis 1876; Roberts 1958; Wegrzynowicz 2002), worldwide distribution. Here follows the classification of Leschen (2003) who divided the subfamily into three tribes: Thallisellini, Languriini, and Hapalipini. All its members live on living plants and feed on decaying plant tissue and fungal hyphae rather than on living plants (Lawrence and Vaurie 1983). Larvae feed on stems of angiosperms. Hapalipini are exclusively phytophagous and they were found in *Opuntia* spp., tree ferns and palms. *Languria mozardi* Latreille, native to Northern Alabama, this species attacks many different plants, mainly rapeseed *(Brassica napus* Linnaeus) and it is considered a serious pest (Ward et al. 2007). Some *Anadastus* are also pests of economic importance. Others eat pollen (Wildermuth and Gates 1920) or fresh flowers (Xenoscelinae), feed on basidiomycetes fungi and the fallen bodies of fungi. Larvae feed on the stems of a variety of plants, particularly Compositae and Legumimosae (Vaurie 1948).

Pharaxonotha is a primitive genus with a very broad distribution comprising Asia, Africa and America. *Pharaxonotha kirschii* Reitter has almost worldwide distribution as a minor pest of stored grains and tubers. Many species of *Pharaxontha* are associated with cicadales species. This genus is considered the most important cycads pollinator. They are associated with *Ceratozamia, Dioon, Microcycas* and *Zamia* (Chaves & Genaro 2005; Leschen 2003). Many languriids adults can be collected on the leaves of angiosperms, by beating and sweeping vegetation, in the dense foliage which lies on the banks of the forest, in trees and shrubs. According to Ulke (1902) in North America *Languria* spp. *"are more frequent in swampy meadows."*

---Subfamily Xenoscelinae: With 9 genera (http://insects.tamu.edu). According to Sen Gupta (1968) this genus is widely distributed in both the New and Old Worlds, *Hapalips filum* Reitter has been recorded from Cuba in corn (maize) stalks, and *Hapalips annulosus* Grouvelle from Guadeloupe in flowers of the cactus (*Cereus triangularis* (Linnaeus)); *Hapalips prolixus* (Sharp) has been recorded from New Zealand in tree ferns, and has been collected both larvae and adults of this species under the sheathing bases of dead palm leaves; Also, according to Sen Gupta (1968) *Hapalips championi* (Grouvelle) and *Hapalips scotti* Grouvelle are recorded from the Seychelles, both from the bases of palm leaves. This genus is widely distributed in both the New and Old Worlds, mainly in tropical and subtropical regions; a few species extend into warm temperate areas. No species are yet recorded from Australia, and onlv one species. And *Xenoscelis* adults occur under leaf bases of palms (Sen Gupta 1968). Adults and larvae of *Bolerus angulosus* (Arrow) have been found on a lichen-covered rock in a cave in Malaya (Sen Gupta 1968).

Adults of Erotylidae are often found on the host fungus in large numbers with the larvae. However, adults are also found away from the host (Skelley 1999) but when the conditions are unfavorable for the host fungi to fruit, erotylids adults often congregate under bark or in other hiding places (Skelley 1999).

The best method to collect Erotylidae is to review the soft or hard mushrooms, including its vicinity and do it mainly at night. The trunks containing mushrooms can also be fumigated. It is preferable not to destroy the mushrooms while checking them, so more specimens of Erotylidae can be captured later (pers. observ.). Also check the loose bark of stumps, fallen trees or alive and the hollows of the wood. Many are collected by beating vegetation and with Malaise trap (Goodrich 1997). A few species come close to the lights. Skelley (1999) cited the next fungi record for some genera: *Megalodacne* spp., found in the harder bracket fungi (*Ganoderma* spp.) found on dead trees and stumps. *Triplax* spp.: On oyster mushrooms (Pleurotus spp.) which grow on dead logs. *Tritoma* spp.: On mushrooms growing from dead roots and logs; *Pseudischyrus* and other *Tritoma* spp.: Feed on fungi that have mycorrhyzal associations with living tree roots. According to Skelley these mushrooms include brittlegill mushrooms (*Russula* spp.) and deathcap mushrooms (*Amanita* spp.). Hawkeswood (2003) has collected *Thallis erichsoni* Crotch from the undersurface of a whitish *Pseudotrametes gibbosa* (Pers) Fries (Polyporaceae) growing on a fallen branch of *Eucalyptus propinqua* Deane & Maiden at Rungli Downs, via Helidon, South-Eastern Queensland (Australia).

Many methods can be used to collect erotylids in the fungi. Also the reader should study the article on "Beetles in fungi", Chapter III.

MONOTOMIDAE

Carlos Aguilar Julio and Josef Jelínek

This family comprises 33 genera and ca.246 species (www.insects.tamu.edu). Bousquet (2010) divides it in two subfamilies: **Rhizophaginae** (*Rhizophagus* only) and **Monotominae**. The latter subfamily is divided into four tribes: Monotomini (*Monotoma*), Thionini (genera *Thione* and *Shoguna*), Lenacini (*Lenax*) and Europini (containing the rest of the family), with worldwide distribution and mainly terrestrial species.

Larvae of Monotomidae are associated with fruiting bodies of Pyrenomycetes fungi (Никитский 1986) or decomposed plant material, living under bark, in association with bark beetles. Some Larvae may be predators (Никитский 1986).

---Subfamily Monotominae

------Tribu Monotomi*:* species of *Monotoma* are mycosaprophagous, living in various decaying vegetable matters, but species of the subgenus *Gyrocecis* live as commensals in the nests of ants. *Monotoma longicollis* Gyllenhal (from Europe, US, Canada and New Zealand) is occasionally found in association with stored products and *Monotoma productum LeConte (from New Brunswick, Florida)* are found under beach wrack (Bousquet 1990, 2002). *Monotoma bicolor* Villa & Villa (Cosmopolitan species) is found in decaying vegetable matter such as grass piles and barnyard litter (Bousquet and Laplante 1999). *Monotoma picipes* Herbst (Cosmopolitan) has been found in carrion in the Hawaiian Islands (Goff et al. 1986).

------Tribu Europini

Europs marinus (LeConte) are under driftwood in the upper intertidal area of the California coast (McCafferty 1981; Bousquet 2010). In U.S.A. *Pycnotomina cavicolle* (Horn) can be collected in large numbers in forests with molasses traps (White 1983). *Crowsonius parensis* Pakaluk & Slipinski and *Crowsonius similis* Pakaluk & Slipinski (both from Brazil) and *Crowsonius meliponae* Pakaluk & Slipinski (Costa Rica) lives in the nests of the bees such as *Trigona sp.* (Pakaluk and Slipinski 1993; Bousquet 2010). According to Bousquet (. 2010) many species live either under bark of trees (e.g.Europs, Malinica,Monotomopsis, Tarunius, Renuka, Leptipsius and others), other feed on decaying vegetal substrates or in fungi (e.g. Bactriidium, Hesperobaenus). Unidentified species of *Mimemodes* have been collected from haystacks and the flowers of *Boga medula* in India (Pakaluk and Slipinski 1993).

------Tribu Thionini: 2 genera are known: *(Shoguna)* distributed in, South Asia, Madagascar, Australasia and Japan. Live either under bark of trees (Bousquet 2010), often in decaying plant debris, in ascomycetous fungi, etc (Никитский 1986) and *Thione* genus with 3 species distributed in Panama, Nicaragua, Guatemala and Australia.

------Tribu Lenacini: endemic to New Zealand. *Lenax* occurs under bark of trees in the New Zealand, but no details on its bionomy are known (Bousquet 2010).

---Subfamily Rhizophaginae: Species of *Rhizophagus* occur mostly under bark of trees primarily feed on some mycels of subcortical fungi, also in those habitats they prey scolytine larvae (Majka and Bousquet 2010). For example, the European species, *Rhizophagus grandis* Gyllenhal, is predator of scolytids of the genus *Dendroctonus.* Also they are found in piles of grass, in heaps of compound plant groups and other decaying plant material and on fermenting sap (Farkač et al 2005). *Rhizophagus aeneus* Richter has been found on submerged wood in streams and rivers (Peacock 1978). According to Bousquet (1990) *Rhizophagus parallelocollis* Gyllenhal (from Canadá and Europe) is frequently found in graves, in the tombs (on the bodies in coffins) feeding on Diptera larvae, is also found in fungi, soil, mammal nests, on mould, plant refuse, old bones, and at sap. According to Majka and Bousquet (2010), in Canada, the following species: *Rhizophagus brunneus brunneus* Horn, has been collected under the bark of *Pinus strobus* Linnaeus, *Pinus palustris* Miller and *Picea glauca* (Moench) (Pinaceae); *Rhizophagus dimidiatus* Mannerheim has been found under bark of oldest Pinaceae (*Picea rubens* Sargent, *Picea mariana* (Miller), *Pinus strobus* Linnaeus, *Pinus palustris Miller, Picea glauca (*Moench*))* and *Betula papyrifera* Linnaeus (Betulaceae); *Rhizophagus remotus* LeConte on bracket fungi growing on white birch. In the same country, Bousquet (1990) has collected *Rhizophagus minutus rotundicollis* Bousquet on Pinaceae (*Abies balsamea* (Linnaeus) and *Picea glauca* Moench). Among European species, *Rhizophagus depressus (*Fabricius*)*, *Rhizophagus grandis* Gyllenhal and *Rhizophagus ferrugineus* (Paykull) are associated with conifers (*Picea abies* Karst and/or *Pinus sylvestris* Linnaeus), the others prefer the deciduous trees, such as *Quercus* spp., *Fagus sylvatica* Linnaeus, *Carpinus betulus* Linnaeus and *Betula* spp.

Monotomidae also can be collected with Berlese o Winkler eclector.

They can be collected by sifting the forest litter, compost and similar substrates, under bark of trees and in/on fruitng bodies of some fungi. *Rhizophagus* species can be also collected in traps baited with some fermenting substrates (bread dough, wine, beer etc.).

Carlos Aguilar Julio and Josef Jelínek *(Department of Entomology, National Museum, CZ-148 00 Praha 4 – Kunratice 1, Czech Republic; E-mail: JJ.Nitidula@seznam.cz)*

HOBARTIIDAE: 6 species are known, they are distributed in Australia, Tasmania, Chile and Argentina. *Hobartius chilensis* Tomaszewska & Slipinski (from Chile), *Hobartius eucalypti* (Blackburn) from Tasmania, The following species are distribuited by Australia: *Hobartius newtonorum* Tomaszewska & Slipinski, *Hobartius niger* Tomaszewska & Slipinski, *Hydnobioides lawrencei* Tomaszewska & Slipinski (New South Walles) and *Hydnobioides pubescens* Sen Gupta & Crowson. They eat soft bodies of rotting Polyporaceae and According to Lawrence (1991) and Tomaszewska and Slipinski (1995) larvae of *Hobartius* are developing in soft, rotting fruiting bodies of basidiomycete fungi *Grifolia* (Polyporaceae). *Hydnobioides pubescens* is a pest of *Pleurotus* fungus (OECD 2006)

Species of Hobartiidae can be collected with Berlese funnel, Winkler/Moczarski eclector and hand-collecting.

CRYPTOPHAGIDAE: With worldwide distribution, 53 genera and over 430 species (www.insects.tamu.edu). Two subfamilies: **Cryptophaginae** and **Atomariinae** (incl. Hypocoprinae, Alfieriellinae) (Lawrence et al. 2010). They can be found in fallen leaves, piles of decaying vegetation, under bark, in leaf litter, grass litter, and in habitats which promote mycelial fungal growth on which adults and larvae feed (Leschen and Skelley 2002). Some species are found in nests of bees and wasps. According to Majka et al. (2010) *Atomaria ephippiata* Zimmermann a Neartic species, it has been reported on fungi and molds in litter and nests (LaBonte 1998). In Atlantic Canada it has been found in mixed forests, coniferous forests, along forest roads, in raised bogs, salt marshes, and pastures. *Cryptophagus* can live in carrion.

Sometimes cryptophagids are found on stored products but they only feed on mold present in these products. Larvae and adults are mycophagous; they feed on pollen, mold, fungus spores of those ball shaped (puffballs) and its hyphas. Many species of *Cryptophagus* and *Antherophagus* develop in the nests of bees and wasps, and *Catopochroius* in termite nests.

Johnson (1993) reported that *"it occurs in a range of habitats – grassland, wetland, and woodland – especially in damp situations, and has been collected in litter, moss, fl ood refuse, cut vegetation and compost heaps."*

According to Majka et al. (2010) *Atomaria apicalis* Erichson is widely distributed in the Palaearctic region being found throughout Europe and North Africa, east across Siberia to the Russian Far East, across the Middle East, to Central Asia, Mongolia, China and North America and Johnson (1993) reported that it species and *Atomaria pusilla* (Paykull), *Atomaria testacea* Stephens: *"They are grassland species which has been recorded especially around farms, gardens and parks but often occurs in other habitats. It has been collected in man-made heaps of refuse, especially cut vegetation, grass, compost and dung heaps, fl ood debris, and more rarely on carrion and rotting fungi."* *Atomaria ferruginea* Sahlberg and *Atomaria fimetarii* (Fabricius) both from Yorkshire, US, have been collected shaking moss at the roots of ash-trees (in: The Entomologist's Monthly Magazine. Vol.V. 1868-9, pag: 143).

To collect cryptophagids beetles the logs can be sprayed with mold and mildew. Also to collect them it is necessary to look at the moldy places where they live, and under dead bark, rotting wood, decomposed plant material and place these materials in a Berlese funnel or Winkler/Moczarski eclector. Animal burrows can also be checked. Sometimes they are collected by beating or sweeping or with pitfall traps baited with a simple mixture of water, detergent and ethylene glycol (Sackmann 2006). Other collection methods are explained in the article "Beetles in fungi", Chapter III.

AGAPYTHIDAE: Family with only one known species: *Agapytho foveicollis* Broun from South Island of New Zealand (Sen Gupta & Crowson 1969; Leschen et al. 2005). Apparently they feed on fungi and have also been found in the mold growing on the bark of some *Nothofagus* (Leschen et al. 2005). *Agapytho* are also associated with sooty mould (Ascomycetes, Dothideales), which form thick black "wefts" of mycelia of up to six different species (Hughes 1972; Leschen et al. 2005) growing generally on the surfaces of foliage, branches, tree trunks and, often surrounding rocks and soil (Gimmel et al. 2009). These growths are associated with sugary exudates of leaves or with the "honeydew" produced by sternorrhynchous (Hemiptera), primarily in association with *Nothofagus*, but other plants as well (Klimaszewski & Watt 1997; Carlton & Leschen 2007; Gimmel et al. 2009). Other collection methods are explained in the article "Beetles in fungi", Chapter III.

PRIASILPHIDAE: Until recently they were considered a subgroup of the Phloeostichidae family. They live only in the Southern hemisphere, Australia, Tasmania, New Zealand and Chile and include 11 species described in 3 genera (www.insects.tamu.edu). These beetles are found in rotting wood and moss and they are probably mycophagous. Larvae and adults are often covered with small particles of trash (Lesch et al. 2005). The New Zealand endemic genus *Priasilpha* are forest dwelling and feed mainly on fungal spores, including those of sooty moulds, and *Chileosilpha elguetai* Leschen, Lawrence & Ślipiński is a flight-capable monotypic genus restricted to Chile (Leschen and Bernard 2005) and *Priastichus* (from Australia and Tasmania), *Telmatophilus olivascens* Broun and *Telmatophilus vestitus* Broun fro New Zealand (Sen Gupta & Crowson, 1969), the species of this last genus are found on plants near water (LeConte 1862). According to Leschen et al. (2005) *Priasilpha* and *Priastichus* species have been collected mainly in leaf litter, but also in decayed wood or among moss on rock or log surfaces. Also, according to Leschen and his colleagues, *Chileosilpha* *elguetai* Leschen, Lawrence & Ślipiński was collected in a forest of Valdivia (Chile) where the dominant trees were "coihue or roble" (*Nothofagus dombeyi* (Mirbel)) and "Tepa" (*Laurelia philippiana* Looser) and the larvae (and adults) of *Priasilpha* and *Priastichus* are often found covered by debris and exudate.

PHLOEOSTICHIDAE: 5 genera and Ca. 15 species are known (www.insects.tamu.edu). Two subfamilies: **Phloeostichinae** and **Hymaeinae** (Leschen et al. 2005; Lawrence 2010).

Phloeostichides eat live and dead fungi, under bark, in leaf litter and moss. Some species live in forests of Fagaceae (Leschen et al. 2005). This family includes 5 genera: *Hymaea* from Australia and Tasmania, *Priasilpha* endemic to New Zealand; *Phloeostichus denticollis* Redtenbacher (only from Central Europe) and *Bunyastichus monteithi* Leschen, Lawrence & Ślipiński (known from Australia) and two species of *Rhopalobrachium: One* of Australia *(Rhopalobrachium crowsoni* Lawrence) and other from Chile and Argentina *(Rhopalobrachium penai* Lawrence) both residents of Fagaceae forest areas, feed hyphae, conidiophores and conidia of various hyphomycetes (Lawrence 1995; Pakaluk et al. 1995).

Little is known about the biology of these beetles. The European species, *Phloeostichus denticollis* Redtenbacher, has been collected under the loose bark of dead trees (Bosselmann 1876; Leschen et al. 2005), the larvae also have been found under mossy bark of old maple trees (Weise 1897).

Some Phloeostichidae are predators, for example, larvae of *Hymaea succinifera* Pascoe, have been collected in the galleries of *Platypus subgranosus* Schedl (Candy 1990), on the other hand, the hindgut of a male of *Hymaea succinifera Pascoe* was packed with very dark, elliptical fungal spores lacking an apiculus (Leschen et al. 2005).

Rhopalobrachium clavipes Boheman has been collected in "Ñirre" (*Nothofagus antarctica* (Forster)) in Chile (Pakaluk et al. 1995).

Bunyastichus monteithi Leschen, were collected by pyrethrum on "Bunya pine" trunks (*Araucaria bidwillii* Hook) (Leschen et al. 2005). According to Leschen et al. (2005) larvae were collected in mossy bark of old maple trees (*Acer* sp). Also, according Leschen and his colleagues the larvae of *Hymaea* and *Rhopalobrachium* are usually found under bark of decaying logs.

Adults of Phloeostichidae are usually collected with Berlese funnel, Winkler/Moczarski eclector and by hand-collecting.

SILVANIDAE: Family with worldwide distribution. According to M.C. Thomas (www.insects.tamu.edu) 61 genera and 502 species have been described in 2 subfamilies: **Silvaninae** and **Brontinae** (Lawrence et al. 2010). The ancestral habitat seems to be the saproxylic sub corticolous with micophytophagous diet (Booth et al. 1990).

---Subfamily Brontinae, they have functional mandibular cavities for transporting fungal spores (Ratti 2007). Other present taxa have strayed a little from the ancestral habitat and food, although the whole group can be considered a member of the saproxylic fauna complex. Some species of Psammoecini tribe are found primarily on plants, especially withered, pendant leaves, and in plant debris, where they probably feed predominantly on fungi. Brontini are are found primarily under bark, where both adults and larvae probably feed on ascomycete and other fungi (Crowson and Ellis 1969; Thomas 2002).

---Subfamily Silvaninae, they are under bark (Halstead 1973) or in leaf litter or soil (Thomas 2002); probably they also eat fungi (Thomas in Arnett et al. 2002). The genera *Ahasverus, Cathartus, Oryzaephilus* and *Nausibius* are pests of stored grains, grain products, nuts and spices (Arnett 2000; Evans and Hogue 2006). Australian *Silvanus lateritius* (Broun) has been collected beneath *Eucalyptus*.

Several Australian species are associated with ants (Sen Gupta and Crowson 1966; Thomas 2002) and two genera from South America have been reported to milk honeydew from mealybugs (Thomas 2002). Some are facultatively predaceous (Thomas 2002). According to Halstead (1973) European and North American species are found beneath the bark of oak (*Quercus* spp.) and sweet chestnut (*Castanea* spp.), less frequently under the bark of pine (*Pinus* spp.); African and oriental species have been found under the bark of *Ficus*. Fungal spores have been observed in the gut of specimens dissected for genitalia, and the Australian *Silvanus lateritius* (Broun) beneath that bark of *Eucalyptus (*Halstead 1973). *Telephanus nodicornis* Nevermann, from Dominica, Guadeloupe, Montserrat, this species was taken at D'leau Gommier, in dead and decaying leaves (Peck 2006).

To collect Silvanidae it is necessary to check under the loose bark, decaying plant material and especially rotten fruit.

CUCUJIDAE: 4 genera with 44 species (M. C. Thomas in: www.insects.tamu.edu), distributed throughout the world, but with very few species in the Neotropical region. *Cucujus* is a Holarctic and Neartic genus, *Pediacus* is a Holarctic and Neotropical genus, *Palaestes (*Neotropical) and *Platisus* are restricted to Australia. Little is known about the biology of these beetles, *Cucujus* are predators (Thomas 2002; Evans and Hogue 2006). They are found under bark, in storage products and in leaf litter and humus.

MYRABOLIIDAE: Known only from South-Eastern part of Australia and Tasmania (Leschen et al. 2005; www.insects.tamu.edu). The family has 14 species in the genus *Myrabolia* (Leschen et al. 2005; Tomaszewska and Slipinski 2008). Adults and also a putative larva have been collected under the bark of several species of *Eucalyptus* and in litter (Blackburn, 1892, 1903; Leschen et al. 2005).

CAVOGNATHIDAE: With 5 genera and 11 species (www.insects.tamu.edu). This family includes *Cavognatha* (from Australia), *Taphropiestes* (from South America) and *Neocercus* and *Zeonidicola* which are restricted to New Zealand and Chatham Islands. Adults and larvae of *Cavognatha pullivora* Crowson, *Zeonidicola chathamensis* Watt and, adults of *Zeonidicola dumbletoni* Crowson and *Taphropiestes* sp., have been found in the nests debris of various seabird species such as cormorants and passeriformes (Crowson 1964a, 1973; Sen Gupta and Crowson 1966, 1969a; Watt 1980).. Larvae of *Cavognatha pullivera* Crowson, were found on nestlings of the magpie *Gymnorhina tibicem* (Latham) at Gungahlin (Australia) and larvae of *Zeonidicola dumbletoni* Crowson were found in nests of the spotted shag *Stictocarbo punctatus* Turbott on Banks Peninsula (Watt 1980). It is unclear how they feed, but it is suspected that they ectoparasite the young pigeons of the birds mentioned above (Di Iorio et al. 2008).

LAMINGTONIIDAE: This family occurs from Southern Queensland to Tasmania and is represented by a single described species *Lamingtonium binnaburrense* Sen Gupta & Crowson (Sen Gupta & Crowson 1969; Lawrence & Britton 1991, 1994; Lawrence et al. 1999) which feeds on polypore fungi (Lawrence & Britton 1994). They live under the bark of standing trees. They are collected by lifting the bark and placing a beating sheet under the tree to be revised, or by hand-collecting.

PASSANDRIDAE: Recently Thomas M. C. (in: www.insects.tamu.edu) has listed 109 species and 9 genera, worldwide distribution (except for the Western Palaeartic and New Zealand), however they are predominantly tropical family (Thomas 2002). Adults and larvae are found under bark (Thomas 2002). Larvae are ectoparasites of other wood infesting beetles especially pupae of Scolytinae and Cerambycidae (Thomas 2002). *Psammoecus desjardinsi* (Guerin-Meneville) has been found in carrion in the Hawaiian Islands (Goff et al. 1986).

PHALACRIDAE: Worldwide distribution. 50 genera and 551species are known (www.insects.tamu.edu). According to Lawrence (2010) there are two subfamilies: **Phaenocephalinae** and **Phalacrinae**. Adults are found in decaying vegetation (Arnett 2000) and larvae live on flowers, especially those of Compositae, and in the spikes of grass that has been attacked by grain rust (Arnett 2000), but according to Wheeler (1984) although the phalacrids are know as the "shining flower beetles" studies have show that the majority of them are asscoiated with fungi, some of which cause diseases of plants. They eat spores of molds and the mold that grows on seeds; others eat fungi spores or mildews growing on dead or dying leaves. For information on host plants of some larvae of phalacrids the reader may consult the work of Wheeler (1984) which is available (in part) online (see bibliographics references).

Phalacridae can be collected by sweeping and beating, with Berlese funnel or Winkler/Moczarski eclector, because many phalacrids are found among the leaf litter, the foliage of trees and composed flowers. Light trap or by fumigating dead logs is other sampling methods. Look on smutty plants, moldy vegetation, and on branches that have blown down. Many species hibernate as adults and they can be sefted from ground litter in spring (White 1983).

PROPALTICIDAE: Found in Old World tropics and Pacific islands, including Japan. According to Matthew Gimmel (www.insects.tamu.edu) two genera: *Discogenia* (13 spp. from Afrika) and *Propalticus* (26 spp. from Afrika, Madagascar, New Caledonia, Southeast Asia, Key Island, some island in Indian Ocean and Japan). They are found under loose bark of dead or dying wood, or logs (Hans 1960). For example, Crowson (cited in Leschen et al. 2005) has collected a larva of *Propalticus* sp. in Lamington National Park, Queensland, under bark of dead *Casuarina.*

LAEMOPHLOEIDAE: This family is represented in all forests areas in the world and the greatest diversity is in the tropics. 37 genera and more than 400 species are known (Thomas 2002). They are found under bark and usually together with Scolytinae, which probably are taken as a prey (Thomas 2002). Also, two species of *Cryptolestes* have been reported as predators of Scale insects, although apparently they are fungivorous rather than carnivores (Thomas 2002). They are also found in leaf litter. Some adults and larvae occur under bark where mostly feed on fungi (Thomas 2002) of the groups Pyrenomycetes, Deuteromycetes and Zygomycetes and sometimes Basidiomycetes. *Cryptolestes* are important pests of stored grain (Thomas 2002), especially those attacked by mold (Halstead 1993). They can be collected with Berlese funnel, Winkler/Moczarski and by hand-collecting.

TASMOSALPINGIDAE: There are 2 known species: *Tasmosalpingus promiscuus* Lea and *Tasmosalpingus quadrispilotus* Lea (Leschen et al. 2005; Lawrence et al. 2010). They have been found in Tasmania and Victoria, Southeastern Australia (Lawrence & Britton 1991). Little is known about their biology. They probably feed on fungi. in this respect and according Leschen et al. (2005) *"adults of Tasmosalpingus have been collected in Malaise traps, and a putative larva was collected by pyrethrum fogging the ark of Phyllocladus aspleniifolius* (Labilladière) Hooker (Podocarpaceae*). The hindgut of a dissected male was packed with fungal hyphae and possibly spores of a* hyphomycete.*"*

CYCLAXYRIDAE: Endemic family to New Zealand, with only one known genus *Cyclaxira* which contains 2 species: *Cyclaxyra jelineki* Gimmel and *Cyclaxyra politula* (Broun) (Gimmel et al. 2009). According to Gimmel et al. (2009) they occurr on the North, South, and Stewart Islands in New Zealand; *Cyclaxyra politula* (Broun) appears to be absent from northeastern South Island. Adults and larvae are strictly associated with sooty moulds Ascomycota (Klimaszewski & Watt 1997; Leschen et al. 2010). Gimmel et al. (2009) in his revision of the family explained: *"Which in south temperate regions form thick black "wefts" of mycelia of up to six different species (*Hughes 1972*), generally on the surfaces of foliage, branches, tree trunks, and often surrounding rocks and soil".* These growths are associated with sugary exudates of leaves or with the "honeydew" produced by sternorrhynchous Hemiptera, in association with *Nothofagus*, but other plants as well (Klimaszewski & Watt 1997; Carlton & Leschen 2007). According Gimmel et al. (2009) they are active at night and may be collected in large numbers on trunks of trees and larger shrubs where sooty moulds are growing. Specimens of *Cyclaxyra* also have been collected by sweeping and beating *Nothofagus* branchs (Gimmel et al. 2009). They were collected also with pyrethrin-fogging *Nothofagus menziesii* (Hook.) bark, with Malaise trap, from leaf litter rotten logs (with Winkler/Moczarski eclector or Berlese funnel) and with yellow pan trap in *Nothofagus* forest and also in *Leptospermum* and podocarp forests (Gimmel et al. 2009).

KATERETIDAE

Josef Jelínek

They were always classified as subfamily of Nitidulidae, only recently they were recognized as a sister group of Nitidulidae and hence the distinct family. 13 genera and ca.90 species are known worldwide (Audisio 1993; Majka et al. 2008; Jelinek and Cline 2010), however, exact numbers of species are changeable, because new species may be described or other synonymized soon. They are distributed mostly in the temperate zones of both hemispheres and throughout Neotropical region; only a few species occur also in the Afrotropica l and Oriental regions. Both larvae and adults live in flowers of plants, feeding on pollen and reproductive structures of flowers. As far as known, particular genera of Kateretidae are associated with specific groups (genus or family) of plants: *Kateretes* with *Carex*, *Heterhelus* with *Sambucus* (Caprifoliaceae), *Brachypterus* with Urticaceae, *Brachyleptus*, *Amartus* and *Anamartus* with Papaveraceae, and *Brachypterolus* with Scrophulariaceae (*Linaria* and *Antirrhinum*). Larvae of Kateretidae are always mono- or oligophagous, but adults may feed on different flowering plants according to season. Kateretidae can be collected by sweeping or beating the relevant flowering plants.

Josef Jelínek
Department of Entomology, National Museum,
CZ-148 00 Praha 4 – Kunratice 1, Czech Republic.
E-mail: JJ.Nitidula@seznam.cz

NITIDULIDAE

Josef Jelínek

About 4000 species known (Jelínek et al. 2010), worldwide distribution, but according to Kirejtshuk (2008) the distribution ranges of its groups have some restrictions. For example, Carpophilin and Nitidulin lineages are asymmetrically distributed. The first group is best represented in the Eastern Hemisphere (not endemic in South America and Hawaii). The second group is much more diverse in the Western Hemisphere and there are more species in the Neotropical region and Hawaii (Kirejtshuk 2008).

According to Arnett et al. (2010) Nitidulidae currently includes the following subfamilies: **Calonecrinae**, **Maynipeplinae**, **Epuraeinae**, **Carpophilinae**, **Amphicrossinae**, **Meligethinae**, **Nitidulinae**, **Cillaeinae**, **Cryptarchinae** and **Cybocephalinae**. The latter subfamily is sometimes classified as a distinct family (Endrödy-Younga 1968, Smith and Cave 2006).

Biology of Nitidulidae is diverse, but with certain simplification the following basic groups can be recognized:

Saprophagous or **mycosaprophagous:** these species in the wide sense, i.e. species developing on various decaying or fermenting organic substrates, like tree sap, fruits, bast, other moulded vegetal substrates, and carrions(*Omosita* and *Nitidula*). This mycosaprophagy is probably an original feeding habit and is represented in all subfamilies of Nitidulidae except for Meligethinae. Fungi like moulds and yeasts represent certainly an important compound of the food of these species. Some species may be facultative predators (e.g. some *Glischrochilus*). Other species (especially some *Carpophilus*) may become pests of stored products (Dobson 1960), Hinton 1945). Those living in tree sap or on decaying fruits may be vectors of fungal and bacterial diseases of trees, or of fruits in orchards and plantations (Dorsey and Leach 1956; Bruck and Lewis 2002, Audisio et al.1990). Some *Glischrochilus, Carpophilus, Colopterus,* are transmitters of *Ceratocystis (*Ascomycetes*)* or *Fusarium* (Hyphomycetes) (Kirejtshuk 2008). Saprophagous species are also fitted to the transportation over long distances with various comodities and some of them can be acclimatized in the territories remote from their original areal (Jelínek 2007).

Mycetophagous: These species develop in the fruiting bodies of various higher fungi like Gasteromycetes (e.g. Nitidulinae: *Pocadius*, *Physoronia* **subgen** *Pocadites*), *Phalales* (some Nitidulinae: Cyllodini), Polyporales (some *Epuraea*) and others. *Pocadius*, *Thalycrodes*, *Thalycra* and *Quadrifrons* genera are associated with forest underground mushrooms (Kirejtshuk and Leschen1998).

Phytophagous: Almost all live, both as larvae and adults, in flowers, feeding on pollen and reproductive parts of flowers: all Meligethinae, Nitidulinae: Mystropini, Epuraeinae: Amystrops and some Epuraea, some other genera of Nitidulinae (e.g. *Aethina* subgen. *Idaethina*), some *Carpophilus* and some genera of Cillaeinae (*Conotelus, Macrostola, Xanthopepus, Lipa-*

ropeplus and probably others); they may be associated with palms (Mystropini, some tropical Meligethinae, *Epuraea* (*Apria*)), *Pandanus* (*Amystrops*) and various families of shrubs and herbs (Kuschel 1990). Only three genera of Nitidulinae have larvae mining in the leaves of some Brassicaceae (*Xenostrongylus, Oxystrongylus* and *Anister*). At least in *Meligethes*, only the larvae are as a rule monophagous or oligophagous, associated with specific host plants, but adults may feed in flowers of various other plants, which are available according to season. Many species living on herbaceous plants prefer to live in communities of herbs found in forests (Kirejtshuk & Viklund 2002.).

Predators: All members of the subfamily Cybocephalinae are predators of scale insects which are employed in some projects of the biological control of scale insects. Otherwise, carnivory is rare among Nitidulidae. Only one genus of Nitidulinae, *Cychramptodes murrayi* Reitter from Australia, is known to prey on scale insects (Kirejtshuk and Lawrence 1992). Facultative predation was described in *Glischrochilus quadrisignatus* (McCoy and Brindley 1961) and in the larvae of *Amphicrossus japonicus* (Kovac et al. 2007). It is supposed to occur also in other species of *Glischrochilus*, *Pityophagus* and *Epuraea*, but the evidence is still doubtful.

Associated with insects: Some species are associated with social Hymenoptera; the Holarctic *Amphotis* and Australian Lawrecerosini (Kirejtshuk 1990c) are associated with ants, Afrotropical species of *Aethina* (according to Lundie 1940) and Australian Onicotis live in the nests of bees and Aethina tumida Murray, originally from Africa, is an important pest of apiculture in North America (e.g. Hood 2000; Toumanoff 1939; Caron 1978; Hepburn and Radloff 1998; Neumann and Ritter 2004). Also larvae of European *Epuraea aestiva* (Linnaeus) and *Epuraea melina* (Erichson) are known to develop in the nests of bumble-bees (*Bombus) (Scott 1920).*

Nitidulidae have a short larval development and adults are relatively long-lived (Kirejtshuk 2008).

Most species of Nitidulidae are regularly collected in temperate and subtropical climate; nevertheless, the phenology may change according to the coordinates and altitude as well as other factors. Even in temperate zones the beetles occur in fact throughout the years (they overwinter as adults somewhere in leaf litter or under bark). The necrophagous species are mostly found outside forest, but the open landscape is biotope of other species, too, e.g. of many *Meligethes*, etc.

Many methods can be used to obtain Nitidulidae, among them the Berlese funnel and sweeping and beating the flowering plants, the bottles in the trees and corrugated card boards traps, baits included: whole wheat bread dough, fermenting brown sugar, cantaloupe, carrion and banana. To collect them, look under loose bark of trees and stumps, in decaying vegetable material, also it is usefull fumigating logs, stumps and trees. Check carefully the mushrooms on the logs and tree sap. Some are caught with lights.

The subfamily **Cybocephalinae** is considered by many to be a separate family (see above). They are distributed in all continents, with 8 genera and 213 species known. They are parasitoid and myrmecophilous (Kirejtshuk 1997). Almost all Cybocephalinae are predators of Coccidae and Aleirodidae (Kirejtshuk and Fallahzadeh 2008), *Cybocephalus* spp. larvae also prey on coccids (Habeck 2002).

In Europe, young adults and larvae of *Cybocephalus* are found in plants infected by scale insects, especially native species of fruit trees, but they are also found in oak, ash, mountain ash and shrubs.

Josef Jelínek
Department of Entomology, National Museum,
CZ-148 00 Praha 4 – Kunratice 1, Czech Republic.
E-mail: JJ.Nitidula@seznam.cz

SMICRIPIDAE: This family, resembling to Nitidulidae, Price (2002) cites 6 species: *Smicrips chontalena* (Sharp) (Guatemala and Nicaragua), *Smicrips distans* (Sharp) (Guatemala), *Smicrips Mexi-*

cana (Sharp) (Mexico), *Smicrips palmicola* LeConte (Cuba, Puerto Rico and U.S.A.), *Smicrips texana* (Casey) (from Mexico, Guatemala, Panama and West Indies) and *Smicrips exilis* Murray (Hispaniola, Guadeloupe, St. Vincent and Grenada).

Smicrips, both larvae and adults are found in fallen leaves and under bark (Lawrence 1982). *Smicrips palmicola* LeConte is associated with *Sabal palmetto* (Walter), but also that species and *Smicrips texana* (Casey) adults and larvae were found among rotting flowers of *Pseudobombax* sp. (Lawrence et al. 1999b; Price 2002).

They can be collected with flight intercept traps and pitfall barriers (Price 2002).

BOTHRIDERIDAE: About 38 genera and 400 species (www.insects.tamu.edu). According to Lawrence et al. (2010) there are 4 subfamilies: **Teredinae**, **Xylariophilinae**, **Anommatinae** and **Bothriderinae**. They are cosmopolitan and the greatest diversity occurs in the Old World tropics. They are found under bark of dead or dying trees, or among the rubbish in the vicinity; they are commonly associated with other wood boring insects. Many species have morphological adaptations to fill the tunnels and galleries of wood-eating beetles. But Anommatinae are wingless and live on organic soils, humus, grass cutting, etc.; they are rare, but sometimes they can be collect in large quantities. Larvae of the Bothriderinae subfamily are ectoparasites of larvae and pupae of wood boring beetles. In its early stages Bothriderinae larvae are triungulines and when they find a host, the larva develops into a grub-like larva that attaches to the host to feed. Before pupation, larvae construct a cocoon or capsule in which they pupate. The "cocoons" may be attached to the old host, or the larvae may migrate a short distance away within the gallery to pupate (Nathan 2009). They are generally predators of Anobiinae, Bostrichidae, Curculionidae, Brentidae, Buprestidae, and Mycteridae, just as several Hymenoptera Apidae *(Xylocopa* sp.*)* and Xiphydriidae *(Xiphydria* sp.*)* (Balduf 1935; Clausen 1940; Pollock and Ivie 2004). *Dastarcus helophoroides* (Fairmaire) attacks Cerambycidae. Those in the *Sosylus* genus are predators specialized in attacking Platypodinae and Scolytinae; other genera are more generalist or specialist predators (Browne 1962; Roberts 1969). *Deretaphrus oregonensis* Horn is a major predator of Buprestidae (Nathan 2009). Larvae of Teredinae subfamily feed on fungi growing in the galleries of Scolytinae. Larvae of Xylariophilinae feed on ascomycete fungi (Craighead 1920).

To collect them, lights can be used. *Sosylus* members may be caught breaking into pieces the wood of trees recently infested with Platypodinae or Scolytinae. Teredinae can also be collected by reviewing or fumigating the wood infested by Scolytinae. Many others are under bark of dead or dying trees.

CERYLONIDAE: Occur in forest litter and under bark. At present, there are about 40 genera and over 300 described species known from all the major zoogeographic regions. 5 subfamilies are known (Lawrence and Stephan 1975; Lawrence et al. 2010): **Euxestinae**, **Loebliorylоninae**, **Ostomopsinae**, **Murmidiinae** and **Ceryloninae**. Little is known about feeding habits and biology in general. They are found under bark and decomposed plant material and feed on mushroom hyphas and spores. Some authors (Besuchet 1972; Lawrence and Stephan 1975) believe that some of them may be predators of small Arthropoda and Nematoda. Their preferred habitats are litter and decaying wood (especially cambium). Species of *Cerylon* spp. *and Mychocerus* spp. are usually collected under the bark, while *Lapethus* spp. appear to be common among forest litter and waste (Lawrence and Stephan 1975). According to Lawrence and Stephan (1975) several species have been recorded in ant nests: *Hypodacne punctata* Leconte is known in *Camponotus* galleries. *Aculagnathus mirabilis* Oke, has been found in the nests of *Amblyopone* (Australian ants), and two species of *Lapethus* have been collected in large numbers in trash storage of ants (*Atta mexicana* Smith). *Elytrotetrantus chappuisi* (Jeannel and Paulian) is known in mole-rat nest, while *Euxestus erithacus* Chevrolat was collected on bat guano in a cave in Jamaica (Lawrence and Stephan 1975). Adults and larvae of certain tropical *Philothermus* live in Polypore fungi (Reitter 1876). *Murmidius ovalis* (Beck) attacks stored products and also have been collected in dead leaves and cut grass in Europe and Northern Asia (excluding China) and North America (Hinton 1945; Halstead 1968; Lawrence and Stephan 1975). *Cerylon castaneum* Say from North America has been collected on dried mushrooms living in trees (Lawrence and Stephan 1975).

Due to its small size, most cerylonids are collected with various techniques, but in general they can be captured checking under bark of various conifers trees and others, or placing plant material in a Berlese funnel. Check carefully under the bark of trees, stumps and fallen logs.

ALEXIIDAE: According to Floyd Shockley (2008) it comprises the *Sphaerosoma* genus (an ambiguous genus also attributed to a fungi genus), with 53 species known from the Palearctic region and North Africa (Apfelbeck 1916). They are mycophagous, found in rotting fungi, on bark eating mosses, on the ground of forests and sometimes in decaying plant debris. They are collected with Berlese or Winkler/Moczarski eclector and by hand-collecting. Other collection methods are explained in the article "Beetles in fungi", Chapter III.

DISCOLOMATIDAE: With 16 genera and 400 species (www.insects.tamu.edu), arranged within 5 subfamilies: **Aphanocephalinae**, **Cephalophaninae**, **Discolomatinae**, **Notiophyginae** and **Pondonatinae**.

They are distributed in all tropical and subtropical regions of the world, with a great diversity in the African tropics and southeastern Asia.

Adults and larvae are collected under bark (Lawrence 1982). However, some species have been collected in leaf litter. Species of *Aphanocephalus* and *Falli* genera have been found eating on the surface of fungi, and species of *Notiophygus* and *Discoloma* have been found in lichens (Lawrence 1991; Shockley and Cline 2009). Larvae feed on fungi. Several species of *Notiophygus, Holophygus and Katoporus* genera live associated with ants (Shockley and Cline 2009; Cline and Slipinski 2009).

They can be collected with Berlese funnel, Winkler/ Moczarski eclector, by hand-collecting and also spraying the trees.

ENDOMYCHIDAE: The University of Texas on its website (www.insects.tamu.edu) cites 131 genera with 1551 species, however the checklist of Shockley et al. (2009) included approximately 130 genera and 1782 species and subspecies arranged among 12 subfamilies. They are distributed in all zoogeographical regions with the highest diversity in tropical and subtropical areas of the world (Tomaszewska 2005; Shockley et al. 2009).

They feed on many types of fungi and molds growing on rotting wood and feed on fermented sap from cuts in the trees; many are found under bark. Larvae in general are mycophagous eating the inner parts of puffballs. (According to Shockley et al. (2009) mycophagy generally refers to feeding on the softer hyphae and fruiting bodies of fungi. Most endomychids feed primarily on macro-Basidiomycetes (Hymenomycetes and Aphyllophorales). Shockley et al. (2009) also explains that *"numerous endomychids species exhibit a special form of mycophagy known as sporophagy. In the case of sporophagy, the beetles have shifted from feeding on the hyphae of their fungal hosts to feeding on the spores."*

Endomychidae has the following subfamilies:

---Subfamily Merophysiinae: With 12 genera and about a hundred of known species. Almost all of them are mycetophagous and myrmecophagous, partial cosmopolitan the genus *Holoparamecus* with 48 species worldwide. For example, according to Wolfgang Rücker (pers. comm. 2010) *Cholovocera* with worldwide distribution that live in nests of ants *(Messor, Aphaenogaster* and *Pheidole), Reitteria,* with 3 species so far found only in the Mediterranean and *Merophysia* with 25 species only in the Mediterranean Region in Europe, North Africa and Asia Minor, distributed almost exclusively to the Mediterranean area, they are myrmecophilous species that live in nests of ants *(Messor, Aphaenogaster* and *Pheidole). Holoparamecus gabrielae* from Mexico was found in bat guano inside a cave in Veracruz State (Rücker 2003). *Lycoperdinella* from Central American seem to live in leaf litter feeding on microfungi or mold and are only collected in berlese funnel samples (Arriaga-Varela et al. 2007). Merophysinae also can be collected with pitfall traps or sifting litter and lights (Wolfgang Rücker, pers. comm. 2010).

---Subfamily Pleganophorinae: With 3 genera worldwide (Tomaszewska 2005). *Trochoideus, Pleganophorus* have been collected with ants and termites (Arriaga-Varela et al. 2007) *Dadocerus* distributed in Oriental region (Tomaszewska 2005). All 26 species of *Saula* are presumed to be entirely predaceous (Leschen 2000). Both adults and larvae of *Saula japonica* Gorham prey on scale insects and phytophagous mites (Sasaji 1978). According to Arrow (1925) *Saula ferruginea* Gerst., was found (from Bombay) upon dry leaves of Palmyra palm (*Borassus*) at Shoranur and upon diseased

Coconut leaves at Pattalam and, *Trochoideus desjardinsi* Guérin-Ménèville, from south Asia, New Guinea, Samoa, Fiji Islands, Mauritius, Madagascar, Philippines, is associated with ants and termites (Arrow 1925).

---Subfamily Anamorphinae: With 163 species known, some Mexican species have been collected from rotting logs and logs infested by hard bracket fungi (Arriaga-Varela et al. 2007). Hoffman et al. (2004) collected an unidentified species of *Rhymbus* in a cave in Yucatan Peninsula (Mexico). *Catapotia laevissima* Thomson was collected in orchids in Mexico (Arriaga-Varela et al. 2007). Most specimens of *Rhymbomicrus* have been collected in forest litter or under bark (Pakaluk 1987). Some specimens of *Exysma laevigata* Gorham (from Mexico) were collected on *Tabebuia* (Bignoniaceae) and in log and on pineapple (Arriaga-Varela et al. 2007). According to Leschen & Carlton (1993) larvae of *Bystus decorator* Leschen & Carlton, from Peru, seem to cloak themselves with debris collected from their surrounding microhabitats and both larvae and adults occur together on rotting logs and feed on fungal spores.

---Subfamily Leiestinae: With 6 genera; distributed mainly in Europe, Japan and the United States. Price and Young (2007) recently reported *Phymaphora pulchella* Newman from damp treehole litter (Blatchley 1910). *Rhanidea unicolor* (Ziegler) from US, it occurs in rotting wodd, beneath bark and sometimes on flowers of thoroughwort in fall (White and Peterson1998) this species has also been collected on flowers (Asteraceae). *Aphorista morosa* LeConte was found in association with a yellow plasmodium of a slime mould (Myxomycetes) (Tomaszewska 2002).

---Subfamily Mycetaeinae: With *Agaricophilus reflexus* Motschulsky and 5 species of *Mycetaea*. In Canada *Mycetaea subterranea* Fabricius has been found in association with stored products in mills, granaries, warehouses and cellars where it feeds on *Zygomycete hyphae* (Majka 2009). It has also been found in old tree trunks, caves, and bee hives (Bousquet 1990).

---Subfamily Eupsilobiinae: With 6 genera distributed mainly in Central and South America and in South Africa. *Eidoreus are* known from Mexico, USA, and Belize but mainly from widely scattered islands around the globe: Antilles, Hawaii, Galapagos, Seychelles, Southern Asia, Samoa, Sri Lanka, Japan (Pakaluk & Ślipiński 1995). The most frequent habitat of Eupsilobiinae seems to be forest debris. *Eidoreus* and *Microxenus* have been collected in nests of ants and termites and their most frequent habitat seems to be forest debris (Arriaga-Varela et al. 2007).

---Subfamily Xenomycetinae: *Xenomycetes laversi* Hatch and *Xenomycetes morrisoni* Horn, both species are endemic to Northwestern North America (Shockley 2009). *Xenomycetes laversi* Hatch appears to be restricted to *Paxillus atrotomentosus* (Batsch) (Agaricales), many larvae and adults have been collected from sporocarps or mycelia of this fungus, found associated with rotten wood (Tomaszewska 2002). *Xenomycetes morrisoni* preferentially has been collected with pitfall traps baited with cantharidin (Dettner 1987; Hemp and Dettner 2001; Price and Young 2007; Young 1989; Shockley et al. 2009).

---Subfamily Danascelinae: With only *Danascelis elongata* Tomaszewska (from Pakistan) and two species of *Hadromychus* (Canada: from New Hampshire, Nova Scotia, Ontario, and Québec). The species appears to by mycophagous, *Hadromychus chandleri* Bousquet &. Leschen has been collected at Campbell Hill were found in a mature red spruce forest while the specimen from Lorne is from an old coniferous forest with flight intercept trap (Bousquet & Leschen 2002; Majka 2007).

---Subfamily Endomychinae: 5 genera with 95 species know (www.insects.tamu.edu). *Endomychus* spp., appear to have obligate relationships with lignicolous Basidiomycetes (Tomaszewska 2000). *Endomychus biguttatus* Say from Canada feeds and reproduces on *Schizophyllum communae* (Leschen & Carlton) (Schizophyllaceae) although it is also found on *Auricularia* spp. (Auriculariaceae) and *Polyporus betulinus* Bull. (Leschen & Carlton 1988; Shockley, cited in: Majka 2007) has been collected with flight intercept trap. *E. biguttatus* in USA has been found feeding directly on the putrefying flesh of pig carcasses during the advanced decay stage of decomposition (Payne and King 1970). Other genus is *Cyclotoma* with 15 species distributed in the Oriental Region (Tomaszewska 2002).

---Subfamily Epipocinae: they have been found in rotting logs or logs infested by lignicolous fungus like *Sirobasidium sanguineum* Lagerheim & Pat. (Sirobasidiaceae); others seem to feed on Boletaceae mushrooms (Navarrete-Heredia 1996). Many have been collected in bromeliads. *Epipocus* has been collected from rotting logs, under bark and feeding on lignicolous fungi of the families Auriculariaceae, Sirobasidiaceae and Polyporaceae and on sporophores of Boletaceae mushrooms (Navarrete-Heredia 1996, 1997), in México *Epipocus* is present in almost all kinds of forests and in plantations of coffee and bananas; collected through the year but more abundant in the rainy season (Arriaga-Varela et al. 2007).

Endomychidae are collected also with bottle traps in trees, by hand- collecting or with Winkler/ Moczarski eclector and Berlese funnel. Their biology is little known, a great amount of work still needs to be done.

---Subfamily Stenotarsinae: worldwide distribution, according to Shockley (2009) "*the largest genus in the entire family, Stenotarsus (with 250, species), is a member of this subfamily and is found throughout the tropical regions of the world (although a few species extend into the Nearctic and Palearctic regions)",* specimens can be collected with Malaise, flight intercept and carrion baited traps. According to Arriaga-Varela et al. (2007) *Stenotarsus circumdatus* Gerstaecker, *Stenotarsus latipes* Arrow, *Stenotarsus marginalis* Arrow, and another *Stenotarsus* spp. have been collected in Russulaceae mushrooms (*Lactarius deliciosus* Fr. and *Russula*) and from the mycelium of *Lactarius* sp in leaf litter. It is also occur under bark and have been collected in high numbers on a trunk infested by *Sirobasidium sanguineum* Lagerh. & Pat. (Sirobasidiaceae) (Arriaga-Varela et al. 2007). Both adults and larvae of *Saula japonica* Gorham (from Africa, Southern Asia, Japan, and Philippines) prey on scale insects and phytophagous mites (Sasaji 1978). Thornton (1957) collected larvae, pupae and adults of a species of *Danae* sp. (extended over a large part of Africa, Southern Asia and Japan) from inside the umbels of papyrus sedge (paper reed) and *Cyperus papyrus* Linnaeus (Cyperaceae).

---Subfamily Lycoperdininae: Species of *Lycoperdina* spp. have internally feeding larvae, which occur in puffballs of the genus *Lycoperdon* (Lawrence 1991). *Eumorphus quadriguttatus* (Illiger) from Southeast Asia, which was reported damaging betel pepper plants, *Piper betle* Linnaeus (Piperaceae) (Mondal et al. 2003; Shockley et al. 2009). In Canada (Nova Scotia) *Lycoperdina ferruginea* LeConte has been collected in coniferous forest with pitfall trap (Majka 2007) this species have been collected also from forest litter or debris (Pakaluk 1984). *Mycetina perpulchra* (Newman) (from Canada) is associated with *Coniophora arida* (Fr.) P. Karst. (Coniophraceae), boletes associated with decaying pine and other soft species in the Agaricales (Shockley, F. cited in: Majka 2007). *Trycherus* (from Africa) larvae seem to eat lichens (Strohecker 1953) and adults are presumed to be truly mycophagous (Shockley et al. 2009). A few species of *Lycoperdina* found in Europe and North America are necrophagous and have been collected with pitfall traps (Shockley et al. 2009). *Aphorista vittata* (Fabricius) (from USA) has been found feeding directly on the putrefying flesh of pig carcasses during the advanced decay stage of decomposition (Payne and King 1970).

According to Shockley (2009) endomychids rarely come to light traps and are most often collected directly from the surface of their host fungi. Several endomychid species has been associated with epiphytes (Stuntz 2001 cited in: Shockley et al. 2009). Many species are frequently collected by beating foliage. They can be also collected with bottle-interceptor traps on the trunks of trees setting traps in the vicinity of fungi or with. Other collection methods are explained in the article "Beetles in fungi", Chapter III.

COCCINELLIDAE

Guillermo González

It is a cosmopolitan family with approximately 6,000 species described to date. There is no consensus on the structure of subfamilies, but most would recognize eight: **Microweiseinae**, **Stilochotidinae**, **Coccidulinae**, **Scymninae**, **Chilocorinae**, **Ortaliinae**, **Coccinellinae** and **Epilachninae**. This structure is derived from Sasaji (1968) with contributions made by Fürsch (1990), Kovar (1996),

Vandenberg (2002), Duverger (2003), Slipinski (2007) and others. The last two authors have risen as different classifications: Duverger (2003) showed a separation in 18 subfamilies and Slipinsky (2007) only two, but these classifications have been followed by others partially only. On the other hand, phylogenetic studies based on DNA, used classifications based on the eight subfamilies previously mentioned (Gregory et al. 2003; Hunt et al. 2007; Giorgi et al. 2009). Microweiseinae is considered the sister group to all other Coccinellidae (Slipinsky, 2007), while the phylogenetic relationships of the other families are not yet elucidated. Specifically Scymninae and Coccidulinae not appear as monophyletic groups, bringing together a group of tribes whose relationship has yet to be clarified.

The bulk of the Coccinellidae are diurnal predators and they are associated in a large proportion with Coccidae and Aphidae. Others species prey on Acari (Sthetorini), Psylloidea (Coccinellini, Ortaliinae), Aleyrodidae (Serangiinii), and others, such as *Neocalvia* ssp., prey on larvae of *Psyllobora* species. By these characteristics they are beneficial for agricultural control pests and many species are bred and sold for that purpose. Members of Epilachninae subfamily are phytophagous and they feed on the leaves of many crops, especially cucurbits and legumes, and as such they are important agricultural pests (*Epilachna paenulata* (Germar)). Species of the tribe Halyziini feed on fungi Erysiphales (Vandenberg, 2002).

Many species hibernate and are in the tops of mountains and hills. Also they are found grouped by hundreds in other places such as trees, rocks, etc., to which they return each year. Other species are congregated in barns, houses, fences, etc. Edwards (1957) has been able to suggest a probable cycle of annual migratory activity in these insects. In the Alpine areas of Washington, Oregon, and California, this cycle includes a general summer migration in July to various summits for the apparent purpose of amassing the local populations prior to mating.

The life cycle of Coccinellidae larvae includes a spring or summer larval period of approximately one month, followed by a short pupation period and adult life period of many months, including a period of hibernation in cold and temperate areas.

The most effective way to collect Coccinellidae is by beating branches of trees and shrubs, as well as sweeping on grass tall. Some species are found at water's edge and many other species can be found floating on the edges of lakes and lagoons after stormy nights. Some species come to lights. During hibernation period many species can be found on fallen branches and dry leaves, also under rocks, logs, under bark, and even inside homes and other buildings, sometimes in large groups. Direct search of specimens of plants infested with aphids or Coccidae can also yield good results. It is important to record the preys associated with collected specimens, verifying that predation is reality and not a casual contact. Larvae usually have the same food as adults and they are in the same places.

In times of migration they are occasionally found hundreds flying on certain days, and can be collected with a net. But I've seen this interesting phenomenon only twice in my life.

Guillermo González
www.coccinellidae.cl
Email: willogonzalez@yahoo.com
Santiago – La Reina
Chile.

CORYLOPHIDAE: Cosmopolytan family with 42 genera (Slipinski et al. 2009) and 284 species known (Lawrence and Thomas 2002) with two subfamilies (Slipinski et al. 2009; Lawrence et al. 2010): Adults and larvae are saprophagous or predators. Some eat spores from the pores of the polyphorous tubes.

---Subfamily Periptyctinae: According to Slipinski et al. (2009) 3 genera are know and som especies are: *Pakalukodes bimaculatus* Ślipiński, Lawrence & Tomaszewska, from Queensland (Australia), has been colleted spraying pyrethrum trees and logs, larvae are associated with adults.

Weirus tozer Ślipiński, Tomaszewska & Lawrence, has been found in open forest litter in Australia (Queensland). Some species of *Periptyctus* (in Australia also) have been colected with flight Intercept traps in *Eucalyptus* forests (Tomaszewska and Slipinski 2002) this genus occurr also in cooler, mountainous regions (Lawrence & Britton 1994).

---Subfamily Corylophinae: According to Slipinski et al. (2009) 39 genera are know, worldwide distribution. *Stanus bowesteadi* Ślipiński, Tomaszewska & Lawrence (from New Zealand) has been collected by sifted litter and rotten wood, also in lichens (with Berlese), in sooty mould and moss on *Nothofagus*, with Malaise trap and window intercept trap. According to (Slipinski et al. 2009) *Stanus tasmanicus* Ślipiński, Tomaszewska & Lawrence from Australia has been collected from *Eucalyptus* litter. *Cleidostethus meliponae* Arrow (from Africa) live in the nests of the stingless bee *Melipona alinderi* Alfken (Bowestead et al. 2001).

Corylophids are found on flowers or between plant litter and piles of decomposing plants, on moist under bark of trees, on mold that occurs on the underside of leaves with a hard consistency. Can be found in carrion or dung (White 1983) and some (*Orthoperus* and *Sericoderus*) live in vertebrate nests (e.g. nests of birds) and with caterpillars (Arnett and Thomas 2002). They can also be collected with Berlese funnel, Winkler/Moczarski eclector, by beating vegetation or spraying logs. A few species come to the lights.

LATRIDIIDAE: Two subfamilies are know: **Latridiinae** and **Corticariinae** (Lawrence et al. 2010) and according to Wolfgang Rücker (pers. comm. 2010) *"32 Genus and up to now at the most 800 valide species. Latridiids show a cosmopolitic distribution, are there is virtually known from all region on land. From Arctic regions like greenland to the tropics and up to 3200-4600 m alpine and paramo zones of the Andes in the flowerheads of Espeletia-species, Mt. Kilimanjaro in Tanzania or on Mt. Wilhelm in Papua-Newguinea has been found."* Many of which are now cosmopolitan (Andrew 2005; Audisio et al. 2003). Associated with stored products and foodstuffs, they are phyto-detriticolous and sub-corticolous. The Latridiidae and their larvae feed on molds, spores or mycelia and may be classified as mycetophag (Wolfgang Rücker, pers. comm. 2010) restricted exclusively to slime molds (mold); they feed on conidies of Pyrenomycetes and Myxomycetes fungi. They are found in plant litter, under rusty bark, among the leaf litter and on the surface of leaves where they feed on fungal spores (Corticariinae); recorded from New Zealand on crops, pastures or stored products and haystack refuse, or on the foliage of trees and shrubs (Watt 1969). *Corticaria ferruginea* Marsham has previously been recorded from Europe, Asia, America and Australian Region, where it has been found in vegetable refuse, manure heaps, moss, under pine bark, in buildings and in flour (Watt, 1969). Some Latridiidae species are found in nests of birds and mammals. The next species: *Lathridius minutus* (Linnaeus) and *Cartodere apicalis (Blackburn),* both, larval and imaginal stages to feed on hyphae ad spores of moulds. Frequently can be extremely numerous on new, incomplettely dried, plaster rendering of walls (Balfour-Browne and Tilbrook 1966). According to Wolfgang Rücker (pers. comm. 2010) *Cartodere delamarei* Dajoz, which spread to Europe through vessels, this species has been collected by sifting litter and sieving Vine branches (Rücker 1998).

According to Balfour-Browne and Tilbrook (1966) "*There is a tendency to regard some species as synanthropic, because they are so commonly met with in habitats connected with man. This is not always a true synanthropism, however, but it may be due to their obtrusiveness in such habitats. Because of their simple requirements of a moist habitat and fungal spores and hyphae, species of Latridiidae are easily transported and many have become established far from their know points of origin.*" Many of these species are common on any damp surface where moulds have developed (in grass tufts, haystacks or flood refuse) and in nature such is their more usual habitat (Balfour-Browne and Tilbrook 1966). Latridiidae can be collected by beating or with pitfall traps using a simple mixture of water, detergent and ethylene glycol (Sackmann 2006).

According to Wolgang Rücker (pers. comm. 2010) Latridiidae can be found in forested environments where decomposing vegetation, in leaf litter, on herbaceous vegetation, on trees, in decomposing wood, in bird, mammal and Hymenoptera nests.

The best sampling method for Latridiids is the sifting from old grass, hay or straw, wilts leaves at trees and all mouldy vegetables (Wolfgang R. pers. comm. 2010). Also they can be collected with

Berlese funnel, Malaise trap or fumigating logs. Wolfgang explains that colleagues collected in Africa many Latridiidae with pyrethrum fog in the treetops. So he thinks that there are probably many species that live only in the canopy. He is convinced, in South America discovering another gigantic number of new species of minute insects will give. Other collection methods are explained in the article "Beetles in fungi", Chapter III.

Superfamily TENEBRIONOIDEA

MYCETOPHAGIDAE: Family with worldwide distribution, 25 genera and 145 species are known (www.insects.tamu.edu). With 3 subfamilies: **Esarcinae**, **Mycetophaginae** and **Bergininae** (Lawrence et al. 2010).

They feed on Hymenomicetes or ascomycetes fungi. They can be found in decaying vegetation and humus, eating on poliphores soft tissues that grow into fruiting and decaying bodies of Xylariaceae fungi. *Typhaea stercorea* (Linnaeus) has been found in carrion in the Hawaiian Islands (Goff et al. 1986) although it is most often referred to as a pest of stored products from warmer climates (Hinton 1945, Brower and Press 1992, Dowdy and McGaughey 1994, Hagstrum et al. 1994, Throne and Cline 1994). *Triphyllus* are associated with sooty mould (*Ascomycota*), which form thick black "wefts" of mycelia of up to six different species (Hughes 1972: 226), growing generally on the surfaces of foliage, branches, tree trunks, and often surrounding rocks and soil (Gimmel et al. 2009). These growths are associated with sugary exudates of leaves or with the "honeydew" produced by sternorrhynchous Hemiptera, primarily in association with *Nothofagus*, but other plants as well (Klimaszewski & Watt 1997, Carlton & Leschen 2007; Gimmel et al. 2009). According to Hinton (1945) *Berginus tamarisci* Wollaston feed on pollen of *Pinus maritima* Lam., and on stamens and pollen of *Cereus grandiflorus* (Linnaeus). A numbre of species undergo their entire life-cycle inside fungi, particularly fleshy hymenomycetes, others species occur under bark and in haystack and other vegetable materials (Hinton1945).

To collect mycethpophagids beetles a Berlese funnel or pitfall trap can be used checking fallen and decayed mushrooms, for example (in Calabria, Italia) according to Fratelli (1913) collected several *Esarcus fiorii* Reitter, by sifting leaves and debris at the foot of an Oak. Some come to the lights. Other collection methods are explained in the article "Beetles in fungi", Chapter III.

ARCHEOCRYPTICIDAE: 10 genera with 50 species are known (Lawrence 1994b; Young 2002; and www.insects.tamu.edu). They are distributed in South and Central America, the Southern parts of North America, the Indian Subcontinent, Southern China and Australia. This distribution pattern suggests that the family is a classical Gondwanan element. Two species are distributed from Eastern Africa to Southern Africa, one extending its area to the south of the rainforests up to the south-west of Angola. These species apparently avoid the dry regions. All African species of the genus live in rainforests, coastal forests, gallery forests or forest islands. Therefore their occurrence depends directly on the conservation of these forests (Lillig 2004). The family comprises *Enneboeus* (Southern USA, South and Central America, Eastern and Southern Australia), *Pseudenneboeus* (Brazil), *Sivacrypticus* (throughout Africa and Asia, one specie in Southern Australia) and *Archeocrypticus* (South America, New Zealand, Tasmania). The remaining six genera (*Australenneboeus, Enneboeopis, Falsoplatydema, Gondwanenneboeus, Nothenneboeus* and *Wattianus*) are restricted to Australia (Lawrence 1994b). They feed on the body of Polyphoros mushrooms. Larvae of the genus *Enneboeus* are known to live in leaf litter and consume rotting plants (Lawrence 1977). The *Enneboeus* genus has been found in the topsoil of the forest floor, in decaying vegetable material, and in accumulations of fallen flowers. The species *Enneboeus ovalis (Casey)* has been collected with light in San José (Costa Rica) and North America. Besides, other undescribed species are known in Southern Guanacaste (Costa Rica). *Enneboeus barrocolorado* Merkl was collected in Panamá. No one knows almost anything about the biology of these rare beetles.

They have been most commonly collected by extraction from litter using Berlese funnels or Winkler/Moczarski eclector, or with pitfall traps. Occasionally with light too (Matthews 1987) and also fumigating logs, but, some Australian species have been associated with the fruiting bodies of Polyporaceae (Young 2002).

PTEROGENIIDAE: Indo-Australian distribution. 7 genus and 26 species are known (Löbl 2005). The majority of the species is restricted to insular tropical Asia and Oceania but 4 of them extend their range onto the Malayan Peninsula. *Histanocerus,* with 14 species from Philippines, Vietnam, Papua New Guinea, Java, Indonesia and Malaysia and 4 species are known only from continental Asia, 2 species of Southern India, one on each side of the Malayan Peninsula and Vietnam respectively. Species of *Kryptogenius* (5 spp.), *Pterogenius* (2 spp.), *Katagenius* and *Tychogenius* (2 spp.) are highly endemic and could therefore potentially be useful for analyzing areas of endemism (Burckhardt and Löbl 2008). *Anogenius clypealis* Löbl & Leschen from Malaysia and *Laenagenius apterus* Löbl in China, both adults and larvae feed on fruiting bodies of macrofungi, especially polyporales (Lawrence 1991; Burckhardt and Löbl 1992). They can be collected by reviewing the mushrooms where they live.

CIIDAE: With worldwide distribution. 42 genera and 617 species are known (www.insects.tamu.edu) in two subfamilies: **Sphindociinae** and **Ciinae** (Lawrence et al. 2010). They have mycophagous habits (Thayer and Lawrence 2002). Mainly found in the timber or shelf fungi (wood ears) usually after they have released the spores. They can be found from a few to thousands of specimens. Their presence is noticed immediately by the holes left by their galleries on the surface of fungi. The complete life cycle takes place in the fungus and lasts from 8 to 10 weeks. According to Thayer and Lawrence (2002) they are occasionally found under bark of logs, in rotting wood (logs, vines, or branches), or in the galleries of Scolytinae, but them are probably associated with fungi found there (Lawrence 1973; Thayer and Lawrence 2002). For example, Hawkeswood (2003) has observed large numbers of *Cis victoriensis* Blackburn, feeding and breeding in the dead, weathered basidiocarps of *Polystictus cinnabarinus* (Jacq. ex Fries) Cooke (Polyporaceae) growing on a fallen log of a *Casuarina* species (Casuarinaceae) in Brisbane, Queensland (Australia), also Hawkeswood collected in dead *Polystictus cinnabarinus* on a dead branch of the Narrow-leaved Ironbark, *Eucalyptus crebra* Muell., at Rungli Downs, via Helidon, South-Eastern Queensland.

They can be collected fumigating the fallen logs with fungi, or placing fungi in a Berlese funnel. They can also be manually collected breaking mushrooms in which they live. Other collection methods are explained in the article "Beetles in fungi", Chapter III.

TETRATOMIDAE: 23 genera and 218 species (www.insects.tamu.edu), worldwide distribution. According to Lawrence et al. (2010) five subfamilies are known: **Tetratominae**, **Piseninae**, **Penthinae**, **Hallomeninae** and **Eustrophinae**. Some genera are: *Allopterus* (from New Zealand), *Ctenoplectron* (Tasmania and New Zealand), *Eustrophopsis* (Neartic, Neotropical and Afrika), *Eucinetomorphus* (from Asturias, Cameroon, Ivory Coast and Algeria), *Mycetoma* (from Russia and Japan), *Cyanopenthe* (from Thailand and India), *Notopisenus* (from Chile), *Abstrulia* (from Europe, North America and Japan), *Tetratoma* (Palaeartic region). All they are saproxylic beetles dependent during some portion of their life cycle upon the dead or dying wood of senescent or dead trees (standing or fallen) or on the presence of other saproxylics beetles (Speight 1989; Pagola 2007). They feed on Hymenomycetes fungi, adults feed on the surface of fungi and larvae within them. Some larvae of the subfamilies Hallomeninae, Eustrophinae are found in polyphores, sporophores, porifera and bracket fungi. To collect these beetles should use the techniques explained in the article "Beetles in fungi", Chapter III.

MELANDRYIDAE: With global distribution. 65 genera and 430 species are known (www.insects.tamu.edu). In two subfamilies: **Melandryinae** and **Osphyinae** (Lawrence et al. 2010). They feed on himenomicetos fungi, ear fungus, and some white fungus in rotting wood. Some adults live on foliage or flowers. *Doxozilora* (endemic to New Zealand) are associated with sooty mould (Ascomycota), which form thick black "wefts" of mycelia of up to six different species (Hughes 1972: 226), growing generally on the surfaces of foliage, branches, tree trunks, and often surrounding rocks and soil (Gimmel et al. 2009). These growths are associated with sugary exudates of leaves or with the "honeydew" produced by sternorrhynchous Hemiptera, primarily in association with

Nothofagus, but other plants as well (Klimaszewski & Watt 1997, Carlton & Leschen 2007; Gimmel et al. 2009). Some larvae of Orchesiini are found in polyphores, sporophores, porifera and bracket fungi. Adults mostly nocturnal and crawl over dead wood and fungi at night (Pollock in: Arnett et al. 2002) which is the best time to collect them. Larvae of Melandryidae life under bark in dry wood, fungi or decaying logs or stumps, a few live in forest litter a few are *carnivorous* others eat plant material (White 1983). They may be collected with pitfall traps. Very little is known of the habits of this species. *Phloeotrya mexicana* (Champion) from Dominica, Central America, Mexico, South America, larvae was found in rotted stump in area being cleared at 1900 feet (Peck 2006). Others are often found on dead trees, under bark or in logs and stumps at night (White 1983); others species are found on carrion. Some come to the lights. Lawrence (1991) provided a detailed discussion of the feeding habits of various groups of Melandryidae. Other collection methods are explained in the article "Beetles in fungi", Chapter III.

MORDELLIDAE: About 1,500 species (Jackman and Lu 2002), but, now there are 115 genera and 2,308 species known (www.insects.tamu.edu), cosmopolitan family wit two subfamilies: **Mordellinae** and **Ctenidiinae** (Lawrence et al. 2001). Adults are found in plant tissues and feeding on pollen of many flowers, especially in umbelliferous (Apiaceae) and composite (Asteraceae) (Jackman and Lu, in: Arnett et al. 2002), sometimes on dead trees or logs; larvae feed on rotten wood; in pith of weeds; some are cecidogenous and others are fungivorous. Some small species can be collected by sweep from leaves or flowers, should be reviewed mainly leafy shrubs with many small flowers especially at the edges of the forest and more open areas such as savannas, grasslands, etc. (pers. observ.); larger species are often found under bark or in rotting logs and stumps at night, other species can be found in carrion (With 1983). Beating differents plants is also very effective method to get many species. When they are captured, a shadow must be produced on the insect body, because with sunlight they are very agile to escape. Adults are occasionally attracted to lights. Some can be reared from dead wood. These beetles must be handled carefully, because they have a body quite fragile.

RHIPIPHORIDAE: 41 genera (www.insects.tamu.edu) with. 400 described species are known (Batelka 2005, 2007), worldwide distribution and according to Falin (2004) and Lawrence et al. (2010) there are five subfamilies know: **Ptilophorinae**, **Pelecotominae** (incl. Micholaeminae), **Hemirhipidiinae**, **Ripidiinae** and **Ripiphorinae**. Adults are among the foliage and flowers. Some larvae are predators and others are ectoparasites of insect larvae. Larvae are triunguline and are found in flowers; many are attached to the body of Hymenoptera of the families Andrenidae, Scoliidae and Tiphiidae to be carried to their nests, where they eat the eggs, larvae and contents of the cells; others attack young cockroaches for example, according to Batelka (2007) as host are given without further detail potter wasps (Vespidae) of the genera *Odynerus* and *Eumenes* and Megachilidae (*Megachile*) (Bétis 1912). *Pelecotoma fennica* (Paykull), from Europe, is confined to old deciduous trees devoid of bark, namely poplars and willows with borings of the *Ptilinus fuscus* Geoffroy (Anobiinae) in whose larvae it develops (Falin 2002). In Europe *Ripidius quadriceps* Abeille is parasitic of larvae of forest cockroaches of the genus *Ectobius* (Besuchet 1956) and *Ripidius pectinicornis* Thunberg, spreads to harbours through shipping with its host cockroach *Blatella germanica* (Linnaeus) (Falin 2002). According to Hudson (1934) in New Zealand *Rhipistena cryptarthra* Broun is parasitic of *Prionoplus reticularis* White (Cerambycidae). According to Hudson (1934) in New Zealand *Rhipistena cryptarthra* Broun it is parasitic of *Prionoplus reticularis* White (Cerambycidae: Prioninae). Nothing is known of the biology of three genera of Pelecotominae (*Ancholaemus*, *Clinops*, and *Micholaemus*), nevertheless, they are likely closely related to *Pelecotoma*, *Allocinops* and *Rhipistena*, and if so they may also have beetles-wood-boring larval hosts (Falin 2002). But, *Ancholaemus acuminatus* Fairmaire, from Dominica, Brazil, Ecuador (Galapagos Islands), Panama, adults are usually collected at light and larvae are endoparasitic in wood-boring beetle larvae (Peck 2006). *Macrosiagon octomaculatum* (Gerstäcker), from: Barbados, Dominica, Guadeloupe, St., Vincent. Central America, South America and United States, adults can be collected by beating vegetation and larvae, are probably hypermetamorphic and parasitic on larvae of wasps (Bembicidae, Tiphiidae, and Scoliidae) (Peck 2006).

When you found parasitized hosts, these must be maintain live and put into a box with a grid (5 mm mesh), so that the box is separated into two compartments, one above (with the wasp or cockroach) and a lower one where the rhipiphorid will fall when escaping the body of the Hymenoptera; if you don't make this way, then the wasp will attack the beetle when it arises from its body. These beetles are rarely collected because they have short lifespan (White 1983). Nevertheless, many species are collected by beating and sweeping vegetation, mainly flowers. Males are infrequently collected however females are more likely to be collected because of their habit of laying eggs on flowers (White 1983).

ZOPHERIDAE: According to Lawrence et al. (2010) there are two subfamilies: **Colydiinae** and **Zopherinae:**

---Subfamily Colydiinae: With 180 genera and 1,300 species are known (Costa 2000), worldwide distribution, and according to Lawrence et al. (2010) most are mycophagous and feed on decaying plants. They are found in rotting wood, under bark, animal material, spores and hyphae of the fruiting bodies of pyrenomycetes fungi. Some larvae are predators, often these species are thin, subcylindrical, allowing them to penetrate the caves formed by their victims; other Larvae are mostly mycetophagous; they are found in decaying plant material, in rotting wood, in or under bark, and in fungi (Sharp 1876). Others larvae attack collections of insects and animal skins.

Syncalus (with 3 species) are found among the dead leaves and decaying vegetable matter of the New Zealand woods and forests (Sharp 1876). *Euxestes erithricus* (Chevrolat) has been found in carrion in the Hawaiian Islands (Goff et al. 1986). *Tarphius* with about 68 species distributed in Spain, the Canary Islands, Azores Islands, Bhutan and India, many of these species are found in dead leaves and decaying vegetable matter in woods (Sharp 1876). Also according to Sharp *Colydium elongatum* (Fabricius) (from Europe) live in burrows in wood found on tree ferns (*Cyathea dealbata* (Forst.)).

---Subfamily Zopherinae: With worldwide distribution, 179 genera and 803 species (Michael Ivie in: www.insects.tamu.edu). Adults feed on fungi. Larvae feed under the bark of dead trees (especially dead white mushrooms). Frequently specimens of the tribe Zopherini are encrusted with a greasy exudate as well as accumulated environmenal debris (Foley & Ivie 2008). They always live under rusty bark in lignicolous fungi, lichens or rotting leaves (Harde & Severa 1984; Zahradník 1990). *Zopherosis georgei* White occurs in higher elevation rainforest and wet sclerophyll forest in Australia (Foley & Ivie 2008).

Adults feed on fungal bodies (Lawrence & Britton, 1994) and vegetable material (Williams 1993) Hawkeswood (2003) has found feeding it on the fresh fruiting bodies of *Ganoderma applanatum* (Fries) Karsten (Polyporaceae). *Pteroderes tuberosus* Germain had been collected only in scanty altered habitats of Chilean forest (Valdivia), in localities of the IX and X Regions (Grez et al. 2003). According to Lafer (2002) in the forests in Southern Primorsky Krai, solitary *Phellopsis* has been found and appear to be very abundant. Lafer observed colonies from 2 to 3 dozen beetles on one tree, in the area of dark coniferous taiga in Sikhote Alin, which includes mainly 2 types: pines and birches and other deciduous trees. The same author has observed *Phelopsis* sp. at the base of these trees in late summer, apparently this genera is one of the relicts of the Southern conifer taiga. Although Lafer found colonies *of Phellopsis amurensis* Heyden were widely dispersed along the trunk and did not gather in groups, and appear to feed on Poriaceae tissues that fructified under the juicy fruit (Lafer 2002). By other hand, García-París et al. (2001, 2006) had collected two species of tribe Zopherini in Australia: *Sesapis doyeni* García, Parra & Coca, from under loose bark of dead pine stumps, and *Sesapis sylvatica* García, Coca & Olea from rotten fungus covered logs in cloud forests; some species of *Sesapis* have been collected under bark of *Quercus* sp. in Cuba and In the Sierra Maestra (Cuba) had been collected *Nosoderma turquinense* (Garrido) under bark (Foley & Ivie 2008). Those of the Monommatini tribe eat decaying vegetable material and are found in decaying wood; they are collected in pastures, in foliage, dead cactus, plant litter, dead twigs and under bark of stumps. *Syrphetodes* species are found under loose bark of standing dead trees or logs and in rotten branches, adults are also found on the underside of logs (Klimaszewski & Watt 1997).

Zopherids species are collected by beating and sweeping vegetation, also with Berlese funnel or Winkler/Moczarski eclector and with pitfall traps; fumigating logs; some species are attracted to lights. Other collection methods are explained in the article "Beetles in fungi", Chapter III.

ULODIDAE: According to Leschen and Rhode (2002) until that year were 30 species known in this family, but now are 41 species in 13 genera known (www.insects.tamu.edu).

According to Leschen and Rhode (2002) they are distribuited in Australia (*Dipsaconia*, *Ganyme*, *Meryx*, *Notocerastes*, *Phaennis* and *Ulodes*), Chile (*Trachyderas*), New Caledonia (*Trachyderastes*) and New Zealand (*Arthopus*, *Brouniphylax*, *Exohadrus* and *Syrphetodes*). New Zealand species (except *Arthopus*) are flightless (Leschen and Rhode 2002). They has been collected and rared from polypore fungi and from rotten wood (Doyen & Lawrence 1979; Klimaszewski & Watt 1997; Leschen and Rhode 2002). *Brouniphylax* have been reared from *Gonoderma* (and possibly *Fomes*), other New Zealand species have been collected from rotting wood or fungi (Leschen and Rhode 2002). *Syrphetodes* (Larvae and adults) has been collected from dead terminal branches of *Coprosma*, *Aristotelia* and *Olearia*, few adults has been collected from fungi, including *Gonoderma* (Doyen and Lawrence 1979). *Exohadrus* (in New Zealand) has been collected from dead wood, including *Kunzea* (Leschen and Rhode 2002). *Ganyme sapphira* (Newman) has been collected in big logs-rolling of *Eucalyptus* in Tasmania (Grove et al. 2006). *Brouniphylax* sp. has been collected with pitfall traps in Maungatautari (Corinne 2007).

PROMECHEILIDAE: 20 species are known (www.insects.tamu.edu), mainly in the Southern hemisphere (Lawrence 1994). Some species like *Perimylops antarcticus* (Müller) and *Hydromedion sparsutum* (Müller) can withstand extremely cold temperatures until - 4 degrees Celsius (Worland et al. 2004). They feed on epiphytic lichens and algae growing on the rocks, and in a second place they prefer mosses, herbs and other plants of lower height (Block et al. 1988).

They can be found between live and dead plant material (Watt 1967a).

CHALCODRYIDAE: Distributed in Tasmania and New Zealand. There are 15 species in 3 genera known (Watt 1974). They eat plants and decaying plant material. Adults of *Chalcodrya, Phylpottia* and *Onysius* have been collected by beating in branches covered with lichens and moss, especially from *Nothofagus* wet forest. Larvae of some species have been collected especially at night, in branches covered with moss, and they are quite agile when they leave their galleries to feed on lichens growing on the trunks of trees (Wat 1974).

TRACHELOSTENIDAE: With 2 species from Chile (Elgueta & Arriagada 1989) and 2 from Perú (www.insects.tamu.edu): *Trachelostenus inaequalis* Solier (Southern Chile, Valdivia) have been found under bark of *Nothofagus* (according to L. Friederichsen & Co., 1907 pag. 90). *Trachelostenus fasciculiferus* Philippi & Philippi can be found under bark in Valdivia (Chile) (Fairmaire et Germain 1862). Elizabeth Arias (http://essigdb.berkeley.edu) collected some specimens of *Trachelostenus* sp., by fogging in the tops of trees in Southern Chile. .Another species are: *Lagriola denticulata* Kirsch and *Lagriola operosa* Kirsch, both species from Peru.

TENEBRIONIDAE: Cosmopolitan family, with over 18,000 described species, with 10 subfamilies (Lawrence et al. 2010): **Lagriinae**, **Nilioninae**, **Cossyphodinae**, **Phrenapatinae**, **Zolodininae**, **Pimeliinae**, **Tenebrioninae**, **Alleculinae**, **Diaperinae** and **Stenochiinae**.

They are popularly known as Darkling beetles. There are micetophagous, coprophagous, predators, mirmecophagous, poliphagous or omnivorous species. Predaceous tenebrionids exist mainly as larvae. These are found in bark beetle galleries (Hypophloeini), in nests of birds (Alphitobiini) or in palms (larvae of Triboliini) fronds where they prey on other beetle (Triplehorn 1990; Aalbu et al. 2002). Saprophagous feed on dead plants, including leaf litter and rotting wood. Thre are diurnal or nocturnal species. Most are specialized to live in very arid habitats but many species can live in swamps or very humid places. Other species attack stored plant material. According to Matthews and Bouchard (2004) very few Darkling beetles pupae have been described because of its short duration and sedentary nature of the pupal stage.

Some species of Eleodini, Pimeliinae and Opatrinae are associated with nests of vertebrates or insects (ants and termites) (Navarrete-Heredia 2001; Aalbu et al. 2002; Steiner 2004 and other authors). These include rodents, some Alphitobiini, Tenebrionini. *Myrmechixenus* spp. (Diaperini) from Europe, are found in nests of ants and birds, in burrows of various rodents, in compost heaps and in the caves (Stebnicka 1999). Some a few Triboliini and Amargymini are also associated with termites. There are species of Triboliini associated with bees. Some Helopini adults have been observed feeding on lichens on tree trunks at night (Steiner 1999); Helopini may also hide under loose bark of dead standing wood. Burlap bands around trunks (for monitoring and control of gypsy moths) often harbor clusters of adult beetles, theirs larvae and pupae inhabit the soil (Steiner 2009).

Following subfamilies are sometimes considered as separate families Tenebrionidae:

---Subfamily Lagriinae: Adults feed on the foliage and they have nocturnal habits. Larvae feed on leaf litter or foliage and are found in stumps and dead bark. To collect them it is necessary to look, during the day, under dead bark and rotting wood. Occasionally they can be collected by beating vegetation and at night with a sweeping bag. Many come to the lights.

---Subfamily Alleculinae: Adults are typically crepuscular or nocturnal (Harde & Severa 1984; Zahradník 1990) but other are diurnal and feed on pollen as adults. Many are walking on the trunks of living or dead trees, especially on hot nights. Their larvae can develop in the soil, based on various plant debris, while another group of species are authentic sapro-xylophagous, with adults and larvae living and feeding on decaying wood of deciduous trees. Many lives and develops under bark, on the trunks, on their lignicolous fungi of recently dead trees and/or attacked by Scolytinae (Zahradník 1990). Some are found in rotting hollow trees. Other are associated with nests of insects (ants and Termites), include nests of rodents and birds. *Lobopoda ebenina* Champion Adults feed in Grenada Island they are found on lichens or algae on tree trunks, and larvae feed on humus (peck 2006).

They can be collected by hand-collecting, by beating and sweeping vegetation, also with Malaise trap, placing bottle intercept traps in trees or fumigating logs, many come to the lights.

How to collect Tenebrionidae:

To collect different species of Tenebrionidae all types of ecosystems must be visited, through the jungles, deserts and high mountains. In almost every imaginable place there are representatives of this family.

In the arid places it is better to walk with a flashlight at night and examine the ground.

Wherever there are trees or shrubs, the logs must be examined during the night because there these beetles are looking for fungus. In these ecosystems there are also many species under the bark. Many species feed on detritus along the line of high-tide, algae touching the stones and wood surfaces. In these cases Darkling beetles can be collected by hand-collecting, or removing algae and organic detritus and immersing them in water to trap the beetles as they flee.

The methods used to trap Carabidae in the same situations (see above) can be used to collect Darkling beetles living on aquatic vegetation in rivers, streams, ponds, etc.

Darkling beetles are manually collected or with beating sheet, Berlese funnel, Winkler/Moczarski eclector, light traps, purple trap, Malaise traps and other intercept traps (e.g. bottle intercept traps) can be used on the trunks of living or dead trees. The pitfall traps, baited or not, are also a good sampling method.

Sweeping vegetation is a good method to collect them also, especially where there are flowers. During the day and night, it is necessary to check the fungi that cling to the trunks of dead or alive trees, on the superficial moss of wood, bark, stone surfaces, some fruits of hard consistency like chestnuts or similar, but rotten.

Darkling beetles are also under sticks, inside logs (or under these), under stones and under any object in almost all available habitats. Generally, the beetles living in all these habitats can be found by hand-collectig. It is also very productive to spray trees, stumps or logs. Careful inspection of various organic matters will often yield tenebrionid larvae. In synthesis, to collect these beetles must be use most of the methods explained in this book. It is also suggested to study the work of Steiner (2005) where this author shows several techniques to collect Tenebrionidae.

PROSTOMIDAE: Young (1991, 2002) cited two genera and about 20 species, but now 32 species are known (www.insects.tamu.edu). They are found almost everywhere except South America.

Prostomis are known from Western North America, Eastern Asia, Turkey, Himalayas, Japan, Australia, New Caledonia, and South Africa (Fleischer 1919; Schawaller 1991, 1992, 1993, 1994, Young 2002). In Europe there is only one species, *Prostomis mandibularis* (Fabricius), which is also in the Caucasus, in Elburs Mountains and North America (Fleischer 1919; Schawaller 1993; Hatch 1961). *Prostomis mandibularis* (Fabricius) is considered a rare species in Central Europe, in the Czech Republic occurs locally in well conserved primeval forest biotopes, in rather moist wood of deciduous trees decomposed by the red putrefaction (Farkač et al. 2005), this specie is more common in Northern and Eastern Europe. *Prostomis atkinsoni* (Waterhouse) were collected in logs from *Eucalyptus obliqua* L'Herit, in wet eucalypt forests in Southern Tasmania (Yee et al. 2006). In Japan *Prostomis latoris* Reitter, has been collected from decaying coniferous wood (Nodoka 1969). *Prostomis mordax* Reitter, are associated with fungus and destroyers of brown rot wood in the Far East (Krivosheina 1991).

Dryocora are known from Australia, New Zealand, Tasmania and Sri Lanka (Young 2002).

Prostomids beetles live in decaying wood of coniferous and others trees. Larvae and adults are found generally in dead logs inside a characteristic clay-like material between layers of the decaying wood fibers (Young 2002).

They can be captured with Berlese funnel, Winkler/Moczarski eclector and by hand-collecting. Larvae were collected from infested trees and fixed in bouin. Pupae were obtained from larvae that were reared in the laboratory in petri dishes filled with decaying wood from the infested tree (Schunger et al. 2003). In Tasmania, Grove et al. (2006) have collected *Prostomis atkinsoni* Waterhouse from big rolling-logs.

SYNCHROIDAE: Young (1991, 2002) cited 3 genera with 8 species, but now are 10 species known (www.insects.tamu.edu), occur in Indonesia, Japan and Eastern North America (Young 2002).

Know species feed largely upon fungal material and decaying wood, and they are commonly found under bark of dead deciduous trees (Payne 1931). According to Harned (1906), Towwnsend (1863) and other authors, larvae live under bark of dead trees (e.g. willow, oak, elm, etc.) and feed on rotting wood and fungi. Adults and larvae of *Synchroa punctata* Newman (from Canada: Nova Scotia and Prince Edward Island) feed largely upon fungal material and decaying wood, commonly under bark of dead deciduous trees (Payne 1931). In Maine, has been collected from *Abies* and *Picea* (Dearborn and Donahue 1993; Majka & Pollock 2006). Other saproxylic species from US (Ohio) is *Mallodrya subaenea* Horn (Young 2002; Majka and Pollock 2006).

Synchroids can be collected at light or as they run over the surface of logs (Young 2002), also can be collected at Malaise, with Winkler/Moczarski eclector or Berlese funnel, pitffal traps either unbaited or baited with ethanol (young 2002) and by hand-collecting.

STENOTRACHELIDAE: Family distributed in U.S.A., Canada, Japan and East Asia, with 10 genera and 29 species (www.insects.tamu.edu). According to Lawrence et al. (2010) four subfamilies are known: **Stenotrachelinae**, **Cephaloinae**, **Nematoplinae** and **Stoliinae**.

They are probably short lived (Young 2002). *Nematoplus* are associated with logs infested with Brown rot fungi (wood in the red rot stage of decay) (Krivosheina 1991; Young 2002). Some larvae live in decaying coniferous wood and others in tree woods inside the jungle. Some larvae have been found in decaying wood of coniferous and hardwood trees in forests (White 1983).

Stenotrachelides are rarely collected and they are found in flowers and foliage by sweeping and beating on different plants or with Malaise traps and flight intercept traps (Young 2002). Species of *Cephaloon* can be collected on flowers or foliage. *Stolius vagepictus* Lewis sometime can be found in Japan by beating vegetation and in grapevines in spring and autumn (see: http://tskk.exblog.jp/). *Nematoplus* species are associated with fungi (Krivosheina 1991).

OEDEMERIDAE: Family with 115 genera and approximately 1,500 species (Britton 1970; Kriska 2002) with worldwide distribution (Lawrence et al. 1999a; Kriska 2002). According to Lawrence et al. (2010) there are 3 subfamilies: **Polypriinae**, **Calopodinae** and **Oedemerinae**.

Adults feed on pollen or nectar; larvae are lignivorous feeding on sticks and rotten wood. A few species can be agricultural pests because they build galleries within the stems of some crop plants (Yerba mate, etc). In France, adults of widespread species *Chrysanthi viridissima* (Linnaeus) are found in flowers while larvae, which are sapro-xylophagous, live in coniferous. Oedemerids are most abundant near oceanic coastlines, particularly on Islands (Hudson 1975).

Polypria cruxrufa Champ. One specimen beaten from *Mimosa* sp. in palmetto thicket at Santo Tomas (cited in the page 90: Vol. 5-11 of Tex. Acad. of Sci. 1903.).

Also, according to López-Colón (2001) the following European species can be found in:

Chrysanthia viridissima (Linnaeus): in *Pinus sylvestris* Linnaeus, *Pinus nigra* Arnold and *Pinus pinaster* Ait, from Eurosiberia Region and Spain. *Chrysanthia reitteri* Seidlitz: in *Pinus halepensis* Miller and *Pinus pinaster* Ait, from Portugal and Spain.

Nacerdes melanura (Linnaeus), it is a cosmopolitan species, infest pilings and is capable of withstanding immersion by the tide (Arnett 1951; Kriska 2002). Larvae of *Xanthochroina bicolor* (LeConte), from US, have been found in wooden flooring and railroad ties (Arnett 1951).

Adults of *Asclera* genus, from North America, are found on flowers and foliage of various plants in US (Kriska 2002). Arnett (1951) cited host plant for each species.

Anogcodes seladonius seladonius (Fabricius): on several flowers, always near from wet places, from Spain and Portugal (Vázquez 1993). *Probosca unicolor* (Küster): in inflorescences of *Thapsia villosa* Linnaeus, from Portugal, Spain and Marruecos (Vázquez 1993, 1995). *Oedemera marmorata* Erichson: larvae are found in twigs (falls and rotten) of *Quercus ilex* Linnaeus and *Quercus rotundifolia* Lam., and possibly *Quercus suber* Linnaueus, where it develops and pupation occurs. Adults come to light (Vázquez 1993). It is known from Marruecos, Argelia, Túnez, Portugal and Spain.

Stenostoma rostratum (Fabricius): distributed by the Atlantic-Mediterranean from the coast of Brittany (France) to the Greek islands of Europe and the Maghreb countries of North Africa, found only on sandy coasts and especially in the dunes. Its natural habitat is sandy beaches with coastal dunes (Vázquez 1995).

Oedemerids found on islands can be particularly abundant and attracted to light in large numbers (Arnett 1968; Kriska in: Arnett et al. 2002) in other places also come to lights, but in smaller quantities. Adults are found on flowers and foliage, under driftwood or in moist to wet rotten logs (Arnett 1968; Hudson 1975).

MELOIDAE: Family of worldwide distribution, with 120 known genera and 2,500 species (Bologna and Pinto 2001; Selander and Fasulo 2003). According to Lawrence et al. (2010) there are 4 subfamilies: **Eleticinae, Meloinae, Tetraonycinae** and **Nemognathinae**.

Larvae are predatory and attack the masses of eggs on leaves and flowers found in decomposed plant material. Adults feed mainly on foliage of various plants, and they can be diurnal or nocturnal. Most eat sweet potatoes, tomatoes or other Solanaceae (Borror and Delong 1969). Some larvae gain access to the bee nests and feed on their eggs. *Epicauta* larvae eat the eggs of grasshoppers (White 1983). According to Pinto and Bologna (in: Arnett et al. 2002) larvae all bee parasitoids, reach their host nest by phoresy (except in Eleticinae), attached to the adult Bees visiting flowers oviposition occurr on vegetation, or in *Cissites*, *Hornia* and *Tricrania*, in or adjacent to nests of host bees. Some rare cases of parasitism on sphecid and masarid wasps also occur in certain meloids species (Arnet et al. 2002). Many larvae of Eleticinae probably are subcorticolous predators (Pinto et al. 1996). First instar larvae of Nemognathinae found in flowers or attached to the hairs of bees are sometimes mistaken for those of Rhipiphoridae (Selander and Fasulo 2003). Oil beetles (Meloinae) secrete an oily substance that protects them from predators. According to Genaro (1996) in Cuba island *Nemognatha punctulata* LeConte is a parasitic of *Megachile poeyi* Guérin and *Cissites maculata* (Swederus) has been collected in nests of *Xylocopa cubaecola* Lucas (Hymenoptera: Anthophoridae).

According to Selander and Fasulo (2003) eggs are laid in masses in the ground or under stones (Meloinae) or on the food plants of adults (Nemognathinae) and most species pass the winter or dry season as coarctate phase larvae, while a few do so as diapausing eggs, triungulin larvae, or adults. Selander and fasulo (2003) also explain *"adults commonly live three months or more. Females typically mate and oviposit periodically throughout their adult lives."* A rare case is *Epicauta atrata* (Fabricius) larvae that prey on other primary stages of *Epicauta* larvae (Selander 1981). Poisonings have been traced to several species. Blister beetles pose a potential threat if horse owners use alfalfa as a source of hay (Selander and Fastulo 2003).

Selander and Fastulo (2003) give the following list of plants that are attacked by Meloidae in the United States:

Epicauta cinerea (Forster): on *Clematis* spp.; *Epicauta excavatifrons* Maydell is recorded from grass; *Epicauta fabricii* (LeConte): on Leguminosae, including alfalfa, *Baptisia*, bean, pea, and sweetclover, potato and glandless cotton. Often taken at lights; *Epicauta floridensis* Werner: on *Ipomoea* (Convolvulaceae), *Schrankia* (Leguminosae), and (in captivity) *Solanum* (Solanaceae); *Epicauta funebris* Horn: on many Leguminosae and Solanaceae, alfalfa, beet, eggplant, potato, soybean, sugar beet, and tomato, on Amaranthus, and in *Cynanchum nigrum* (Linnaeus); *Epicauta heterodera* Horn: on *Helenium* and other Compositae; *Epicauta obesa* (Chevrolat): on *Clematis, Amaranthus*, alfalfa, *Tribulas* and tomato; *Epicauta pensylvanica* (De Geer): attach wide variety of plants, including many Compositae, such crops as alfalfa, beet, and potato, It is most commonly taken on inflorensences of *Solidago, and in Mexico on Amaranthus hypocondriacus* Linnaeus (Amaranthaceae) (Salas-Araiza and Boradonenko 2006); *Epicauta sanguinicollis* (LeConte): on Compositae, *Schrankia* (Leguminosae), Asteraceae, and cotton; *Epicauta strigosa* (Gyllenhal): principally on cotton, okra, Compositae, *Opuntia* (Cactaceae), *Ipomoea* (Convolvulaceae), and *Vigna* (Leguminosae); *Epicauta torsa* (LeConte): on *Ilex* (Aquifoliaceae), *Sapindus* (Sapindaceae), *Albizzia*, *Amorpha*, *Robinia* (Leguminosae); *Epicauta vittata* (Fabricius): on wide variety of plants, including Amaranthaceae (*Amaranthus*), Solanaceae (*Solanum*) and Fabaceae (*Medicago*, alfalfa), crops as bean, beet, cotton, potato, and tomato, attracted to lights; *Pyrota lineata* (Olivier): on Compositae and *Gerardia* (Scrophulariaceae); *Pyrota mutata* (Gemminger): on *Cicuta, Daucus, Eryngium*, and several other Umbelliferae; *Pyrota sinuata* (Olivier): on *Gerardia* (Scrophulariaceae); *Nemognatha nemorensis* Hentz: can be found in several Compositae, including *Bidens*, *Erigeron*, *Heterotheca*, and, particularly, *Rudbeckia*; *Nemognatha piazata* (Fabricius): on *Cirsium* and *Tetraognotheca* (Compositae); and *Nemognatha punctulata* LeConte: on *Bidens* and "thistle" (Compositae).

Genaro (1996) cited to Cuba:

Nemognatha punctulata LeConte: on flowers of *Spilanthes urens* Jack. and *Viguiera dentata* (Cavann.) (Asteraceae), *Tetraonyx quadrimaculata* (Fabricius): on Convolvulaceae (*Ipomoea*) and Leguminosae (*Bradburya, Coelosia*) these families and Bignoniaceae, Euphorbiaceae, and Verbenaceae in the West Indies. Reported damaging grapefruit flowers in Puerto Rico; *Zonitis cribricollis* (LeConte): on *Achillea*, *Coreopsis*, *Helianthus*, and *Rudbeckia* (Compositae); *Zonitis vittigera* (LeConte): on numerous Compositae and *Psoralea* (Leguminosae).

Martínez (1992) cited for Argentina and neighboring countries:

Acrolytta colon (Burmeister): on flowers of *Portulaca* sp. (Portulacheae); *Pyrota vittigera* Blanchard on flowers of *Portulma* sp. (Portulacacea); *Pseudopyrota riojanensis (*Pic*)* on *Bulnesia* y *Lama* (Zygophyllaceae); *Wagneronota aratae (*Berg*)*: on "mistol" *Zizyphus mistol* Grisebach (Rharnnaceae), salad "lechuga" (Compositae) and potato (Solanaceae); *Pseudomeloe andensis* (Guerin-Meneville): found in the Puna between 3,200 - 3,700 MASL eating on flowers Leguminosae that creeps along the ground, from December to January; *Pseudomeloe gracilior* Fairmaire: found on yellow flowers of a Leguminosae that creeps along the ground. The area is xerophytic, with cacti and leguminosae, and with places rather large lacking of vegetation; *Epicauta atomaria* (Ger.): on Solanaceae and Amarantaceae, 'fumo bravo" (Solanaceae); *Epicauta bruchi* Borchmann: on potato (*Solanum* spp.) and wild Solanaceae; *Epicauta excavata (*Klug*)*: on wild Solanaceae; *Epicauta fulvicornis (*Burmeister*)*: on Mimosoidaes; *Epicauta leopardina* (Haag-Rutenb.): on *Portulaca* sp. (Portulacaceae) and Amarantacea; *Epicauta lizeri* Denier: on wild Solanaceae; *Epicauta monachica (*Berg*)*: on 'girasolillo" (*Helianthus* sp.: Compositae); *Epicauta nigripes* Borchmann: on "palán palán" (Nicotiana glauca Graham: Solanaceae); *Epicauta pluvialis* Borchmann: on ornamental crops(Solanaceae), Chard "Acelga" (Quenopodiaceae) and *Capsicum* sp. (Solanaceae); *Epicauta rosilloi* Martinez: on tobacco (*Nicotiana tabacum* Linnaeus), on *Capsicum* spp. (Solanaceae) and tomato (*Lycopersicum* sp.); *Epicauta rubella* Denier: on Mimosoidea; *Epicauta zischkai* Martinez:

on potato (*Solanum* spp.); *Epicauta tristis* (Maklin): on potato (Solanum spp.); *Epicauta zebra* (Dohrn): on *Cassia aphylla* Cavanilles (Leguminosae); *Tetraonyx brunnescens* Haag: on *Urvillea* sp. (Sapindaceae); *Tetraonyx kirschi* Haag-Rutenberg: on *Baccharis* sp. (Compositae); *Tetraonyx lampyroides* Burmeister: on *Sphaeralcea bonariensis* (Cavann.)(Malvaceae); *Tetraonyx propinqua* Burmeister: on Iridaceae, Malvaceae, Solanaceae, Convolvulaceae and Cactis; *Tetraonyx sericea* Selander & Martínez: on flowers of *Abutilon* sp. (Malvaceae); *Nemognatha coeruleipennis* Perty: on flowers of *Convolvulus* sp. (Convolvulaceae); *Nemognatha nigronotata* (Pic): on flowers of *Helianthus* sp. (Compositae); *Nemognatha nigrotarsata* (Fairmaire & Germain): on flowers of 'girasolillo" (*Helianthus* sp.); *Nemognatha weiseri* Pic: on flowewrs of *Helianthus* sp. (Compositae).

In the arid and hot regions, many different meloids species can be found with lights at night can be captured also by beating and sweeping vegetation.

MYCTERIDAE: With 30 genera (Costa 2000) and, according to Darren Pollock (2009) until today the number of known species has risen to 165. There are 3 subfamilies (Lawrence et al. 2010): **Mycterinae**, **Eurypinae** and **Hemipeplinae**. They are distributed throughout the world, but apparently absent in Europe. Adults feed on pollen or nectar; few larval habits are known for *Mycterus* species (Pollock 2002), the larva of the Palearctic species *Mycterus curculionoides* Fabricius has been found from under dead pine bark by Crowson and de Viedma (1964).

A Western species of *Lacconotus* was collected under bark of dead poplar (Lawrence 1991). According to Pollock (2002) appears that most if not all larvae of Lacconotinae live under dead tree bark. But in general based on other published and unpublished data, larvae are lignivorous and feed on decaying wood or rotten sticks. A few species can be agricultural pests, because they build galleries within the stems of some crop plants (Yerba mate, etc). Typical *Hemipeplus* spp. (Neotropical, Neartic, Etiopic and Australian Region) occur in grasses and species live together with their larvae under the leaf bases of palms (Sen Gupta 1968; Thomas 1985; Willemstein 1987), where they feed on unopened vegetative material (Pollock 2002). Adults of most species of *Mycterus* are mainly collected at flowers of *Ceanothus*, *Daucus* and *Yucca* (Pollock 2002). *Holopeplus* are known from palms and grasses in warmer areas world-wide (Pollock 2002).

BORIDAE: This family includes two subfamilies **Borinae** and **Synercticinae** (Lawrence et al. 2002):

---Subfamily Borinae: with 2 genera and 3 species, with distribution in Mittelleuropa Mountains, Pyrenees, Transylvania, Siberia and the region of Amur. It has been found in bark of hardwoods and it is very rare. Also, they are widespread in North America to Canada, predominantly in the east. Immature stages of *Boros unicolor* Say (from USA) and the Palearctic species *Boros schneideri* Panz is widespread across central, Northern and Eastern Europe and lives in coniferous and mixed forests under the bark of dead conifer trees pines *(Pinus* spp.) and other conifers trees (St. George 1931; Young 1991a; Pollock 2002) is endanger specie in Appendix 1 of Red List status of European saproxylic beetles (Nieto and Keith 2010). *Lecontia discicollis* (Le Conte) are found in Canada, US, and Northern Mexico. Larvae can be found in the subcortical region of fired-killed coniferous trees (Young et. al. 1996; Arnett and Thomas 2002).

---Subfamily Synercticinae: *Synercticus heteromerus* Newman restricted to Australia and Papua New Guinea (Lawrence & Pollock 1994) this species has been found in mesic and xeric environments (Williams 2002).

The members of this family are found under bark, under rocks, plant debris and even flowers. They eat rotting wood and cab be collected by beating and sweeping vegetation.

TRICTENOTOMIDAE: A family of 13 described species (Arnett & Thomas 2001 and www.insects.tamu.edu), in two genera: *Autocrates* (3 spp.) from India, Nepal, Myanmar, Malaysia, Thailand, China, and Vietnam. *Trictenotoma* (10 spp.) from: Malaysia, Borneo, Java, Thailand, Indonesia, Vietnam, China, Taiwan, India, Bangladesh, Bhutan, Laos, Myanmar, Sri Lanka and Philippines. *Trictenotoma davidi* Deyrolle (from tropical China and Vietnam) inhabits damp tropical forests (Walters 2008). A few species of trictenotomides have distribution extended into the South Eastern Palearctic. They live in subtropical and tropical forests, in monsoon forests. Although these beetles are very large and showy, little is known about them. They are found under bark and feed on rotting wood. Larvae are only known because they live under bark of dead trees (Telnov 1999). *Autocrates aeneus* Parry from India, Nepal, Myanmar, Malaysia, Thailand, is apple stem borer (Ananthakrishnan 2004).

Trictenotomidae are very similar to Cerambycidae but can be immediately distinguished by having the tarsal formula 5-5-4, opposed to Cerambycidae with tarsal formula 5-5-5 (Nikitsky 1992). These beetles fly well and are attracted to lights, while adults spend most of their life under the bark.

PYTHIDAE: 50 species are known distributed throughout the Americas, Asia, Europe, New Zealand and Australia (Campbell 2009). They live under bark, eat rotting wood, but *Pytho depressus* (Linnaeus) in Europe are predatory of xylophagous beetles. According to Majka (2006) larvae and adults of several species have been found under bark of various trees: *Pinus* spp., *Tsuga* spp., *Laris* spp., *Pseudotsuga* spp., *Abies* spp. and decaying coniferous wood. Larvae of *Priognathus monilicornis* (Randall) feed on tree fermented substances and *Pythos* spp. rotting bark. All nine *Pytho* species are saproxylic, with larvae and adults living under the bark of coniferous trees that have been dead for 3-4 years and which stage the decay is not very advanced (Pollock 1991) and in Eastern North America, individual of two or three species may be present under the bark of one host tree (Pollock 1991). Larvae of *Sphalma* are found in deciduous logs and, *Priognathus* are found in the sapwood of red rotten conifer logs (Pollock 1991; Young 1976, 1991).

Adult pithydes have been collected on dead leaves of bananas also. They are collected by hand-collecting. Adults of *Trimitomerus riversii* Horn, from North America, have been collected several times at ultraviolet lights (Pollock 2002).

PYROCHROIDAE: According to Lawrence and Newton (1995) and Young (2002) approximately 200 species have been described from all major regions of the world; however, recently Young (www.insects.tamu.edu) cited 30 genera and 167 species. According to Lawrence et al. (2010) this family includes five subfamilies: **Tydessinae**, **Pilipalpinae**, **Pyrochroinae**, **Agnathinae** and **Pedilinae**.

Adults are nocturnal (Young 2002) and very rare in collections, are in the sunny and harvested wood, in shrubs and herbaceous plants and flowers (Harde & Severa 1982; Young 2002). Larvae develop under the bark of dead trees where other insects live (Young 2002), but apparently they are more fungivorous than predators feeding on fungi or xylophagous eating decaying wood and are associated in cold and wet subcortical conditions, between the loose bark and decaying wood of deciduous trees and conifers particularly in riparian areas (Majka 2006).

---Subfamily Pilipalpinae: With 12 genera. They are distribuited in Australia, Sumatra, New Zealand, Madagascar, Chile and Argentina and are found on vegetation and on flowers (Pollock 1995). *Exocalopus pectinatus* Broun (endemic to New Zealand) are found on *Podocarpus* sp.and *Weinmanaia* sp., Compositae and Cunoniaceae forest (Pollock 1995). Very little is known about Incollogenius (3 species from Madagascar) due in part to the scarcity of collected material, these were collected in Malaise traps in a rainforest clearing and the larvae is unknown (Pollock 1995).

---Subfamily Tydessinae: with only genera *Tydessa* from US and Japan, adults are more active on vegetation or flowers (Pollock 1995).

---Subfamily Pyrochroinae: *Neopyrochroa* (from US) and *Dendroides* (from North America, Russia and Japan) are often collected at lights or on forest vegetation while collecting at night with a head lamp (Young 2002). Adults Adults are more active on vegetation or flowers (Pollock 1995). Larvae of *Neopyrochroa flabellata* (Fabricius) and *Neopyrochroa sierraensis* Young are generally found beneath bark and in decaying wood (Young 2002).

---SubfamilyAgnathinae: Appear to be adapted to xeric habitats, they have been found under Rocks and in rather dry, decaying vegetation (Young 2002). Many larvae of *Cononotus bryanti* Van Dyke, from United States, were collected from beneath dried cow dung (Doyen 1979).

---Subfamily Pedilinae: Distributed mainly in the Northern hemisphere. Today 3 genera are included: *Anisotria shooki* Young; *Lithomacratria mirabilis* Wickham, both species from U.S.A., and *Pedilus* genus with 41 known species. Adults are found on flowers (young 2002), in foliage, in decaying vegetation, in the soil surface, often in very wet areas. Some species live among the foliage of streams. Others can be common in brome fields feeding on flowers, pollen and nectar. Larvae are phytophagous, omnivorous or fungivorous.

They are often found in moist areas and even in the sand along the beach. To collect them must be check the flowers of False indigo, Osage orange and red haw. They can be collected by hand-collecting or with Winkler/Moczarski eclector, Berlese funnel and pitfall trap using a simple mixture of water, detergent and ethylene glycol (Sackmann 2006); by sweeping foliage of wet areas or near water; even in dry grass fields in early summer, particularly in the foliage of mayapples, horseweeds, willows, maples, buckeyes and alders (White 1983); also placing the sand where they live in a container with water, so they are caught by hand or with a colander when insects float. Other species come at lights. Some species can be collected also with pitfall traps baited with dead beetles (Oedemeridae and Meloidae) (Dettner and Bauer 1997).

Pyrochroides adults can be collected from dead trees also, among the foliage, in flowers, under and on the bark, or other hiding places, some can be attracted with substances such as Cantharidin, molasses and decomposing fruit (Young 2002). Others with Malaise traps and many are caught by beating vegetation. Larvae are much more frequently found than adults.

SALPINGIDAE: A moderately large family with worldwide distribution, with 8 subfamilies (Lawrence & Newton 1995), 45 genera and 300 species in the world (Pollock 2002).

Adults live on flowers, foliage or decaying trees. Although members of this family are mostly terrestrial, those of the Aegialitinae subfamily live in coastal waters in the intertidal zone (Lawrence & Britton 1994). They are brilliant, small and prey bark beetles and other insects living in wood (Hard & Severa 1982; Troukens 2006).

Albert Allen (pers. comm. 2010) has informed me that the following subfamilies are known:

---Subfamily Aegialitinae: Distributed in the Northern Pacific and Subantarctic zone, Russia, British Columbia, USA, Alaska and Robben Island (South Africa), *Antarcticodomus fallai* Brookes (from Campbell Island, Auckland Island) the larvae have been collected under stones in beaches (Spilman 1967), and *Aegialites* (31 spp.). Sugihara (1938) studied *Aegialites stejnegeri* (Kono) in the Kurile Islands (Japan), although he did not observe feeding, he investigated stomach contents and found that the species is phytophagous. Anyway, the predatory habit in Aegialitinae is present in some stage of the life cycle (Leech & Chandler 1956; Spilman 1967).

---Subfamily Salpinginae: With some genera *(Cariderus, Vincenzellus)* widely distributed, including *Cariderus* spp., distributed by Tunisia, Algeria and Armenia, they are associated with *Thilia* sp., *Betula* sp. and *Alnus* sp. The species *Rabocerus gabrieli* (Gerhardt) from Northern and central Europe is found in maple bark, in dead branches of beech, fir trees and wooden fences (Keer 1930). The genus *Sphaeriestes* spp. is associated with *Pinus* spp. (Majka 2006); Sphaeriestes castaneus (Panzer) appears to be restricetd to the Coniferae, with or without bark beetles (Crowson 1984).

The European *Sphaeriestes bimaculatus* (Gyllenhall) feeds on fruiting bodies of Cucurbidothis (Ascomycetes: Pleosporales) (Franz 1955) Adults and larvae of many species are found in galleries in the wood where the beetles attack other borers. Other larvae are found under bark and they are fungivorous or saprophagous. In Japan it has been found in decaying coniferous wood (Nodoka 1969) adults and larvae of *Serrotibia* spp., which feed on cataphylls (Araceae) dead plant tissue and petioles (Arecaceae), respectively, this genus is associated with cloud forests. They can be collected from host plants, with Lindgren funnel traps (Escalona 2008). *Rhinosimus unguiculus* (Broun) from New Zealand is a small, brown, variegated beetle predator of scale insects (Morales et al. 1988). *Rhinosimus planirostris* (Fabricius) and *Rhinosimus* ruficollis (Linnaeus) from Europe lives under the bark of oaks and their dead branches, in beeches and other hardwoods, on the hunt for bark beetles (Crowson 1984). *Lasius fuliginosus* (Latreille) and *Vincenzellus ruficollis* (Panzer) lives under the bark of various deciduous trees, throughout Europe (Harde & Severa 1982).

---Subfamily Dacoderinae: 2 genera: *Dacoderus* (Neotropical) and *Tretothorax* (Australia); they are associated with ants (Pollock 2002), although *Dacoderus striaticeps* LeConte, from Mexico and US, has been collected with pitfall traps under stones and under bark of cottonwood but almost always near ant nests of *Neotoma* sp. (Aalbu et al. 2005). *Dacoderus werneri* Aalbu, distributed from USA to Panama, has been collected in desert wash, sand dunes, cereal bowl, pitfall trap with carrion, in palm leaf litter and rodent feces of *Neotoma* sp., with barrier pitfall traps and ultraviolet light (Andrews and Pollock 2005; Aalbu et al. 2005). Other species of *Dacoderus* from Central and North America have been collected with Malaise traps, with Berlese funnel (of leaf litter in a moist forest), with pitfall trap in clay soil with bushes, and from a riparian oak forest. *Myrmecoderus laevipennis* (Horn) from Mexico and USA has been collected with light late at night. *Tretothorax cleistostoma* Lea of Queensland (Australia) was found in ant nests Ponerinae *(Odontomachus* and *Leptogenys)* (Ward in: Aalbu et al. 2005). *Dacoderus* and *Myrmecoderus* were also found associated with ants, however the Californian species of *Dacoderus sleeperi* Aalbu has been found on islands that had no Ponerinae ants (Andrews & Pollock 2005). Larvae of the Dacoderinae subfamily are unknown and presumed to be specialized to live in ant nests.

---Subfamily Prostominiinae: The subfamily is worldwide in distribution (Lawrence 1977). There are no representatives of this subfamily from North America north of Mexico, however, there are many genera in Mexico and Central America and nothing is known about its biology (Pollock 2002).

---Subfamily Othninae: 51 known species in the genus *Elacatis* (with global distribution) and 2 species in the genus *Paralecatis* from Philippines Islands. Most species live in the tropics of Asia, Africa and South and Central America.

Adults are predators and are found under bark of trees, foliage, fallen leaves, on dead stumps and cactus. Larvae are scavengers, with saprophages habits. In North America larvae have been collected on dead bark pine at high elevations in Arizona (Pollock 2002) and adults are usually found in dead foliage (Young 1991a). In Paraguay, the author has found them running on large-leaved plants (Araceas and *Philodendron)* in high humidity places (pers. observ.).

Species of othnines can be collected by reviewing the bark of falling plants or trees. Also a butterfly net can be used because these beetles quickly run through the foliage when frightened; they are also found in cactus. Some types of attractants can be used to collect the Othninae, including: Frontline, Ipsdienol, Ipsenol, Turpentine of pine and/or Pinene, which can be purchased at stores specialized in products against agricultural pests or laboratories where chemicals are developed (Sanchez et al. 2008).

---Subfamily Inopeplinae: Larvae and adults are found under dead tree bark (Young 1991). For example *Inopeplus praeustus* (Chevrolat), from Cuba, Dominica, Guadeloupe, Martinique, St. Vincent, adults are rarely at lights, adults and larvae mostly collected under bark of fallen trees (Peck 2006).

---Subfamily Eurystethinae: With 2 genera: *Aegialites* and *Antarcticodomus.* They are found in rock crevices on the Pacific coast, and on Campbell and Auckland Island, respectively. These beetles live in the intertidal zone (Spilman 1967; White 1983).

---Subfamily Agleninae: With a single species *Aglenus brunneus* (Gyllenhall) from Europe, USA, Balearic Islands, has been collected in caves and it seems to be a scavenger (Young 1991).

---Subfamily Inopeplinae: The genus *Inopeplus* Smith contains about 60 known species (Nodoka 1969), genera widely distributed. They are found among the bark of dead trees (Young 1991c).

ANTHICIDAE: Family with worldwide distribution. 3,000 species known in the world (Costa 1999; Chandler 2002) In Europe only 40 species are known (Troukens Willy 2006) and according to Lawrence et al. (2010) there are seven subfamilies currently accepted: **Eurygeniinae**, **Macratriinae**, **Steropinae**, **Copobaeninae**, **Lemodinae**, **Tomoderinae** and **Anthicinae**.

Adults are found in decaying vegetation. Limited to specific habitats such as dunes and margins of fresh or salt water, they are omnivorous or predators; however, the diet of adults consists mainly of dead beetles (Hard & Severa 1982). Larvae are omnivorous, mycophagous or predators and are found on decaying tissues of plants. They may be collected with pitfall trap. It is necessary to look carefully at the edges of places with water or very wet areas, since adults are usually active there. Some are gathered, walking on the hard clay on the banks of lake, rivers, streams or ponds of water. These habitats may be within the forest or outside, for example in a savanna. There are diurnal and nocturnal species.

In New Zealand the *Macratria* species may be quite common on riparian vegetation; they are active during the day and can be collected by sweeping (Werner and Chandler 1995).

Many *Anthicus* spp. (Anthicinae) are diurnal and they can are associated with disturbed areas near human habitation (Werner and Chandler 1995), some species can be collected under dung, under stones, under logs, compost, in pastures etc., even in the nests of some birds (Werner and Chandler 1995).

Many species of the *genus Notoxus,* living in xerophytic regions, can be collected with pitfall traps, using dead Meloidae as bait (pers. obs.). *Protoanthicus* are collected with Winkler/Moczarski eclector or Berlese funnel.

Those of the genera *Anthicus, Ischyropalpus, Copobaenus, Chileanthicus, Vacusus and Mitraelabrus* can be collect between the leaves, with Malaise trap while walking on the ground, on the banks of streams, in flowers, with light traps. In South American countries (mainly in the subtropics) either in winter (or dry season), spring and/or summer and at night when the moon is not visible, thousands of specimens belonging to various species are attracted to the lights, in front of white walls or to a white light (or UV) which is located in front of a white sheet near a small forest (better if the forest is large). Several species of Anthicidae reach the lights in places much disturbed.

ADERIDAE: There are about 50 genera with approximately 1,000 species worlwide, but concentrated in the tropics (Chandler in: Arnett et al. 2002). They live in flowers and plant debris and are usually found resting on the under side of leaves of various shrubs and trees, particularly angiosperms with broad leaves (Chandler in: Arnett et al. 2002). Larvae and adults are predators or saprophagous. They are primarily associated with foliage, flowers or tree wounds (Elgueta 1993) and vegetable refuse (Willemstein 1987). They are associated with decaying vegetation and they live in the foliage of plants. Larvae live in rotting wood. They can be collected with Winkler/Moczarski eclector, Berlese funnel, beating and sweeping vegetation. Many aderides are collected with Malaise trap and some have been captured in termite nests (Lawrence et al. 1990).

SCRAPTIIDAE: Family distributed throughout the world. 35 genera with 480 species (Arnett and Thomas 2001; www.insects.tamu.edu), according to Lawrence et al. (2010) there are two subfamilies: **Scraptiinae** and **Anaspidinae**. Adults are found in decaying plants in the soil, in the foliage and flowers. Larvae live in the bark of damaged trees, and usually they are saprophagous.

---Subfamily Scraptiinae: adults of *Canifa pallipes* (Melsheimer) are found on foliage in the vegetation of forested environments in Canada (Nova Scotia) in coniferous (*Picea rubens* Sarg., *Picea mariana* (Mill.), and *Pinus strobus* Linnaeus) and deciduous forests (Majka & Pollock 2006). *Canifa* has been found primarily in coniferous forests (*Picea rubens Sarg., Picea mariana* (Mill.), *Tsuga canadensis* (Linnaeus), *Pinus strobus* Linnaeus and *Abies balsamaea* (Linnaeus) from Nova Scotia (Majka & Pollock 2006). *Canifa* larvae have been found under the bark of dead poplar (Pollock 2002b; Majka & Pollock 2006).

---Subfamily Anaspidinae: 18 genera, *Anaspis* with many species mainly distributed in the Palearctic and Nearctic region. According to Majka & Pollock (2006) in Canada (Nova Scotia and Prince Edward Island) *Anaspis* has been found in coniferous (*Picea rubens Sarg,, Picea mariana* (Mill.)*, Tsuga canadensis* (Linnaeus)*, Pinus strobus* Linnaeus, and *Abies balsamaea* (Linnaeus)) forests and deciduous forests of all types and in many forest edge areas (old field, scrub, bog, streamedge, pine barren, etc.). *Anaspis* adults can be very abundant on flowers (Majka & Pollock 2006) and may be taken in very large numbers from blossoms of Apiaceae and Rosaceae (Young 1991). Larvae occur under loose tree bark (Hatch 1965).

Scraptides beetles can be collected with Winkler/Moczarski eclector and Berlese funnel, also by beating and sweeping vegetation.

INCERTAE SEDIS: The taxonomic position of the following families, subfamilies and genera within the superfamily Tenebrionoidea is unclear (sensu Lawrence et al. 2010):

ISCHALIIDAE: Family distributed by Old World, in Eastern Eurasia, the Oriental region, North American and Southern of Canada (Nikitsky & Egorov 1992; Nikitsky 1996). One genus is known *(Ischalia)* with 34 species. Adults and larvae are found in fungal mycelia associated with decaying logs (Young 1985). They have been collected inside rotting logs.

Rhizonium: According to Watt (1983) in Auckland Islands *Rhizonium antiquum* Sharp live in the dead rachides of various tree ferns especially on *Cyathea dealbata* (Forst.) and *Cyathea medullaris* (Forst.). Adults tend to be long and thin, an adaptation to movment between the longitudinal fibres of the rachides (Watt 1983).

Aprostomis cephalotes Grouvelle, 1912, this species is distributed through the Montserrat (Ivie et al. (undated)) and Guadeloupe (Trois-Rivieres). According to Grouvelle and Raffray (1912) this species was collected under the bark of small branches and creepers.

LAGRIOIDINAE: Found in Australia, New Zealand, and temperate South America (Chandler 2002). According to Champion (1890) they are found on flowers of *Mesembryanthemum* sp. and *Lagrioida obscurella* Fairm and Ger., were found on the coast of Chile under plants on the dunes at Coquimbo and Valparaiso. *Lagrioida brouni* Pascoe, in New Zealand have been captured under a log on the sea shore in coastal dunes (Champion 1890; Stephenson 1999).

AFREMINAE: According to Levey (1984) there are two species known: *Afremus madgei* Levey from Zimbabwe, and *Afremus pickeri* Levey from Namibia. Currently classified with the Anthicidae (Arnett et al. 2010)

Superfamily CHRYSOMELOIDEA

CERAMBYCIDAE: With 4,000 genera and 35,000 known species (Cleide Costa 2000). Adults are phytophagous. Many species live in flowers, fruit or other decomposing sweet substances. Larvae live inside the wood of living trees or dead, others are on the floor, eating the outside of roots of healthy trees. Larvae are somewhat cylindrical, with a well developed thorax, so that your galleries are similar to the shape of their bodies (Rojas and Gallardo 2004). Many different species can be captured with Malaise traps, with bottle traps in the trees and many species are attracted at lights. But the principally method to collect them is the beating and to look logs, tree fallen, branches, etc., see article "Xylophagous beetles", Chapter III. The best places to practice the beating are living trees found along forest paths (natural flyways) and around the periphery of open areas and fields that are in the sun (see article "beating method", Chapter III). In temperate zone, at least for species of Holarctic distribution, hunting by sight on flowers yields excellent results as far as Lepturinae are concerned. Many Cerambycinae are also found on flowers (Clytini, etc). The hunter will quickly learn to recognize the most productive flowers, keeping in mind that almost no type of flower should be overlooked, because many Cerambycidae species are associated with a single feeding nectar source. Another effective method to obtain many rare species is to breed the larvae (see article "rearing larvae, Chapter III"). Some very rare species can be collected with pitfall traps baited with fruits. In South America, some species of the subfamily Prioninae can be found under bark of trees (eg. some *Macrodontia*, *Mallosoma*, *Stenodontes*, etc), other species (*Macrodontia flavipennis* Chevrolat) are in the lower parts of the trees and the entry of its galleries can be detected by having an oval shape. Another effective method to collect many different cerambycids species is to make small cuts with a machete or a hatchet in different branches or trunks of trees (including woody climbing plants), which must be examined some days later.

CHRYSOMELIDAE: This family now includes 11 subfamilies (Reid 2000; Özdikmen and Turgut 2008), one of which has been renamed for nomenclatural reasons (Staines 2002). About 33,000 species have been reported and 10,000 species probably have not yet been described (Özdikmen and Turgut, 2008). Most species are terrestrial, but genera *Donacia*, *Plateumaris, Macroplea* are found exclusively on aquatic or riparian vegetation. The genera *Potamogeton, Ruppia, Phragmites, Sparganium, Typha, Nymphaea, Nuphar, Sagittaria, Alisma, Glyceria, Carex, Butomus,* live above or below the water surface (Jäch 1998). Many times *Galerucella* spp., *Altica* spp., *Agasicles hygrophila* are found on the water surface. These species were introduced in U.S.A. from South America, for biological control of the alligatorweed *(Alternanthera* sp.) and they have proved to be very effective in this regard (Jäch 1998). *Longitarsus nigerrimus* (Gyllenhal) is found on *Utricularia sp.,* and may be submerged under water long enough (Mohr 1960). *Lysathia ludoviciana* (Fall) is found on the plant *Myriophyllum sp.,* and the species *Pseudolampsis guttata* (LeConte) were also reported to feed on aquatic plants (Jäch., M.A. 1998). Some Chrysomelinae *(Hydrothassa* spp.) and one species of *Prasocuris* are aquatic or semiaquatic (White & Brigham1996). Many species of Galerucinae (incl. Alticinae) live on emerging aquatic plants. Larvae of some species of Clytrinae live in ant nests and others between humus and litter (White 1983). Nearly all species feed on live plants and only a few do it on remnants of vegetation. Larvae are terrestrial and feed on leaves, stems and flowers of plants in which they live, but in some species larvae are aquatic and feed on submerged plant stems. Others are found in the soil eating roots. Many larvae feed outdoors on the leaves of plants (Criocerinae, Chrysomelinae, Cassidinae, Chlamisinae and Eumolpinae). Others species undermine plant leaves (Zeugophorinae, Hispinae and Alticinae).

To collect many Chrysomelidae different techniques must be used: sweeping and beating in all type of foliage. Check the leaves, stems, flowers and fruits of various plants, both terrestrial and aquatic, since several species are walking among the clusters of fallen petals and fruits at the base of trees. Many species of Hispinae, Chrysomelinae, Eumolpinae, Cryptocephalinae and Cassidinae are slow moving and generally easy to catch, even just by hand, but there are more agile groups which fly quickly when frightened (Alticinae, Donaciinae, many Galerucinae). Furthermore, many Chrysomelidae stand still or pretend to be dead and become less visible to the amateur collector (Clytrinae, Chlamisinae, Hispinae and Cassidinae). Almost always they are easily collected with beating-sheet, but some times some of them start to roll over the cloth and it is necessary to prevent escape. Many different species can be captured with Malaise traps, and many come to the lights.

Keep in mind that to achieve a variety of species it is necessary to visit different habitat and look carefully at each plant found in our path (including exotics). Chrysomelidae are abundant in all types of weeds, grasses and shrubs; however, they are sparser in trees.

Finally, it is important to keep in mind that in general, but not always, the species of yellows, greens and/or red crystalline color (or semi-transparent) should not be put in alcohol since the liquid will change the color, making them more dark and dull. That happens to many Cassidinae and Galerucinae, thus for these cases, potassium cyanide vapors can be used; however, if using cyanide, newly collected specimens should not remain a long time in these deadly vials because a long-time exposure to the poison also fade them. To avoid this, it is best to use ethyl acetate and have on hand several empty bottles (filled with toilet paper or sawdust) and when the bugs are changed to the new bottle a few drops of ethyl acetate can be added to the flasks.

ORSODACNIDAE: 3 genera are known with 37 species. Almost all species are found in the neotropics and North America, but some live in the Holarctic region, and even Japan. They resemble the Cerambycidae but have short antennae, and other species may be confused with Anthicidae and Pedilinae, but differ from those families not to present a neck between the separation of pronotum and elytra, which is present in these families. Orsodacnidae adults feed on foliage, flowers or fruits of various plants. In the Neotropical region, adults are on fronds of cycads and feed on pollen of various types of angiosperm flowers (Clark and Rileyin in: Arnett et al. 2002). In general, larvae of this family feed on roots. In the Tropical and Subtropical South America, the author has found adults on various plants throughout the year. In North America, adults are active mostly in spring and sometimes are abundant on the snow covering the ground (Clark and Rileyin in: Arnett et al. 2002). Like other Chrysomeloidea, adults are collected by sweeping and beating on foliage of various types of trees, shrubs and plants.

BRUCHIDAE: Cosmopolitan family. J. M. Kingsolver cites 70 genera and 1,632 species for the world fauna (www.insects.tamu.edu). They are found in seeds of various plants (including stored) from which they feed; others eat Dicotyledonous seed capsules. They are quite common in legume seeds.

They are collected reviewing the capsules with seeds or just free seeds of various plants. A good method is selecting and collecting seeds of native trees (both dry and the newest) and transport those to the laboratory in small cloth bags or plastic boxes previously prepared for that purpose. Each set of seeds must be separated and labeled in the collecting site with the sample data (collection site, date, collector, host plant, GPS coordinates, etc). In the lab, each group of seeds will be stored in individual and sealed boxes. These should be sealed, and with a cotton bed (or blotting paper) on the bottom. The author has used various small plastic containers (tupperware type) with snap-on caps. They come in various sizes and are sold in most supermarkets. According to Alex Delobel (pers. comm. 2010) small zipper storage bags aerated with pinholes may also be used. They are cheap and convenient, but care must be taken as some species may eat their way through the plastic film. At the top of each box a label must be pasted containing all data collected, and then it is stored in a cool and bright place, but avoiding direct sunlight. At least twice a week the boxes should be opened to give air for a while, and occasionally spray the seeds with some water, depending on the humidity inside the boxes. Besides, be careful with the appearance of fungus in the seeds, if they are too wet; completely unripe seeds should never be used, as they will mould without reaching maturity, as will beetle larvae (Alex Delobel pers. comm.2010). Bruchids adults are born after a few weeks or months, depending on the species and conditions of the parasitized seeds.

Also, the adult's members of this family are collected with Malaise traps, by beating and sweeping vegetation and many are attracted to lights. According to Alex Delobel (pers. comm. 2010) a good way of collecting large numbers of adults is visiting flowers, especially Convolvulaceae and Cistaceae; adults of many species feed on pollen of these flowers, and may be easily collected with a mouth aspirator. For a more extensive study on the biology of this family, please read the work of Borowiec (1987) that is available online (see bibliographical references).

MEGALOPODIDAE: Distributed in all regions of the world, but most species are found in tropical regions. Over 400 species in 30 genera (Monrós 1960; Seeno & Wilcox 1982; Kuschel & May 1996); adults are found in flowers and fly quickly when frightened.Larvae are mostly leafminer, mainly Solanaceae. According to Flowers (1996), R.W. in Costa Rica the species of Megalopus genus lives on Solanaceae during a few weeks in the rainy season. Palophagoides vargasorum Kuschel (Palophaginae) are associated with Araucaria araucana (Mol.) in Chile and Argentina (Neuquen by the border with Chile), according to Kuschel and May (1996) the diet of adults is presumed to be pollen judging and larvae were rearing from cones of Araucaria araucana and others larvae retrieved from soil under the host tree (Kuschel and May 1996). Palophagus occur in Australia, they can be attracted to light (Kuschel and May 1996). Specimens of Palophugus bunyae Kuschel were reared from cones of bunya-pine (Araucaria bidwillii Hook) (Kuschel and May 1996).

They are collected manually checking various plants, by beating and sweeping vegetation, and also with malaise traps and light traps.

Superfamily CURCULIONOIDEA

Classification after Oberprieler et al. (2007) [2]

NEMONYCHIDAE: 72 described species in 21 genera (Kuschel 1994), a few more awaiting description; distribution highly disjunct, mainly in temperate Northern and Southern regions, most diverse in Australian and Neotropical regions, less so in Nearctic and Palaearctic regions.

They are classified into three subfamilies, **Nemonychinae** (for *Nemonyx* only), **Rhinorhynchinae** and **Cimberidinae**.

Associated predominantly with conifers, especially Araucariaceae, which host half of all species, Pinaceae the common hosts in Northern hemisphere; two genera on angiosperms, *Rhynchitomacer* on *Nothofagus* (Nothofagaceae) in South America and *Nemonyx* on *Consolida* and *Delphinium*

2. Almost all the information about this superfamily was provided by Oberprieler.

(Ranunculaceae) in Western Palaearctic. Adults feeding mainly on pollen; eggs laid between sporangia in strobili or into fruit follicles by means of ovipositor but sometimes aided by elongated rostrum of females; larvae freely (ectophytically) among sporophylls in dehiscing male conifer strobili feeding on pollen in open sporangia, or in follicles of host fruits feeding on developing seeds.

Adults and larvae often common during short coning period; collected either by beating vegetation, fogging the hostplants or by gathering fallen male cones and soil beneath them and rearing larvae and pupae to adults.

ANTHRIBIDAE: About 3,860 described species in 371 genera (Rheinheimer 2004), distribution cosmopolitan, most diverse in tropical regions.

Classified into three subfamilies, **Anthribinae**, **Choraginae** and **Urodontinae**, but monophyly of Anthribinae with respect to Choraginae uncertain and their tribal classification poorly established; Urodontinae different in numerous features and possibly a distinct family.

Anthribinae and Choraginae are associated predominantly with wood-decaying ascomycete fungi. Adults feeding mostly on fungal tissues; eggs laid into fungus-infested wood by means of sclerotised, toothed ovipositor; larvae feeding endophytically on fungal mycelium; some species on wood, lichens, pollen, seeds and even scale insects (Oberprieler et al. 2007); larvae of Urodontinae in stems of Aizoaceae or in fruits of Brassicaceae, Resedaceae and Iridaceae and related groups.

Adults are collected mainly by beating dead branches, also by fogging, light-trapping or rearing from fungus-infested wood; larvae found in the latter; adults and larvae of Urodontinae best extracted from infested stems and fruits of hostplants. Many Australian species are found in *Eucalyptus* forests or forests of Acacia spp.

BELIDAE: About 375 species described in 38 genera (Oberprieler et al. 2007); distribution predominantly in Southern hemisphere; Subfamily **Belinae** restricted to Neotropical and Australo-New Guinean regions, **Oxycoryninae** worldwide but with highly fragmented range.

They are classified into two subfamilies:

---Subfamily Belinae: 155 species in 25 genera restricted to Neotropical and Australo-New Guinean regions, the former with tribes **Pachyurini**, **Agnesiotidini** and **Belini**. They are associated with gymnosperms, angiosperms and possibly ferns. Adults feeding on various plant tissues; Larvae of Belinae boring in stems and logs: Pachyurini and Agnesiotidini mostly in conifers, Belini in angiosperms. Oviposition holes drilled into host tissues by means of rostrum.

---Subfamily Oxycoryninae: About 200 species in 13 genera, worldwide distribution, but with highly fragmented range, with tribes: **Aglycyderini**, **Metrioxenini** and **Oxycorynini**. Larvae are found in reproductive plant organs of gymnosperms (araucarias, cycads) and angiosperms but Aglycyderini also in stems or under bark. There are species spawning in fleshy flower roots of parasitic plants such as those of Hydnoraceae family.

Belidae adults are collected mainly by beating branches of hostplants, also by fogging, light-trapping or rearing from branches or wood; adults and larvae of Metrioxenini and Oxycorynini best extracted from cones and inflorescences of their hostplants.

ATTELABIDAE: About 2,500 described species in 150 genera, distribution cosmopolitan.

Classified into two subfamilies, **Rhynchitinae** and **Attelabinae**, but monophyly of the former uncertain; classification of both into tribes unsettled.

They are associated predominantly with angiosperms, very few with conifers. Adults feeding on various tissues; females generally preparing larval development substrate by severing sap flow of living stems, buds or fruits (Attelabinae rolling leaves into cigar-like nidi), into which they lay eggs and which they can inoculate with fungal spores carried in mycetangia next to hind coxae; larvae feeding on decaying tissues in withering plant organs.

Adults are collected mainly by beating or fogging general or hostplant foliage, also by rearing from leaf rolls or cut flowers or buds of hostplants.

CARIDAE: only 6 described species in 4 genera (but several undescribed ones known); restricted to Southern hemisphere (South America, Australia-New Guinea). Classification: No subdivision.

Associated with Cupressaceae but perhaps also with other conifers. Adult feeding not recorded;

oviposition holes drilled into female cones with rostrum; larvae (only known for Car) in sporophylls and seed chambers of Callitris cones. *Caenominurus* were collected on plants of *Austrocedrus* genus and *Pilgerodendron* (Kuschel 2003).

Adults are collected by beating or fogging hostplant foliage, larvae extracted from immature female cones.

BRENTIDAE: approximately 4,000 described species in 400 genera, about 90% in subfamilies Brentinae and Apioninae; distribution cosmopolitan but smaller subfamilies with more restricted ranges.

In wider family concept 6 subfamilies recognised, **Ithycerinae**, **Microcerinae**, **Eurhynchinae**, **Nanophyinae**, **Brentinae**, **Apioninae**; tribal classification of last two not well established.

Associated almost exclusively with dicotyledonous angiosperms, only a few and unrelated apionines on conifers, none on monocotyledons. Adults feed on various tissues. Classified into six subfamilies:

---Subfamily Microcerinae: Larvae are found in soil on stems of Malvaceae. Oviposition holes drilled into host tissues by means of rostrum but eggs laid into soil with ovipositor. They are collected under rocks or with pitfall traps.

---Subfamily Ithycerinae: oviposition holes drilled into host tissues by means of rostrum but eggs laid into soil with ovipositor. Larvae are found in soil on roots of Fagaceae and related families.

---Subfamily Eurhynchinae: There are only 3 extant genera (2 fossil ones, from Africa and South America). Most are rare insects. The larvae bore in stems of Proteaceae and other families.

---Subfamily Nanophyinae: This small family has about 70 species described. Distribution: Paleartic, Afrotropical, Oriental and Australo-Pacific regions. The members of this family are found in forests, but also in pastures. They feed on several plants. Adults and larvae are found in fruits and stems.

---Subfamily Apioninae: 126 genera are known at least, cosmopolitan. They feed on a wide range of dicotyledonous plant families mostly, especially legumes. Adults feed on foliage, producing small holes in the leaves. Larvae are found mostly in young stems and inflorescences, fruits or seed pods. They can be collected with beating and sweeping vegetation and also manually.

---Subfamily Brentinae: Adults are found under or on the bark of stumps and trees. Some are predaceous on other xylophagous beetles. Larvae feed on dead and dying trees (some predaceous on other xylophagous beetles). They can also be caught spraying logs and reviewing at night stumps and fallen trees; some species can be collected with beating from dead branches; other with pitfall traps, or with sawdust traps or with some attractants (ethyl acetate or phenol). Some species come to the lights.

In general, brentids adults can be collected by beating method, fogging hostplant (foliage or wood), Light traps, Malaise intercept trap and rearing from stems, trunks, buds or fruits; larvae best extracted from infested stems and fruits of hostplants.

CURCULIONIDAE: With about 4,600 genera and with more than 52,000 described species. They live all over the world, from the Arctic Zone, in the North, to the Subantarctic Islands, in the South, beaches to mountain tops, from deserts to rainforests. They feed on virtually all plants, mainly angiosperms but also gymnosperms, pteridophytes, bryophytes and lichens and occasionally they even browse on algae and cyanobacteria, they also make extensive use of monocotyledons as hosts (Oberprieler et al. 2007).

---Subfamily Dryophthorinae: approximately 1,200 described species in 150 genera; distribution cosmopolitan, most diverse in tropical regions. They are generally divided into 5 tribes, Dryophthorini, Stromboscerini, Orthognathini, Cryptodermatini and Rhynchophorini.

Associated predominantly with woody monocotyledons, especially palms and pandans; several species of considerable economic importance. Adults feeding on various host tissues, often in wounds; oviposition holes drilled into host tissues by means of rostrum; larvae mostly in living or dying stems, often in large numbers, but also in infructescences or seeds (e.g. stored grain).

Adults are collected mainly by beating or fogging hostplants, also by light- and malaise-trapping or rearing from stems, trunks, fruits or seeds; larvae best extracted from infested hostplant tissues.

---Subfamily Platypodinae: about 1,500 described species in 38 genera; distribution cosmopolitan but mainly in Tropical, well-forested regions.

They are divided into 4 tribes, Mecopelmini, Schedlariini, Tesserocerini and Platypodini, the first 2 monotypic.

They are specialised mycetophagous woodborers attacking dead or dying trees of a wide range of families. Adults constructing galleries in dead or dying wood and inoculating them with spores of ambrosia fungi carried in special mycangia, caring for brood; larvae feeding on fungi in parental galleries.

Adults are collected mainly by light-trapping or rearing from infested wood. They can be collected in spring when they go flying by thousands in the afternoon and before sunset. Also, the sawdust traps are very productive.

- Subfamily Brachycerinae: about 1.200 described species in 113 genera; distribution cosmopolitan, diverse in arid regions but also in aquatic and semi-aquatic habitats.

Concept of subfamily not settled, in wider sense including tribes Brachycerini, Byrsopini, Cryptolaryngini, Ocladiini, Erirhinini, Tanysphyrini (Stenopelmini), Raymondionymini and Myrtonomini.

They are associated predominantly associated with monocotyledons, Tanysphyrini with aquatic Alismatales and Commelinales (some also with aquatic grasses and ferns), Erirhinini and Ocladiini with Poales (a few with dicotyledons or mosses), Cryptolaryngini, Byrsopini and Brachycerini with geophytic Asparagales and Liliales. Adults feeding on various host tissues; oviposition holes drilled into host tissues by means of rostrum but in terricolous groups (Brachycerini and Byrsopini) eggs laid into soil by means of sclerotised ovipositor; larvae endophytic in stems, corms, bulbs and perhaps roots.

Adults are collected mainly by hand-picking, beating or fogging from hostplant foliage, some also by light-trapping, Brachycerini and Byrsopini by hand-collecting under rocks or pitfall-trapping, or by rearing from infested plant organs; larvae best extracted from infested stems and bulbs of hostplants.

---Subfamily Cyclominae: approximately 1,500 described species in 150 genera; distribution predominantly in Southern hemisphere and in arid regions, some also at high altitudes in mountains.

Classified into 8 tribes, Cyclomini, Amycterini, Hipporhinini, Rhythirrinini, Listroderini, Notiomimetini, Dichotrachelini and Aterpini, but concept of some remaining unclear and monophyly of subfamily not established.

Host associations poorly known, Cyclomini and Amycterini associated with monocotyledons, others largely with dicotyledons. Adults feeding on various host tissues; eggs mostly laid in soil by means of specialised, sclerotised ovipositor but rostrum used for preparing oviposition sites in some Aterpini; larvae generally in soil feeding on roots and underground stems, some crawling up onto shrubs at night to feed on leaves, of Aterpini mostly in roots, trunks, stems, shoots and inflorescences.

Adults are collected mostly flightless, taken mainly by hand-collecting under rocks or plants or by pitfall-trapping, some also by beating or fogging hostplant foliage and by light-trapping; larvae dug up from soil around roots of hostplants.

- Subfamily Entiminae: approximately 12,000 species described in 1340 genera; distribution cosmopolitan, including subantarctic islands.

55 tribes listed in latest catalogue but most of them poorly defined and constituted.

Feeding on virtually all plant groups but rarely closely associated with any one; many serious agricultural pests. Adult feeding mostly on leaves and young shoots, often at night; eggs generally laid in soil by means of unspecialised ovipositor but sometimes inserted into natural spaces on hostplants; larvae generally free-living in soil feeding on roots, rarely endophytic in stems.

Adults are collected mainly by beating or fogging hostplant foliage, flightless species also by hand-collecting under rocks or shrubs or by pitfall-trapping; larvae dug up from soil around roots of hostplants.

---Subfamily Molytinae: approximately 10,000 described species in 1,200 genera; distribution cosmopolitan.

Composition and definition unclear, both in narrower, traditional concepts and in the wider one proposed recently.

They are associated mainly with dicotyledons but several taxa also with monocotyledons (especially palms), gymnosperms (conifers, cycads) and ferns. Adults feeding on various tissues; oviposition holes drilled into host tissues (mostly wood) by means of strong rostrum; larvae generally boring in woody tissues of dying trees but several also in living stems, trunks, roots and fruits.

Adults are collected mainly by beating or fogging hostplant foliage or wood, also by light-trapping and baiting with felled trunks and stems, or by rearing from sections of stems; larvae best extracted from wood.

---Subfamily Cossoninae: about 1,700 described species in 275 genera; distribution cosmopolitan and many species with large ranges (often aided by human trade).

Tribal classification unsettled, tribes varying from 18 to 5 but their concepts largely unclear.

Associated mostly with dicotyledons but without great degree of host specificity; some taxa more narrowly associated with monocotyledons, conifers and ferns. Adults are typically feeding on decaying wood, tunnelling into it; larvae in dying or dead wood in often advanced stages of decay but sometimes in branches, fruits or seeds.

Adults and larvae are collected mainly by extraction from decaying wood, occasionally by light-trapping.

---Subfamily Scolytinae: about 6,000 described species in 235 genera; distribution cosmopolitan but mainly in tropical, well-forested regions, several species widely dispersed by human trade activities. Currently 28 tribes recognised.

They are specialised phloeo - or mycetophagous woodborers - attacking dead or dying trees of a wide range of families, some also in leaf stalks, fruits and seeds. Adults construct galleries in wood and mostly inoculating them with spores of ambrosia fungi carried in special mycangia, often with complex mating and brood care systems; larvae feeding on fungi in parental galleries.

Adults are collected mainly by light-trapping and rearing from infested wood or similar plant organs, or by baiting with felled trunks and stems; larvae best extracted from infested wood or similar plant organs. They can be collected in spring when they go flying by thousands during the afternoon and before sunset. Sawdust traps are very productive, like those of pheromones. Many are attracted with ethanol (one of the by-products of trees discomposure).

---Subfamily Baridinae: about 8,000 described species in 900 genera; distribution cosmopolitan. They are unsettled, in wider sense including Baridini, Conoderini, Ceutorhynchini, Trigonocolini and Orobitini.

They are associated mainly with angiosperms including many monocotyledons, a few groups also with gymnosperms. Adult feeding on various tissues; oviposition holes drilled into host tissues by means of rostrum; larvae in wood, stems, flower buds, fruits and seeds.

Adults are collected mainly by beating or fogging hostplant foliage or wood, also by light- and malaise-trapping or rearing from plants; larvae best extracted from infested plant organs.

---Subfamily Curculioninae: about 9,000 described species in 450 genera; distribution cosmopolitan.

Unsettled, combining 3.5¬40 tribes but their precise concepts and status often unclear.

They are associated predominantly with angiosperms, often in relatively close associations with particular genera or families. Adult feeding mostly on reproductive plant organs; oviposition holes drilled into host tissues by means of rostrum; larvae also mostly inside reproductive plant organs (buds, flowers, fruits, seeds) but a few ectophytic on leaves, coprophagous or predeaceous.

Adults are collected mainly by beating or fogging hostplant foliage, also by light- and malaise-trapping or rearing from buds or fruits; larvae best extracted from infested stems and fruits of hostplants. Some Cryptorhinchini can also be collected from leaf litter.

ORDER: STRESIPTERA

This order is quoted here because some entomologists have included it as a family of Coleoptera and other authors have separated it from Coleoptera

Stylopidae: Some authors consider it a separate order from Coleoptera (Whiting et al. 1997; Jiron 2006). 8 genera and approximately 600 species are known, worldwide distribution (Grimaldi and Engel 2005). They are endoparasites of hymenoptera, grassopphers, planthoppers, leafhoppers, spittlebugs, treehoppers. The sexes have different shapes and different behaviors. Males are free living, have wings and leave the bodies of their victims to be able to reproduce. In contrast, females are wingless, usually without legs and live a completely parasitic life; they do not transform into pupae and do not leave their hosts. However, there are some free-living females with legs. Their hosts generally do not die (although many do), but they cause diseases and deformities in the abdomen. Stylopidae generally produce a high number of eggs and between 1,000 to 5,000 larvae with worm shape, which are dedicated to find a host to be parasitized, but only a small percentage survives. Males are collected only occasionally, mainly by sweeping vegetation. To collect females (and sometimes males) it is necessary to look for insects parasitized by them. When catching a wasp, bee, etc. which could contain Stylopidae, look at the abdominal segments which are deformed, and sometimes the parasite can be seen (with a magnifying glass) between the abdominal segments. The host can be sacrificed to extract the Stylopidae female, but it is better to put the host inside a transparent plastic box (with pressure lid) containing a grid. Once inside the infested insect must be watched carefully for several days until the parasitoid (if a male) leave the host (Jiron 2006) and quickly move away (especially if its victim was a wasp) because, as they discover the parasite, they try to kill it immediately. Therefore it is advisable to put a very fine grid at the bottom of the box, through which only Stylopidae could pass. Finally, remember to write in a label all data collection including, if possible, the host name. Many males have been collected with Malaise trap (Grimaldi and Engel 2005).

CHAPTER V

KILLING
TEMPORARY STORAGE
PRESERVATION
LABELLING
FORMULAS
REARING

HOW TO KILL, STORE AND LABELLING THE COLLECTED BEETLES

HOW TO KILL

Potassium cyanide jars

To kill the collected beetles, a wide-mouth deadly bottle can be used. At the bottom potassium cyanide (KCN) is placed, covered with a cotton layer; then put some sawdust and above a plaster layer. Plaster will dry for a few minutes, and then the inside will be carefully cleaned, removing the excess of plaster on the interior walls. On the plaster put a corrugated cardboard lid which will prevent the beetles to get in contact with the plaster. Inside the bottle put some toilet paper, so that insects can walk on it and will not become dirty. The lid must be hermetic, don't forget to glue a label saying "poison". The potassium cyanide is dangerous and must be carefully manipulated.

When using Potassium cyanide to kill the beetles, it is important to remember that these should not be more than 1 hour in the killing Jar. Must be transferred to another jar containing toilet paper soaked in ethyl acetate (if you want to transport dry) because sometimes the insects come to life if they have been little time to poison, but If they were long periods in contact with Potassium cyanide, the internal tissues are destroyed and after a while they fall the legs and antennae. See below.

Bottle with ethyl acetate

It is easy to use a plastic bottle like those used to store vitamins. Inside this bottle put piece of wrinkled toilet paper and soak it with ethyl acetate. This bottle is easier to use than the one with cyanide jars, because ethyl acetate is not so much poisonous. As this liquid is very volatile, a dropper filled with this product should be always at hand to moisten the paper into the bottle.

Plastic dropper

It is also helpful to use a plastic eyedropper (like the ones used in ophthalmology) filled with ethyl acetate. Before filling the dropper with ethyl acetate, take care don't destroy the plastic with the liquid. This chemical product can't dissolve some plastic eyedropper. When evaporate ethyl acetate, inside the "killing jar", with the dropper can be renew the deadly liquid.

Ammonia injection

Ammonia can be used to kill large beetles. With a very fine needle syringe a few drops of ammonia are injected between the pygidium and the tip of the elytra of the beetle. Thus, the beetle is killed instantly.

Ammonium carbonate

This product can also be used as a poison to kill beetles. The powder is not poisonous or flammable, but can be irritating to the lungs and corrosive to mucous membranes. Ammonia gas is liberated when a small amount of powder is exposed to oxygen. The release of fumes is enhanced by wetting the powder with water. The powder can also be used in other collecting traps as a killing agent. Ammonium carbonate is light sensitive and must be stored in an airtight container or packed in a polypropylene jar, there are no shipping restrictions for this product. This product should be used in glass bottles because may dissolve the plastic.

Ethanol bottles

It is convenient to have on hand several bottles with different measures, filled with alcohol ethanol at 75% and others with higher graduation which may also serve to kill beetles.

HOW TO STORE

Cotton bed envelopes

These envelopes are good for storing dead beetles. They are made from newspaper sheets, because this paper has fungicide ink. To make them, cut two strips of newspaper, stick them crosswise and, in the middle of the cross is placed a loose layer of pressed cotton. Dead beetles will remain on the cotton bed. (Fig. 131)

Fig. 131 - Beetles in the center of cotton-bed envelope. *(Photo: Carlos Aguilar)*

The fins on the sides of the cross are fold upward, covering the cotton with beetles. When there are insects inside the envelope, some crushed naphthalene should be added to repel other insects which could attack the collection. All collecting data are written on the paper. Then, the envelopes can be exposed directly to sunlight to dry the insects, with closed envelopes' lids because dead beetles should not be exposed to sunlight, which fades them. These envelopes with insects must be stored in a locked box, which will be placed inside a plastic bag; then store it in a place free from moisture.

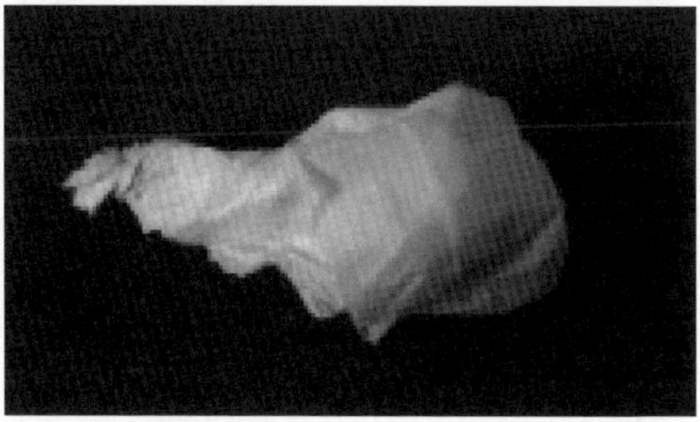

Fig. 132 - Tissue paper envelope with beetles inside. *(Photo: Carlos Aguilar)*

Fig. 133 - Plastic container's to store envelopes with beetles. *(Photo: Carlos Aguilar)*

A moist chamber is used at the laboratory to soften the dried insects and remove them from the cotton envelopes (to see below).

Tissue paper envelopes

This envelops are useful to save for short time the beetles. Cross two pieces of tissue or toilet paper, then in the center should be placed a small group of beetles (with full date). Then the wings of paper must be folded and twist until it is firm in that position and will form a small bag of beetles (Fig. 132). These envelopes are stored in tightly covered plastic containers (Fig. 133) with thymol or some naphthalene balls. Don't keep long time the insects inside the envelopes to prevent fungal growth.

Bottles filled with sawdust

Bottles filled with sawdust can be used instead of cotton envelopes to store the dead insects.

Moist chamber

After a while the beetles are dried and are difficult to handle without breaking legs or antennae. So to softened them it may be useful a moist chamber. This chamber can be made with an airtight plastic box (Tupper) or a wide-mouth jar. At the bottom there is some wet sand, or wet newspaper with some drops of carbolic acid or acetic acid (vinegar) to prevent fungal growing. On the sand or wet paper, wooden bars are placed, and above a plastic grid is located to avoid direct contact of the insects with water (Fig. 134). The dry beetles are put inside the chamber, which must be subsequently closed. After 1 to 2 days, the beetles will be soft. To find out when they are ready to be removed from the moist chamber, move their appendages (antennae and legs) with a tweezers. If the insects are still stiff, leave them for

Fig. 134 - Plastic moist chamber used to softening beetles. *(Drawing: Carlos Aguilar)*

a while in the moist chamber. To soften the beetles, others prefer to dip them in hot water or expose them to steam. Some large species such as Dynastinae can be softened by soaking them with isopropyl alcohol, but it is always better to use the moist chamber.

Labelling

All specimens must be accompanied by labels of 8 mm by 13 mm which can be done on cardboard or acid-free paper.

The labels must include all information relating to the collection: country, department, state or province. Specific locality (better with coordinates), date of collection (the day must be written in Arabic numerals and the month in Roman numerals, then the full year in Arabic numerals) and finally the collector's name.

Ecological data are recorded on other label: height above sea level, habitat, host plant, etc.

And finally, when you know the species or genus, it is recorded on another label with the scientific name of the beetle (in italics), who identified it, and the year.

These labels are placed beneath the insect, prodding them with the same pin, but the label should not touch the insect, not only for aesthetic reasons, but to prevent damage to any part of the insect.

Try to keep in the same bottle the beetles have been collected from the same plant species and the same place, identified by a label with all the data collection. It is best to try to identify plants where each beetle was collected, for that you can collect samples of flowers and other plant parts and put them between sheets of newspaper to herbarium and identified later with the help of a book or Better yet, with the help of a botanist. Floral accurate records are very important to understand the eating habits of the beetles

The letters, size 5, can be printed with a computer, but it is always better to handwriting with India ink.

Examples of labels' contents

Capture location:

Paraguay: Dpto Central:
Capiatá
24 III. 2006
Leg. Antonio Varela

Label with ecological data:

collected on leaves
of Palm (Copernicia
alba) in savannah

Identification label:

Rhynchophorus
palmarum Linnaeus

Det. P. Franco L. 2007

BASIC COLLECTION KIT

Tools

The tools required are several vials with preservative liquid, one entomological forceps, a small knife or small machete, two brushes (big and small size), one aspirator hose and a normal aspirator.

Aspirator

It is an indispensable tool for collecting beetles which can not be taken by hand due to their small size. It consists of a bottle with a pressurized lid, with two holes on top. A long hose and a shorter one are inserted in each hole. The longer hose has a stuck piece of cloth inside the bottle, since this material prevents the insects to be swallowed by the collector. The beetles come in through the shorter hose and then they are sucked through the longer hose. It is convenient to put a piece of toilet paper in the jar, so that the beetles could walk and do not get dirty with each other. Afterwards, the beetles must be transferred to a killer jar.

Fig.135 - Aspirator (picture courtesy of BioQuip 2010, www.bioquip.com)

Aspirator hose

It consists of a single hose which has at the end a smaller glass tube and a small piece of cloth between them. This aspirator operation is the same as described above, except that in this case keep the insect inside the aspirator for a few seconds aspirating continuously, until the insect has been transferred to the vial with ethanol. The advantage with this aspirator is that the insect falls directly into the bottle with ethanol, eliminating several necessary steps with a normal aspirator. The disadvantage is that it is useful only with very tiny insects and too slow to flee (Ptiliidae, Latridiidae, Pselaphidae, etc). Besides, it is difficult to use on a moving surface, for example the beating sheet, in that case use the common aspirator.

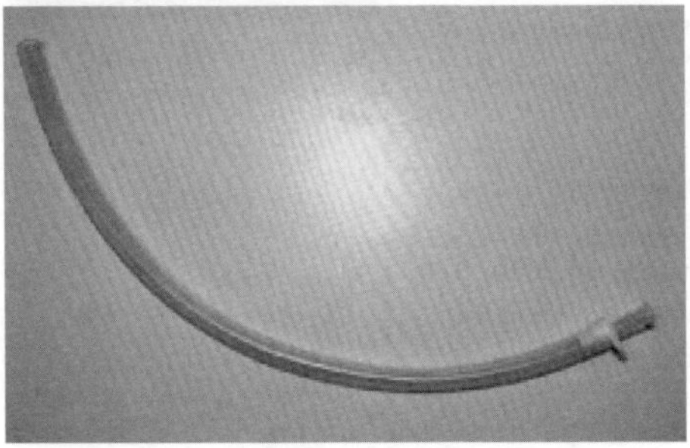

Fig 136 - Aspirator hose for small beetles

Entomological tweezers

The entomological tweezers are essential tools for every entomologist, there are many types (Fig. 137) shows some of these models. These devices can be purchased in stores distributing dental tools, but also there are shops specializing in entomological equipment.

Aerial nets

Aerial nets are very useful for collecting beetles flying or that cannot be collected by hand. Made with tulle very fine, soft and resistant. There are several models of these aerial nets here are two of them:

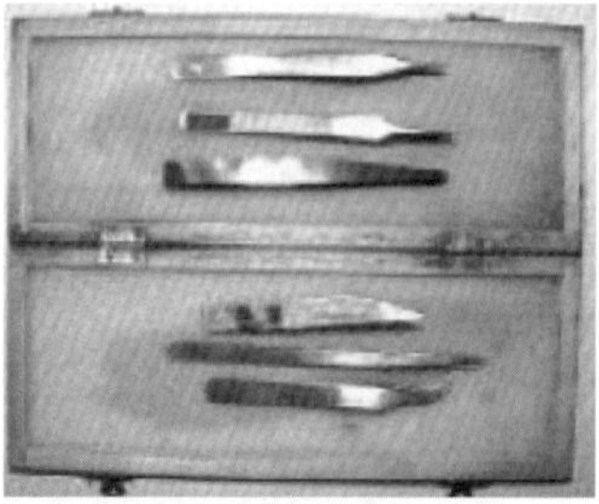

Fig. 137 - Box with tweezers. *(Photo: Carlos Aguilar)*

1) Telescopic aerial net

This net has an extendable telescopic handle, used to catch beetles that fly high in the foliage.

2) Coleopterist's net bag

It is an aerial net that prevents the beetles could escape when they are captured. The outside looks like a normal aerial net, but inside it has a bag that allows the beetles to enter but not to go out.

Fig. 138 - Telescopic aerial net. *(Picture courtesy BioQuip Products. http://www.bioquip.com)*

Plastic shower bottle

This shower bottle is useful to get beetles living in organic materials of the aquatic habitats.

The artificial shower is made with a plastic bottle (1 lt. or more), to which many small holes have been practiced at the bottom. The bottle is filled with water and serves as a shower to wash the organic materials gathered on the banks of watercourses. Sometimes it is inconvenient to carry a large container to the collection site. However, it is advisable to have on hand the shower, container, strainers and grid, to be able to use all techniques of collecting and obtain a greater representation of species present at each site.

List of equipment to collect

Every collector needs a basic collection kit. Which may consist of the following materials, but for more specific collections, other materials are required (specific traps, etc.).

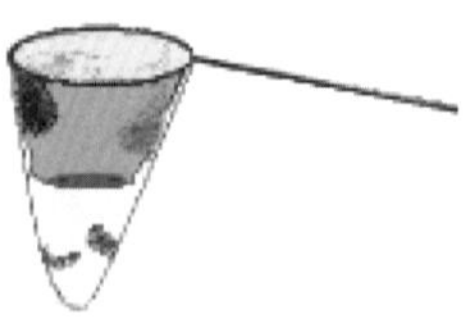

Fig. 139 - Aerial net with a bag in their interior so the beetles do not escape. *(Drawing: Carlos Aguilar)*

Basic Collection Kit

The basic equipment for any collector should consist of:
Collection nets (aerial net - sweeping bag – aquatic net)
Beating sheet
Aspirator
Fine-bristle brushes
Scissors

Maps (with trails, roads, longitude, latitude and altitude)
Collection bottles (various sizes)
Ethanol bottles (various sizes)
Plastic bags (various sizes)
Cotton beds or bottles filled with sawdust (to transport the specimens)
Plastic box with pressure lid (Tupperware) to save the bed envelopes
Permanent markers
Box to keep insects or larvae alive
Knife or penknife
Machete
A hand saw to cut tree branches with larvae
Linen or cotton bags to keep living fungi
Resistant bags to keep branches or pieces of material containing larvae or adult beetles
Flashlight with extra batteries
Insect repellent
Preservatives (Thymol, Naphthalene, etc)
Sufficient quantity of methylated spirits 40%, mixed with wine vinegar 50% and water 10%, to set traps, and ethanol to preserve the beetles during some time.
Ethanol: To keep many specimens in the same bottle: 80% to 90% (in vials 200 ml. approximately). To keep a few specimens 75% 1Lt.
Tweezers (one with fine tip and the other with a soft broad tip)
White paper sheets.
Rapidograph with India ink or pencil
Notebook.
First aid kit

Other important materials[1]

This list can be modified according to the place you are going to visit and what groups of beetles you are going to collect.

Insecticide with base of pyrethroids (Permethrin is usable, but it is not the best).
Plastic containers for live specimens
White plastic trays (for removal of dried specimens)
Grids - sieves
Beating sheets
Replacement aerial nets
Small saucepan
GPS device
Portable stove
Two white sheets and long rope
Light traps with accessories
Winkler strainer and berlese (with accessories)
Malaise trap and accessories
Flight intercept traps and accessories

1- See also "Beetles in fungi", Chapter III.

Bottle traps for collecting in water and trees (with accessories)
Pitfall trap materials
Bucket or large white containers for aquatic collection
Small shovel
Materials and traps for fungi collection
Cotton bed envelopes (with a box for storage)
Baits for pitfall traps (meat, cheese, fruit, etc.). They can be transported to the collection site in a container with ice, depending on how many hours you are traveling
Strong rope and thin rope
Clothespins
Plastic bags with an airtight lid
Scissors
Camera
Magnifying glass
Toilet paper or tissue paper

FORMULAS

Quimical attractants and pheromones

There are many chemicals attractants that are used to attract beetles to the traps (Toth and Imrei 2003; Donaldson et al. 1986; Yasuda 1996; Leather et al. 2005). These compounds can be combined or can be used separately in the different traps that are used with bait (of which many are described in this book), here only mentioning some of these attractants, which are available in laboratories specialized:

- Anisyl acetone,
- Benzoic/acetic acid
- Beta-ionone,
- Chloral hydrate,
- Cinnamyl acetate,
- Cinnamyl alcohol,
- Ethyl acetate,
- Ethylene glycol 25–50% solution (freely available as car antifreeze, toxic to birds and mammals),
- Eugenol,
- Fenol,
- Geraniol,
- Hexanoic acid,
- Isosafrol,
- Phenethyl propionate,
- Valeric acid+ hexanoic,
- …etc.

Other types of baits may be highly specific and are used more frequently in pest monitoring programs (Leather et al. 2005). Such baits have included pheromones (Yasuda 1996) and alpha- and beta-pinenes (Rieske & Raffa 1993).

Cantharidin has recently come into use as an extremely effective attractant for various beetles, such as pedilids beetles (Schauf, undate).

Traps will attract more insects if they are positioned upwind. Positioning of the traps must be carefully planned as insects from surrounding vegetation may be attracted to the trap (Colleen 2007).

Substances naturally produced by insects to attract others of their own kind are known as pheromones. They are often used in traps to aid in controlling pest species (Barclay 1984). Many pheromones types can be purchased in specialized laboratories.

According to Dowdy and Mullen (1998) when mixed into a trap several Pheromones three things can happen:

1) A beetle can be attracted to the trap by pheromones of their own species and this is not affected by the presence of other pheromones.
2) A combination of some pheromones can attract strongly to trap a species of beetle.
3) The attraction to the trap can be reduced by the presence of certain pheromones that act as repellents for certain species of beetles. More information on pheromones can be found in Barclay (1984).

For put in the pitfall traps

10% acetic acid
Brine
Glycol (in very small amounts)

To preserve beetles temporarily

The most useful field preservation technique for adult insects like the author of this book and many others entomologists have successfully used for years involves the use of chlorocresol (more correctly 4-Chloro-m-cresol).
Some coleopterists preserve small beetles in a 4% solution of glacial acetic acid until they are dried and set. If kept cool the material should be fine for many months. But there are also other methods:

1ª formula

50 parts good quality vinegar
40 parts water
1 tablespoon salt
10 parts ethanol at 75%

2ª formula

50 parts ethanol at 80°
50 parts ethyl acetate

This formula is useful to prevent the beetle's hardening. It is a semi-preserving formula and after a few days the beetles must be moved to ethanol at 75% where they can be kept forever, or they can be dried.

Also, the author used the following formula to preserve, for a few days, the beetles collected

White vinegar (good quality)	20 parts
Ethanol (75%)	50 parts
Ethyl acetate	30 parts

The volume of this depends on the size of samples.

To keep beetles permanently in liquid mediums

When large numbers of beetles caught and are kept permanently in ethanol it is better 90% ethanol. What is explained here, although it contradicts most of the books that touch this subject, because the beetles (and all insects) have water in their bodies and after a time, this water leaves her body and mixed with ethanol from the bottle, thus lowering graduation, in addition, Osvaldo Diorio of Argentina (expert on Neotropical Cerambycidae), has informed me in 2003 that 90% ethanol fixation[1] and preserves the tissues and muscles of insects, whereas the lower-ranking (75%) only preserves the tissues and muscles, but not fixed those tissues. In any case, the ethanol of 75% is good when they are few specimens preserved in a jar. For ensuring that the total mass of insects are kept in optimal conditions in a jar with ethanol, is better than the volume of ethanol (inside the jar) occupies a third more than the volume of insects into the bottle. So, it will prevent insects from rotting.

The live beetles can be placed in the following solutions

Hadden's solution (1933):

75% to 80% Ethanol	200 cc
Acetic acid	1 cc

Hangay and Dingley (1985) recommend the following formulas to preserve the material collected

Van Emden's fluid

To kill and fix the larvae of beetles:

Glacial acetic acid	2 ml
Distilled water	30 ml
Formalin	5 ml
Ethanol (95%)	15 ml

1.- When the animal dies, (e.g. beetles) it is necessary to stop its vital processes before an autolysis of its materials. This process is known as steadfast. It preserves the cells and tissues in a state as close as possible in morphology and chemical composition lived.

Bles' fluid

Designed to kill and fix beetles larvae.

Glacial acetic acid	3 ml
Formalin	7 ml
Ethanol (70%)	15 ml

To preserve dry beetles permanently

It is best to store the dry insects in cotton envelopes, as described above or in bottles with sawdust. But if you want to do DNA studies they should be stored in almost pure alcohol (ethanol).

Conservation at low temperature

Another form of conservation is to keep the specimens within the vials in a refrigerator, but not freezing. With this method the author has preserved specimens in perfect condition for many years and, when removed from the refrigerator, they are still in good condition and soft as the first day when they were collected.

The jars must be sterilized; they should be washed with hot water and then finish the cleaning with pure alcohol (to avoid mold). Then you need to dry them thoroughly. After cleaning and drying the jars, dive the beetles into ethanol for a few moments to kill bacteria and then put them on blotting paper or newsprint. Once the specimens have no traces of moisture out of their bodies (but not dry inside), they can be placed into the sterilized and dry jar, which is deposited in the refrigerator with a temperature of + 4° Celsius or so. The jar must be labeled with all data collection.

GLOSSARY

Acid soil: Soil that has a pH less than 7.0. Soil is regarded as 'very acid' when the reaction is less than pH 5.0. The USDA lists five standard ranges of soil: acidity less than pH 4.5, is extremely acid; very strongly acid (4.5-5.0); strongly acid (5.1-5.5); medium acid (5.6-6.0); and slightly acid (6.1-6.5). The surface soil horizons of acid brown earths have a reaction of pH 5.0 or less.

Actinomycete: A mycelium-forming bacterium of the order Actinomycetales.

Ambimorphs: Organisms with certain morphological modifications for life in caves, which have retained most of the features of epigean organisms.

Anophthalmos: Lacking eyes

Anthophilous: Attracted by flowers or feeding on flowers.

Argillaceous: Terms referring to sediment clay rocks with grain diameter less than 0.0625mm. They represent more than 50 percent of the sedimentary rocks and most have a high mineral content. Many contain a high percentage of organic matter and can be considered as a potential source of hydrocarbons.

Arctic desert (Arctic Tundra): Arctic tundra is located in the Northern Hemisphere, encircling the North Pole and extending south to the coniferous forests of the taiga. The arctic is known for its cold, desert-like conditions. Tundra is the coldest of all the biomes. Tundra comes from the Finnish word "tunturi", meaning treeless plain. It is noted for its frost-molded landscapes, extremely low temperatures, little precipitation, poor nutrients, and short growing seasons.

Asymptote: A straight line associated with a curve such that as a point moves along an infinite branch of the curve the distance from the point to the line approaches zero and the slope of the curve at the point approaches the slope of the line.

Azonal: Soils without definite horizons.

Basin River: The space defined by the union of all headers that form a major river, or the territory drained by a single natural drainage system, which drains its waters into the sea through a single river.

Biodiversity: A term which gained popularity in the late 1980s, used to describe all aspects of biological diversity, including species richness, ecosystem complexity, and genetic variation.

Biogeography: It is a branch of biology that studies the geographical distribution of plants and animals.

Biomass: Represents the total mass, total weight, total volume or energy of all the organisms of given specie of in a given area.

Biome: A biological subdivision that reflects the ecological and physiognomic characteristics of vegetation. They are the largest geographical biotic communities that correspond with climatic regions and extend to all living organisms (not just plants) and their interaction with the environment.

Biota: Represents the total of plants and animals occupying a place together (e.g. forest biota, marine biota).

Biotope: An environmental region characterized by certain conditions and populated by a characteristic biota.

Boreal: Pertaining to the north (derive from from "Boreas", the Greek god of the north wind).

Brachypterous: adj. Having very short or rudimentary wings, for example some Carabidae, Staphylinidae, Scarabaeidae, etc.

Bracket fungus: Is the name of a group fungi with many species (Aphyllophorales), many of which produce fruiting bodies found in the trunks of trees. Some are parasites of conifers and cause serious disease in these trees.

Bromeliads: The name given to plants of the family Bromeliaceae. As bromeliads are found in the same places as the orchids, the latter introduced its roots in the leaves of bromeliads to supply water and food. Also, there are other plants epiphytes that depend on aquatic environment provided by the bromeliads.

Caatinga: A thorn forest of arid, tropical, of northeastern Brazil and similar types of vegetation, including thorn woods and thorn bush. It is differentiated from savannah woodlands by the absence or paucity of grass, with prolonged dry period and smaller amounts of rainfall. The name is also applied to similar vegetation that occurs in patches through the Amazonian rain forest.

Cerrado: A tree-covered savannah, which is one type of Brazilian campo.

Cecidogenous: A term applied to insects that produce galls (tumors) in branches and leaves of living plants.

Cleptoparasite: An organism that exploits the food reserves of another species to the latter's detriment to provide food for its own young.

Coccidophagous: That feed on scale insects (Homoptera).

Colluvial deposits: When a alluvium, a variety of materials consisting of fine particles of silt and clay and larger particles of sand and gravel, which are deposited as sand or cement in a lithological unit or lithified, it is called alluvial deposit.

Congeneric species: Species belonging to the same genus.
Conidia: Plural of conidium: asexual spores of fungi produced by constricting one end of a hypha or sterigma and are not enclosed in a sporangium.
Coprophagous: Animals that feed on fecal matter
Deciduous: Applied to parts of a plant or animal that fall seasonally (e.g. the leaves of certain plants or the horns of a deer). Or the trees that lose their leaves in a particular season.
Detritus: Litter consists of fragments of dead material (e.g. the litter leaves, manure, feathers and carcasses).
Ecological indicator: Any organism or group of organisms indicative of a particular environment or set of environmental conditions.
Ecological niches: Is a term describing the relational position of a species or population in its ecosystem to each other. The role of an organism in a community in terms of the habitat it occupies, its interactions with other organisms and its effect on the environment.
Ecology: The scientific study of the interrelationships among organisms, and between them and all aspects (living and non-living) of their environment.
Ecosystem: Community of different species that depend on each other and relate with their nonliving environment. Something autonomous in terms of energy flow and it is different from the surrounding communities.
Ectoparasites: Parasite that lives on the surface of an organism.
Edaphic: Belonging or influenced by conditions of soil or substratum.
Endemic: Restricted to a certain region or part of region.
Endogeic: Feed below the soil surface on mineral soil mixed with organic matter in various stages of decay.
Environment: All external conditions, physical and biological, in which an organism evolves. Environment also includes the social environment in which they develop in humans, along with their economic, cultural and political as well as more general characteristics such as soil, climate and foods.
Epigeal (epigean): Growing or occurring above ground.
Epiphyte: A plant that uses another plant, typically a tree, for its physical support, but which does not draw nourishment from it.
Etiopical faunal region: Zoogeographic area that includes the entire African continent south of the Sahara and the Southern edge of the Arabian Peninsula, with warm or very warm weather.
Facultatively predaceous: That has the ability to feed on different prey.
Generalist: Species with a very broad ecological niche, which can tolerate a wide range of environmental conditions and eat a variety of foods.
Genome: The complete set of genes carried by an organism or virus.
Habitats: The environment within which an organism is normally found.
Hepatic plants: A group of green plants commonly called liverworts, Which with the hornworts and mosses, are the bryophytes. They are small and usually found in groups under the ground or on rocks or tree bark.
Hygropetric: Belonging to the fauna of submerged rocks.
Hyphas: Tubular filament, which is the basic form of growth of the vegetative phase of a fungus. Hyphal extension and branching produces the fungal mycelium.
Imago: A term applied to the adult stage of an insect.
Incertae sedis: From Latin for "of uncertain placement", abbreviated "inc. sed.", is a term used to define a taxonomic group where its broader relationships are unknown or undefined.
Instar: The stage in an insect's life history between two moults.
Interstitial fauna: Occurring in interstices or spaces. Flora and fauna is living between sand grains or soil particles.
Intertidal zone: The area that is exposed to the air at low tide and underwater at high tide (for example, the area between tide marks). Also known as the foreshore and seashore and sometimes referred to as the littoral zone.
Introduced specie: It is non-native species, alien, exotic, is a species living outside its native distributional range, which has arrived there by human activity, either deliberate or accidental.
Larva: The newly hatched, wingless, often wormlike form of many insects before metamorphosis. The newly hatched, earliest stage of any of various animals that undergo metamorphosis, differing markedly in form and appearance from the adult.
Lignívorous: Feeding on wood.
Meanders: In general is a bend in a sinuous watercourse. It forms when water moves in a river eroding the Outer Banks and expand its valley. Meanders are curves in a stream (or river). Where a stream flows across a flood plain there may be many meanders.
Micro-habitats: A precise location within a habitat where an individual species is normally found (e.g. within a forest, a small waterfall can be a microhabitat for some species of Torridincolidae).

Middle hypogeum: Animals that live on land and feed on plant debris, and predators
Monotypic: Applied to any taxon that has only one immediately subordinate taxon.
Morphospecies: A group of biological organisms that differs in some morphological respect from all other groups.
Mucilaginous: Containing, or composed of mucilage. Mucilage: complex substances composed of various types of polysaccharides, becoming viscous and slimy when wet, widely occurring in plants, and secreted by bacteria and by plant roots.
Mycophilia: Means mushroom lover.
Mycorrhiza: A close physical association between a fungus and the roots of a plant, from which both fungus and plant appear to benefit; a mycorrhizal root takes up nutrients more efficiently than does an uninfected root.
Myrmecophlle: Applied to a species which relies on ants for food or protection in order to complete its life cycle.
National Park: As defined by the IUCN in 1975, as a long area of land with ecosystems that have not been altered by human activities, including plant and animal species, with landscape features, and with habitats of great scientific interest, and of beauty, recreational or educational interest. Is controlled by the state, and the public is allowed to visit, cultural, and recreational purposes. Is an area for the perpetual conservation, within which such public recreational activity is permitted as is compatible with the conservation objectives.
Native: Applied to a species that occurs naturally in an area, and therefore one that has not been introduced by humans either accidentally or intentionally.
Natural woodland: Comprising trees that have not been planted by humans, and where no human interference has occurred.
Nature reserve: It is an area of land for nature conservation and scientific research related activities. Usually, they are well protected legally against other operating uses. Public access may be restricted partially or completely. The status and definition of reserves varies from one country to another.
Nearctic faunal region: The fauna from North America to south to Mexico.
Neotropical faunal region: Includes South and Central America, including Southern Mexico, the West Indies, and the Galápagos Islands.
Oasis: In an arid region, an isolated area that supports water-loving plants throughout the all year. Oases most commonly occur in depressions, where the water-table lies close enough to the surface to be within the reach of plant roots. Also, it is a small area of vegetation of a single type, which is surrounded on all sides by a much larger area with vegetation of a very different type.
Omnivore: That feeds on both plants and animals, and thus operates at a range of trophic levels.
Oriental faunal region: The area that includes India and Asia south of the Himalayan-Tibetan mountain barrier, and the Australasian archipelago, excluding New Guinea and the Sulawesi.
Palaearctic faunal region: Includes the region that includes Europe, Asia north of the Himalayan-Tibet. Also, North Africa and much of Arabia.
Parasitism: A special case of symbiosis in which one parasite receives advantage to the detriment of the other (the host).
Parasitoid: Organism alternately parasitic and free-living and whose parasitism ultimately kills its host.
Perennial: A plant that normally lives for more than two seasons and which, after an initial period, produces flowers annually.
Permafrost: A layer of soil or bedrock that has been continuously frozen for at least two years and as long as tens of thousands of years. Permafrost can reach depths of up to 1,524 m (4,999 ft). It is found throughout most of the Polar Regions and underlies about one fifth of the Earth's land surface. Most permafrost is located in high latitudes (i.e. land in close proximity to the North and South poles), but alpine permafrost may exist at high altitudes in much lower latitudes.
Pest: An animal that causes economic damage to eat grains, fruits, fiber, or other materials used by humans.
Pheromone: A chemical substance produced and released into the environment by an animal, affecting the behaviour or physiology of others of its species.
Phototropism: Growth movement of animal or plants in response to stimulus of light.
Phytophagous: Feeding on plants.
Polyphagous: They are organisms that feed on many types of plants or animals.
Polyporous: Includes a group of basidiomycete fungi that are commonly known as the bracket fungi, where the fruit body is typically leathery, papery or woody with tubes or gills on the underside. many of these fungi grow on wood. This group also includes fungi with smooth, ridged, warty or spiny surfaces bearing the hymenium, and includes the coral fungi and cantharelles (agarics).
Population: A group of organisms of the same species living in a determined area.

Prairie: It is for temperate grasslands in North America, dominated by grasses and with a progressive decrease in rainfall. Several broad-leaved herbaceous plants, annuals and perennials are mixed with herbs.
Predation: It is the interaction between populations of species in which one organism, the predator, to get energy (food), through consumption, by killing the prey. A predator is an animal which captures kills and devours its prey, but also includes the feeding of predation by insectivorous plants and grazing interactions (e.g., consumption of leaves by beetles).
Pupa: The third stage in the life cycle of a beetle while it undergoes a complete metamorphosis.
Pyrethroid: A synthetic chemical compound similar to the natural chemical pyrethrins produced by the flowers of pyrethrums. Pyrethroids constitute a major proportion of the synthetic insecticide market and are common in commercial products such as household insecticides. In the concentrations used in such products, they may also have insect repellent properties and are generally harmless to human beings in low doses but can harm sensitive individuals. They are usually broken apart by sunlight and the atmosphere in one or two days, and do not significantly affect groundwater quality. But the Pyrethroids are toxic to fish and other aquatic organisms.
Relict (relic): The term refers to species that previously had a much wider distribution than at present and have been able to survive locally through unfavorable periods (glaciations or subsidence of the earth). There are regions called "refuges", while they are in danger of extinction elsewhere (eg, some beetles from the Andean forests).
Salt lake: It is a lake that has highly concentrated mineral salts, about 100 parts per thousand or more.
Saprophage: An organism that feeds on other, dead, organisms. The Saprophages form part of the heterotrophs (organisms that feed on other organisms) and consist mainly of bacteria and fungi, but also some invertebrate animals, such as insect larvae.
Saproxylic beetles: Are species that are involved or depend on the decomposition process produced by fungi of the wood, or products of this degradation and are associated with live or dead trees.
Savanna: Tropical vegetation dominated by grasses with different mixed tall shrubs and / or trees in open order. Has a long dry season and with plants adapted to fire.
Stygobiontic: Species live only in groundwater, they are almost never found at the surface.
Subcortical: Is the part beneath the cortex or cortical layer. The Cortex in vascular plants, is the tissue in stem and root surrounding but not part of the vascular tissue; also, in eukaryotic cells, the region just under the cell membrane.
Synanthropic: Species ecologically associated with humans.
Taiga: Name used for all boreal forests, but by some only to the more open, otherwise known as lichen woodland.
Taxon, pl. taxa: a group of organisms of any taxonomic rank (e.g. family, genus, or species).
Taxonomy: The scientific classification of organisms. This term is also used in a broader sense (e.g. soil taxonomy). The term derives from "Tasso" Greek word meaning "arrange".
Temperate rain forest: Forest which extends into the temperate regions, where rainfall is high (typically 1500-3000mm) or the fog is common. The trees have broad leaves, the evergreen conifers are common. They are found in coastal areas of Southeastern United States and Northwestern North America, in Southern Chile, in some parts of Australia and New Zealand, and Southern Japan and China.
Termitophile: Concerning the species that live with termites
Monoterpenes (Terpenoids): Natural plant products, based on isoprenoids. Include aromatic compounds such as menthol, pigments, such as β-carotene, and gibberellins.
Pygidium: The terminal exposed and sclerotized tergite at the tip of abdomen; this may be tergite 7 or 8. (the tergite is the sclerite forming dorsal surface of any body segment).
Testaceous: Of, pertaining to, or derived from shells. Protected by a shell-like outer shell, or like material made of shell or carapace.
Triungulinas: It is the first instar of a mobile larva (that has three claws at the end of each of its legs) and of an insect that has hypermetamorphosis. The term Triungulin applies only to the Meloidae. Sometimes is used in a broader sense, for example, to any similar mobile larva (planidium).
Troglobiont: Organism that lives in caves.
Univoltine: Species has one generation per year.
Vascular plants: It is the common name given to all plants that contain xylem and phloem and include mosses, ferns, cycads, gymnosperms and angiosperms.
Xeric: A dry or intermediate environment.
Xerophyte: A plant adapted to living in a dry arid habitat; a desert plant.
Xilophagous: Wood-eating.

USEFUL BOOKS FOR STUDY

Hangay, G. & Dingley, M.O., 1985 Biological Museum Methods. Vol. 2. Plants, Invertebrates and Techniques. Academic Press. Australia. 323 pages.

Helga Blanco Metzler 2004. Las feromonas y sus usos en el manejo integrado de plagas. Manejo Integrado de Plagas y Agroecología (Costa Rica) No. 71 p.112-118 (in spanish version, but very useful to know the pheromones traps). Available online (2010) in: http://orton.catie.ac.cr/repdoc/A1933E/A1933E.PDF

Johnson, Paul J.; Roughley, Rob; Pollock, Darren A. and Solís, Angel. version 2009. Biodiversity Sampling for Coleoptera in Cosat Rica: A Parataxonomist's Manual for collectinng, rearing and preparing Beetles. Available online (2010) in: www.inbio.ac.cr/papers/manual_coleoptera/manual.htm

Lowman, M.D., Moffett, M. & Rinker, H.B. 1993. A new technique for taxonomic and ecological sampling in rainforest canopies. Selbyana. Vol. 14, pp: 75 - 79. Available online (2010) in: www.canopymeg.com/PDFs/papers/0031.pdf

Lowman, M. D. and Moffett, M. 1993. The Ecology of Tropical Rain Forest Canopies. TREE. Vol. 8, N°. 3, pp: 104- 107. Available online (2010) in: http://www.canopymeg.com/PDFs/papers/0030.pdf

Maree, Hunt and Ian Millar 2001. Cave invertebrate Collecting Guide. Department of Conservation Technical. Series 26, pp: 1 - 28. Available online (2010) in: http://www.doc.govt.nz/upload/documents/science-and-technical/docts26.pdf

Martin, J.E.H. 1977. Collecting, preparing and preserving insects, mites and spiders. The Insects and Arachnids of Canada. Part 1. Publication 1643. Canada Department of Agriculture, Ottawa. 182 pp.

McGavin, George C 2007. Expedition Field Techniques INSECTS and other terrestrial arthropods. Published by Geography Outdoors: the centre supporting field research, exploration and outdoor learning. 93 pags.

Pablo Bahillo De la Puebla, 1996. Sobre la recolección, tratamiento y conservación de los Cerambycidae. Bol. SEA. N° 14, pp: 55- 62. Available online (2010) in: www.sea-entomologia.org/PDF/BOLETIN_14/B14-013-055.pdf

Ribera, Ignacio y Aguilera, Pedro 1995. Metodos de recolección y estudio de coleópteros acuáticos. Bol. SEA. N° 12, pp: 43-48. Available online (2010) in: www.sea-entomologia.org/PDF/BOLETIN_12/B12-016-043.pdf

Schauf, M. E. (Undated). Collecting and preserving insect and mites: techniques and tools. U.S. Department of Agriculture, Systematic Entomology Laboratory. version of the USDA Misc. Publication no. 1443 published by the Agricultural Research service in 1986 and Edited by George C. Steyskal, William L. Murphy, and Edna M. Hoover. Available online (2010) in: http://www.ars.usda.gov/SP2UserFiles/ad_hoc/12754100CollectingandPreservingInsectsandMites/collpres.pdf

REFERENCES

Aalbu, R.L., C.A. Triplehorn, J. M. Campbell, K. W. Brown, R. A. Somerby, and D. B. Thomas. 2002. Family 106. Tenebrionidae Latreille 1802, pp: 463- 509. In: American Beetles. Volume 2. Polyphaga: Scarabaeoidea through Curculionoidea (R. H. Arnett, Jr.; M. C. Thomas; P. E. Skelley and Howard, J. Frank, eds.). CRC Press, Boca Raton, FL.

Aalbu, R. L.; Andrews, F. G. and Darren A. Pollock 2005. Revision of Western Hemisphere Dacoderinae with a phylogenetic analysis of world taxa Coleoptera: Salpingidae. Coleopterists Society Monograph. N° 4, pp: 1– 40.

Abdullah, M. 1969. The natural classification of the family Anthicidae with some ecological and ethological observations. Deutche Entomolo-gische. Zeitschrift, N.F. N° 16, Helft IV-V pp: 323-366.

Aguilella, Antoni y Segundo Rios 2004. Bosques, sotos y herbazales: quintaesencia de la ribera. Monográfico Anuario. (España). N° 92. 9 pages.

Aguilera, C. H. P and Ribera, I. 1998. *Bothriophorus atomus* Mulsant & Rey, 1852 nuevo para la península Ibérica (Coleoptera, Limnichidae). Orsis. Vol. 14, pp: 39-41.

Aiken, R. B. and Roughley, R. E. 1985. An Effective Trapping and Marking Method for Aquatic Beetles. Source: Proceedings of the Academy of Natural Sciences of Philadelphia, Vol. 137, N° 1. Proceedings of the First International Conference on Classification, Phylogeny, and Natural History of Hydradephaga, pp: 5-7. Available online (2010) in: http://www.mta.ca/~raiken/Pubs/Trap.pdf

Alarie Yves and Bilton David T. 2005. Larval morphology of aspidytidae (Coleoptera: Adephaga) and its phylogenetic implications. Annals of the Entomological Society of America. Vol. 98, N° 4, pp: 417-430.

Aldemaro. Romero 2009. Cave Biology. Life in Darkness. Cambridge University Press. 319 Pages.

Alexander, K. (coord.). 2002. The invertebrates of living and decaying timber in Britain and Ireland – A provisional annotated checklist. English Nature Research Reports (number 467). Peterborough, UK.

Allaby, Michael 2008. Temperate Forests. Facts On File, Inc. 552 pags.

Alistair, S. Ramsdale 2007. A new species of *Calochromus* Guérin-Méneville (Coleoptera: Lycidae) from Fiji. Zootaxa N° 1625, pp: 43–46.

Allemand, R. Aberlenc, H-P. 1991. Une méthode efficace d'échantillonnage de l'entomofaune des frondaisons: le piège attractif aérien. Bulletin de la Société Entomologique Suisse Vol. 64, pp: 293-305.

Allemand, R. & H. P. Aberlenc 1996. Un método eficaz de muestreo para la entomofauna de las zonas frondosas: la trampa atrayente aérea. Bol. S.E.A., 14: 23-30. Available online (2010) in : www.sea-entomologia.org/PDF/BOLETIN_14/B14-004-023.pdf

Allsopp, P. G., 1984. Checklist of the Hybosorinae (Coleoptera: Scarabaeidae). The Coleopterists Bulletin. Vol. 38, pp: 105–117.

Allsopp, P. G., 1989. *Schizognathus apricagger* sp. nov. (Coleoptera: Scarabaeidae: Rutelinae) from Southeast Queensland and new locality records for other Schizognathus spp. Journal of the Australian Entomological Society. Vol. 28, pp: 187–190.

Allsopp, P. G., 1995. Biogeography of Australian Dynastinae, Rutelinae, Scarabaeinae, Melolonthini, Scitalini and Geotrupidae (Coleoptera: Scarabaeoidea). Journal of Biogeography.Vol. 22, pp: 31 – 48.

Almeida, L. M.; C. S. Ribeiro-Costa & L. Marinoni. 1998. Manual de coleta, conservação, montagem e identificação de insetos. Ribeirão Preto, Holos, 78 Pages.

Alonso, R. N. 1998. Los boratos de la Puna. Editorial Cámara de la Minería de Salta, Salta. 196 pages.

Alonso-Zarazaga, Á.; Moreno, I. P. & Grijalba, F. M. 2002. Presencia de dos especies raras de Anthribidae (Coleoptera) en la Península Ibérica Miguel. Aracnet. 10 - Bull. S.E.A., N° 31, pp: 141 – 143.

Alonso-Zarazaga, M. A. 2002. Lista preliminar de los Coleoptera Curculionoidea del área ibero-balear, con descripción de Melicius gen. nov. y nuevas citas. Boletín de la S.E.A. Vol. 31, pp: 9-33.

Alonso-Zarazaga, M. l. Á.; M. Sánchez-Ruiz & A. Sánchez - Ruiz. 2003. Una nueva familia de Coleoptera para España: Acanthocnemidae. ARACNET 11. Bull. S.E.A. N° 32, pp: 179 - 180.

Ananthakrishnan, T. N. 2004. General and applied entomology. Tata McGraw-Hill. 1184 pages. ISBN 0070434352, 9780070434356.

Anderson, R. S. 1982. Resource partitioning in the carrion beetle (Coleoptera: Silphidae) fauna of Southern Ontario: ecological and evolutionaiy considerations. Canadian Journal of Zoology. Vol. 60, pp: 1314 - 1325.

Anderson, R.S. ; Roughley, R.E.; Behan-Pelletier, V. and Danks, H.V. 1994. Terrestrial Arthropod Biodiversity: Planning a Study and Recommended Sampling Techniques (Reprint 2007). Biological Survey of Canada. Entomological Society of Canada Supplement to the Bulletin. Vol. 26. Fasc.1. 32 pages.

Andreas Herrmann 2001. Car Catcbers - Improved methods for collecting beetles and other insects!. The Bulletin of the Amateur Entomologist's Society. Vol. 60. N° 439, pp: 210 - 214.

Andrew, A. Calder 1996. Family Cantharidae, Soldier Beetles. CSIRO Entomology, Canberra, Australian. Australian Biological Resources Study: http://www.environment.gov.au/biodiversity/abrs/online (2010) -resources/fauna/afd/taxa/CANTHARIDAE

Andrew, D. A. David K Y., *Gerasimos* C., Murray J F., John La Salle, John F Lawrence; P. B McQuillan; L. A Mound; D. J Bickel; P. J Gullan; D. F Hales and G. S Taylor. 2004. Insects 'Down Under' – Diversity, endemism and evolution of the Australian insect fauna: examples from select orders. Australian Journal of Entomology. Vol. 43, pp: 216 – 234.

Andrew, Edward, Z. Short 2004. A New Genus and Species of Sphaeridiinae from Costa Rica (Coleoptera: Hydrophilidae) The Coleopterists Bulletin. Vol 58. Fasc.4, pp: 534-537.

Androw, R. A. 1992. An Unusual Method for Collecting Buprestids. The Ohio Coleopterist. Vol. I, N° 1. Available online (2010) in: www.ohiocoleo.org/members/issues/iss1~1.asp#three

Anwar, Maun, M. 2009. The Biology of Coastal Sand Dunes. Oxford University Press. 265 pages.

Apfelbeck, V. 1916. Zur Kenntnis der Gattung *Sphaerosoma* Leach (Col.). Revision der Arten von der Balkanhalbinseln. Ann. Mus. Nat. Hung. Vol. 14, pp: 471 - 500.

Aponte, O. I. C. 1996. Arthropodes associated with colonies of stingless bees (Apidae: Meliponinae). Pegone. Vol. 4, pp: 3-6.

Araujo, Pablo; Fabián Bersosa; Rubén Carranco; Valeria Granda; Paulo Guerra; Nelson Miranda; Ana María Ortega; Paulina Rosero 2005. Evaluación preliminar de la diversidad de escarabajos (Insecta: Coleoptera) del Chocó ecuatoriano. Politécnica. Biología 6. Vol. 26. Fasc. 1: pp: 120 - 140.

Arbea, J. I. & Baena, M. 2003. Colémbolos cavernicolous de Andalucía (Insecta: Collembola) Cavernicolous springtails of Andalusia (Insecta: Collembola). Zool. Baetica. Vol. 13/14, pp: 71 – 84.

Arce-Pérez, R. Y R. Novelo-Gutiérrez 1988. Primer registro de *Lepicerus bufo* (Hinton 1934) (Coleoptera: Lepiceridae) para el estado de Morelos, Mexico. Folia Entomológica Mexicana.Vol. 75, pp: 156- 158.

Arce Pérez, R. y M. Morón, A. 2008. Primer registro de species del suborden Myxophaga (Iinsecta: Coleoptera) en la Província del Golfo de México. Acta Zoologica Mexicana (nueva serie). Vol. 24, N° 002. Pp. 225-226

Archangelsky, M. 1997. Studies on the biology, ecology, and systematics of the immature stages of New World Hydrophiloidea (Coleoptera: Staphyliniformia). Bull. Ohio Biol. Survey, New Series, 12: 1-207.

Archangelsky, M. 2001. Coleoptera. En: Fernández, H. R. & E. Domínguez. (eds.). Guía para la determinación de los Artrópodos Bentónicos Sudamericanos. Universidad Nacional de Tucumán, Facultad de Ciencias Naturales e Instituto M. Lillo, pp. 131-153.

Archangelsky, M., Beutel, R. G. & Komarek, A. 2005. Hydrophiloidea. 10, 10.1. Hydrophiloidea, Hydrophilidae. En: Beutel, R. G. & A. B. Leschen (eds.), Coleoptera, Volume 1: Morphology and Systematics (Archostemata, Adephaga, Myxophaga, Polyphaga partim). Handbuch der Zoologie, De Gruyter, Berlin, pp: 157-183.

Archangelsky, M. 2007. Escenarios evolutivos en larvas y adultos de Hydrophiloidea (Coleoptera) analizados en función de diferentes hipótesis filogenéticas. Rev. Soc. Entomol. Argent., Vol..66. N° 3-4, pp: 103-111.

Arias, D. M. et al. 2007. Distribución Espacialy Temporal de los coleópteros acuáticos En la cuenca del Río Coello (Tolima, Colombia). Caldasia Vol. 9 Fasc.1, pp: 177- 194.

Arnett, R. H. Jr. 1963. The Beetles of the United States: A Manual for Identification. Catholic University press. Washington DC. XII. + 1112 Pages.

Arnett, R. H. Jr. 1968. The Beetles of the United States: A Manual for Identification. The American Entomological Institute, Ann Arbor, MI. XII+ 1112 Pages.

Arnett, R. H. Jr.; N. M. Downie and H. E. Jaques 1980. How to know the bettles. Second Edition. The pictures Key Nature Series. 416 Pages.

Arnett, Ross H. 2000. American insects: a handbook of the insects of America north of Mexico. CRC Press. (2ª ed.). 1003 pages.

Arnett, R. H. Jr. & Thomas, M. C. Eds. 2001. American Beetles. Vol. 1 CRC Press, Florida, USA. 443 pags.

Arnett, R. H. Jr.; M. C. Thomas; P. E. Skelley and Howard, J. Frank (Eds.). 2002. American Beetles. Vol. 2: Polyphaga: Scarabaeoidea through Curculionoidea. CRC Press, Boca Raton. XVI + 861 Pages.

Arriaga-Varela, Emmanuel; K. Wioletta Tomaszewska & José Luis Navarrete-Heredia 2007. A synopsis of the Endomychidae (Coleoptera: Cucujoidea) of México. Zootaxa. N° 1594, pp: 1 – 38.

Arrow, G. J. 1912. Scarabaeidae: Pachypodinae, Pleocominae, Aclopinae, Glaphyrinae, Orphninae, Idiostominae, Hybosorinae, Dynamopinae, Acanthocerinae, Troginae. Coleoperorum Catalogus. Vol.19. Pp : 1-66.

Arrow G J Arrow 1925. The Fauna of British India Including Ceylon and Burma - Coleoptera; Clavicornia; Erotylidae, Languriidae, and Endomychidae. Edit.: Read Book. 436 pages.

Ashe J. S. and Timm R. M. 1987. Activity by and activity patterns of parasitic beetles of the genus *Amblyopinus* (Coleoptera: Staphylinidae). Journal of Zoology (London). Vol. 212, pp: 429 - 437.

Ashe J. S. 1987. Egg chamber production, egg protection, and clutch size among fungivorous beetles of the genus

Eumicrota Casey (Coleoptera: Staphylinidae) and their evolutionary implications. Zoological Journal of the Linnean Society. Vol. 90, pp: 255 - 273.

Assing, V. 2003. On the identity of *Trichophya huttoni* Wollaston (Coleoptera: Staphylinidae, Trichophyinae). Linzer biol. Beitr. 35/1, pp: 515 - 518.

Atkins, M. D. 1958. Observations on the flight, wing movements and wing structure of male *Priacma serrata* (Lec.) (Coleoptera: Cupedidae). Canadian Entomology. Vol. 90, pp: 339 – 347.

Audisio, P.; Capolupo, P.; Parisella, R. and Santorelli, S. 1990. Associazione Nitidulidae (Coleoptera) – *Rhizopus*: probabile flagello dell´agricoltura degli anni ´90. L´Informatore Agrario 46. Fasc. 21, pp: 69-71.

Audisio, Paolo; C. Canepari; A. De Biase; R. Poggi: E. Ratti e M. Franco Z.i 2003. Check List delle specie della fauna Italiana. COLEOPTERA POLYPHAGA XI. (CLAVICORNIA II) Fascicolo 56. 19 Pages.

Auffenberg, K. 1986. A commensal relationship between *Ryssota* (Pulmonata, Helicarionidae) and a beetle in the Philippines. Malacological Review. Vol. 19, pp: 113- 114.

Ausden, M. 1996. Invertebrates. En: Sutherland, W. J. (ed.) Ecological census techniques: a handbook: 139-177. Cambridge University Press, Cambridge.

Bahillo de la Puebla, P. & J. Ignacio L Colón 2003. La familia Prionoceridae Lacordaire, 1857 en la Península Ibérica (Coleoptera: Cleroidea).Bull. S.E.A. N° 33, pp: 161 – 165.

Bahillo de la Puebla, P. & J. I. López-Colón 2005. Los Drilidae Lacordaire, 1857 de la Península Ibérica e Islas Baleares (Coleoptera). Boletín Sociedad Entomológica Aragonesa. N° 37, pp: 119 - 128.

Balachowsky, A. 1949. Famille des Platypodidae (Pp.: 275-279). En: Coléoptères Scolytides. Faune de France 50. Fédération Française des Sociétés de Sciences Andorra (Insecta). Monografías S.E.A. vol. 8. Sociedad Entomológica Aragonesa. Zaragoza.

Balasubramanyan, K.; Srinivasan, M. and Kathiresan, K. 1979. Insects. Centre of Advanced Study in Marine Biology, Annamalai University.pp: 359-363.

Balduf, W. V. 1926. On the habits and development of a checkered beetle (*Cymatodera undulata* Say). Transactions of the American Entomological Society. Vol.52, pp: 29-37.

Balduf, W. V. 1935. The Bionomics of Entomophagous Coleoptera. E. W. Classey Ltd., Hampton.220 Pages.

Balfour-Browne, F. 1940. British water beetles 1: xx + 375 pp: 89 figs. Roy Soc. London. General account of origin, respiration, locomotion life-cycle, and morphology of adults and larvae with keys to families, genera, and British species.

Balfour-Browne, F. 1958. British Water Beetles III. Ray Society, London, Bernard Quaritch Ltd. 210 Pages.

Balfour-Browne and Tilbrook, P. J. 1966. Coleoptera collected in the South Orkney and South Shetland Islands. Br. Antarct. Surv. Bull. N° 9, pp: 41-43. Available online (2010) in: http://www.antarctica.ac.uk/documents/bas_bulletins/bulletin09_04.pdf

Balke, M., 1995. The Hydroporini (Coleoptera: Dytiscidae: Hydroporinae) of New Guinea: systematics, distribution and origin of the fauna. Invertebrate Taxonomy. Vol. 9, pp: 1009 – 1019.

Barr, William F. 2005. New species and distribution records for mexican *Pelonium* (Coleoptera: Cleridae: Enopliinae). Folia Entomológica Mexicana. Vol. 44. Supl. N° 1, pp: 55-62.

Barbalat, S. 1995. Efficacité comparée de quelques méthodes de piégeage sur certains coléopteres et influence de l'antophilie sur le résultat des captures. Bulletin de la Societé neuchâteloise des Sciences naturelles. Vol.118, pp: 39 - 52.

Barber H.S. 1913. The remarkable life-history of a new family (Micromalthidae) of beetles. P. Biol. Soc. Wash. Vol. 26, pp: 185 – 190.

Barber, H.S. 1931. Traps for cave inhabiting insects. Journal of the Elisha Michell Scientific Society. Vol ,46, pp: 259–266.

Barclay,Hugh J. 1984. Pheromone trapping models for pest control effects of mating patterns and inmigration. Res. Popul. Ecol. 26, pp: 303-311. Available oline in:http://meme.biology.tohoku.ac.jp/POPECOL/RP%20 PDF/26(2)/pp.303.pdf

Barranco, P. ; Ruiz-Portero, C. ; Fernández-Cortés, A.; Bellés, X. y Tinaut, A. 2003. Ptínidos de cavidades en yeso de Almería (Coleoptera, Ptinidae). Boletín Asoc. esp. Ent. Vol.27. Fasc.1-4, pp: 53-69.

Barrera, A. 1974. Las Colecciones Científicas y su problemática en un país subdesarrollado: México. Biología. Vol. 4. Fasc.1, pp: 12-19.

Barrera, R. 1996. Species concurrence and the structure of a community of aquatic insects in tree holes. *Journal of Vector Ecology. Vol.* 21, pp: 66–80.

Bartlett, J. S. 2009. The Cleridae of Lord Howe Island, with descriptions of two new species (Coleoptera: Cleroidea). Records of the Australian Museum. Vol. 61, pp: 225-228.

Bartlett, J. S. 2009. *Scrobiger splendidus* (Newman) (Coleoptera: Cleridae) associated with *Hylaeus* sp. (Hymenoptera: Colletidae) in Southeastern Queensland. Australian Ent

Bartolozzi L. & G. Onore 1993 Observations on the biology and behaviour of *Sphaenognathus oberon* Kriesche (Coleoptera: Lucanidae). Coleopterists Bulletin. Vol. 47. Fasc. 2, pp: 126 – 128.

Bartolozzi L. & G. Onore 2006. *Sphaenognathus (Chiasognathinus) xerophilus* sp.n. from Peru (Coleoptera: Lucanidae). Koleopterologische Rundschau. Vol. 76, pp: 361– 365.

Basset, Yves; Kitajima, K. & Wright, J. S. (no date). The International Canopy Crane Network: key findings in canopy science, pp: 159- 196.: Available online (2010) in: www.stri.org/english/research/facilities/terrestrial/cranes/PDFs/Part4UNEPBook.pdf

Basset, Y. 1985. Comparaison de quelques méthods de piégeage de la faune dendrobie. Bulletin Romand d'Entomologie. Vol. 3, pp: 1–14.

Basset, Y., Springate, N.D., Aberlene, H.P. & Delvare, G. 1997. A review of methods for sampling arthropods in tree canopies. (In) Stork, N.E., Adis, J. & Didham, R.K. (eds.), Canopy arthropods. Chapman & Hall, London, pp: 27- 52.

Batelka, Jan 2007. Coleoptera: Ripiphoridae. Folia Heyrovskyana. Series B, 7 pp: 1-7.

Bedick, J. C., Ratcliffe, B. C., and L. G. Higley. 2004. A new sampling protocol for the endangered American burying beetle, *Nicrophorus americanus* Olivier (Coleoptera: Silphidae). The Coleopterists Bulletin. Vol.58, pp: 57-70.

Behan-Pelletier, V.M., Tomlin, A., Winchester, N., & Fox, C. 1996. Sampling protocols for microarthropods. In A Workshop Report on Terrestrial Arthropod Sampling Protocols for Graminoid Ecosystems (ed. A.T. Finnamore). Available online in: http://www.eman-rese.

ca/eman/reports/ publications/sage.

Beiriger R. and Robert W. 1996. The Bostrichidae (Coleoptera) of Missouri.Journal of the Kansas Entomological Society, Vol. 69, No. 1, pp: 45 – 68.

Belcher, D.W. 1989. Influence of cropping systems on the number of wireworms (Coleoptera: Elateridae) collected in baits in Missouri cornfields. Journal of the Kansas Entomological Society. Vol. 62, pp: 590–592.

Bellés, X., 1976. Ptínidos recogidos en cavidades subterráneas ibéricas (Col. Ptinidae).Speleon. Vol. 22, pp: 145 - 147.

Bellés, X., 1985. Sistemática, filogenia y biogeografía de la subfamilia Gibbiinae (Coleoptera, Ptinidae). Treballs del Museu de Zoología. Barcelona. 94 pages.

Bellés, X., 1991: Survival, opportunism and convenience in the processes of cave colonization by terrestrial faunas. In: Ros, J.D. & N. Prat (eds.) Homage to Ramon Margalef; or, Why there is such pleasure in studying nature. – Oecol. Aquat. Vol. 10.pp: 325-335.

Benesh, B. 1960. Coleopterorum Catalogus Supplementa, pars 8: Lucanidae. W. Junk. Gravenhage, Netherlands.

Berenguer, José Manuel 2007. Luci, Sin nombre y sin memoria. Reflejos electrónicos de un manglar lejano. Edit: Nau Coclea De Camallera. Impressio: Arts Grafiques Trayter. 83 pages.

Bernhard, D.; Schmidt, C.; Korte, A.; Fritzsch, G.; Beutel, RG 2006. From terrestrial to aquatic habitats and back again — molecular insights into the evolution and phylogeny of Hydrophiloidea (Coleoptera) using multigene analyses. Zoologica Scripta. Vol. 35. Fasc. 6, pp: 597–606.

Besuchet, C. 1956. Biologie, morphologie et systématique des *Rhipidius* (Col. Rhipiphoridae). Bulletin de la Société Entomologique Suisse. Vol.29. Fac.2, pp: 74-144.

Besuchet, C. 1972 Les Coleopteres Aculagnathides. Revue Suisse Zool. Vol. 79. Pp : 99 - 145.

Besuchet, C. 1981. Description of *Scydmaenus aelleni*, a new species of cavernicolous scydmaenid beetle from New Caledonia. Revue Suisse de Zoologie. Vol.88, pp: 459–62.

Besuchet, C., Burckhardt, D.H. & Lobl, I. 1987. The "Winkler/Moczarski" eclector as an efficient extractor for fungus and litter Coleoptera. The Coleopterists Bulletin Vol. 41. Fasc. 4, pp: 392-394.

Bentancourt, Carlos M. y Scatoni, Iris B. 1999. Guía de Insectos y Acaros de importancia agrícola y forestal en el Uruguay. Universidad de la República, Fac. de Agronomía, Montevideo.

Benton AH, Wilcox J. 1955. On the habits of beetles of the genus *Catops*. Coleopterists Bulletin. Vol.9, pp: 29.

Bétis, L. 1912. Les rhipiphorides Gallo-Rhénans. Annales de la Société d'Histoire Naturelle de Toulon. 3, pp: 125-155.

Beutel, R. G. & Vanin S. A. 2005. 6.2. Torridincolidae Steffan, 1964. Pp. 46-48. In: Beutel R. G. & Leschen R. A. B. (eds.): Handbook of zoology. A natural history of the phyla of the animal kingdom. Volume IV. Arthropoda: Insecta. Part 38. Coleoptera, Beetles. Volume 1: Morphology and systematics (Archostemata, Adephaga, Myxophaga, Polyphaga partim.). Walter de Gruyter, Berlin, New York, xi + 567 pp.

Beutel, Rolf G. 2005. Coleoptera, Beetles: Handbuch der Zoologie/begr. von Willy Kükenthal. Fortgef. von M. Beier.Bd. 4. Arthropoda. Morphology and systematics, (Archostemata, Adephaga, Myxophaga, Polyphga partim).Volumen 1. Parte 38.de Gruyter. 567 pags.

Bickhardt, H., 1913. H. Sauter's Formosa-Ausbeute. Histeridae II (Col.) (16. Beitrag zur Kenntnis der Histeriden). – Ent. Mitt. 2, pp: 166-177.

Bickardt, H. 1916a. Biologische Notizen über puläarktische Histeriden.- Entomologische Blätter. 12, pp: 49-54.

Bickhardt, H., 1916. Histeridae. In: P. Wytsman. Genera Insectorum, fasc. 166a, pp: 1-112. — La Haye.

Biggs J.; Williams P.; Whitfield M.; Nicolet P. and Weatherby, A. 2005. 15 years of pond assessment in Britain: results and lessons learned from the work of Pond Conservation. Aquatic Conservation: Marine and Freshwater Ecosystems 15, pp: 693-714.

Binckley, C. A. & Resetarits, W. J. 2005. Habitat selection determines abundance,richness and species composition of beetles in aquatic communities.Biol. Lett. 1, pp: 370–374.

Birch, M., ed. 1974. Pheromones. 495 pgs. North-Holland Publ. Co., Amsterdam.

Birtele, D. 2003. The succession of saproxylic insects in dead wood: a new research method. In: Mason, F.; Nardi, G.; Tisato, M. (eds.) Proceedings of the International Symposium "Dead wood: a key to biodiversity", Mantova, May 29-31 2003: 91-93. Sherwood 95, Suppl. 2.

Blackburn, T. 1892. Further notes on Australian Coleoptera, with descriptions of new genera and species. Transactions of the Royal Society of South Australia. Vol. 15, pp: 20 – 73.

Blackburn, T. 1903. Further notes on Australian Coleoptera, with descriptions of new genera and species. Transactions of the Royal Society of South Australia. Vol. 28, pp: 91 – 182.

Blakely, Tanya J. 2008. A thesis submitted for the degree of Doctor of Philosophy in Ecology at the University of Canterbury. 202 pages.

Blatchley, W. S. 1910. Family XVII. Endomychidae. The handsome fungus beetles. p. 533-539. In: W. S. Blatchley (ed.). An illustrated descriptive catalogue of the Coleoptera or beetles (exclusive of the Rhynchophora) known to occur in Indiana. The Nature Publishing Co.; Indianapolis. 1385 pages.

Block, W. L. Somme; Ring, J. P. and M. R. Worland 1988. Reports on Antartic Fieldwork adaptations of Arthropods the Sub-Anatarctic environment British Antarctic Survey, Natural Environment Research Council, Department of Biology, Division of Zoology, University of Oslo and Department of Biology. University of Victoria.Canadá. Reports on Antarctic Fieldwork. Vol. 67, pp: 1-3.

Blouin-Demers, G. & Weatherhead, P. J. 2000. A novel association between and beetle and snake: parastirism of *Elaphe obsoleta* by *NIcrophorus pustulatus*. Ecoscience. Vol. 7. Fasc. 4, pp: 395 - 397.

Bocak, Ladislav; Brlik, Milan 2008. Revision of the family Omalisidae (Coleoptera, Elateroidea). Insect Systematics & Evolution. Vol. 39, N° 2, pp: 189- 212.

Bocakova, M.; Bocak, L.; Hunt, T. & Vogler, A. P. 2007. Molecular phylogenetics of Elateriformia (Coleoptera): evolution of bioluminescence and neoteny. – Cladistics. 23, pp: 477–496.

Bocakova, Milada & Janisova, Kristyna 2010. A new genus and species of ototretine firefly from Borneo (Coleoptera: Lampyridae). Zootaxa. 2347, pp: 59–63.

Bohác, J. 1978. Description of the larva and pupa of *Thoracophorus brevicristatus* (Coleoptera, Staphylinidae). Acta Entomologica Bohemoslovaca. Vol. 75, pp: 394 - 399.

Boix, D. & Sala, J. 2002. Riqueza y rareza de los insectos acuáticos de la laguna temporal de Espolla (Pla de l'Estany, Cataluña). Bull. Asoc. Esp. Entomol. Vol. 26. Fasc.1-2, pp: 45 - 57.

Bologna, M. A. and Pinto J. D.2001. Phylogentic studies of Meloidae (Coleoptera), with emphasis on the evolution of phorsey. Systematic Entomology. Vol. 26, pp: 33-72.

Booth, R. G.; Cox, M. L.; Madge, R. B. 1990. IIE guides to insects of importance to man: Coleoptera. London: International Institute of Entomology / The Natural History Museum, 384 Pages.

Bordy, B. 2000. Coléoptères Chrysomelidae. Volume 3. Hispinae et Cassidinae. En: Faune de France N° 85. Fédération Francaise des Societés de Sciences Naturelles. Paris. 250 Pags. XXVI Planches.

Borgemier, T. 1937. *Canthon dives* Harold (Col. Copridae), predador das femeas de *Atta laevigata* Smith (Hym. Formicidae). Revista de Entomología. N° 7. Fasc.1, pp: 117-118.

Borowiec, Lech 1987. The genera of seed-beetles (Coleoptera, Bruchidae). Bulletin Entomologique de Pologne. Tomo 57, pp: 3-207. Available online (2010) in: http://www.biologybrowser.org/weblinks/goto/1156927

Borror, D. J. and Delong, D. M. 1969. Introdução ao estudo dos insetos. São Paulo: E. Blücher. 653 Pages.

Borror, D. J., DeLong, D. M., Triplehorn, C. A. 1976. cuarta edición. An introduction to the study of insects. Holt, Rinehart and Winston. New York, Chicago.

Bosselmann, G. 1876. Zeitschrift für die Gesammten Naturwissenschaften. Naturwissenschaftlicher Verein für Sachen und Thüringen in Halle. Volume 48. Pag. 396.

Bouchard, P; Wheeler, T. A.; Goulet, H. 2000. Design for a low-cost, covered, ramp pitfall trap. Canadian Entomologist 132, pp: 387-389.

Bournaud, M.; Keck, G. & Richoux, P. 1980. Les prélèvements des macroinvertébrés benthiques en tant que révélateurs de la physionomie d'une rivière. Ann. Limnol. Vol. 16, pp: 55 – 75.

Bousquet, Y. 1990. A review of the North American species of *Rhizophagus* Herbst and a revision of the Nearctic members of the subgenus *Anomophagus* Reitter (Coleoptera: Rhizophagidae). The Canadian Entomologist. N° 122, pp: 131–171.

Bousquet, Y. 2002. Monotomidae Laporte 1840. In American Beetles, Volume 2: Polyphaga: Scarabaeoidea through Curculionoidea. Edited by Arnett, R. H. Jr.; Thomas, M. C.; Skelley, P. E. and Howard, J. Frank. CRC Press, Boca Raton, USA. Pp. 319–321.

Bousquet, Y. & Leschen, R. A. B. 2002. Description of a new genus and species of Endomychidae (Coleoptera: Cucujoidea) from northeastern North America. Coleopterists Bulletin. Vol.56. Fasc.2, pp: 291-298.

Bousquet, Y. 2010. Monotomidae Laporte, 1840, Pp.319-324. In: Leschen R.A.B., Beutel R.G. and Lawrence J.F. (eds.): Handbook of Zoology. Arthropoda: Insecta Coleoptera, Beetles. Volume 2: Morphology and Systematics (Elateroidea, Bostrychiformia, Cucujiformia partim.). Berlin-New York, Walter de Gruyter, xiii+786 pp.

Bøving, A. G., and A. B. Champlain. 1920. Larvae of the North American beetles of the family Cleridae. Proceedings of the United States National Museum. Vol. 57, pp: 576-649.

Bowestead, S., Booth, R.D., Slipinski, A. & Lawrence, J. F. 2001. The genus *Cleidostethus* Arrow, 1929 reappraisal and transfer from Coccinellidae to Corylophidae (Coleoptera: Cucujoidea). Annales Zoologici, 51, pp: 319–323.

Brandmayr, P. 1991. The reduction of metathoracic alae and of dispersal power of carabid beetles along the evolutionary pathway into the montains. Form and funtion in zoology, pp: 363 – 378. in: G., Lanzavecchia, R. Valvassori (eds.) Selected Symposia and Monographs V.Z.I., 5, Mucchi, Modena.

Branham, M. & Wenzel, J. W. 2001. The evolution of bioluminescence in cantharoids (Coleoptera: Elateroidea). Florida Entomologist. Vol.84. Fasc. 4, pp: 565–586.

Branham, M. & Wenzel, J. W. 2003. The origin of photic behavior and the evolution of sexual communication in fireflies (Coleoptera: Lampyridae). – Cladistics 19, pp: 1–22.

Branham, M. A. 2003. A new technique for collecting glowworm fireflies (Coleoptera: Lampyridae). The Coleopterists' Bulletin. Vol.57, pp: 114-115.

Bridges, E. 1952. El último confín de la tierra. Buenos Aires - Argentina: Emecé Editores S.A. 530 pags.

Brokaw, N. V. L. and Lent, R. A. 1999. Vertical structure. In Maintaining biodiversity in forest ecosystems, ed. M. L. Hunter, pp: 373–399. Cambridge University Press, Cambridge.

Broun, T. 1921. Descriptions of new genera and species of Coleoptera. Part VII. New Zealand Institute bulletin. Vol. 1. Fasc. 7, pp: 591 - 665.

Broun, T. 1921: Descriptions of new genera and species of Coleoptera. Part VI. New Zealand Institute bulletin, Vol. 1. Fasc. 6, pp: 475 - 590.

Brower, J. H., and J. W. Press. 1992. Suppression of residual populations of stored-product pests in empty corn bins by releasing the predator *Xylocoris flavipes* (Reuter). Biol. Control. Vol. 2, pp: 66-72.

Browne, F.G. 1962. *Sosylus spectabilis* Grouvelle (Coleoptera, Colydiidae), a predator and parasite of African ambrosia beetles. 5th Rep. W. Afr. Timber Borer Res. Unit, pp: 91 - 96.

Brown, H. P. 1973. Survival records for elmid beetles, with notes on laboratory rearing of various dryopoids (Coleoptera). Entomological News 84, pp: 278-284.

Bruch, C. 1918 - Nuevos huéspedes de formigas procedentes de Córdoba. Physis. Vol. 4. Fasc. 17. Pages: 186-195, figs.

Bruch, C. 1919 Un nuevo coleoptero ecitófilo. PHYSIS. Vol. 4, pp: 579- 582.

Bruck, D. J. and Lewis, L. C. 2002. *Carpophilus freemani* (Coleoptera, Nitidulidae) as a vector of Beauveria bassiana. Journal of Invertebrate Pathology. Vol. 80, pp: 188-190.

Bruhl, C., Gunsalam, G. & Linsenmair, K.E. 1998. Stratification of ants (Hymenoptera, Formicidae) in a primary rainforest in Sabah, Borneo. Journal of Tropical Ecology. Vol. 14, pp: 285 - 297.

Bruns, T. D. 1984. Insect mycophagy in the Boletales: fungivore diversity and the mushroom habitat. In: Q. D. Wheeler and M. Blackwell, eds., Fungus/insect relationships: perspectives in ecology and evolution. Columbia University Press. New York, pp: 91 - 129.

Bryant, In Champion, 1921, Ann. Mag. Nat. Rist. Tomo IX. Fasc.7. Page 325.

Bryant, In Champion 1921, Ann. Mag. Nat. Rist. Tomo IX, Fasc. 8. Page 195.

Buffington, M.L. & Redak, R.A.1998. A comparison of vacuum sampling versus sweepnetting for arthropod biodiversity measurements in California coastal sage scrub. Journal of Insect Conservation. Vol. 2, pp: 99–106.

Bullard, Donald R. and Peterson, Marilyn 1982. APTI course SI:409: basic air pollution meteorology : student guidebook. Air Pollution Training Institute, Northrop

Services, inc, Environmental Research Center (Research Triangle Park, N.C.), United States. Environmental Protection Agency. Office of Air Quality Planning and Standards. U.S. Environmental Protection Agency, Air Pollution. 125 pages. Training Institute.

Bullians, Mark. S. & Leschen, Richard. A. B. 2004. *Noteucinetus* new genus from New Zealand and Chile and notes on *Eucinetus stewarti* (Broun) (Coleoptera: Eucinetidae). New Zealand Entomologist. Vol. 27, pp: 29-38.

Burakowski B. 1975 Development, distribution and habits pof *Trixagus dermestoides* (L.) with notes on the Throcidae and Lissomidae (Coleoptera: Elateroidea) Annales Zool. Vol. 32, pp: 375 – 405.

Burckhardt, D.; Löbl, I. 1992: A review of the Pterogeniidae (Coleoptera). Zoological journal of the Linnean Society. Vol. 104, pp: 243 - 291.

Bustos-Gómez, Fernanda Lida & Lopera Toro, Alejandro 2003. Preferencia por cebo de los escarabajos coprófagos (Coleoptera: Scarabaeidae: Scarabaeinae) de un remanente de bosque seco tropical al norte del Tolima (Colombia). Sociedad Entomológica Aragonesa. Monografías Tercer Milenio. Vol. 3. Pp.59-65.

Byrne, K. J.; Swigar, A. A.; Silverstein, R. M.; Borden, J. H. and Stokkink, E. 1974. Sulcatol: population aggregation pheromone in the scolytid beetle, Gnathotrichus sulcatus. Journal of Insect Physiology. Vol. 20, pp: 1895 - 1900.

Calder, A. A. 1996. Click Beetles. Genera of Australian Elateridae (Coleoptera). Monography of Invertebrate Taxonomy. Vol. 2 (i-x), pp: 1- 401.

Cambefort, Yves, 1991. Biogeography and Evolution. Chapter 4, pp: 51-67, in: Hanski, Ilkka and Cambefort, Yves, 1991. Dung beetle ecology. Princeton University Press (Princeton, N.J). xii, 481 pags.

Cambefort, Yves and Walter, Philippe 1991. Dung Beetles in Tropical Forests in Africa. Chapter 11, pp: 198-210, in: Hanski, Ilkka and Cambefort, Yves, 1991. Dung beetle ecology. Princeton University Press (Princeton, N.J). xii, 481 pags.

Campbell, J.M. 2009. Family Limulodidae (Horse-shoe crab beetles): Available online (2010) in: www.canacoll.org

Canaday, C. L. 1987. Comparison of insect fauna captured in six different trap types in a Douglas fir-forest. Canadian Entomologist. Vol. 119, pp: 1101-1108.

Canevari M. y Decarre, J. 2007. Edit. por M. E. Zaccagnini, y R. Suárez. Monitoreo de reptiles. INTA. 32 Pages.

Canio, B. Enio 1998. *Deltochilum valgum acropyge* Bates (Coleoptera: Scarabaeidae: Scarabaeinae): Habits and Distribution. The Coleopterists Bulletin. N°52. Fasc.2, pp: 174-178.

Cannings, R. A. and Scudder, G. G. E. 2006. Families of Coleoptera of British Columbia. Klinkenberg, Brian. (Editor) 2010. E-Fauna BC: Electronic Atlas of the Fauna of British Columbia [www.efauna.bc.ca]. Lab for Advanced Spatial Analysis, Department of Geography, University of British Columbia, Vancouver.

Cano, E.B. 1993. Pasálidos. In: Evaluación Ecológica Rápida de la Reserva de la Biosfera "Sierra de las Minas". CDC-CECON. USAC. Guatemala, pp: 36-39.

Carabajal, E. M.; Carrillo, J. G. & Rodríguez, F. F. 1996. El medio subterráneo: Adaptaciones de los invertebrados terrestres. Bull. SEA. Vol. 15, pp: 55-60.

Carlson, D. C. and P. O. Ritcher. 1974. A new genus of Ochodaeinae and a description of the larva of *Pseudochodaeus estriatus* (Schaeffer) (Coleoptera: Scarabaeidae). Pan Pacific Entomologist. Vol. 50, pp: 99- 110.

Carlson, D. C. 1975. Taxonomic characters of the genus *Ochodaeus* Serville with descriptions of two new species in the *O. pectoralis* LeConte species complex (Coleoptera: Scarabaeidae). Bulletin of the Southern California Academy of Sciences Vol. 74, pp: 49-65.

Carlson, D. C. 1977. Taxonomic revision of *Lichnanthe* Burmeister with studies on the biology of *L. rathvoni* (LeConte) (Coleoptera: Scarabaeidae). Ph.D. Thesis, Oregon State University, Corvallis.

Carlson, D. C. 1980. Taxonomic revision of *Lichnanthe* Burmeister (Coleoptera: Scarabaeidae). Coleopterists Bulletin. Vol. 34, pp: 177-208.

Carlson, D. C. 2004. 30.OCHODAEIDAE Mulsant and Rey 1870. Pp.28- 31.

Carlton, C. E.; Leschen, R. A. B. & Kovarik, P. W. 1996. Predation on adult blow flies by a Chilean hister beetle, *Euspilotus bisignatus* (Erichson). The Coleopterists Bulletin. 50. Page 154.

Carlton Chris and Richard A. B. Leschen 2007. Descriptions of *Soronia* complex (Coleóptera: Nitidulidae: Nitidulinae) larvae of New Zealand with comments on life history and taxonomy. New Zealand Entomologist. Vol. 30, pp: 41 - 51.

Caron D. M. 1978. Other insects, in: Honey bee pests, predators, and diseases, Morse R.A. (Ed.), Cornell University Press, London, pp: 158–196.

Carrera, C.J. y A.E. Lugo 1978. Los Sistemas de Mangles de Puerto Rico. Programa de la Zona Costanera, Departamento de Recursos Naturales y Ambientales. Page 102.

Carrero, Planes 2008. Plagas del campo. Mundi Prensa Libros. 775 págs.

Carter RWG, Nordstrom KF and Psuty NP 1990b. The study of coastal dunes. In K Nordstrom, Putsy and B Carter, eds, Coastal dunes: form and process. Wiley, Chichester. Pp. 1–14.

Carter, D. J. y Hargreaves B. 1987. Guía de campo de las orugas de las mariposas y polillas de España y de Europa. Omega. Barcelona. 310 pages.

Casale, A., P. M. Giachino, & B. Jalzic. 2004. Three new species and one new genus of ultraspecialized cave dwelling Leptodirinae from Croatia (Coleoptera, Cholevidae). Natura Croatica. Vol. 13, pp: 301 – 17.

Cassola, Fabio 1997. Studies on Tiger Beetles. Revision of the neotropical genus *Pseudoxycheila* Guérin, 1839 (Coleóptera, Cicindelidae). Fragmenta Entomologica. Roma. Vol. 29. Fasc. 11, pp: 1 – 121.

Castro, A.; J. M. Hidalgo & A. M. Cárdenas. 2003. Nuevos datos sobre los coleópteros acuáticos del Parque Nacional de oñana (España): Capturas realizadas mediante trampas de luz y técnicas de muestreo para fauna edáfica (Coleoptera). Bull. S.E.A., n° 33, pp: 153 – 159. Available online (2010) in: http://www.sea-entomologia.org/PDF/BOLETIN_33/B33-024-153.pdf

Caterino, M. S., and A. P. Vogler. 2002.The phylogeny of the Histeroidea. Cladistics. Vol. 18. Fasc.4, pp: 394-415.

Caterino. M. S. and Dégallier, N. 2007. A review of the biology and systematics of Chlamydopsinae. Invertebrate Systematics 21:1-28.

Caterino, Michael S. 2009 Histeridae in: Tree of life web project. Available online (2010) in: http://tolweb.org/Histeridae

Cazier, Mont A. 1961. A New Species of Myrmecophilous Scarabaeidae, with Notes on Other Species (Coleoptera, Cremastocheilini). American Museum Novitates. N°2033, pp: 1-12.

Céréghino, R.; J. Biggs, B. Oertli, and S. Declerck. 2008. The ecology of European ponds: Defining the characteristics of a neglected freshwater habitat. Hydrobiologia 597, pp: 1-6.

Champion, G. C. 1890. On the true affinities of the Heteromerous genus Lagrioida, Fairmaire and Germain. The Entomologist's Monthly Magazine, pp: 121-122.

Chandler, D. S., and S. B. Peck. 1992. Diversity and seasonality of leiodid beetles (Coleóptera: Leiodidae) in an old-growth and a 40-year-old forest in New Hampshire. Environmental Entomololgy. Vol. 21, pp: 1283-1293.

Chandler, Donald S. 2002. Family 117. Anthicidae Latreille 1819, pp: 549-558 in: Arnett, H. and Thomas, M. 2002. American Beetles: Polyphaga: Scarabaeoidea through Curculionoidea. Vol. 2. CRC Press. 880 pags.

Chapin, E. A. 1917. Miscellaneous notes on Coleoptera. Bulletin of the Brooklyn Entomological Society. Vol. 12, pp: 29-31.

Chapin, E. A. 1938. The nomenclature and taxonomy of the genera of the scarabaeid subfamily Glaphyrinae. Proceedings of the Biological Society of Washington Vol. 51, pp: 79-86.

Chapman, J. A.; Kinghorn, J. M. 1955. Window flight traps for insects. Can. Ent. 87: 46-47.

Chapuis, M.F. 1876. Famille des Érotliens, des Endomychides et des Coccinellides. Histoire Naturelle des Insectes. Genera des Coléopteres ou Exposé Méthodique et Critique de tous les Genres Proposés jusqu'ici dans cet Ordre d'Insectes. (ed. by T. Lacordaire and F. Chapuis) Librairie Encyclopédique de Roret, París. 424 Pages.

Chaves, R.; Genaro, J.A. 2005. «A new species of *Pharaxonotha* (Coleoptera : Erotylidae), probable pollinator of the endangered Cuban cycad, *Microcycas calocoma* (Zamiaceae)» Insecta Mundi. Vol. 19. N.° 3, pp: 143-150.

Chénier, J. V. R.; Philogène, B. J. R. 1989. Evaluation of three trap designs for the capture of conifer-feeding beetles and other forest Coleoptera. Can. Ent. 121: 159-167.

Chernov, Yu. I.; Makarov, K. V., and Eremin, P. K. 2000. Family of Ground Beetles (Coleoptera, Carabidae) in the Arctic Fauna: Communication 1. Entomological Review, Vol. 80, No. 8, 2000, pp. 1009–1021. Translated from Zoologicheskii Zhurnal, Vol. 79, No. 12. Available online (2010) in : http://www.zin.ru/Animalia/Coleoptera/pdf/chernov_makarov_1.pdf

Chernov, Yu. I.; K. V. Makarov & P. K. Eremin 2001. Family of Ground Beetles (Coleoptera, Carabidae) in the Arctic Fauna: Communication 2. Entomological Review. Vol. 81, No. 1, pp: 108 – 117. Available in: www.eurocarabidae.de/images/up_files/library/files/chernov_makarov_2.pdf

Chevrolat, A. 1874. Révision des Cébrionides. Ann. Soc. Entomol. France, ser. 5, fasc.4, pp: 503-540, lámina 1.

Chiesa, E. 1959. Hydrophilidae Europa. Arnaldo Forni, Bologna. Page 199 .

Chittenden FH. 1915. The violet rove-beetle (*Apocellus sphaericollis* Say). USDA Bulletin. Vol. 264, pp: 1 - 41.

Christiansen, K. 1962. Proposition pour la classification des animaux cavernicoles. Spelunca. Vol. 2, pp: 76 - 78.

Cicero, Joseph. 1982. The genus Bicellonycha in the United Stated with descriptions of a new species and subspecies (Coleoptera: lampyridae, Photurinae). The Coleopterists Bulletin. Vol. 36. Fasc. 2, pp: 270-278.

Clausen, C.P. 1940. Entomophagous Insects. McGraw-Hill Book Company. New York. 688 Pages.

Cline A. R. 2005. Revision of *Pocadius* Erichson (Coleoptera: Nitidulidae). Ph.D. Dissertation: Louisiana State University, i-vii + 376 pp. ttp://etd.lsu.edu/docs/available/etd-11042005-122348/

Cline, A.R. and A. Ślipiński. 2009. Discolomatidae Horn, 1878. In R.G. Beutel and R.A.B. Leschen (eds.). Handbuch der Zoologie. Band IV, Teilband 38, vol. 2. De Gruyter, Berlin. In Press.

Cobos S. J. M. y E. Martín Bernal. 1987. Métodos para la determinación del periodo de vuelo de los coleópteros escolítidos. Itea. Vol. 73, pp: 65 - 72.

Colin Johnson 1982. An Introduction to the Ptiliidae (Coleoptera) of New Zealand. New Zealand Journal of Zoology. Vol. 9, pp: 333-376.

Colleen Hepburn 2007. Composition and phenology of insect pests of *Capsicum* (Solanaceae) cultivated in the Makana District, Eastern Cape Province, South Africa. 218 pags. Available online (2010) in: http://eprints.ru.ac.za/1167/1/hepburn-msc.pdf

Collinson, N. A.; Biggs, J.; Corfield, A.;Hodson, M. J.; Walker, D.: Whitfield, M. / Williams, P. J. 1995. Temporary and permanent ponds: An assessment of the effects of drying out on the conservation value of aquatic macroinvertebrate communities. Biol. Conserv. 74, pp: 125-133.

Commonwealth of Australia 2004. Import Risk Analysis for Sawn Coniferous Timber from Canada, New Zealand and the United States. Draft Pest Categorisation Tables, pp: 1- 73.

Compte Sart A. 2008. Odonata. En: El reino animal en la Península Ibérica y las Islas Baleares. Fauna Ibérica. CSIC. Available online (2010) : http://www.fauna-iberica.mncn.csic.es

Constantin, R. 1990. Descriptions des larves d'*Enicopus pyrenaeus* Fairmaire et de *Danacaea pallipes* (Panzer). Contribution à l'étude de la biologie et de la systématique larvaire des Melyridae (Coleoptera). Nouvelle Revue d'Entomologie (N.S.), 6 (1989) pp: 387–405

Cooper, K. W. 1991. Artematopidae In: F. W. Stehr, ed. Immature Insects. Volume 2. Kendall/Hunt. Dubuque, IA, pp: 407- 409.

Cooter, J. 1991. A Coleopterist´s Handbook. The Amateur Entomolgists' Society. General Editor: P. W. CRIB. Portsmouth. 294 pp.

Cooter, Jonathan and Barclay, M. V. L. 2006. A coleopterist's handbook. The Amateur Entomologist' Society. Volumen 11 de Amateur entomologist. N°4, 439 pags.

Corinne Watts 2007. Beetle Community Responses to Mammal Eradication in the Southern Exclosure on Maungatautari. Edit.: Landcare Research New Zealand Ltd. Landcare Research Contract Report: LC0607/170, pp: 3-26.

Corporaal, J. B. 1939. Revision of the Thaneroclerinae (Cleridae: Coleoptera). Bijdragen tot de Dierkunde. Vol. 27, pp: 347-359.

Correia, Ch., Maria L. 1999. Monitorização biológica de lagoas temporárias mediterrânicas- Contribuição para a definição de estados de rêferencia. Dpto de Zoología e Antropología. Faculdade de Ciencias da Universidade de Lisboa. Relatorio apresentado para obtenção da Licenciatura en Biología Aplicada aos Recursos Animais- Ramo Terrestres. N°1/PRODEP/98. 60 pages. Available online (2010) in: http://portal.icnb.pt/NR/rdonlyres/19F5E6EA-E120-4674-9454-2A203A3CFD42/0/PNSACVLagunasTemporarias_Monitorizacao_1999.pdf

Costa Lima 1952. Insetos do Brasil. Coleópteros (1ª parte).

7° Tomo, 2ª parte, capítulo XXIX. Escola Nacional de Agronomía. Serie Didáctica N° 9, pp: 1- 372.
Costa Lima 1953. Insetos do Brasil. 8° Tomo. Coleópteros (2ª parte). Escola nacional de Agronomia. Rio de Janeiro. 323 pgs.
Costa, Cleide, Casari–Chen, S. A. & Vanin, S. A. 1992. On the larvae of Tetralobini (Coleoptera, Elateridae). – Revista Brasileira de Entomologia. Vol.36. Fasc.4, pp: 879–888.
Costa, Cleide 1994. Cladistic Analysis and Systematics of the Tetralobini sensu Stibick, 1979 (Coleoptera, Elateridae, Pyrophorinae). – Arquivos de Zoologia. Vol.32. Fasc. 3, pp: 111–157.
Costa, Cleide 1999. Systematics and bionomics of Cneoglossidae with cladistic analysis of Byrrhoidea Sensu Lawrence & Newton 1995 (Coleoptera, Elateriformia). Arquivos de zoologia. Vol. 35. Fasc. 3, pp: 231- 300.
Costa, Cleide 2000. Estado de conocimiento de los Coleoptera neotropicales. Neotropical. PrIBES. Proyecto Iberoamericano de Biogeografía y Entomología Sistemática: PRIBES 2000 : trabajos del 1er taller iberoamericano de entomología sistemática / coord. por Fermín Martín Piera, Juan J. Morrone, Antonio Melic, 2000, ISBN 84-922495-1-X, pages. 99 - 114.
Costa, Cleide; Vanin S. A. John F. Lawrence, Sergio Ide, And Marc A. B.. 2006. Review of the Family Brachypsectridae (Coleoptera: Elateroidea). Annals of the Entomlogical Society of America. Vol. 99. N° 3, pp: 409- 432.
Coutin, R., 1966 Un insecte fort curieux. Bull. des Naturalistes des Yvelines. Vol. 23. Fasc. 1, pp: 8 - 10.
Craighead, F.C. 1920. Biology of Some Coleoptera of the Families Colydiidae and Bothrideridae. Proceedings of the Entomological Society of Washington. Vol. 22: pp: 1 - 13.
Crotch, G. R. 1873. «Synopsis of the Erotylidae of boreal America» Transactions of the American Entomological Society. Vol. 4, pp: 349 - 358.
Crowson, R. A. 1955. The natural classification of the families of Coleoptera. Nathaniel Loyd and Co., London.187 pages.
Crowson, R. A. 1955. The Natural Classification of the Families of Coleoptera. Nathaniel Lloyd, London (seen as 1967 reprint, E.W. Classey Ldt., Hampton. 214 pp.)
Crowson, R. A. 1961: Considerations on the genera *Endecatomus* Mellié and *Euderia* Broun (Coleoptera: Bostrichidae), with descriptions of their larvae. Proc. R. Ent. Soc. Lond. (B), Vol.30. Fasc. 9 – 10, pp: 113 - 120. Available online (2010) in: http://www3.interscience.wiley.com/journal/122202741/abstract
Crowson, R. A. 1964. A review of the classification of Cleroidea (Coleoptera), with descriptions of two new genera of Peltidae and of several new larval types. Transactions of the Royal Entomological Society of London. Vol.116, pp: 275 – 327.
Crowson, R. A. 1964. The habits and life cycle of *Phloiophilus edwardsi* Steph. (Coleoptera: Phloiophilidae). Proceedings of the Royal Entomological Society of London. Series A, General Entomology. Vol. 39 Issue: 10-12, pp: 151 – 152.
Crowson, R. A. and M. G. De Viedma. 1964. Observations on the relationships of the genera *Circaeus* Yablok. and *Mycterus* Clairv. with a description of the presumed larva of *Mycterus*. Eos. 40, pp: 99-107.
Crowson, R. A. 1964a. Observations on British Tetratomidae (Col.), with a key to the larvae. Entomologist's Monthly Magazine. Vol. 99.
Crowson, R. A. 1966a. Observations on the constitution and subfamilies of the family Melandryidae (Coleoptera). Eos-Revista Espanola de Entomologia. Vol 41.
Crowson, R.A. and T. SenGupta. 1969. The systematic position of Propalticidae and of *Carinophloeus* Lefkovitch (Coleoptera: Clavicornia) with description of a new species of *Propalticus* and of its supposed larva. Proceedings of the Royal Entomological Society of London (B) 38, pp: 132-140.
Crowson, R. A. 1971, Observations on the superfamily Dascilloidea (Coleoptera: Polyphaga), with the inclusion of Karumiidae and Rhipiceridae. Zoology Journal of the Linnean Society. 50, pp: 11-19.
Crowson, R. A. 1972. A review of classification of Cantharoidea. Coleoptera, with the definition of two new families, Cneoglossidae and Omethidae. Revista Univ. Madrid. Vol. 21. Fasc. 82, pp: 35 - 77.
Crowson, R. A. 1973. Further observations on Phloeostichidae and Cavognathidae, with definitions of new genera from Australia and New Zealand. Coleopterists Bulletin 27, pp: 54-62. Available (in part) online (2010) in: http://www.jstor.org/pss/3999431
Crowson, R. A. 1981. The Biology of the Coleoptera. Academic Press, London. 802 Pages.
Crowson, R. A. 1984. The Associations of Coleoptera with Ascomycetes. 11, pp: 256-285, in: Wheeler, Quentin and Blackwell, Meredith (edits) 1984. Fungus-insect relationships: perspectives in ecology and evolution. Columbia University Press Hardcover, 514 pages.
Crowson, R. A. 1990. A new genus of Boganiidae (Coleoptera) from Australia, with Observations on glandular openings, Cycad associations and geographical distribution in the family. J. Austr. ent. Soc. Vol. 29, pp: 91 - 99.
Pakaluk, James; Ślipiński,Stanisław Adam and Crowson, Roy Albert 1995. Biology, phylogeny, and classification of Coleoptera: papers celebrating the 80th birthday of Roy A. Crowson, Volumen 1. Muzeum i Instytut Zoologii PAN. 1092 páginas.
Culver, D. C., L. M. Lawrence, M. C. Christman & H. H. Hobbs III. 2000. Obligate cave fauna of the 48 contiguous United States. Conservation Biology. Vol. 14, pp: 386 – 401.

Dajoz, R. 2000. Insects and forests: the role and diversity of insects in the forest environment. Editions Tec & Doc, París.
Dajoz, R. 1967. Ecologie et biologie des coléoptères xylophages de la hêtraie. Vie et Milieu. Vol. 17, pp: 525-636.
Dalla Torre, K W von 1911. Birrhydae, Chelonariinae. Coleopterorum catalogus. Vol. 14. Fasc. 33, pp: 11- 12.
Dallara, P. L.; Seybold S. J.; Meyer H.; Tolasch T.; Francke W.; Wood D. L. 2000. Semiochemicals from three species of *Pityophthorus* (Coleoptera: Scolytidae): Identification and field response. Canadian Entomologist. Vol. 132, pp: 889 - 906.
Dave Gardiner 2009. Microscale meteorology and atmospheric hazards. Rev. 34b — page content was last changed September 22, 2009 consequent to editing by RA-Aus member Dave Gardiner. Available online (2010) in: http://www.auf.asn.au/groundschool/umodule21.html#layer
David Boukal's website: Collecting water beetles in tropical streams: beginner's field guide http://www.entu.cas.cz/boukal/guide.html.
Davis, A.J., Huijbregts, J., Kirk-Spriggs, A.H., Krikken, J.

& Sutton, S.L. 1997. The ecology and behaviour of arboreal dung beetles in Borneo. (In) Stork, N.E., Adis, J. & Didham, R.K. (eds.), Canopy arthropods. Chapman & Hall, London, pp: 417 - 432.

Day J. C. 2009, in: http://www.bioluminescentbeetles.com/evolution.html

Daly, D. C. & J. D. Mitchell 2000. Lowland vegetation of tropical South America – an overview. Pages 391-454. In: D. Lentz, ed. Imperfect Balance: Landscape Transformations in the pre-Columbian Americas. Columbia University Press, New York. Available online (2010) in: http://www.nybg.org/bsci/acre/pdfs/Lowlandvegetation.pdf

Darlington and Ball 1985. In: Ball, George (ed) 1985.Taxonomy, phylogeny and biogeography of beetles and ants: Dr. W. Junk Publisher, The Hague, The Netherlands, pp: 437- 488.

Dearborn, R.G. & Donahue, C.P. 1993. An annotated list of insects collected and recorded by the Maine Forest Service: Order Coleoptera, Beetles. Maine Forest Service, Technical Report 32. Augusta, Maine. 101 pages.

Decu, V., Juberthie, C. & Nitzu, E.Coleoptera (Varia) 1998. In Encydopaedia Biospeologica, vol. 2, edited by C.Juberthie & V. Decu, Moulis and Bucharest: Société de Biospéologie, pp: 1163 - 1173.

Deloya, C. 1988. Coleopteros lamelicornios asociados a depositos de detritos de *Atta mexicana* (Smith) (Hymenoptera): Formicidae) en el sur del estado de Morelos, Mexico. Folia Entomologica Mexicana. Vol. 75, pp: 77- 91.

Demillio, R., 1994. How Weather Works. Emeryville, California: Ziff David Press. Stern, A.C. y otros, 1984. Fundamentals of Air Pollution. 2da. Edición. Nueva York: Academic Press.

Dettner, K. 1987. Chemosystematics and evolution of beetle chemical defenses. Annual Review of Entomology 32, pp: 17-48.

Deuve, T. 2001. Le genre Eustra Schmidt-Goebel, 1846, insectes (Coleoptera, Paussidae, Ozaeninae) à genitalia femelles orthotopiques. Zoosystema. Vol. 23. Fasc. 3, pp: 547- 578.

Deuve, T. 2004. Nouveaux ozaenini néotropicaux [Coleoptera, Paussidae] New neotropical Ozaenini [Coleoptera, Paussidae]. Revue française d'entomologie. Vol. 26, no3, pp: 117- 130

Dettner, Konrad and Bauer, G. Wolfgang V. 1997. Vertical food web interactions: evolutionary patterns and driving forces. Volumen 130 of Ecological studies. 390 pags.

Di Giulio, A. and W. Moore. 2004. The first-instar larva of the genus *Arthropterus* (Coleoptera: Carabidae: Paussinae): implications for the evolution of myrmecophily and phylogenetic relationships within the subfamily. Invertebrate Systematics. Vol. 18. Fasc. 2, pp: 101- 115.

Di Iorio, Osvaldo; Bulit, F.; Aldatz, F. & Masson, V. 2008. Insects found in birds'nests from Argentina: *Tachycineta leucorrhoa leucorrhoa* (Vieillot, 1817) [Aves: Hirundinidae], a new host for *Acanthocrios furnarii* (Cordero & Vogelsang, 1928) [Hemiptera: Cimicidae]. Zootaxa. N° 1896, pp: 1 – 24.

Diamond, J.M. 1976. Island Biogeography and conservation: strategy and limitations. Science 193, pp: 1027-1029.

Dietrick, E.J. 1961. An improved backpack motor fan for suction sampling of insect populations. Journal of Economic Entomology. Vol. 54, pp: 394–395.

Dobson R.M. 1954. The species of *Carpophilus* Stephens (Col., Nitidulidae) associated with stored products. Bulletin of Entomological Research London 45, pp: 389-402.

Donaldson, J. M.; Ladd, T.L. and McGovern, T. P. 1986. Trapping techniques and attractants for Cetoniinae and Rutelinae (Coleoptera: Scarabaeidae). J. Econom. Entom. Vol. 79, pp: 374-377.

Dondale, C.D., Nicholls, C.F., Redner, J.H., Semple, R.B., & Turnbull, A.L. 1971. An improved Berlese–Tullgren funnel and flotation separator for extracting grassland arthropods. *Canadian Entomologist. Vol.* 103, pp: 1549–1552.

Dorsey C.K. and Leach J.G. 1956. The bionomics of certain insects associated with oak wilt with particular reference to the Nitidulidae. Journal of Economic Entomology 49, pp: 219-230.

Douglas, James Ellis Jr. 2003. The Ecology and Control of Small Hive Beetles (Aethina tumida Murray). Volume I. Submitted in fulfillment of the requirements for the degree of Doctor of Philosophy at Rhodes University. 251 pags. Availabale online (2010) in: http://www.pir.sa.gov.au/__data/assets/pdf_file/0015/41262/apiary_shb_fact_sheet_2006.pdf

Dourojeanni, M.J. 1990. Entomology and Biodiversity Conservation in Latin America. American Entomologist. Vol. 36. Fasc. 2, pp: 88 - 93.

Dowdy, A. K., and W. H. McGaughey. 1994. Seasonal activity of stored-product insects in and around farm-stored wheat. J. Econ. Entomol. 87, pp: 1351-1358.

Dowdy, A. K. and Mullen, M.A. 1998 Multiple stored-product insect pheromone use in pitfall traps. Journal of Stored Products Research Vol. 34. N° 1, pp: 75-80.

Downie, N. M., And R. H. Arnett, Jr. 1996. The beetles of Northeastern North America, Vol. 1. The Sandhill Crane Press, Gainesville, FL, 880 pgs.

Doyen, J. T. 1979. The larva and relationships of *Cononotus* LeConte (Coleoptera: Heteromera). Coleopterists Bulletin. Vol.33, pp: 33-39.

Doyen, J. T, Lawrence, J. E. 1979. Relationships and higher classification of some Tenebrionidae and Zoperidae (Coleoptera). Systematic Entomology. 4, pp: 333-377.

Dudley, t., and N. H. Anderson. 1982. A survey of invertebrates associated with wood debris in aquatic habitats. Melanderia. Vol. 39, pp: l - 21.

Dybas, S. Henry. 1980. The smallest beetles (Coleoptera: Ptiliidae) Fla. Dpt. Agricult. & Consummer Serv. Entomology circular N° 218, pp: 1- 4.

Dyer, L. A., and D. K. Letourneau. 1999. Trophic cascades in a complex terrestrial community. Proceedings of the National Academy of Science, USA. Vol. 96, pp: 5072-5076.

Dyer, L. A., and D. K. Letourneau. 2003. Top-down and bottom-up diversity cascades in detrital vs. living food webs. Ecology Letters. Vol. 6, pp: 60-68.

Eaton, Eric R. and Kaufman, Kenn 2007. Field guide to insects of North America. Houghton Mifflin Harcourt (edit.). 391 pages.

Edmonds, W.D. 2000. Revision of the Neotroipical dung beetle genus *Sulcophanaeus* (Coleoptera: Scarabaeidae: Scarabaeinae). Folia eyrovskyana. Supplementum. Vol. 6, pp: 1- 60.

Edwards, J. G. 1957. Entomology above Timberline: II The Attraction of Ladybird Beetles to Mountain Tops, Coleopt. Bull. II, pp: 41-46.

Eikenbary, R. D., and R. C. Fox. 1968. Arthropod predators of the Nantucket pine tip moth, *Rhyacionia frustrana*

(Lepidoptera: Tortricidae). Annals of the Entomological Society of America. Vol. 61. Fasc.5, pp: 1218-1221.

Einar, K. E.; Byrkjedal, I. & Atle, K. J. 1989 Resource partitioning among seven carabid species on Hardangervidda, Southern Norway. Ann. Zool. Fennici. 26, pp: 113-120.

Eisner, T., M. Eisner, A. B. Attygalle, M. Deyrup and J. Meinwald. 1998. Rendering the inedible edible: Circumvention of a millipede's chemical defense by a predaceous beetle larva (Phengodidae). Proceedings of the National Academy of Sciences, 95, pp: 1108 - 1113.

Eisner, Thomas; Wiemer, David; Haynes, Leroy and Meinwald, Jerrold 1978. Lucibufagins: Defensive steroids from the fireflies *Photinus ignitus* and *P. marginellus* (Coleoptera: Lampyridae), The National Academy of Sciences of the USA.

Elgueta, M & Arriagada, G.,1989. Estado actual del conocimiento de los coleópteros de Chile (Insecta: Coleoptera). Revista Chilena de Entomología. Vol. 17, pp: 5-60.

Elgueta, M., 1993. Invertebrados asociados a suelo en bosque de *Nothofagus pumilio* (Poepp. et Endl.) Krasser, XII Region - Chile, con especial referencia a Insecta. Revista Chilena de Entomología. Vol. 20, pp: 49-60.

Eliason, E. A., and D. A. Potter. 2000. Biology of *Callirhytis cornigera* (Hymenoptera: Cynipidae) and the arthropod community inhabiting its galls. Environmental Entomology. Vol. 29, pp: 551-559.

Ellertson, F.E. & Ritcher, P.O. 1959. Biology of rain beetles, *Pleocoma* spp., associated with fruit trees in Wasco and Hood River Counties. Oregon Agricultural Experiment Station Bulletin. Vol. 44, pp: 1- 42.

Elzen P.J., Baxter J.R., Westervelt D., Randall C., Delaplane K.S., Cutts L., Wilson W.T. 1999. Field control and biology studies of a new pest species, *Aethina tumida* Murray (Coleoptera, Nitidulidae), attacking European honey bees in the Western Hemisphere. Apidologie. Vol. 30, pp: 361-366.

Elzinga, R.J. 1977. Observations on *Sandalus niger* Koch (Coleoptera: Sandalidae) with a description of the triungulin larva. Journal of the Kansas Entomological Society. Vol. 50, pp: 324–328.

Encyclopedia of Caves and Karst Science 2004. Taylor and Francis Insecta: Coleoptera: Available online (2010) in: www.bookrags.com/tandf/insecta-coleoptera-tf/

Endrödy-Younga, S. 1959. Systematischer Überblick über die Familie Clambidae (Col.). Opuscula Entomologica. Lund. Vol. 24, pp: 81–116

Endrödy-Younga, S. 1960a. Monographie der Paläarktischen Arten der Gattung *Clambus* (Col. Clambidae). Acta Zoologica Academiae Scientiarum Hungaricae. Vol. 6, pp: 257–303

Endrödy-Younga, S. 1960b. Neue Angaben zur Klärung des Systems der Familie Clambidae under Beschreibung einer neuen Liodiden-Gattung (Coleoptera). Annales Historico-Naturales Musei Nationalis Hungarici. Vol. 52, pp: 239–245

Endrödy-Younga, S. 1961a. Revision der Gattung *Calyptomerus* Redtb. (Col. Clambidae). Acta Zoologica Academiae Scientiarum Hungaricae. Vol. 7, pp: 401–412

Endrödy-Younga, S. 1961b. Eine neue *Clambus*-Art aus Südamerika (Col. Clambidae). Annales Historico-Naturales Musei Nationalis Hungarici. Vol. 54, pp: 325–326

Endrödy-Younga, S. 1961c. Revision der aethiopischen Arten der Gattung Clambus Fischer von Waldheim (Coleoptera, Clambidae). Annales Historico-Naturales Musei Nationalis Hungarici. Vol. 53, pp: 313–323

Endrödy-Younga, S. 1965. Clambiden-Studien (Coleoptera). Annales Historico-Naturales Musei Nationalis Hungarici. Vol. 57, pp: 259–265

Endrödy-Younga S. 1968: Monographie der Paläarktischen Arten der Familie Cybocephalidae (Coleoptera: Clavicornia). Acta Zoologica Ascademiae Scientiarum Hungaricae 14: 27-115.

Endrödy-Younga, S. 1974. A revision of the described Australian and New Zealand species of the family Clambidae (Coleoptera) with description of a new species. Records of the South Australian Museum Vol. 17. Fasc. 1, pp: 1–10

Endrödy-Younga, S. 1978. A revision of the Oriental Clambidae with references to the Papua and Hawaiian species (Coleoptera). Pacific Insects. Vol. 18. N°1-2, pp: 67–84

Endrödy-Younga, S. 1981. The American species of the family Clambidae (Coleoptera: Eucinetoidea). Entomologia Generalis. Vol. 7. Fasc. 1, pp: 33–67

Endrödy-Younga, S.; Crowson, R.A. 1986: Boganiidae, a new beetle family for the African fauna (Coleoptera: Cucujoidea). Annals of the Transvaal Museum. Vol. 34. Fasc. 12, pp: 253 - 273.

Endrödy-Younga, S. 1990. A revision of the Australian Clambidae (Coleoptera: Eucinetoidea). Invertebrate Taxonomy. Vol. 4. Fasc. 2, pp: 247–280

Endrödy-Younga, S. 1991: Boganiidae (Coleoptera: Cucujoidea) associated with cycads in South Africa: two new species and a new synonym. Annals of the Transvaal Museum. Vol. 35. Fasc. 20, pp: 285 - 293.

Endrödy-Younga, S. 1997. Active extraction of water-dissolved oxygen and descriptions of new taxa of Torridincolidae (Coleoptera: Myxophaga). Annals of the Transvaal Museum. Vol.36. Fasc.24, pp: 313-332.

Erichson, W. F. 1845-1847. Natureschichte der Insecten Deutschlands. Erste Abtheilung, Coleoptera, Vol. 3, pp: 1-800. Verlag der Nicolaischen Buchhandlung, Berlin.

Erwin, T. L. 1983. Beetles and other insects of tropical forest canopies at Manaus, Brazil, sampled by insecticidal fogging. Reprinted from Tropical Rain Forest: Ecology and Management. Blackwell Special Publications Number 2 of the British Ecological Society. Boston Melboume, pp: 59 - 75.

Erwin, T.L. 1985. The taxon pulse: a general pattern of lineage radiation and extinction among carabid beetles. In: Ball GE (ed) Taxonomy, phylogeny and biogeography of beetles and ants: Dr. W. Junk Publisher, The Hague, The Netherlands, pp: 437- 488.

Esat, T. Topkara & Süleyman Balık 2008. Brève note – Short note First record of the family Hygrobiidae (Coleoptera: Adephaga) from Turkey. Ann. soc. entomol. Fr. (n.s.), Vol. 44. Fasc. 3, pp: 315 – 316.

Escalona, H. E. 2008. Revisión del genus neotropical *Serrotibia* Reitter (Coleoptera: Salpingidae, Prostominiinae). Entomotropica. Vol. 23. Fasc. 2, pp: 105-166.

Evans, Howard E. and Rubink, William L. 1978. Observations on the prey and nests of seven species of *Cerceris* (Hymenoptera: Sphecidae). Great Basin Naturalist. Vol 38. No 1.pp: 59-63.

Evans A., and James N. Hogues 2006. Field Guide to Beetles of California. California Natural History Guides. Phyllis M. Faber and Bruce M. Pavlik (Edit.). UNiversity California Press. 377 Pages.

Eyre, M. D. & G. N. Foster. 1989. A comparison of aquatic Heteroptera and Coleoptera communities as a basis for environmental and conservation assessments in static water sites. Journal of Applied Entomology. Vol. 108, pp: 355 - 362.

Eyre, M. D., S. G. Ball & G. N. Foster. 1986. An initial classification of the habitats of aquatic Coleoptera in North-East England. Journal of Applied Ecology. Vol. 23. Fasc. 3, pp: 841-852.

Fabio Penati 2009. An updated catalogue of the Histeridae (Coleoptera) of Sardinia, with faunistic, zoogeographical, ecological and conservation remarks. Zootaxa. N° 2318, pp: 197–280.

Fairmaire, L. et P. Germain 1862. Révision des Coléoptères du Chili. Annales Societe entomologique de France, Paris. Suite 1. Vol 32, pp: 225-284.

Falin, H. 2002. Family 102. Ripiphoridae, pp: 431-444, in: Arnett, H. and Thomas, M. 2002. American Beetles: Polyphaga: Scarabaeoidea through Curculionoidea. Vol. 2. CRC Press. 880 pags.

Falin, Z. 2004. Phylogenetic analysis and revision of the genera and subfamilies of the Ripiphoridae (Coleoptera). PhD Thesis, University of Kansas. 560 pags. University Microfi lms International, Ann Arbor, Michigan.

Fall, H. C. 1901. Two new species of Lucanidae from California. Canadian Entomologist 33, pp: 289-292.

Fall, H. C. 1932 *Diphyllostoma* a third species. Pan-Pacific entomologists. Vol. 8, pp: 159-161.

Fall, H. C. 1909. List of the Coleoptera of Southern California, with notes on habits and distribution and descriptions of new species. In: Occasional Papers of the California Academy of Sciences. (2009) Edit. BiblioBazaar, LLC. 290 pags.

Farkač J.; Král D. & Škorpík, M. [eds.] 2005. Červený seznam ohrožených druhů České republiky. Bezobratlí. List of threatened species in the Czech Republic. Invertebrates. – Agentura ochrany přírody a krajiny ČR, Praha, 760 pp.

Fauna Europaea: Coleoptera. Fauna Europaea versión 1.3. /www. faunaeur.org

Fellin, D.G. 1981. *Pleocoma* spp. in Eastern Oregon coniferous forests: observations on adult flight habits and on egg and larval biology. Pan-Pacific Entomologist. Vol. 57, pp: 461-484.

Fender, K. M. 1969. Notes on *Matheteus theveneti* LeConte (Coleoptera: Lampyridae). Coleopterists Bulletin. Vol. 23. Page 52.

Fernández, L. A. and R. E. Campos. 2005. Developmental biology of the aquatic beetle Berosus alternans Brullé (Coleoptera: Hydrophilidae) in the laboratory. Coleopterists Bulletin 59, pp: 1-6.

Fikáček, Martin & Trávníček, Dušan 2009. Order Coleoptera, family Georissidae. Arthropod fauna of the UAE. N° 2, pp: 145–148.

Fikáček, Martin & Šípková, Hana 2009. New Asian *Hydroscapha*, with comments on male-female association of co-occuring species (Coleoptera, Myxophaga, Hydroscaphidae). Zootaxa. N° 2286.pp: 31–48.

Finch, S. 1990. The effectiveness of traps used currently for monitoring populations of the cabbage root fly (*Delia radicum*). Annals of Applied Biology. Vol. 116, pp: 447–454.

Fincke, O.M. 1992a Interspecific competition for tree holes: consequences for mating systems and coexistence in neotropical damselflies. *The American Naturalist. Vol.* 139, pp: 80–101.

Fincke, O.M. 1998. The population ecology of *Megaloprepus coerulatus* and its effect on species assemblages in water-filled tree holes. In Insect Populations: in Theory and in Practice (ed. J.P. Dempster & I.F.G. McLean). Pp. 391–416. Kluwer, Dordrecht.

Fish, D. 1983. Phytotelmata: flora and fauna. In *Phytotelmata: Terrestrial Plants as Hosts for Aquatic Insect Communities* (ed. Howard, J. Frank & Lounibos, L. P.). Plexus, Medford, NJ, pp: 1–27.

Fisher F.W.; Moore I.; Legner, E. F. and Orth, R. F. 1976. *Ocypus olens*: a predator of brown garden snail. California Agriculture. Vol. 30, pp: 20- 21.

Fleenor, S. B. and S. W. Taber. 1999. Review of *Brachypsectra* LeConte with a new record of the Texas Beetle (B. fulva LeConte; Coleoptera: Brachypsectridae). Coleopterists Bulletin. Vol. 53. PP: 359 - 364.

Flegg, J. 1993. Deserts: a miracle of life. Facts on File inc., New York. 159 pages.

Fleischer, 1919. Eine neue Prostomis-Art aus dem Elbursgebirge. Entomol. Bl. Vol. 15. Page 211.

Flores, G.E. 2000. Systematics of the Andean genera *Falsopraocis* Kulzer and *Antofagapraocis* gen. n. (Coleoptera: Tenebrionidae), with descriptions of two new species. Journal of the New York Entomological Society. Vol. 108, pp: 52−75.

Flores, G.E. 2001. Taxonomic placement of the Andean genera *Praocidia* Fairmaire and *Pilobaloderes* Kulzer (Coleoptera: Tenebrionidae: Praocini). Journal of the New York Entomological Society. Vol. 109, pp: 171−178. Lucas

Flores, G. E.; Susana. J. Lagos y Sergio Roig-Juñent. 2004. Arthropodos epigeos que viven bajo la copa del Algarrobo (*Prosopis flexuosa*) en la reserva Telteca (Mendoza, Argentina). Multequina. Vol. 13, pp: 71- 90. Available in: http://redalyc.uaemex.mx/redalyc/pdf/428/42801308.pdf

Flowers, R. W. 1996. La subfamilia Eumolpinae (Chrysomelidae: Coleoptera) en America Central. Revista de Biología Tropical, Publicación Especial 2, pp: 1-60.

Foley, I. A. & Ivie, M. A. 2008. A phylogenetic analysis of the tribe Zopherini with a review of the species and generic classification (Coleoptera: Zopheridae). Zootaxa. Edit. by Magnolia Press Auckland, New Zealand. N° 1928, pp: 1–72. Available online (2010) in: http://www.mapress.com/zootaxa/2008/f/zt01928p072.pdf

Fontanarrosa, M. S.; Patricia L. M. T. y Michat, M. C. 2004. Comunidades de insectos acuáticos de charcos temporarios y lagunas en la ciudad de Buenos Aires (Argentina). Rev. Soc. Entomol. Argent. vol.63 n.3-4 p.55-68.

Foster, D. E. 1976. Revision of North American *Trichodes* (Herbst) (Coleoptera: Cleridae). Special Publications of the Museum of Texas Tech University. Vol.11, pp: 1-86.

Foster, D. E., and W. F. Barr. 1972. Notes on the distribution and bionomics of some North American Cleridae. Journal of the Kansas Entomological Society. Vol. 45. Fasc.1.pp: 122-125.

Francke, W. and Heemann,V. 1974. Tests on the attractiveness of 3-hydroxy-3- methylbutan-2-one to Xyloterus domesticus L. and X. lineatus Oliv. (Coleoptera: Scolytidae). Zeitschrift fur Angewandte Entomologie. Vol. 75, pp: 67 – 72.

Francke, W.; Heemann, V.; Gerken, B.; Renwick, J. A.; Vité, J. P. 1977. 2- ethyl-1,6-dioxaspiro[4.4]-nonane, principal aggregation pheromone of Pityogenes chalcographus (L.). Naturwissen-schaften. Vol. 64, pp: 590 – 591.

Franz, J. 1955. Tannenstammlause (*Adelges piceae* Ratz.) unter einer pelzdecke von *Curcurbitaria pithyophila* (Kze.

et Schm.) De Not., nebst Beobachtungen an *Aphidoletes thompsoni* Mohn (Dipt., Itonididae) und *Rabocerus mutilatus* Beck (Col., Pythidae) als Tannenlausfeinde. Ztschr, f. Planzenkrankheiten u. Pflanzenschutz. Vol. 62. Fasc. 2. Pp : 49-61, 8 figs.

Fratelli Pagano (edit.) 1913. Bollettino della Società entomologica italiana, Tipografi Editori. Volúmenes 44-45. Page: 160.

French, M. Hugh 2007. The Periglacial Environment. John Wiley & Sons Ltd. 458 pages.

Fukuda A. 1943. Description of the larva of *Helota gemmata* Gorham. Trans. Nat. Hist. Soc. Taiwan 33: 15-18. (in Japanese).

Gadagkar, R., Chandrashekara, K. & Nair, P. 1990. Insect species diversity in the tropics: sampling methods and a case study. Journal of the Bombay Natural History Society. 87, pp: 337–353.

Galan, Carlos 2004. Fauna Cavernicolous de la Sierra de Aralar. Ecología Taxonomía y Evolución. Sociedad de Cincias Aranzadi, pp: 1 – 20.

Galan, Carlos y Herrera, Francisco F. 2006. Fauna Cavernicolous de Venezuela: Una Revisión. Bull. Soc. Venezolana Espel. Vol. 40. Pp : 39-57.

Garcete, Bolivar 2001. Las species de *Brachygnathus* (Coleoptera: Carabidae) Del Parque Nacional Defensores del Chaco. Boletín del Museo Nacional de Historia Natural del Paraguay. N° 13, pp: 30 – 37.

García-París, M.; Parra-Olea, G. & Coca-Abia, M. M. 2001. A new species of *Noserus* (Coleoptera: Zopheridae) from Mexico. Pan-Pacific Entomologist. Vol. 77, pp: 144–155.

García-París, M.; Coca-Abia, M. M. & Parra-Olea, G. 2006. Re-evaluation of the genera *Phloeodes*, *Noserus*, and *Nosoderma* (Coleoptera: Zopheridae) with a description of a new species of Nosoderma from Northern Mexico. Annales Société Entomologique de France. Vol. 42, pp: 215–230.

Garcia, Rocio C. and Chacón de Ulloa, Patricia 2005. Rove beeties (Coleoptera Staphyiinidae) in dry forest fragments of the Cauca river valley. Rev. Colomb. Entomol. Vol. 31. N°1, pp: 43 – 50.

Gardner, J. C. M. 1929. Immature stages of Indian Coleoptera (6). Indian Forest Records. Vol. 14, pp: 103 – 132, pls. 1–6.

Gardner, J. C. M. 1930. The early stages of Niponius andrewsi. Bulletin of Entomological Research. Vol. 21, pp: 15-17 +1 plate.

Gardner, J.C.M. 1936. A larva of the subfamily Balginae (Col., Elateridae). Proceedings of the Entomological Society of London B 5, pp: 3–5.

Gardner, J. C. M. 1937. Immature stages of Indian Coleoptera, part 21, Cleridae. Indian Forest Records, Entomology. Vol.3. Fasc.2, pp: 31-47.

Genaro, J. A. 1996. Resumen del conocimiento sobre los meloidos de Cuba (Insecta: Coleoptera) Caribbean Journal of Science. Vol. 32, No. 4, pp: 382-386.

Genaro, Julio A. y Gutiérrez, Esteban 2001. La familia Ptiliidae en Cuba (Insecta: Coleoptera). Cocullo, pp: 9-11.

Geoff Monteith. 2008. a bit about beetles in Brisbane. Paper Available online (2010) in: www.saveourwaterwaysnow.com.au/_dbase_upl/BrisbaneBeetlesMonteith.pdf

Gigli, Maurizio 1998. Where can I find Beetles?. Available online (2010) in: http://bup.xoom.it/where.htm

Gimmel, Matthew L.; Leschen, Richard A. B. & Ślipiński, S. Adam 2009. Review of the New Zealand endemic family Cyclaxyridae, new family (Coleoptera: Polyphaga). Acta Entomologica Musei Nationalis Pragae. Vol. 49. Fasc.2, pp: 511–528. Available online (2010) in: http://www.aemnp.eu/PDF/49_2/Gimmel.pdf

Glaister, A. 1985. Laboratory rearing of Australian elmid larvae (Elmidae: Coleoptera). Australian Society for Limnology Bulletin 10, pp: 51-58.

Goddard D. G. 1999. *Hydrochus elongatus* (Schaller) en Warndon Worcestershire. A Red Data Book Water Beetles. Record N°. 6. p. 26.

Goff, M. Lee; Early, Marianne; Odom, Charles B. and Tullis, K. A. 1986. Preliminary Checklist of Arthropods Associated with Exposed Carrion in the Hawaiian Islands. Proceedings, Hawaiian Entomological Society. Vol. 26, pp: 53-57. Available online (2010) in: scholarspace.manoa.hawaii.edu/bitstream/10125/11192/1/26_53-57.pdf

Gonzáles del Tágano, M., & D. García de Jalón. 1998. Restauración de ríos y riberas. Coedit. Fundación Conde del Valle de Salazar y Ediciones Mundi-Prensa. Madrid. 318 pages.

Gonzalez, G.,2006. Los Coccinellidae de Chile [online (2010)]. In: http://www.coccinellidae.cl

Goodrich M. A. 1997. The Pleasing Fungus Beetles (Coleoptera: Erotylidae) of Beall WoodsState Park, Wabash Co., Illinois. Transactions of the Illinois State Academy of Science. Vol. 90. Fasc. 3-4. Pp. 139 - 145.

Goodrich M. A. 2002. Family 88. Byturidae, pp: 354- 355, in: Arnett, Ross H. and Thomas, Michael 2002. American Beetles: Scarabaeoidea through Curculionoidea. Charles CRC Press. 880 pags.

Gordon, R. D. and Cartwright, O. L. 1980. The Wstern Hemisphere Species of Rhyssemus and Triciorhyssemus (Coleoptera: Scarabaeidae). Smithsonian Contributions to Zoology. N°317. 31 pgs.

Gordon S. C.; Woodfords, J.A.T. & Birch A.N.E. 1997. Arthropods pests of Rubus in Europe: Pests status, current and future control and strategies. Review article. J. of Hort. Sc. Vol. 72. Fasc. 6, pp: 831- 862.

Gordon, Robert D. and Barbero, Enrico 2008. Dung beetles (Coleoptera: Scarabaeidae: Scarabaeinae) of the Mpala Research Centre and Environs, Laikipia district, Kenya. Journal of East African Natural History. Vol. 97. Fasc.2, pp: 135–164.

Grebennikov, Vasily V. 2005. Older-instar larvae of Pseudopsinae (Coleoptera: Staphylinidae): morphological description of three genera and phylogenetic placement of the subfamily. Eur. J. Entomol. 102, pp: 713–724. Available online (2010) in: http://www.eje.cz/pdfarticles/1061/eje_102_4_713_Grebennikov.pdf

Grebennikov, Vasily V. 2004. Review of larval morphology of beetles of the suborder Archostemata (Insecta: Coleoptera), including first-instar chaetotaxy. Eur. J. Entomol.Vol. 101, pp: 273–292.

Green, E. E. 1912. On some luminous Coleoptera from Ceylon. Transactions of the Entomological Society of London 1912, pp: 717–719, pl. 86.

Grez, A. A.; Moreno, Paloma y Mario Elgueta 2003. Coleópteros (Insecta: Coleoptera) epigeos Asociados al bosque Maulino y plantaciones de pino aledañas. Revista Chilena de Entomología. Vol. 29, pp: 9-18.

Grimaldi, D. and Engel M. 2005. Evolution of the Insects. Cambridge Univ. Press. 755 Pages.

Grouvelle, A. H. 1889. Voyage de M. E. Simon au Venezuela (Decembre 1887 – Avril 1888). Coléopteres, 2e mémoire. Cucujidae, Rhysodidae, Dryopidae, Cyathoceridae, Heteroceridae. Ann. Soc. Entomol. France. Vol. 6. Fasc. 9, pp: 157- 166, 10 figs. And lists *Cyathocerus horni* (now *Lepicerus*).

Grouvelle, A. H. and Achille Raffray 1912. Supplement a la liste des Coleopteres de la Guadeloupe. 3e supplement. Ann. Soc. entom. France Vol. 81, pp: 289 - 312. (14 nn. spp. in: *Scalenarthrus, Mystrops, Carpophilus, Ithris, Pycnomerus, Metacerylon, Botrodus, Cathartus* 2, *Laemophloeus* 2, *Aprostomis* n. g., *Telmatoscius, Derolathrus*).

Grove, S.J. & Stork, N.E 2000. An inordinate fondness for beetles. Invertebrate Taxonomy. Vol. 14. Fasc.6, pp: 733–739.

Grove, S. J. 2000. Trunk window trapping: an effective technique for sampling tropical saproxylic beetles. Memoirs of the Queensland Museum. Vol. 46, pp: 149-160.

Grove, S. J. 2002. Saproxylic insect ecology and the sustainable management of forests. Annu. Rev. Ecol. Syst. 33: 1-23.

Grove, S. J. and Bashford, R. 2003. Beetle assemblages from the Warra log-decay project: insights from the first year of sampling. Tasforests Vol. 14, pp: 117-129.

Grove, Simon; Richards, K.; Spencer, C. and Yaxley, B. 2006. What lives under large logs in Tasmanian eucalypt forest?. The Tasmanian Naturalist. 128, pp: 86-93. Available online (2010) in: http://www.tasfieldnats.org.au/TasNatArticles/TasNat128/TN128-GROVE2.pdf

GTLI 2007. Métodos de muestreo para insectos saproxílicos. Working Group on Iberian Lucanidae (GTLI). Available online (2010) in: http://entomologia.rediris.es/gtli/engl/zero/saproxyl.htm

Gustavo, E. Flores; Susana. J. Lagos y Sergio Roig-Juñent. 2004. Arthropodos epigeos que viven bajo la copa del Algarrobo (Prosopis flexuosa) en la reserva Telteca (Mendoza, Argentina). MULTEQUINA. Vol. 13, pp: 71-90. Available online (2010) in: http://redalyc.uaemex.mx/pdf/428/42801308.pdf

Habeck, D. H. 2002a. Brachypteridae Erichson 1845. In: Arnett RH Jr.; Thomas M. C,; Skelley P. E.; Howard, F. J. (Eds) American Beetles, Volume 2: Polyphaga: Scarabaeoidea through Curculionoidea. CRC Press, Boca Raton, pp: 309 - 310.

Habeck, Dale H. 2002b. Family 77. Nitidulidae, pp: 311-315, in: Arnett, R. H. Jr.; M. C. Thomas; P. E. Skelley and Howard, J. Frank (Eds.). American Beetles. Vol. 2: Polyphaga: Scarabaeoidea through Curculionoidea. CRC Press, Boca Raton. XVI + 861 Pages.Hadden, Fred. C. 1933 .Collecting Beetles in the tropics. The Hong Kong Naturalist, pp: 38-47.

Hadley, N. F. 1970. Water relations of the desert scorpion, *Hadrurus arizonensis*. Journal of Experimental Biology. Vol. 53, pp: 547- 58.

Halffter y Mathews 1966 La Historia natural de los escarabajos peloteros de la subfamilia Scarabaeinae (Coleopt. Scarabaeidae). Folia Entomológia Mexicana. Vol. 12. Fasc. 14, pp: 1 – 312.

Hall, E. 2002. Ptiliidae Erichson, 1845. In: R. H. Arnett, Jr. and M. C. Thomas, eds. American Beetles. vol. I. CRC Press, Florida, pp: 233-246.

Hall, W. Eugene & Andrew E. Z. Short 2010. A new species of *Hydroscapha* LeConte from northwestern Venezuela. Zootaxa. N° 2358, pp: 49–56.

Hangay, George and Zborowski, Paul 2010. A Guide to the Beetles of Australia. Publisher: CSIRO publishing. 248 pages.

Halstead, D. G. H. 1968. Observations on the biology of *Murmidius ovails* (Beck) (Coleoptera : Cerylonidae). Jour. Stored Prod. Res. Vol. 4, pp: 13-21.

Halstead, D. G. H. 1973. A revision of the genus *Silvanus* Latreille (s.l.) (Coleoptera: Silvanidae). Bulletin of the British Museum of Natural History (Entomology). Vol. 29, pp: 39-112.

Halstead, D. G. H. 1993. Keys for the identification of beetles associated with stored products II. Laemophloeidae, Passandridae and Silvanidae. Journal of Stored Products Research. Vol.29, pp: 99-197.

Halstead, D. G. H. 1973. A revision of the genus Silvanus Latreille (s.l.) (Coleoptera: Silvanidae). Bulletin of the British Museum of Natural History (Entomology) Vol. 29, pp: 37–112. Available online (2010) in: http://hbs.bishopmuseum.org/fiji/pdf/halstead1973.pdf

Hammond, H. E. J. 1997. Arthropod biodiversity from Populus coarse woody material in Nort-central Alberta: a review of taxa and collection methods. Can. Ent. 129: 1009-1033.

Hammond, H.E.J., Langor, D.W., and Spence, J.R. 2004. Saproxylic beetles (Coleoptera) using Populus in boreal aspen stands of Eastern Canada: spatiotemporal variation and conservation of assemblages. Canadian Journal of Forest Research. Vol. 34, pp: 1–19.

Hagstrum, D. W., A. K. Dowdy, and G. E. Lippert. 1994. Early detection of insects in stored wheat using sticky traps in bin headspace and prediction of infestation level. Environ. Entomol. Vol. 23 pp: 1241-244.

Hanley, R. S. 2001. Generic revision and pliylogeny of tlie aleocharine beetle tribe Hoplandriini Casey 1910 (Coleóptera: Staphylinidae). Ph.D. Dissertation, University of Kansas, Lawrence. 537 pp.

Hans John 1960. Eine Jebersicht uber die familie Propalticidae (Coleoptera). Pacific Insects Vol. 2. Fasc. 2, pp: 149 – 170.

Hansen, M. 1991. A review of the genera of the beetle family Hydraenidae (Coleoptera). Steenstrupia. Vol. 17. Fasc.1, pp: 1- 52.

Hansen, M. 1991. The Hydrophiloid Beetles. Biologiske Skrifter. Vol.40, pp: 1- 367.

Hansen, M. 1997. A new subfamily for a remarkable new genus and species of Hydrophilidae from New Guinea (Coleoptera: Hydrophilidae). Annales zoologici. Vol.47, pp: 107–110.

Hansen, M. 1997 Phylogeny and classification of the staphyliniform beetle families (Coleoptera). Biologiske Skrifter 48, Copenhagen.

Hansen, M. 1998: The systematic position of *Kiransus* Makhan, 1994, with the description of three new species of *Hydrochus* Leach, 1817 (Coleoptera: Hydrochidae). Entomol. Scandinavica. Vol. 29, pp: 223 - 232.

Hansen, M. and Richardson B. A. 1998. A new species of *Omicrus* Sharp (Coleoptera: Hydrophilidae) from Puerto Rico and its larva, the first known larva of Omicrini. Systematic Entomology. Vol. 23, N° 1, pp: 1-8 (16).

Hansen, M. 1999. World Catalogue of Insects. Volume 2. Hydrophiloidea (s.str.)(Coleoptera). Apollo Books, Stenstrup, 416 peg.

Hansen, M. 2000. Observations on the immature stages of Georissidae (Coleoptera: Hydrophiloidea) with remarks on the evolution of the hydrophiloid egg cocoon. Invertebrate Taxonomy. Vol. 14, pp: 907–916.

Hansen M. & Richardson B. A. 2002. A new species of *Omicrus* Sharp (Coleoptera: Hydrophilidae) from Puerto Rico and its larva, the first known larva of Omicrini. Systematic Entomology. Volume 23 Issue 1, Pages 1 – 8.

Hansen, Victor and Larsson, Sv. G.1973 (2. utgave). Biller X. Blødvinger, klannere m. m.: (*Malacodermata*, *Fossipedes*, *Marcrodactylia* og *Brachymera*) Danmarks Fauna N° 44. 344 sider.

Hanski, Ilkka, 1989. Dung Beetles: Tropical Rain Forest Ecosystems, ed. by H. Lieth and M.J.A. Werger. Elsevier Science Publishers B.V., Amsterdam — Printed in The Netherlands Chapter 28, pp: 490 - 511.

Hanski, Ilkka and Cambefort, Yves, 1991. Dung beetle ecology. Princeton University Press (Princeton, N.J). xii, 481 pags.

Hanski, Ilkka and Krikken, Jan 1991. Dung Beetles in Tropical Forests in South-East Asia. Chapter 10, pp: 179-197, in: Hanski, Ilkka and Cambefort, Yves, 1991. Dung beetle ecology. Princeton University Press (Princeton, N.J). xii, 481 pags.

Harde, K. W. & Severa, F. (vert. J. Huizenga) 1982. Thieme's kevergids. — Thieme & Cie, Zutphen.200 - 205

Harned, R. W. 1906. Insects Affecting Willows. in: Annual Report of the Columbus Horticultural Society for the Year Ending December 31, 1906. Proceedings of the Columbus Horticultural Society. Page108.

Harris, A.C. 1970a. Coastal beetles of the Wanganui-Manawatu area. Part 1. Environmental factors and zonation. Bulletin of Natural Sciences 1, pp: 45-58. Victoria University of Wellington Biological Society, Wellington.

Harris, A.C. 1970b. Coastal beetles of the Wanganui-Manawatu area. Part 2. Morphological, physiological and behavioural adaptations of psammophilous burrowing beetles. Bulletin of Natural Sciences,1, pp: 59-87. Victoria University of Wellington Biological Society, Wellington.

Hatch, M. H. 1929. The genera and subgenera o f Leiodidae and Clambidae. Journal of New York Entomological Society. Vol. 37, pp: 1-6.

Hatch, M. H. 1933 - Studies on the Leptodiridae (Catopidae) with descriptions of new species. Journal of New York Entomological Society. 41:187-238, 1 est.

Hatch, M. H. 1961. Beetles of the Pacific Northwest. Part III: Pselaphidae and Diversicornia I. Vol.16, University of Washington Publications in Biology. University of Washington Press. Seattle. 503 pages.

Háva, Jiří & Löbl, Ivan 2005. A World catalogue of the family Jacobsoniidae (Coleoptera). Studies and reports of District Museum Prague-East. Taxonomical Series 1. Fasc. 1-2, pp: 89-94.

Háva, Jirí 2006. A World catalogue of the family Derodontidae (Coleoptera). Polish Journal of Entomology. Vol. 75. Pp : 29 – 38.

Háva, Jirí & Alain Lenoir 2008. *Thorictus martinezi* sp. n. from Burkina Faso (Coleoptera: Dermestidae:
Thorictini). Calodema Supplementary Paper No. 77, pp: 1-5.

Hawkeswood, Trevor J. 2003. New host records for adults of some fungus-feeding beetles (Coleoptera) from New South Wales and Queensland, Australia. Calodema. Vol.1, pp: 7-11. Available online (2010) in: http://www.calodema.com/freefiles/263.pdf

Hawkeswood, Trevor J. 2006. Review of literature and the biology of the Australian family Rhinorhipidae (Insecta: Coleoptera). Calodema. Vol. 8, pp: 39 – 41.

Hayashi, M. 2007. Distributional records and ecological notes on aquatic Coleoptera of Shimane Prefecture. Bulletin of the Hoshizaki Green Foundation 10. (in Japanese, English abstract), pp: 77-113

Hayashi, M. & Kadowaki, H. 2007. Records on aquatic Coleoptera from rivers in and around Mt. Daisen, Tottori Prefecture, Japan. Bulletin of the Hoshizaki Green Foundation, 10, pp: 149-168.

Hayek, L.-A.; Buzas, M. A. 1997. Surveying natural populations. Columbia University Press, New York.

Hebauer, F. 1997. Revision der Arten der Familie Spercheidae Erichson, 1837 (Coleoptera: Hydrophiloidea).- Entomologische Blätter. Vol. 93, pp: 9-42. Jena.

Hebauer, F. & Wang, L.-J. 1998: New species of the genus *Oocyclus* Sharp, 1882 from India, Sri Lanka and Taiwan with a key to all known species (Coleoptera: Hydrophilidae). – Acta Coleopterologica. Tomo XIV. Fasc.1, pp: 37 - 46.

Hebauer, F. 1998. Teil A: Imagines. Pp. 1-90 in: F. Hebauer, & B. Klausnitzer: Insecta: Coleoptera: Hydrophilidae (exkl. Helophorus). Süsswasserfauna von Mitteleuropa, 20, part 7, 8, 9, 10-1. Gustav Fischer, Stuttgart, Jena, Lübeck, Ulm, xii + 134 pp.

Helava, J. V. T. 1978. A revision of the Nearctic species of the genus *Onthophilus* (Histeridae). Contributions of the American Entomological Institute. Vol.15, pp: 1-43.

Helfrich,Louis A. and Parkhurst, James 2009. Sustaining America's Aquatic Biodiversity Aquatic Habitats: Homes for Aquatic Animals. Virginbia Cooperative Extension. Publication 420-522- pp: 1-4. Available online (2010) in: http://pubs.ext.vt.edu/420/420-522/420-522.pdf

Heliövaara, K. and Väisänen, R. A. 1984. Effects of modern forestry on northwestern European forest invertebrates: a synthesis. Acta Forestalia Fennica. N° 189, pp: 1–3.

Helava, J.V.T., Howden, H.F. & Ritchie, A.J. 1985. A review of the New World genera of the myrmecophilous and termitophilous subfamily Hetaeriinae. Sociobiology. 10, pp: 127–386.

Hemp, C., and K. Dettner. 2001. Compilation of canthariphilous insects. Beiträge zur Entomologie Vol. 51, pp: 231-245.

Henderson, I.F. & Whittaker, T.M. 1977. The efficiency of an insect suction sampler in grassland. *Ecological Entomology. Vol* 2, pp: 57–60.

Hepburn H.R., Radloff S. E. 1998. Honeybees of Africa, Springer Verlag, Berlin, Heidelberg, New York.

Herman, Lee H. 1983. *Pseudopsis* : two new species from India (Coleoptera, Staphylinidae, Pseudopsinae). American Museum novitates. N°. 2755. 6 Pages.

Herman, Lee. H. 1986. Revision of *Bledius*. Part IV. Classification of species groups, phylogeny, natural history, and catalogue (Coleoptera, Staphylinidae, Oxytelinae). Bulletin of the American Museum of Natural History. Vol. 184, pp: 1 - 368.

Herms, D.A., Neilsen, D.G., & Davis Snydor, T. 1990. Comparison of two methods for sampling arboreal insect populations. Journal of Economic Entomology. Vol. 83, pp: 869–874.

Hernando, C. 2002. *Phloeocharis* (*Scotodytes*) *montnegrensis* sp. nov., un nuevo estafilínido endógeoidel noreste de la Peninsula Ibérica (Coleoptera: Staphylinidae: Phloeocharinae). Heteropterus Revista de Entomología. Vol. 2, pp: 1 - 5.

Hernando, C. & Ribera, I. 2003. Phalacrichus semicaecus sp.nov., the first micropterous and microphthalmic forest-litter species of Limnichidae (Coleoptera). Entomological Problems. Vol. 33. Fasc.1–2, pp: 1-5.

Hertel, F. and Colli, R. 1998. The Use of Leaf-cutter Ants, *Atta laevigata* (Smith) (Hymenoptera: Formicidae), as a

Substrate for Oviposition by the Dung Beetle *Canthon virens* Mannerheim.The Coleopterists Bulletin. Vol.52. Fasc.2, pp: 105-108.

Hertz, M. 1927. Huomioita petokuoriaisten olinpaikoista. Luonnon Ystävä. Vol. 31, pp: 218–222.

Hervé Brustel and Roge Jean 2006. Présence en aveyron de *Phloiophilus edwardsii* stephens, 1830 (Coleoptera Phloiophilidae). L' Entomologiste. Vol. 62. N° 3-4, pp: 87- 90.

Hickman, V. V. 1974. Notes on the biology of Ptinus exulans Erichson (Coleoptera: Ptinidae). – Journal of the Entomological Society of Australia. Vol. 8, pp: 7–14.

Hidalgo-Galiana, A.; Jäch, M. A. & Ribera, I. 2010. *Hydrochus farsicus* sp.n. from Iran and notes on other Palearctic species of the genus (Coleoptera: Hydrophiloidea: Hydrochidae). Zootaxa 2344, pp: 61–64. Available online (2010) in: http://www.mapress.com/zootaxa/2010/f/zt02344p064.pdf

Hilsenhoff, W. L. and B. H. Tracy. 1985. Techniques for collecting water beetles from lentic habitats. Proceedings of the Academy of Natural Sciences of Philadelphia, 137: 8-11.

Hinton, H. E. 1933. Two new coleopterous families new to mexico. Pan-Pacific Entomologist,. Vol. 9. Fasc. 4, pp: 160 - 162. (este paper ha sido publicado en 1934).

Hinton, H. E. 1945 A Monograph of the Beetles associated with Stored Products. Vol. 1. London, British Museum (Natural History), viii + 443 Pages.

Hisamatsu, S., 1985a. Notes on some Japanese Coleoptera, I. Trans. Shikoku ent. Soc. Vol. 17. Fasc: 1- 2, pp: 5 - 13.

Hlstead, D. G. H. 1968 Observations on the biology of *Murmidius owails* (Beck) (Coleoptera: Cerylonidae). Jour. Stored Prod. Res., Vol. 4, pp: 13- 21.

Hockberger, P. E. 2002. "A history of ultraviolet photobiology for humans, animals and microorganisms", en Photochem. Photobiol., vol. 76. 561-579

Hoebeke ER, Wheeler QD, Gilbertson RL. 1987. Second Eucinetidae Coniopharaceae association (Coleoptera; Basidiomyctes), with notes on the biology of *Eucinetus oviformis* LeConte (Eucinetidae) and two species of Endomychidae. Proceedings of the Entomological Society of Washington. Vol. 89, pp: 215-218.

Hoffman, C. H. 1940. Additions to annotated lists of insects reared from elm bark and wood. Bulletin of the Brooklyn Entomological Society. Vol. 35, pp: 54-63.

Hoffman, A., A. G. López-Campos, and I. M. Vázquez-Rojas. 2004. Los artrópodos de las cavernas de México. p. 229-328. In: J. E. Llorente, J. J. Morrone, O. Yánez, and I. Vargas (eds.). Biodiversidad, taxonomía y biogeografia de artrópodos de México: Hacia una síntesis de su conocimiento. Universidad Nacional Autónoma de México; México. 660 pages.

Hogarth, Peter 2007. The Biology of Mangroves and Seagrasses. Oxford University Press. 281 pages.

Holger, H. D. 2003. Wirbellose Tiere. Lehrbuch der Spexiellen Zoologie. Spektrum Akademischer Verlag, Based 1: Wirbellose Tiere.

Hölldobler B. 1971. Communication between ants and their guests. Scientific American. Vol. 224, pp: 86-93.

Hood, W. M. 2000. Overview of the small hive beetle, *Aethina tumida*, in North America. Bee World 81, pp: 129-137.

Hörnschemeyer, T.; Goebbels, J.; Weidemann, G.; Faber, C. and Haase, A. 2006. The head morphology of *Ascioplaga mimeta* (Coleoptera: Archostemata) and the phylogeny of Archostemata. Eur. J. Entomol. N° 103, pp: 409 – 423.

Hörnschemeyer T. 2009. The species-level phylogeny of archostematan beetles—where do *Micromalthus debilis* and *Crowsoniella relicta* belong? Systematic Entomology. Vol. 34. Fasc. 3, pp: 533-558.

Hovore, F. T. 1972. Three new sympatric Pleocoma from the Southern Sierra Nevada mountains of California. Bulletin of the Southern California Academy of Science. Vol. 71, pp: 69-80.

Hovore, F. T. 1979. Rain beetles: small things wet and wonderful. Terra Magazine. Vol. 17: 10-14.

Howard, J. Frank and Thomas, Michael C. 1999. EENY-114.Staphylinidae. Available online (2010) in: http://entnemdept.ufl.edu/creatures/misc/beetles/rove_beetles.htm

Howard, Frank, J. and Thomas, M. C. 2008. Rove Beetles of Florida, Staphylinidae (Insecta: Coleoptera: Staphylinidae). DPI Entomology. Circular No. 343, pp: 1- 13.

Howard, Frank, J.; Mizell, R. F. 2007. III Ladybirds, Ladybird beetles, Lady Beetles, Ladybugs of Florida, Coleoptera: Coccinellidae. 12 Pages.

Howden H. 1955a. The biology and taxonomy of the North American beetles of the subfamily Geotrupinae with revisions of the genera *Bolbocerosoma*, *Eucanthus*, *Geotrupes* and *Peltotrupes* (Scarabaeidae). Proc. U.S. Nat. Mus. 104, pp: 151-319.

Howden H. 1968 A review of the Trichiinae of North and Central America Memoirs of the entomological Society of Canada. Ottawa. Vol. 54, pp: 5-77.

Howden, H. F. and B. D. Gill. 1988a. *Xenocanthus*, a new genus of inquiline Scarabaeidae from Southeastern Venezuela (Coleoptera). Canadian Journal of Zoology. Vol. 66, pp: 2071 - 2076.

Howden, H. F. and B. D. Gill. 1988b. A fourth genus of unusually modified Ceratocanthinae (Coleoptera: Scarabaeidae) from South America. Canadian Journal of Zoology. Vol. 66, pp: 2077-2079.

Howden, H. F. and B. D. Gill. 1995. *Trachycrusus*, a new genus of Ceratocanthinae (Coleoptera: Scarabaeidae) with two new species from Peru. The Canadian Entomologist. Vol. 127, pp: 587-593.

Howden, H. F. and B. D. Gill. 2000. Tribes of New World Ceratocanthinae, with keys to genera and descriptions of new species (Coleoptera: Scarabaeidae). Sociobiology. Vol. 35. Fasc. 2B, pp: 281-329.

Howden H. 2008 Association of Cetoniinae with *Thrincopyge* LeConte in North America. The Coleopterists Bulletin. Vol. 62. Fasc. 2, pp: 181-184.

Hu Gy and Howard, Frank J. 1995b. Biology of *Neohypnus pusillus* (Sachse) and its predation on immature horn flies in the laboratory. Coleopterists Bulletin. Vol. 49, pp: 43- 52.

Hudson, G. V. 1934. New Zealand beetles and their larvae. Edit: Furgeson and Osborn. Wellington. 236 pags.

Hudson, L. 1975. A systematic revision of the New Zealand Oedemeridae (Coleoptera, Insecta). Journal of the Royal Society of New Zealand, 5, pp: 227-274.

Humble, Leland M.; Winchester,Neville N. and Ring, Richard A. 1999.The Potentially Rare and Endangered Terrestrial Arthropods in British Columbia: Revisiting British Columbia's Biodiversity. Proc. Biology and Management of Species and Habitats at Risk, Kamloops, B.C., 15–19

Hyvärinen, E. 2006. Green-tree retention and controlled burning in restoration and conservation of beetle diversity in boreal forests. Dissertationes Forestales 21. Finnish Society of Forest Science, Helsinki, Finland. 55

pags.

Ignacio, José López-Colón y Sánchez-Ruiz Antonio 2002. La Familia Cebrionidae Latreille, 1802 En la fauna Ibérica (Coleoptera, Elateroidea). Bol. S.E.A., N° 31, pp: 133–139.

Incekara, U; Mart, A. and Erman, O. 2004a. Distribution of Turkish *Helophorus* Fabricius, 1775 (Coleoptera, Helophoridae) I. Subgenus *Rhopalhelophorus*, with two new records. Gazi Entomological Research Society. Vol. 6. Fasc. 2, pp: 51-62.

Incekara, Omit and Abdullah Mart 2004. Two New Records of Hydrochidae (Coleoptera) Species from Turkey, with Some Ecological Notes. Turk J Zool. 28, pp: 213-216.

Incekara, U., Mart, A. and Erman, O. 2004b. *Helophorus erzurumicus* sp. n. A New Species from Turkey and Some Notes on *H. Ponticus* Angus, 1988 (Coleoptera: Hydrophilidae). Aquatic Insects (in press).

Inoda, T., M. Hasegawa, S. Kamimura, and M. Hori. 2009. Ditary program for rearing the larvae of a diving beetle, Dytiscus sharpi (Wehncke), in the laboratory (Coleoptera: Dytiscidae). Coleopterists Bulletin 63, pp: 340-350.

Inoda, T. and S. Kamimura. 2004. New open aquarium system to breed larvae of water beetles (Coleoptera: Dytiscidae). Coleopterists Bulletin 58, pp: 37-43.

Instituto de Investigaciones de Recursos Biologicos Alexander Von Humboldt. 1998. Informe Nacional sobre el estado de la biodiversiclad 1997- Colombia. Chávez. M. E.; Arango. N. (eds.). Santafé de Bogotá: Instituto Humboldt, PNUMA, Ministerio del Medio Ambiente. Vol. 3.

Ireland, R. R., Jr. 1982. Moss Flora of the Maritime Provinces. National Museum of Canada, Publications in Botany, 13, pp: 1- 738.

Ireland, R. R., Jr. 2007. Paraleucobryum (Lindberg ex Limpricht) Loeske. In Flora of North America north of Mexico. Vol. 27, Bryophyta, part 1. Edited by Flora of North America Editorial Committee. Oxford University Press, New York & Oxford, pp: 425–428.

Ivie, M. A.; Marske, K. A.; Foley, I. A. & Ivie, L. L. (undated). The beetles of Montserrat: an annotated checkllist. Appendix 2. Species lists of the beetles, non-beetle hexapods and non-hexapod invertebrates of Montserrat. A Biodiversity Assessment of the Centre Hills, Montserrat 81. pages.

Jäch & Koda 1995. Elmidae: 1. Check list and bibliography of the Elmidae of China (Coleoptera). - In Jäch, M.A. & Ji, L. (eds.): Water Beetles of China, pp. 289-298. - Wien: Zoologisch-Botanische Gesellschaft in Österreich and Wiener Coleopterologenverein, 410 Pages.

Jäch, M. A. 1995c: Hydraenidae (Coleoptera). - In Jäch, M. A. & Ji, L. (eds.): Water Beetles of China, pp. 173-180. Wien: Zoologisch-Botanische Gesellschaft in Österreich and Wiener Coleopterologenverein. 410 Pages.

Jäch, M. A. 1995d: Eulichadidae: Synopsis of the species of the genus *Eulichas* Jacobson from China, Laos and Vietnam (Coleoptera). - In Jäch, M.A. & Ji, L. (eds.): Water Beetles of China, pp.359-388. - Wien: Zoologisch-Botanische Gesellschaft in Österreich and Wiener Coleopterologenverein. 410 Pages.

Jäch, M. A. 1998. Water Beetles of China, Vol. II, Annotated check list of aquatic and riparian/littoral beetle families of the world (Coleoptera), pp: 25 – 42, in: M. A. Jäch & L. Ji (eds.): Water Beetles of China, Vol. II. Jäch, M. A. and M. Balke 2008. Global diversity of water beetles (Coleoptera) in freshwater. Freshwater Animal Diversity Assessment. Vol.198, pp: 19 – 442.

Jackman, John and Lu, Wenhua 2002. Family 101. Mordellidae Latreille 1802, pp: 423-430. In: Arnett, R.H., Jr., Thomas, M.C., S ., Skelley, P.E. & Howard, Frank, J. (Eds.). American Beetles, Volume 2: Polyphagha: Scarabaeoidea through Curculionoidea. CRC Press, Boca Raton, USA. 861 pages.

Jansen, H. D. 1991. How to save tropical biodiversity. American Entomologist. Vol.37. Fasc. 3. PP: 159 - 178.

Jeannel, R. G. & M. H. Hatch 1914 -1928 - Silphidae. Coleopt. Catal. Vol. 7 Fasc.60 e 95. 246 Pages.

Jeannel, R. G. 1922 - Silphidae Catopinae (Coléoptères, 2ª série) avec une étude phylogénique et paléogeographique de la sous- -famile. Arch. Zool. Exper. Vol. 61, pp: 1-98.

Jeannel, R. G. 1926. Faune cavernicole de la France avec une étude des conditions d'existence dans le domaine souterrain. p. 300. In: P. Lechevalier (ed.). Encyclopedie entomologique. Paul Lechevalier; Paris. 334 pages.

Jeannel, R. G. 1934. Coléoptères cavernicoles de la grotte de Fersine, en Asie Mineure. Annales de la Société Entomologique de France. Vol.103, pp: 159-174.

Jeannel, R. G. 1935 - Sur la distribution géographique des Catopidae (Coleoptera). 6. a Congr. Int. Entom. 1940, pp: 483-499, 3 figs.

Jeannel, R. G. 1936 - Monographie des Catopidae. Mem. Mus. Hist. Nat. (n. s.) Vol. 1. Fasc.1. 433 Pages.

Jeannel, R. G. & N. Leleup, 1952. L'évolution souterraine dans la region méditerranéenne et sur les Montagnes du Kivu.-Notes Biospéologiques, pp: 7 7-13.

Jelinek, J. 2007. Sphindidae. En: Audisio P. (Ed.). Fauna Europaea: Coleoptera. Fauna Europaea versión 1.3. http://www.faunaeur.org.

Jelínek, J. 2007. Adventivarten der Nitidulidae und Kateretidae (Coleoptera: Cucujoidea) in Mitteleuropa. Entomologica Romanica 12, pp: 147-150.

Jelínek, J.; Carlton, C.; Cline, A. and Leschen, R. 2010. Nitidulidae Latreille, 1802. Pp. 390-407. In: Leschen R.A.B., Beutel R.G. and Lawrence J.F. (eds.): Handbook of Zoology. Arthropoda: Insecta Coleoptera, Beetles. Volume 2: Morphology and Systematics (Elateroidea, Bostrychiformia, Cucujiformia partim.). Berlin-New York, Walter de Gruyter, xiii+786 pp.

Jelinek, J. and Cline, A. 2010. Kateretidae Erichson in Agassiz, 1846.Pp. 386-290. In: Leschen R.A.B., Beutel R.G. and Lawrence J.F. (eds.): Handbook of Zoology. Arthropoda: Insecta Coleoptera, Beetles. Volume 2: Morphology and Systematics (Elateroidea, Bostrychiformia, Cucujiformia partim.). Berlin-New York, Walter de Gruyter, xiii+786 pp.

Jenkins, D. W. & Carpenter, S.J. 1946. Ecology of the tree hole breeding mosquitoes of nearctic North America. *Ecological Monographs. Vol.* 16, pp: 31–47.

Ji, L. & Jäch, M. A. 1998b. Epimetopidae: Synopsis of the genus *Eumetopus* Balfour-Browne (Coleoptera), pp. 195 - 205. - In Jäch, M.A. & Ji, L. (eds.): Water Beetles of China. Vol. II. - Wien: Zoologisch-Botanische Gesellschaft in Österreich and Wiener Coleopterologenverein, 371 Pages.

Jian-Rong Wei, Zhong-Qi Yang, Therese M. Poland and Jia-Wei Du Parasitism and olfactory responses of *Dastarcus helophoroides* (Coleoptera: Bothrideridae) to different Cerambycid hosts. BioControl. Editor:Springer Netherlands. Volume 54, Number 6 / diciembre de 2009.

Jiří Hájek & Martin Fikáček 2008. A review of the genus *Satonius* (Coleoptera: Myxophaga: Torridincolidae): taxonomic revision, larval morphology, notes on wing polymorphism, and phylogenetic implications. Acta Entomologica Musei Nationalis Pragae. Vol.48. Fasc.2, pp: 655 - 676.

Jirón, J. 2006. Observación de Stresiptera (Stylopidae) sobre avispas papeleras *Polistes erithrocephalus* Latreille en Cali. Boletín del Museo de Entomología de la Universidad del Valle. Nota Científica. Vol. 7. Fasc. 1, pp: 32 – 33.

Jolivet, Pierre and Verma, Krishna K. 2005. Fascinating insects: some aspects of insect life. Pensoft. 310 pgs.

Johnson, C. 1992. A bionomic review of the British Clambidae. Entomologist's Gazette. Vol.42, pp: 67–71.

Johnson, Paul J. 2002. Family 55. Cerophytidae Latreille 183, pp: 150-151. In: Arnett, R.H., Jr., Thomas, M.C., S ., Skelley, P.E. & Howard, Frank, J. (Eds.). American Beetles, Volume 2: Polyphagha: Scarabaeoidea through Curculionoidea. CRC Press, Boca Raton, USA. 861 pages.

Johnson, P. J. 2002. Family 42. Byrrhidae, pp: 113-116, in: Arnett, R.H., Jr., Thomas, M.C., S ., Skelley, P.E. & Howard, Frank, J. (Eds.). American Beetles, Volume 2: Polyphagha: Scarabaeoidea through Curculionoidea. CRC Press, Boca Raton, USA. 861 pages.

Johnson C, Otero JC, Leschen RAB, 2007. Cryptophagidae Kirby, 1837. In: Löbl I, Smetana A (Eds) Catalogue of Palaearctic Coleoptera, Volume 4: Elateroidea, Derontoidea, Bostrichoidea, Lymexyloidea, Cleroidea, Cucujoidea. Apollo Books, Stenstrup, Denmark. Pp :513–531.

Johnson, J. B.; Emberson, R. M.; Marris J. M. W. 2008. Biology of Metaxina ornata Broun (Coleoptera: Metaxinidae), with Notes on Associated Beetle Taxa.The Coleopterists Bulletin. Vol. 62. Fasc. 2, pp: 215 - 219.

Johnson, P. J. et al 2008. Biodiversity sampling for Coleoptera in Costa Rica: a parataxonomist's manual for collecting, rearing and preparing beetles http://www.inbio.ac.cr/papers/manual_coleoptera/manual.htm#Introduction

Johnson, P. J. 1985. A new species of *Exomella* from Idaho, with notes on the biology of *Exomella pleuralis* (Casey) (Coleoptera: Byrrhidae). Coleopterists Bulletin. Vol. 39. Fasc. 2, pp: 151-157.

Jonsell, M.; Nordlander, G. 1995. Field attraction of Coleoptera to odours of the wood-decaying polypores *Fomitopsis pinicola* and *Fomes fomentarius*. *Ann. Zool. Fennici* 32: 391-402.

Jonsell, M.; Schroeder, M.; Larsson, T. 2003. The saproxylic beetle Bolitophagus reticulatus: its frequency in managed forests, attraction to volatiles and flight period. Ecography 26: 421-428.

Jonsell, M., Hansson, J., 2007. Comparison of methods for sampling saproxylic beetles in fine wood. Entomol. Fenn. Vol. 18, pp: 232–241.

Juan de Ferrer 2005. Confirmación de la presencia en España de *Platylomalus gardineri* (Scott, 1913) (Coleoptera, Histeridae, Dendrophilinae, Paromalini). Boletín Sociedad Entomológica Aragonesa, N° 36, pag: 274.

Juha Siitonen 1994. Decaying wood and saproxylic Coleoptera in two old spruce forests: a comparison based on two sampling methods.Ann. Zool. Fennici.31, pp: 89-95.

Juberthie, C.; Delay, B. et Bouillon, M. 1980. Sur l'existence d'un milieu souterraine superficiel en zone con calcaire. C.R. Acad. Sci. París. N° 290, pp: 49 - 52.

Juberthie, C. et Decu, V. 1994. Encyclopedia Biospeologica. Soc. Biospeol. Moulis (C.N.R.S.). 834 pages.

Jyrki Muona 2000. A Revison of Nearctic Eucnemidae. Acta Zoologica Fennica. N° 212. 106 pages.

Kaesler, R. L.; E. E. Herricks and J. S. Crossman. 1978. "Use of Indices of Diversity and Hierarchical Diversity in Stream Surveys.". Biological Data in Water Pollution Assessment; Quantitative and Statistical Analyses, ASTM STP 652, pp: 92-111.

Kaila L. 1993. A new method for collecting quantitative samples of insects associated with decaying wood or wood fungi. Entomol. Fenn. 4: 21-23.

Kaila, L.; Martikainen, P.; Punttila, P.; Yakovlev, E. 1994. Saproxylic beetles (Coleoptera) on dead birch trunks decayed by different polypore species. Ann. Zool. Fennici 31: 97-107.

Kalshoven, L. G. E. 1955. Additional note on the giant Elaterid, Oxynopterus mucronatus Ol., a predator on termites in Java. Entomol. Ber. Deel. Vol. 1. Fasc. 1, pp: 273-78.

Kanaar, P. 2003. Revision of the genus *Trypeticus* Marseul (Coleoptera: Histeridae). Zool. Verh. 342, pp: 1-318, figs 1-454. Available online (2010) in: www.repository.naturalis.nl/document/46741

Kapler, O. 1999. *Paromalus (Paromalus) picturatus* sp.n. from China, and notes about Hister hanka and *Scapicoelis tibialis* (Coleoptera: Histeridae). Folia Heyrovskyana. Vol. 7, pp: 217–220.

Kathiresan, K. 1992. Insect foliovory in Mangroves. Environment and Ecology, Vol. 10. Pp :988 - 89.

Katovich, Kerry 2002. Family 47. Heteroceridae MacLeay 1825, pp: 127-132. In: Arnett, R.H., Jr., Thomas, M.C., S ., Skelley, P.E. & Howard, Frank, J. (Eds.). American Beetles, Volume 2: Polyphagha: Scarabaeoidea through Curculionoidea. CRC Press, Boca Raton, USA. 861 pages.

Kaufmann, T. and Stansly, P. 1979. Bionomics of *Neoheterocerus pallidus* Say (Coleoptera: Heteroceridae) in Oklahoma. Journal of the Kansas Entomological Society, 52, pp: 565-577.

Kavanaugh, D.H., Erwin, T.L. 1991. The tribe Cicindisini Banniger (Coleoptera: Carabidae): comparative morphology, classification, natural history, and evolution. Proceedings of the Entomological Society of Washington. Vol. 93. Fasc. 2, pp: 356–389.

Kazantsev, Sergey V. 2004. Phylogeny of the tribe Erotini (Coleoptera, Lycidae), with descriptions of new taxa. Zootaxa N° 496, pp: 1-48.

Keer, P. M. 1930. Calwer Keverboek. — Thieme & Cie, Zutphen. Pag. 783.

Keller, W. L. & Heske, E. J. 2001. An observation of parasitism of black rat snake (*Elaphe obsoleta*) eggs by a beetle (*Nicrophorus pustulatus*) in Illinois. Transactions of the Illinois State Academy of Science. Vol. 94. Fasc. 3, pp: 167 - 169.

Khen, C. V.; A. Chung, Y. C. and N. Muslim 2002. Insects sampled by flight intercept trap at a logged forest in Sabah. Malayan Journal of Nature. Vol. 56, pp: 15-22.

Kenneth Charles Neece, B.S. 1980. A faunistic survey of the organisms associated with Ant colonies of West Texas (especially Solenopsis spp.). A thesis in Entomology. Submitted to the graduate Faculty of Texas Tech University in partial fulfillment of the requirements for the degree of Msater of scinece. 105 pgs.

Kerstin Sundseth 2009. In the Alpine Region.European Communities.16 pages. Available online (2010) in:

http://ec.europa.eu/environment/nature/info/pubs/docs/biogeos/Alpine.pdf

Кирейчук А.Г. 1989. Новые палеарктические род и виды сем. Kateretidae (Coleoptera) и замечания по синонимии. Зоологический журнал, Т. 68. вып. 2, pp: 145 - 149.

Kirejtshuk, A. G. and Lawrence, J. F. 1992. Cychramptodini, a new tribe of Nitidulidae (Coleoptera) from Australia. Journal of Australian Entomological Society 31, pp: 29-46.

Kirejtshuk A. G. 1997. On the evolution of anthophilous Nitidulidae (Coleóptera) in tropical and subtropical regions. Bonner Zoologische Beiträge. Vol.47, pp: 111 - 134.

Kirejtshuk A. G. and Leschen R.A.B. 1998. Revue of the *Thalycra* complex (Coleoptera: Nitidulidae) with three new genera and notes on mycophagy. Annales Zoologii. Vol. 48. Fasc. 3/4, pp: 253-273.

Kirejtshuk, A. G. 1998. The position of the subfamily Maynipeplinae subfam. n. (Coleoptera, Nitidulidae) in the classification and notes on the evolution and structural modifications among sap-beetles. Entomologicheskoe obozrenie. Vol. 77. Fasc.3, pp: 540-554, 732.

Kirejtshuk A. G. 1999-2000..*Sikhotealinia zhiltzovae* (Lafer, 1996)-Recent representative of the Jurassic Coleopterous fauna (Coleoptera, Archostemata, Jurodidae). Proceedings of the Zoological Institute Ras. Vol. 281, pp: 21 - 26.

Kirejtshuk A. G. & Viklund B. 2002. Contribution to the knowledge on the subgenus *Meligethes* (*Clypeogethes* Scholtz, 1932) from Kenya (Coleptera, Nitidulidae). Annales Historico-Naturales Musei Nationalis Hungarici, Vol. 94, pp: 9 - 21.

Kirejtshuk A. G. 2008. Essay on the family Nitidulidae from the monograph Family Nitidulidae (incl. Cybocephalinae): http://www.zin.ru/animalia/coleoptera/eng/kirejt4.htm

Kirejtshuk, A.G. and Fallahzadeh, M. 2008. A new species of the genus Cybocephalus from Iran and a note on synonymy (Coleoptera: Nitidulidae). Proceedings of the Zoological Institute RAS. Vol. 312, Nos. 1/2. Pp: 83–86.

Kirk, T. K.; Cowling, E. B. 1984. Biological decomposition of solid wood. En: Rowell, R. M. (ed.) The chemistry of solid wood: 455-487. American Chemical Society, Washington DC.

Kirk-Spriggs A.H. 1996. Pollen Beetles, Coleoptera: Kateretidae and Nitidulidae: Meligethinae.Handbooks for the Identification of British Insects. Vol 5, Part 6a. Royal Entomological Society of London, 157 Pages.

Kiselyova, Tatiana and Mchugha, Joseph V. 2006. Phylogenetic study of Dermestidae (Coleoptera) based on larval morphology. Systematic Entomology. Vol. 31, pp: 469–507.

Kissinger, D.G. 1968. Curculionidae subfamily Apioninae of North and Central America, with reviews of the world genera of Apioninae and world subgenera of *Apion* Herbst (Coleoptera). Taxonomic Publications, South Lancaster, Massachusetts, 559 pages.

Kistner, D. H. 1982. The social Insects'Bestiary, pp: 1-244 in Herman, H. H. (ed.). Social Insects. Vol. III. Academic Press. New York.

Kistner, D. H. and F. A. Abdel-Galil 1986. A new genus and species of termitophilous Throscidae from South Africa (Coleoptera). Sociobiology 12, pp: 305–313.

Kitching, R.L. 1971a. An ecological study of water-filled tree-holes and their position in the woodland ecosystem. *Journal of Animal Ecology. Vol.* 40, pp: 281–302.

Kitching, R. L. 2000. Food Webs and Container Habitats: the Natural History and Ecology of Phytotelmata. Cambridge University Press, Cambridge.

Klass, C.; Abrahamson, L. and Simeone, J. 2009. Powder Post Beetles: Families: Lyctidae, Anobiidae, Bostrichidae. Cornell Cooperative Extension-Suffolok County. Insect and Plant Disease Diagnostic Laboratory.

Klein, M. G., K. O. Lawrence y T. L. Ladd. 1973. A Modified Trap for the Japanese Beetle. Journal of Economic Entomology. 66(1):275-276.

Klima, A. 1934. Curculionidae: Gymnetrinae, Nanophyinae. [in]: S. Schenkling (ed.), Coleopterorum Catalogus auspiciis et auxilio W. JUNK, 135, pp: 1-68 [Gymnetrinae] + 1-26 [Nanophyinae].

Klimaszewski, J. and J.Ch. Watt. 1997. Coleoptera: family-group review and keys to identification (Insecta: Coleoptera). Fauna of New Zealand. Vol. 37, pp: 1 - 198.

Knull, J. N. 1928. Descriptions of two new species of *Acmaeodera* (Buprestidae) with notes on other Coleoptera (Cleridae, Platysomidae, Curculionidae). Entomological News. Vol. 39, pp: 314-316.

Knull, J. N. 1934. Notes on Coleoptera, No. 4. Entomological News. Vol.45, pp: 207-212.

Knull, J. N. 1951. The checkered beetles of Ohio (Coleoptera: Cleridae). Ohio Biological Survey Bulletin. Vol.8. Fasc.42, pp: 268-350.

Kolibac, J. 2004. Metaxinidae fam.nov., a new family of Cleroidea (Coleoptera). Entomologica Basiliensia. 26, pp: 239–268.

Kolibac, J. 2005. A review of the Trogossitidae. Part 1: Morphology of the genera (Coleoptera: Cleroidea). Entomologica Basiliensia et Collectionis Frey. 27, pp: 39-159.

Kopecký Tomáš. 2006. Dasyceridae. Poslední úprava. Available online (2010) in: http://kopido.sweb.cz/DASYCERIDAE.pdf

Korotkevich, E. S. 1972. Poliarnye pustyni / E.S. Korotkevich Gidrometeoizdat, Leningrad. 420 Pages. Polar regions.

Kotze, D. J.; Samways. Ti. J. 1999. Support for the multi-taxa approach in biodiversity assessment, as shown by epigaeic invertebrates in an Afromontane forest archipelago. Journal of insect Conservation. Vol. 3, pp: 125 - 143.

Kovac, D.; Jelínek, J.; Hashim, R. and Wiwatwitaya, D. 2007. Transition from bamboo sap to water: aquatic habits in the Oriental beetle *Amphicrossus japonicus* (Coleoptera: Cucujoidea: Nitidulidae). European Journal of Entomology 104, pp: 635-638.

Kovarik, P. W. & Caterino, M. 2000. Family 15. Histeridae Gyllenhall, 1808, Pp.212-227 in: Arnett, R. H. Jr. & Thomas, M. C. (eds). American Beetles. Vol. 1. Archostemata, Myxophaga, Adephaga, Polyphaga: Staphylliniformia. CRC Press, Boca Raton, London, New York and Washington.

Kovarik, W. P. and Caterino, M. S. 2001. "Histeridae", in Ross H. Arnett, Jr. and Michael C. Thomas, American Beetles (CRC Press). Vol.1.

Kovarik, P. W. & Caterino, M. S. 2005. 10.4. Histeridae Gyllenhal, 1808. Pp. 190-222 in Kristensen, N. P. & Beutel, R. G. (eds.): Handbook of Zoology, Volume IV Arthropoda: Insecta, Part 38 Coleoptera, Beetles, Volume 1: Morphology and Systematics (Archostemata, Adephaga, Myxophaga, Polyphaga partim.). 567 pp. Walter de

Gruyter, Berlin - New York.Krebs, J. R. 1989. *Ecological methodology*. HarperCollins, New York. Leather, S. (ed.). Insect sampling in forest ecosystems. Blackwell, Malden, MA.

Krell, F.T. 2004. East African dung beetles (Scarababaeidae) attracted by defensive secretions of millipedes. Journal of East African Natural History. Vol.93, pp: 69–73.

Kricher John, 2008. Un compañero neotropical. 2ª edición. Edit: Alvaro Jaramillo y Luis Segura. American Birding Assoc. Inc. 437 Pages.

Krivosheina, Nina P. 1991. Relations between wood-inhabiting insects and fungi. (Moscow), pp: 335-346. In: baranchikov, Y. N.; Mattson, W. J.; Hain, F. P., and Payne, T. L., eds. 1991. Forest Insect Guilds: Patterns of Interaction with Host Trees. U.S. Dep. Agric. For. Serv. Gen. Tech. Rep. NE-153. Available online (2010) in: http://nrs.fs.fed.us/pubs/gtr/gtr_ne153/gtr_ne153_335.pdf

Kuijten, P. I. 1978. Revision of the Indo-Australian species of the genus *Phaeochrous* Castelnau, I840 (Coleoptera: Scarabaeidae, Hybosorinae). with notes on the African species. Zool. Verhand. 165, pp: 1- 40.

Kuijten, P. I. 1981. Revision of the genus *Phaeochroops,* Candeze (Coleoptera: Scarabaeidae, Hybosorinae). Zool. Verhand. 183, pp: 1-76.

Kundrata, Robin; Bocák, Ladislav 2007. Una revisión de *Euanoma* y *Pseudeuanoma* (Coleoptera: Drilidae). Annales Zoologici. Vol. 57, N° 3, pp: 427-441.

Kuschel G. 1990. Beetles in a suburban environment: a New Zealand case study. The identity and status of Coleóptera in the natural and modified habitats of Lynfield, Auckland (1974-1989). DSIR Plant Protection Report. Vol. 3. 119 Pages.+ 230 figs.

Kuschel, G. 1994. Nemonychidae of Australia, New Guinea and New Caledonia, pp: 563-637 In Zimerman, E.C. Australian Weevils, Vol. I. (X.I.R.O., Melbourne.

Kuschel G. 1995. A phylogenetic classification of Curculionoidea to families and subfamilies. Memoir Entomol. Soc. Wash. Vol. 14, pp: 5-33.

Kuschel, G. and May, B. M. 1996. Discovery of Palophaginae (Coleoptera: Megalopodidae) on Araucaria araucana in Chile and Argentina. New Zealand Entomologist, 1996, Vol. 19, pp: 1-13. Available online (2010) in: http://www.ento.org.nz/nzentomologist/free_issues/NZEnto19_1_1996/Volume%2019-1-13.pdf

Kuschel, G. 2003. Nemonychidae, Belidae, Brentidae (Insecta:Coleoptera: Curculionoidea) Fauna of New Zealand. Vol. 45. 100 Pages.

Lafer G. Sh. 2002 (http://lod.geospecies.org/ses/2u5RJ.html) Zopheridae beetles *Phellopsis amurensis* in the nature.

Lan-Yu Liu, Klaus Schönitzer & Jeng-Tze Yanga 2008. Review of the literature on the life history of Bostrichidae (Coleoptera). Mitt. Münch. Ent. Ges. München. Vol. 98, pp: 91 – 97.

Лафер, Герман Шлёмович 1996. "Сем. Sikhotealiniidae Lafer, fam. n. // Определитель насекомых Дальнего Восток России. Жесткокрылые или жуки. Том 3. Часть 3. С. 390- 396.

Larson, D. J. & J. R. Labonte. 1994. *Stygoporus oregonensis*, a new genus and species of subterranean water beetle (Coleoptera: Dytiscidae: Hydroporini) from the United States. Coleopterists Bulletin. Vol. 48, pp: 371 – 9.

Larson, D. J. 1997. Habitat and Community Patterns of Tropical Australian Hydradephagan Water Beetles (Coleoptera: Dytiscidae, Gyrinidae, Noteridae). Australian Journal of Entomology. Vol. 36, pp: 269 - 285.

Lawrence, J. F., and Hans Reichardt 1966. Revision of the Genera *Gnostus* and *Fabrasia* (Coleoptera: Ptinidae). Psyche. Vol. 73, pp: 30 - 45.

Lawrence, J. F. 1973. Host preference in ciid beetles (Coleoptera: Ciidae) inhabiting the fruiting bodies of Basidiomycetes in North America. Bull. of the Mus. of Comparative Zool. Vol. 145, pp: 163-212.

Lawrence, J. F. and Stephan, Karl 1975. The North American Cerylonidae (Coleoptera: Clavicornia). Psyche. Vol. 82 N°. 2, pp: :131-166. Available online (2010) in: http://psyche.entclub.org/pdf/82/82-131.pdf

Lawrence, J. F. 1977. The family Pterogeniidae, with notes on the phylogeny of the Heteromera. Coleopterists Bulletin, 31, pp: 25-56.

Lawrence, J. F., and T. F. Hlavac. 1979. Review of the Derodontidae, with new species from North America and Chile. Coleopterists Bulletin. Vol. 33. Fasc. 4, pp: 369–414.

Lawrence J. F.; Newton A. F. 1980. Coleoptera associated with the fruiting bodies of slime moulds (Myxomycetes). Coleopterists Bulletin. Vol. 34, pp: 129-143.

Lawrence, J. F. 1982. A Catalog of the Coleoptera of America North of Mexico. Family: Ciidae. USDA Agr. Handbook. N°.529-105. 18 p.

Lawrence, J. F. 1982. A Catalog of the Coleoptera of America North of Mexico. Family: Cerylonidae. USDA Agr. Hdbk. N° 529-95. 10 p.

Lawrence, J. F. 1982. Lathridiidae, In S. P. Parker (edit.), Synopsis & Classification of Living Organisms. McGraw Hill, Inc., New York. pp. 533-534.

Lawrence, J. F. 1982. Coleoptera, pp: 482–553 in Parker, S.P. (ed.). Synopsis and Classification of Living Organisms. New York: McGraw Hill. Vol. 2 pp. Vii 1232.

Lawrence, J. F. and P. Vaurie. 1983. A Catalog of the Coleoptera of America North of Mexico. Family: Languriidae. United States Department of Agriculture, N°. 529/92. 13 Pages.

Lawrence, J. F. 1985. The genus *Nothoderodontus* (Coleoptera: Derodontidae) with new species from Australiia, New Zealand and Chile. in Ball, G.E. (ed.). Taxonomy, Phylogeny and Zoogeography of Beetles and Ants. A Volume Dedicated to the Memory of Philip Jackson Darlington, Jr 1904–1983 (Series Entomologia 33). Dordrecht : Dr W. Junk, pp: 68 – 83.

Lawrence, J. F. 1988. Rhinorhipidae, a new beetle family from Australia, with comments on the phylogeny of the elateriformia. Invertebrate Taxonomy. CSIRO. Vol. 2. Fasc. 1. PP: 1 - 53.

Lawrence, J. F., D. H. Kistner and J. M. Pasteels.1990.
A new genus and three new species of termitophilous Aderidae (Coleoptera) from Australia, Papua New Guinea and the Philippines, with notes on their biology. Invertebrate Taxonomy. Vol. 4, pp: 643 - 654.

Lawrence, J. F. 1991.Ommatidae (Arcostemata) (=Ommadidae,including Tetraphaleridae), Cupedidae (Archostemata) (=Cupesidae), Micromathidae (Archostemata), Buprestidae (Buprestoidea) (incl. Schizopodidae), Zopheridae (Tenebrionoidea) (incl. Merycidae). Cerambycidae (chrysomeloidea) (incl.Disteniidae, Hypocephalidae, Oxypeltidae, Parandridae, Spondylidae, Vesperiidae). In Stehr F. W. (ed.): Immature Insects, Vol. 2. Kendall/Hunt Publishing Company, Dubuque, Iowa, pp: 298 - 302, 386 - 388, 518 - 519, 556 - 561.

Lawrence, J. F. 1991b. Omalisidae, pp. 422-423. In F. W. Stehr (ed.), Immature Insects. Vol. 2, Kendall Hunt Pub-

lishing Company, Dubuque, Iowa, 974 pages.
Lawrence John F. 1991. Coleoptera in: Immature Insects. Kendall/Hunt Publishing Company. Frederick W. Stehr (edit.) Vol. 2. 50 Pages.
Lawrence, J. F. 1991. Lathridiidae (Cucujoidea), In F. W. Stehr [ed.], Immature Insects. Kendall-Hunt Publishing Company, Dubuque, IA. Vol. 2, pp: 497 - 498.
Lawrence, J. F. 1991. Brachypsectridae In: F. W. Stehr, ed, Immature Insects. Kendall/ Hunt. Dubuque, IA. Vol. 2, pp: 421- 422.
Lawrence, J. F. 1991. Mordellidae. In: F. W. Stehr, ed. Immature Insects. Kendall/Hunt. Dubuque, IA, 975 pages.
Lawrence, J. F. 1991. Eucinetidae (Eucinetoidea). In: Immature InsectsVol 2. (ed. FW Stehr). Pp. 364- 365. Kendall/Hunt Publishing Co, Dubuque, Iowa
Lawrence, J. F., & E.B. Britton, 1991. Coleoptera "beetles". Chap. 35 in The Insects of Australia, ed. I.D. Naumann, pp. 543–683.
Lawrence, J. F. 1991. Ptinidae (Bostrichoidea). Pp. 444–445 in Stehr, F. W. (ed.) Immature Insects, Vol. 2. Kendall/ Hunt Publishing Co., Dubuque, Iowa.
Lawrence, J. F. 1991. Melandryidae (Tenebrionoidea) (=Serropalpidae). Pp. 505-508. In: F. W. Stehr, ed. Immature Insects. Volume 2. Kendall/Hunt. Dubuque, IA. xvi + 975 pp.
Lawrence, J. F. & Viedma, M. G. de 1991: Anobiidae (Bostrichoidea). Pp. 441–444 in Stehr, F. W. (ed.) Immature Insects, Vol. 2. Kendall/Hunt Publishing Co., Dubuque, Iowa.
Lawrence, J. F. & Stribling, J. B. 1992. A new genus of Ptilodactylidae (Coleoptera: Elateriformia) from north Queensland, with description of the presumed larva. Journal of the Australian entomological Society. Vol. 31, pp: 19 - 27.
Lawrence, J. F 1992b. Review of the *Thalycrodes* complex of genera (Coleoptera: Nitidulidae) endemic to the Australian region. Journal of the Australian Entomological Society. Vol. 31, pp: 119–142.
Lawrence, J. F., & D.A. Pollock, 1994. Relationships of the Australian genus *Synerctinus* Newman (Coleoptera: Boridae). Journal of the Australian Entomological Society. Vol. 33, pp: 35–42.
Lawrence, J. F., & E. B. Britton, 1994. Australian Beetles. Melbourne: Melbourne University Press. X. 192 Pages.
Lawrence, J. F., 1994a. The larva of *Sirrhas variegatus*, sp. nov., with notes on the Perimylopidae, Ulodidae (stat. nov.), Zopheridae and Chalcodryidae (Coleoptera: Tenebrionoidea). Invertebrate Taxonomy. Vol.8, pp: 329 – 349.
Lawrence, J. F., 1994b. Review of the Australian Archeocrypticidae (Coleoptera), with descriptions of a new genus and four new species. Invertebrate Taxonomy. Vol.8, pp: 449–470.
Lawrence, J. F. and A. F. Newton. 1995. Families and subfamilies of Coleoptera (with selected genera, notes and references, and data on family-group names), in Biology, Phylogeny, and Classification of Coleoptera: Papers Celebrating the 80th Birthday of Roy A. Crowson. Pakaluk, J. and Slipinski, S. A., eds. Muzeum i Instytut Zoologii PAN, Warsaw, pp: 779-1006.
Lawrence, J. F. 1995. Two new species of *Rhopalobrachium* Boheman (Coleoptera: Phloeostichidae: Hymaeinae) from Australia and Chile. Pp. 433-447 in Pakaluk, J.; Slipinski, S.A. (eds.) Biology, phylogeny, and classification of Coleoptera: papers celebrating the 80th birthday of Roy A. Crowson, 1. Muzeum i Instytut Zoologii PAN, Warsaw, Poland.
Lawrence J. E. 1999. The Australian Ommatidae (Coleoptera: Archostemata): New species, larva and discussion of relationships. Invertebrate Taxonomy. Vol. 13. PP: 369–90.
Lawrence, J. F., Hastings, A. M., Dalwitz, M. J., Paine, T. A., and Zurcher, E.J. 1999a. "Beetles of the World: A Key and Information System for Families and Subfamilies." CD-ROM, Version 1.1 for MS-Windows, CSIRO Publishing, Melbourne.
Lawrence, J. F.; Hastings, A. M.; Dallwitz. M. J.; Paine, T. A. and Zurcher, E. J. 1999b. Beetles of the World: Descriptions, illustrations, and information retrieval for families and subfamilies. CD-ROM, Version 1.0 for MS-Windows. CSIRO Publishing. Melbourne. Zoology, Supplementary Series 4.
Lawrence, J. F., Hastings, A. M., Dallwitz, M. J., Paine, T. A., and Zurcher, E. J. 2000 (onwards) Elateriformia (Coleoptera): descriptions, illustrations, identification, and information retrieval for families and subfamilies. Version: 9th October 2005. http://delta-intkey.com
Laurance, J.. F.; Lovejoy, T. E., 2002. Ecosystem decay of Amazonian forest fragments: a 22-year investigation. Conservation Biology, Boston, Vol.13. N 3, pp: 605 - 618.
Lawrence, J. F. and Leschen, R.A.B. 2003: Review of Lamingtoniidae (Coleoptera, Cucujoidea) with descriptions of two new species. Memoirs on entomology international. Vol. 17, pp: 905-919.
Lawrence, J. F. 2005a. *Brevipogon*, a New Genus of North American Artematopodidae (Coleoptera), with a Key to World Genera. The Coleopterists Bulletin. Vol. 59. Fasc. 2 pp: 223-236.
Lawrence, J. F; Muona, J.; Teravainen, M.; Stahls, G. and Vahtera, V. 2007. Insect Sytematic and Evolution. Vol. 38, pp: 205-239.
Lawrence, J. F. & Yoshitomi, H. 2007. Nipponocyphon, a new genus of Japanese Scirtidae and its phylogenetic significance. Elytra, Tokyo, 35, pp: 507-527.
Lawrence, J. F.; Bruce, S. Archibald, and Adam Ślipiński 2008. A new species of Prionoceridae (Coleoptera: Cleroidea) from the eocene of British Columbia, Canada. Annales Zoologici (Warszawa). Vol. 58. Fasc. 4, pp: 689 – 693.
Lawrence, J. F.; Beutel, R. G.; Leschen, R. A. B.; Ślipiński, A. 2010: 1. Changes in classification and list of families and subfamilies. Pp. 1-7 in: Leschen, R.A.B.; Beutel, R.G.; Lawrence, J.F. (volume eds.) Coleoptera, beetles. Volume 2: Morphology and systematics (Elateroidea, Bostrichiformia, Cucujiformia partim), Kristensen, N.P. & Beutel, R.G. (eds.) Handbook of zoology. Berlin, New York: Walter de Gruyter.
Lawrence, J.F.; Ślipiński, A. 2010. 11.13. Trachelostenidae Lacordaire, 1859. Pp. 571-574 in: Leschen, R.A.B.; Beutel, R.G.; Lawrence, J.F. (volume eds.) Coleoptera, beetles. Volume 2: Morphology and systematics (Elateroidea, Bostrichiformia, Cucujiformia partim). In: Kristensen, N.P. & Beutel, R.G. (eds.) Handbook of zoology. A natural history of the phyla of the animal kingdom. Volume IV. Arthropoda: Insecta. Part 38. Berlin, New York: Walter de Gruyter. ISBN 3110190753 ISBN 9783110190755
Lawson, S. A. and F. D. Morgan 1993. Prey specificity of adult *Temnochila virescens* F. (Col., Trogossitidae), a predator of Ips grandicollis Eichh. (Col., Scolytidae). Journal of Applied Entomology. Vol. 115, Issue: 1-5, pp: 139

– 144.

Lea, A. M. 1901. Descriptions of new species of Australian Coleoptera. Part VI. The Proceedings of the Linnaean Society of New South Wales. Vol. 1901. Fasc.4, pp: 481-513.

Leather, S. R., Bale, J. S., & Walters, K. F. A. 1993. The Ecology of Insect Overwintering. Cambridge University Press, Cambridge.

Leather, S. R.: Lawton, J.H. and Likens, G.E.(edits.) 2005. Insect sampling in forest ecosystems. Blackwell, Malden, MA.. 315 pags.

Leather, S. R. and Watt, Allan D. 2005. Sampling theory and practice Chapter I.pp: 1-15 in: Leather, S.R.: Lawton, J.H. and Likens, G.E.(edits.). Insect sampling in forest ecosystems. Blackwell, Malden, MA.

Leather, S. R.; Watt, A. D. 2005. Sampling theory and practice. In: Leather, S. R. (ed.) Insect sampling in forest ecosystems: 1-15. Blackwell, Malden, MA.

LeConte, John Lawrence 1862. Classification of the Coleoptera of North America, Partes 1-2.Vol. 3; Volumen 11 de Smithsonian miscellaneous collections Classification of the Coleoptera of North America, John Lawrence LeConte. 285 pages.

Lee, Chi-Feng and Satô, M. 2006: The Helotidae of Taiwan (Coleoptera: Cucujoidea). Zoological studies. Vol. 45, pp: 529 - 552. http://zoolstud.sinica.edu.tw/Journals/45.4/529.pdf

Lee, Chi-Feng 2007. Revision of Family Helotidae (Coleoptera: Cucujoidea):I. *Gemmata* Group of Genus *Helota*. Ann. Entomol. Soc. Am. 100. Fasc.5, pp: 623-639. Available online (2010) in: http://digileps.biology.nsysu.edu.tw/TaiEntNews/news20071004/AnnalsHelota.pdf

Lee, Chi-Feng 2009: Revision of the family Helotidae (Coleoptera: Cucujoidea) III: The *Thibetana* group and a checklist of *Helota* MacLeay species. Annals of the Entomological Society of America. Vol. 102, pp: 48 – 59.

Leech, H. B., and H. P. Chandler 1956. Aquatic Coleoptera. In Aquatic insects of California with keys to North American genera and California species. R. L. Usinger, ed. Univ. California Press, Berkeley, pp: 293 – 371.

L. Friederichsen & Co. 1907. Ergebnisse der Hamburger Magalhaensischen Sammelreise 1892/93. Naturhistorisches Museum in Hamburg. Vol.2. Pag: 90+ *Trachelostenus fasciculifer* Philippi

Léonard, J. & Dessart, P. 1994. Avis de recherche: Torridincolidés (Coleoptera) vivant en symbiose avec des podostémacées (Podostemales). Bulletin et Annales de la Société Royale Belge d'Entomologie. N° 130, pp: 71-76.

LeSage, L. 1991. Order Coleoptera. Cantharidae (Cantharoidea). pp. 429–431 in Stehr, F.W. (ed.). Immature Insects. Iowa: Kendall/Hunt Vol. 2. pp. XVI 975.

Leschen, R. A. B. and Allen R. T. 1988. Immature stages, life histories and feeding mechanisms of three *Oxyporus* spp. (Coleoptera: Staphylinidae: Oxyporinae). Coleopterists Bulletin. Vol. 42, pp: 321-333.

Leschen, R. A. B. 1988. Natural history and immatures of *Scaphisoma impunctatum* (Coleoptera: Scaphidiidae). Entomological News. Vol. 99, pp: 225 - 232.

Leschen, R. A. B. & Carlton, C. E. 1988. Immature stages of *Endomychus biguttatus* Say (Coleoptera: Endomychidae) with observations on the alimentary canal. Journal of the Kansas Entomological Society. Vol. 61, pp: 321–327.

Leschen, R. A. B. & Carlton C. E. 1993. Debris cloaking in Endomychidae, a new species from Peru (Coleoptera). Zoological Journal of the Linnean Society, 109, pp: 35-51.

Leschen, R. A. B. 1994. Fungal Host Use in Two Species of *Derodontus* LeConte (Coleoptera: Derodontidae). The Coleopterists Bulletin, Vol. 48, No. 2, pp: 126 – 130.

Leschen, R. A. B. and C. E. Carlton. 1996 Slime production in mycophagous Nitidulidae (Coleoptera) including a new species of *Eusphaerius*. J. of Nat. Hist. Vol. 30, pp: 1861- 1873.

Leschen, R. A. B. 2000. Pseudoliodini (Coleoptera: Leiodidae: Leiodinae) of New Zealand. New Zealand Entomologist. Vol. 22, pp: 33- 44.

Leschen, R. A. B. 2000. Beetles feeding on bugs (Coleoptera, Hemiptera): Repeated shifts from mycophagous ancestors. Invertebrate Taxonomy. Vol.14, pp: 917-929.

Leschen, R. A. R. and Skelley, P. E. 2002. Cryptophagidae Kirby 1837. In: Arnett, R. H. Jr.; Thomas, M. C.; Skelley P. E. and Howard, Frank J. (Eds). American Beetles, Volume 2: Polyphaga: Scarabaeoidea through Curculionoidea. CRC Press, Boca Raton, USA, 338–342.

Leschen, R. A. B. and Rhode, Birgit, E. 2002. A new genus and species of large extinct Ulodidae (Coleoptera) from New Zealand. New Zealand. Entomologist. Vol 25, pp: 57-64.

Leschen, R. A. B. and A. F. Newton 2003. Larval Description, Adult Feeding Behavior, and Phylogenetic Placement of *Megalopinus* (Coleoptera: Staphylinidae). The Coleopterists Bulletin. Vol. 57. Fasc. 4, pp: 469 - 493.

Leschen, R. A. B. 2003. Erotylidae (Insecta: Coleoptera: Cucujoidea): phylogeny and review Part. 1. Fauna of New Zealand. Manaaki Whenua Press, New Zealand. Vol. 47. 108 Pages.

Leschen, R. A. and Early, John 2004. New collection of *Brounia thoracica* (Coleoptera: Chelonariidae). The Weta. Vol.27, pp: 15 – 21.

Leschen, R. A. B. & Beutel, R. G. 2004. Ocellar atavism in Coleoptera: plesiomorphy or apomorphy?. Journal of Zoological Systematics and Evolutionary Research. Vol. 42. Fasc. 1, pp: 63 - 69.

Leschen, R. A. B.; Lawrence, J. F., and Slipinski, S. A. 2005. Classification of basal Cucujoidea (Coleoptera: Polyphaga): cladistic analysis, keys and review of new families. Invertebrate. Systematics. Vol.19. Fasc.1, pp: 17- 73. Available online (2010) in: http://bio150.chass.utoronto.ca/labs/cool-links/lab9/Leschen,_Lawrence_and_Slipinski_2005_lab_9.pdf

Leschen, R. A. B. and Bernard, M. 2005. Biogeography and Evolution of New Zealand Priasilphidae (Coleoptera: Cucujoidea). New Zealand Entomologist. Vol.28, pp: 55 - 64.

Leschen, R. A. B.; Ślipiński, A. 2010: 10. Cucujoidea, Latreille, 1802. Introduction, phylogeny. Pp. 281-282 in: Leschen, R.A.B.; Beutel, R.G.; Lawrence, J.F. (volume eds.) Coleoptera, beetles. Volume 2: Morphology and systematics (Elateroidea, Bostrichiformia, Cucujiformia partim). In: Kristensen, N.P. & Beutel, R.G. (eds.) Handbook of zoology. A natural history of the phyla of the animal kingdom. Volume IV. Arthropoda: Insecta. Part 38. Berlin, New York: Walter de Gruyter.

Lesne, Pierre 1896. Révision des coléoptères de la famille des Bostrychides. 1ª mémoire. Ann. de la Soc. Entom. de France. Vol. 65, pp: 95-127.

Lesne, Pierre 1897. Révision des coléoptères de la famille des Bostrychides. 2ª mémoire. Ann. de la Soc. Entom. de France. Vol. 66, pp: 319-350.

Lesne, Pierre 1899. Révision des coléoptères de la famille des Bostrychides. 3ª mémoire. Ann. de la Soc. Entom. de France. Vol. 67 (1898).pp: 438- 621.

Lesne, Pierre 1901a. Révision des coléoptères de la famille des Bostrychides. 4ª mémoire. Ann. de la Soc. Entom. de France. Vol. 69. (1900), pp: 473-639.
Lesne, Pierre 1901b. Synopsis des Bostrychides paléartiques. L'Abeille. Vol. 30, pp: 73-136.
Lesne, Pierre 1904. Supplèment au synopsis des Bostricides paléartiques. L'Abeille. Vol. 30, pp: 153-168.
Lesne, Pierre 1905. Notes additionnelles es rectificatives sur les Bostrychides paléartiques. L'Abeille. Vol. 30, pp: 248-51.
Lesne, Pierre 1906a. Synopsis des Micrapate de l'Amerique Centrale. L'Abeille. Vol. 30. PP:269-281.
Lesne, Pierre 1906b. Nouvelles notes sur les Bostrychides paleartiques. L'Abeille. Vol.30. Pag. 282.
Lesne, Pierre 1906c. Révision des coléoptères de la famille des Bostrychides. 5ª mémoire. Ann. de la Soc. Entom. de France. Vol. 75, pp: 445-561.
Lesne, Pierre 1907. Un Lyctus africain nouveau (Col.). Ann. de la Soc. Entom. de France. Vol.76. Pp302-303.
Lesne, Pierre 1924. Les coléoptères Bostrycides de L'Afrique tropicale franÇaise. Les presses universitaires de France-Librarie Paul Lechevalier. Paris.
Lesne, Pierre 1934a. Sur le Phonapate frontalis Fahr. (Col. Bostrychidae). Bulletin de la Societé Entomologique de France. Vol. 34, pp: 217-220.
Lesne, Pierre 1934b. Mission J.de Lépiney au Soudan franÇais 1933-34 (deuxième note). Bostrychidae. Bulletin de la Société des Sciences Naturelles du Maroc. Vol.14. Pag.22.
Lesne, Pierre 1938. Bostrychidae (pp.:1-84). En: Junk W. Schenkling S. Coleopterorum Catalogus. Pars 161. W. Junk. Gravenhage.
Letourneau, D. K. 1990. Code of ant-plant mutualism broken by parasite. Science. N° 248, pp: 215-217.
Letourneau, D. K., & L. A. Dyer. 1998. Density patterns of *Piper* ant-plants and associated arthropods: top-predator trophic cascades in a terrestrial system? Biotropica. Vol.30. Fasc.2, pp: 162-169.
Letourneau, D. K., Dyer, L. A., & G. Vega C. 2004. Indirect effects of a top predator on a rainforest understory plant community. Ecology. Vol. 85. Fasc.8, pp: 2144-2152.
Levesque, C., and Levesque, G.-Y. 1993. Abundance and seasonal activity of Eucinetoidea (Coleoptera) in a raspberry plantation and adjacent sites in Southern Québec (Canada). Entomological News. N° 104, pp: 180–186.
Levey, Brian 1984. Afreminae: A new subfamily of Anthicidae (Coleoptera) from Southern Africa. Insect Systematics & Evolution. Vol.15, N° 3, pp: 419 - 422
Lewis, G. 1882. Synteliidae: a family to include Syntelia & Sphaerites, with a note of a new species of the first genus.- Entomologist's. Monthly Magazine. Vol. 19, pp: 137-138.
Lewis, G. 1884. On some Histeridae new to the Japanese fauna, and notes of others.— Ann. Mag. nat. Hist. (5) 13, pp: 131-140.
Lewis, G. 1885. On a new genus of Histeridae. Transactions of the Entomological Society of London. N° 1885, pp: 331-334.
Liebherr, J. and Mahar, J. 1979. The carabid fauna of the upland oak forest in Michigan: survey and analysis. Coleopterists Bulletin. Vol. 33, pp: 183–197.
Liebherr, J. K. and J.V. McHugh. 2003. Coleoptera (Beetles, Weevils, Fireflies). Pp. 209-230, In: Resh, V.H. and R. Carde (eds.). Encyclopedia of Insects. First Edition. Academic Press, San Diego. 1266 pages.
Lillig, Martin 2004. Stuttgarter Beiträge zur Naturkunde Serie A (Biologie). Herausgeber: Staatliches Museum für Naturkunde. Stuttgarter Beitr. Naturk. Ser. A, Nr. 663. 12 S, 6 Abb. Stuttgart, pp: 1 - 12.
Lindgren, B. S. 1983. A multiple funnel trap for scolytid beetles (Coleop.). Can. Entomol. Vol. 115, pp: 299–302.
Lloyd, J. E. 1973: A firefly inhabitant of coastal reefs in New Guinea (Coleoptera: Lampyridae). Biotropica. Vol. 5. Fasc.3, pp: 168 - 174.
Lloyd, J. E. 2002. Lampyridae. In: R. H. Arnett, Jr. & M. C. Thomas [Eds.] American Beetles: Vol. 2, pp: 187-196. CRC Press, Boca Raton, Florida
Löbl, I. 1996. A new species of Sphaerites (Coleoptera Sphaeritidae) from China- Mitteilungen der Schweizerischen Entomologischen Gesellschaft. Vol 69, pp: 195-200.
Löbl, I. & Háva, J. 2002. A new species of Sphaerites (Coleoptera: Sphaeritidae) from China.- Entomological Problems. Vol. 32, pp: 179-181.
Löbl, I. 2005. *Laenagenius apterus* gen. et sp.nov. from China (Coleoptera: Pterogeniidae). Entomological problems, Vol.35, pp: 153 - 155.
Locascio de Mitrovich; Villagra de Gamundi; Juarez, A.; Jorgelina et al. 2005. Limnological characteristics and zooplankton in five shallow lakes, Puna - Argentina. Ecología en Bolivia. Vol.40. N° 1, pp: 10 - 24.
Loebl, I. and Leschen, R. A. B. 2003. Scaphidiinae (Insecta: Coleoptera: Staphylinidae). Fauna of New Zealand. Vol. 48. B5, 94 Pages.
Lohse, G. A. & Lucht, W. H. (eds.), 1992. Biphyllidae. In: Lohse G.A. & Lucht W.H. (eds.), Die Käfer Mitteleuropas. Supplementband mit Katalogteil, 2. Goecke & Evers, Krefeld: 114 Pages.
López, Romero S.; Ochoa, P. R.; Bilbao, J. C. & Lafuente, A. G. 2007. Los escolítidos de las coniferas del País Vasco. Guía práctica para su identificación y control. 1ª ed. Vitoria-Gasteiz: Eusko Jaurlaritzaren Argitalpen Zerbitzu Nagusia = Servicio Central de Publicaciones del Gobierno Vasco. Colección Lur. N° 11. 189 Pages.
López-Colón, J. I. 2001. Nuevos registros de Oedemeridae Latreille, 1810 Ibéricos (Coleoptera). Bol. S.E.A., N° 28, pp: 85-87. Available onlione in: http://www.sea-entomologia.org/PDF/BOLETIN_28/B28-012-085.pdf
López-Colón, J. I. 2001. Rhipiceridae Latreille, 1834, una familia muy poco conocida. Fauna de Marruecos (Coleoptera, Dascilloidea). ARACNET. Bull. S.E.A. Vol. 8. N° 29. Pp : 83 - 84. Notas breves.
López, Heriberto and Oromí, Pedro 2010. A pitfall trap for sampling the mesovoid shallow substratum (MSS) fauna. Speleobiology Notes 2, pp: 7-11.
Louton, J. 1996. The Aquatic Macrofauna of Water Filled Bamboo (Poaceae: Bambusoideae: *Guadua)* Internodes in a Peruvian Lowland Tropical Forest. Biotropica. Vol. 28 Fasc. 2, pp: 228-242.
Louw, S. V. M. 1984. Comparison of the faunistic and ecological patterns of the ground-living Beetle faunas of the Kalahari and Namibia Deserts. Part II. Supplement tyo Koedoe, pp: 157 - 159.
Louw, S. 1993. Systematics of the Urodontidae (Coleoptera: Curculionoidea) of Southern Africa. Entomology Memoir, Department of Agriculture, Republic of South Africa. Vol. 87, pp: 1–92.
Luckenbach, R. A. and R. B. Bury 1983. Effects of off-road vehicles on the biota of the Algodones Dunes, Imperial County, California. J. Appl. Ecol. Vol. 20, pp: 265–286.
Lumaret, Jean-Pierre and Kirk, Alan A. South Termperate

Dung Beetles. Chapter 6, pp: 97-115, in: Hanski, Ilkka and Cambefort, Yves 1991. Dung beetle ecology. Princeton University Press (Princeton, N.J). xii, 481 pags.

Luna, Juan M. 2005. Técnicas de colecta y preservación de insectos. Boletín Sociedad Entomológica Aragonesa. N° 37, pp: 385 – 408.

Lundberg, S. 1984. The beetle fauna of burnt forest in Sweden. Entomologisk Tidskrift 105, pp: 129–141.

Lundie, A. E. 1940. The small hive beetle Aethina tumida. Science Bulletin 220. Dep. Agr. Forestry, Government Printer, Pretoria, South Africa. 30 Pages.

MacArthur, R. H. and Wilson, E. O. 1963. An equilibrium theory of insular zoogeography. Evolution. Vol. 17, pp: 373- 387.

MacArthur, R. H. and Wilson, E. O. 1967. The theory of island Biogeography. Princeton: Princeton University press. 203 Pages.

MacConnell, J. G.; Borden, J. H.; Silverstein, R. M. and Stokkink, E. 1977. Isolation and tentative identification of lineatin, a pheromone from the frass of Trypodendron lineatum (Coleoptera: Scolytidae). Journal of Chemical Ecology. Vol. 3, pp: 549 – 561.

McCoy, C. E. and Brindley, T. A. 1961. Biology of the four-spotted fungus beetle, *Glischrochilus* q. *Quadrisignatus*, and its effect on European corn borer populations. Journal of Economic Entomology 54, pp: 713-717.

Machado, T. 1989. Distribución ecológica e identificación de los coleópteros acuáticos en diferentes pisos altitudinales del departamento de Antioquia. Universidad de Antioquia, Facultad de Ciencias Exactas y Naturales, Departamento de Biología, Medellín. McGavin GC. 2002. Entomología esencial. Ariel Ciencia. Barcelona.

MacLean, S. F. & F. A. Pitelka 1971. Seasonal patterns of abundance of tundra arthropodsnear Barrow. Arctic 24, pp: 19-40. Available online (2010) (only read) in: http://pubs.aina.ucalgary.ca/arctic/Arctic24-1-19.pdf

Maes, Jean-Michel & Ivie, Michael (undated). Familia Bostrichidae. Available online (2010) in: http://www.bio-nica.info/Ento/Coleo/BOSTRICHIDAE.htm

Magro, A.; Lecompte, E.; Magne, F.; Hemptinne, J.-L.; Crouau-Roy, B. 2010: Phylogeny of ladybirds (Coleoptera: Coccinellidae): are the subfamilies monophyletic? Molecular phylogenetics and evolution. Vol. 54, pp: 833-848.

Majer, Karel 1994a. A review of the classification of the Melyridae and related families (Coleoptera, Cleroidea). Entomologica Basiliensia. Vol.17, pp: 319 - 390.

Majer, Karel. 1995. A review of the classification of the Melyridae and related families (Coleoptera, Cleroidea). Entomologica Basiliensia. Vol. 17, pp: 319 - 390.

Majka, C. G. & Klimaszesk, J. 2004. *Phloeocharis subtilissima* Mannerheim (Staphylinidae: Phloeocharinae) and *Cephennium gallicum* Ganglbauer (Scydmaenidae) new to North America: a case study in the introduction of exotic Coleoptera to the port of Halifax, with new records of other species. ZOOTAXA. N° 781, pp: 1–15.

Majka, C. G. 2006. The Mycteridae, Boridae, Pythidae, Pyrochroidae, and Salpingidae (Coleoptera:Tenebrionoidea) of the Maritime me Provinces of Canada. Zootaxa. N° 1250, pp: 37 – 51.

Majka, C. G. & Pollock, D. A. 2006. Understanding saproxylic beetles: new records of Tetratomidae, Melandryidae, Synchroidae, and Scraptiidae from the Maritime Provinces of Canada (Coleoptera: Tenebrionoidea). Zootaxa 1248, pp: 45–68.

Majka, C. G. 2007a. The Erotylidae and Endomychidae (Coleoptera: Cucujoidea) of the Maritime Provinces of Canada: New records, zoogeography, and observations on beetle-fungi relationships and forest health. Zootaxa. 1546, pp: 39–50.

Majka, C. G., Reginald Webster, Andrew R. Cline 2008. New records of Nitidulidae and Kateretidae (Coleoptera) from New Brunswick. Canada ZooKeys. Vol. 2, pp: 337 - 356

Majka, C. G. 2008. Contributions to the knowledge of Atlantic Canadian Histeridae (Coleoptera). ZooKeys. 2. PP: 189-202. Available online (2010) in: http://pensoftonline.net/zookeys/index.php/journal/article/viewFile/3/24

Majka, C. G. and Langor, D. W. 2009. Clambidae (Coleoptera) of Atlantic Canada Journal Acad. Entomol. Soc. Vol. 5, pp: 32-40. Available online (2010) in: http://www.acadianes.org/journal/papers/Majka_Clambidae.pdf

Majka, C. G. and Langor, D. W. 2009, Clambidae (Coleoptera) of Atlantic Canada Journal Acad. Entomol. Soc. Vol. 5, pp: 32-40.

Majka, C. G.; Johnson, C. and Langor, D. W. 2010. Contributions towards an understanding of the Atomariinae (Coleoptera, Cryptophagidae) of Atlantic Canada. Zoo Keys. Pp 35: 37–63.

Majka, C. G. and Bousquet, Yves 2010. Monotomidae (Coleoptera) of the Maritime Provinces of Canada. J. Acad. Entomol. Soc. Vol.6, pp: 1-8. Availbale online (2010) in: http://www.acadianes.org/journal/papers/Majka_Monotomidae.pdf

Majka, C. G. 2010. *Macropogon piceus* (Coleoptera: Artematopodidae) in the Maritime Provinces of Canada. Note of: J. Acad. Entomol. Soc. 6, pp: 9-11. Available online (2010) in: http://www.acadianes.org/journal/papers/majka_artematopodidae.pdf

Makarova, O. L.; Bieńkowski, A. O.; Bulavintsev, V. I. and Sokolov, A. V. 2007. Beetles (Coleoptera) in Polar Deserts of the Severnaya Zemlya Archipelago. Published in Zoologicheskii Zhurnal. Vol. 86. N° 11, pp: 1303 - 1314.

Makhan, Dewanand 1995. Descriptions of ten new species of *Hydrochus* from different parts of the World (Coleoptera: Hydrochidae). Phegea 23: 187-193.

Makhan, Dewanand 2001. A new genus, *Soesilius*, and a new species of Hydrochidae (Coleoptera) from America. Russian Entomol. J. Vol.10. Fasc. 4, pp: 389–393.

Makihara, Hiroshi and Ôhira, Hitoo 2006. Supplementary notes on the elaterid-beetles (Coleoptera, Elateridae) of the Ogasawara Islands held in the collection of the Forestry and Forest Products Research Institute (FFPRI), including a description of a new species. (Bulletin of FFPRI), Vol.5, N°.1 (N°.398), pp: 93 – 97.

Malaise, R. 1937. A new insect trap. Entomologisk Tidskrift. Vol. 58, pp: 148–160.

Mallis, A. 1997. Handbook of Pest Control: the behavior, life history, and control of household pests. Mallis Handbook and Technical Training Company; Cleveland, OH. 1456 pags.

Mamaev, B. M. 1974. The inmature stages of the beetle Syntelia histeroides Lewis (Synteliidae) in comparison with certain Histeridae (Coleoptera).- Entomol. Obozrenie. Vol. 53, pp: 866-871 (in Russian; English translation in Entomol. Review Vol. 53. Fasc. 4, pp: 98-101.

Mandl, Kael W. 1967. Österreichische entomologische Expeditionen nach Persien und Afghanistan. Beiträge zur Coleopterologie. Teil VII : Neue Arten aus der Familie

Karumiidàe. Ann. Naturhist. Mus. Wien. 70, pp: 445-452.

Margaret, K. Thayer 1985. The larva of *Brathinus nitidus* Leconte and the systematic position of the genus (Coleoptera: Staphylinidae). The Coleopterists Bulletin. Vol. 39. Fasc. 2, pp: 174 – 184.

Margaret, K. Thayer 1987. Biology and phylogenetic relationships of *Neophonus bruchi*, an anomalous South Andean staphylinid (Coleoptera). Systematic Entomology. Vol. 12. N°3, pp: 389-404.

Mariconi, F. A. M. 1970. As Saúvas. Editora Agronómica "Ceres" Ltda, São Paulo, xi+167 pgs.

Marinoni, Renato C. y Dutra, R. R. C. 1997. Famílias de Coleoptera capturadas com armadilha malaise em oito localidades do Estado do Paraná, Brasil. Diversidades alfa e beta. Revista Brasileira de Zoologia. vol.14 no.3, pp: 751 – 770.

Marinoni, Renato C. 2001. Os grupos tróficos em Coleoptera. Rev. Bras. Zool. Vol.18. N° 1. Pp.205 - 224.

Marinoni, Renato C.; Ganho, N. G.; Monné, M. L. & Mermudes, J. R. M. 2001. Hábitos alimentares em Coleoptera (Insecta). Ribeirão Preto, Holos. 63 Pages.

Márquez, L. J. & Asiain, J. 2000. La colección de Coleoptera (Insecta) del Museo de Zoología "Alfonso L. Herrera", Facultad de Ciencias, UNAM, México. Acta Zoológica Mexicana, nueva serie. Vol. 79, pp: 241-255.

Márquez, L. J. 2005. Técnicas de colecta y preservación de insectos. Boletín Sociedad Entomológica Aragonesa. N° 1, Fasc. 37, pp: 385 – 408.

Martikainen, P. 2003. Saproxylic beetles in boreal forests: temporal variability and representativeness of samples in beetle inventories. In: Mason, F.; Nardi, G.; Tisato, M. (eds.) Proceedings of the International Symposium "Dead wood: a key to biodiversity", Mantova, May 29-31 2003: 83-85. Sherwood 95, Suppl. 2.

Martinez, Antonio 1992. Los Meloidae de Salta, Argentina (Coleoptera). Insecta Mundi. Vol. 6, No.1, pp: 1-12. Available online (2010) in: http://digitalcommons.unl.edu/cgi/viewcontent.cgi?article=1442&context=insectamundi

Martínez, Antonio 1994. Notas sobre Hybosorinae (Coleoptera: Scarabaeidae), II. Elytron. Vol. 8, pp: 223-239.

Martínez, Claudia y George, E. Ball 2003. Los Platynini (Coleoptera: Carabidae) de Colombia. Biota Colombiana. Vol. 4. Fasc. 2, pp: 175 - 186.

Martínez de Murguía, L.; Lapaza, J.; Salaberria, E. ; Méndez, M. y Molino-Olmedo, F. 2006. Coleópteros saproxílicos (Insecta: Coleoptera) de un hayedo acidófilo en regeneración del norte peninsular. Munibe. Ciencias Naturales – Natur Zientziak. Vol. 55, pp: 167-181.

Martínez de Murguía, L.; Castro, A.; Molino-Olmedo, F. 2007. Artrópodos saproxílicos forestales en los Parques Naturales de Aralar y Aizkorri (Guipúzcoa, España) (Araneae y Coleoptera). Boletín de la S.E.A. 41, pp: 237-250.

Marvaldi, A. E.; Oberprieler, R. G.; Lyal, C. H. C.; Bradbury, T. & Anderson, R. S. 2006. Phylogeny of the Oxycoryninae sensu lato (Coleoptera: Belidae) and evolution of host-plant associations. Invertebrate Systematics. Vol. 20, pp: 447– 476.

Mascagni, A. 1995. Heteroceridae: check list of the Heteroceridae of China and neighbouring countries, and description of two new species (Coleoptera). In Jäch, M.A. & Ji, L. (eds.): Water Beetles of China, pp. 341-348. Wien: Zoologisch-Botanische Gesellschaft in Österreich and Wiener Coleopterologenverein. 410 Pages.

Masner, L. and H. Goulet. 1981. A new model of flight-intercept trap for some hymenopterous insects. Entomological News. Vol. 92, pp: 199-202.

Matile, L. 2000. Les diptères d'Europe occidentale. Vol. 1. Société nouvelle des éditions Boubée. París.

Matthews, E. G. 1972. A revision of the Scarabaeinae dung beetles of Australia. I. Tribe Onthophagini. Australian Journal Zool. Suppl. 9, pp: 1-330.

Matthews, E. G. and P. Bouchard. 2004. Available online (2010) : http://www.environment.gov.au/biodiversity/abrs/online (2010) -resources/fauna/afd/taxa//TENEBRIONIDAE.

Matthew, G. L. 2007. Phylogenetic placement of *Cyclaxyra* Broun, with a revision of the genus (Coleoptera: Cucujoidea: Phalacridae, Cyclaxyridae). ESA Annual Meeting, December, pp: 9- 12.

Matti, Koivula J. and Vermeulen, Hendrik J. W. 2005. Landscape ecology. Springer, Dordrecht, Pays-Bas (Revue). Vol. 20, No 8, pp: 911-926.

Mavnea, W. 1968. A general account of the fauna and flora of mangrove swamps and forests in Indo-west Pacific region. Adv. Mar. Biol. Vol. 6, pp: 27-270.

Mawdsley, J. R. 1999. New records and biological notes on species of Cleridae (Coleoptera) from the Adirondack Park, New York. Great Lakes Entomologist. Vol. 32. Fasc.1-2, pp: 39-45.

Mazur, S. 1997. A world catalogue of Histeridae. Genus (Supplement), pp: 1- 373.

McAlpine, J. F. 1964. Arthropods of the bleakest barren lands: composition and distribution of the arthropod fauna of the northwestern Queen Elizabeth Islands, Can. Ent. V. 96. P. 127 -129. 1965. Insect and related terrestrial invertebrates of Ellef Ringnes Island. Arctic. Vol.18, pp: 73 -103.

Mccafferty, W. P. 1981. Aquatic Entomology. Boston: Science Books International. 448 Pages.

McClain, E. S; Seely, M. K.; Hadley, N. F and Gray, V. 1985. Wax blooms in tenebrionid beetles of the Namib Desert: correlations with environment. Ecology. Vol. 66, pp: 112 –19.

Mcdermott, F. A. 1966. Lampyridae. In Steel, W.O. (ed.): Coleopterorum Catalogus, Supplementa, pars 9 (second ed.). 's-Gravenhage: W. Junk, 149 Pages.

McGeoch, M. A.; Chown, S. L. 1998. Scaling up the value of bioindicators. Tree. Vol. 3. Fasc. 2, pp: 46 - 47.

McLain, D. K. 1982. Density dependent sexual selection and positive phenotypic assortative mating in natural populations of the soldier beetle, *Chauliognathus pennsylvanicus*. Evolution. Vol.36, pp: 1227 – 1235.

Medri, I. M. and Lopes, J. 2001. Forest and pasture coleopterofauna in North of Paraná, Brazil, collected using pitfull traps. Rev. Bras. Zool. vol.18 supl.1, pp: 125- 133.

Menier, J. J. 1991. Les Chaetosomatidae malgaches. 3. Description d'un nouveau *Malgassochaetus* et de des especes du genre (Coleoptera, Cleroidea). Revue Francaise d'Entomologie (nouvelle serie). Vol. 13. Fasc. 4, pp: 159 - 161.

Mezquita, F. 2000. Vivir a oscuras. Mètode. Universitat de València, anuario. N°. 2000, pp: 30-35.

Miller, R. S. 1991. A revision of the Leptolycini (Coleoptera: Lycidae) with a discussion of paedomorphosis., Dissertation [unpublished]. The Ohio State University; Columbus. 403 p.

Miller, R. S. 1997. Female *Phengodes* feeding and an associated risk (Coleoptera: Phengodidae). Entomological News. Vol. 108, pp: 213-214.

Ming-Luen, Jeng; Ping-Shih, Yang and Masataka, Satô 2006. Synopsis of *Cyphonocerus* (Coleoptera: Lampyridae) with the Description of Four New Species and a Key to the Genus. Zoological Studies. Vol. 45. Fasc.2, pp: 157-167.

Mitchell, R. W. 1971a. Distribution and dispersion of the troglobitic carabid beetle Rhadine subterranea. Intnl. J. Speleol. Vol. 3, pp: 271- 288, pls. 88-90.

Mitchell, R. W. 1971b. Food and feeding habits of the troglobitic carabid beetle Rhadine subterranea. Intnl. J. Speleol. Vol. 3, pp: 249- 270, pls. 83-87.

Mitchell, R. W. 1971c. Preference responses and tolerances of the troglobitic carabid beetle Rhadine subterranea. Intnl. J. Speleol.Vol. 3, pp: 289-304, pls. 91-96.

Mizunuma, T. and Nagai, S. 1994. The Lucanid Beetles of the World. Mushisha. Tokyo. 337 Pages.

Moffett, M. W. 1993. The high frontier: Exploring the tropical rainforest Canopy. Harvard University Press, Cambridge, MA. 192 pages.

Mohr, K. H. 1960. Erdflöhe. Wittenberg: Neue Brehm Bücherei, 261 Pages.

Mondaca, J. y Paulsen, M. J. 2004. Ceratognathus argentinense Martínez, nuevo registro de Aesalinae (Coleoptera: Lucanidae) para Chile. Revista Chile Entomología. Vol. 30. Fasc. 2, pp: 21 -25.

Mondaca, J. & Smith, A. 2008. A revision of the Southern South American genus *Bolborhinum* Boucomont (Coleoptera: Geotrupidae: Bolboceratinae). Zootaxa. N° 1794, pp: 1–48.

Mondal, S.; Ray, R. and Ghorai, M. 2003. Observation on the destruction of piper betel plant by *Eumorphus quadriguttatus* (Coleoptera: Endomychidae). Bionotes 5. P. 51.

Monrós, F. 1960. Los genera de Chrysomelidae (Coleoptera). Opera lilloana. Vol.3 [1959], pp: 5 - 337.

Moore, I. and Legner, E. F. 1976. Intertidal rove beetles (Coleoptera: Staphylinidae). Pp. 521-552, In Cheng L (ed.) Marine Insects. North Holland, New York.

Moore, T. & Vidal, P. 2005. Nuevo genus y dos species nuevas de Anthicidae para Chile (Coleoptera). Gayana. Vol. 69. N°2, pp: 411 - 416.

Moore, Peter D. 2008. Tropical Forests. Facts On File, Inc. An imprint of Infobase Publishing. 265 pages.

Morales, C. F.; Hill, M. G. and Walker, A. K. 1988. Life history of the sooty beech scale (*Ultracoelostoma assimile)* (Maskell), (Hemiptera: Margarodidae) in New Zealand *Nothofagus* forests. New Zealand Entomologist. Vol. 11. 15 Pages.

Moret, Pierre 1998. Les *Dyscolus* de la zone périglaciaire des Andes équatoriennes (Coleoptera, Harpalidae, Platyninae). Bull. Soc. Ent. Fr. Vol. 103. Fasc.1, pp: 11-28.

Moret, Pierre 2001. El género *Bradycellus* Erichson, 1837 en los Andes de Ecuador (Coleoptera: Carabidae, Harpalini). Revta. Aragon. Ent. Vol. 9, pp: 25– 29.

Moret, Pierre 2006. Los coleopteros Carabidae del páramo en los Andes del Ecuador. 306 Pages, with 398 Figs.

Moretti, M.; Barbalat, S. 2004. The effects of wildfires on wood-eating beetles in deciduous forests on the Southern slope of the Swiss Alps. Forest Ecol. Manag. 187: 85-103.

Moron, M. .A. & Krikken, J. 1990. A new Mesoamerican genus of Trichiinae Folia entomologica mexicana. Mexico 78:71-84.

Moron, M. A. 1991. Larva and pupa of *Aegidium cribratum* Bates (Coleoptera: Scarabaeidae: Orphninae). Coleopterists Bulletin. Vol. 45. Fasc.4, pp: 360-367.

Moron, M. .A. 1997. Inventarios faunisticos de los Coleoptera Melolonthidae Neotropicales con potencial como bioindicadores. Giornale Italiano di Entomologia. Vol. 8, pp: 265 – 274.

Moulds, T. 2004. Review of Australian cave guano ecosystems with a checklist of guano invertebrates. Proceedings of the Linnean Society of New South Wales. 125, pp: 1-42.

Mueller, Gregory M.; Bills, G. F. and Foster M. S. 2007. Biodiversity of fungi, Inventory and Monitoring Methods. 760 Pages.

Muir, R. C. and R. G. Gambrill. 1960. A note on the knock-down method for estimating numbers of insect predators on fruit trees. Annual Report of the East Malling Research Station 1959, pp: 109-111.

Müller, J. 2005. Wie beeinflusst Forstwirtschaft die Biodiversität in Wäldern? Eine Analyse anhand der xylobionten Kafërn (Insecta: Coleoptera). Beitr. Bayer. Entomofaunistik. Vol.7, pp: 1 - 8.

Müller, J.; Bussler, H.; Bense, U.; Brustel, H.; Flechtner, G.; Fowles, A.; Kahlen, M.; Möller, G.; Mühle, H.; Schmidl, J. and Zabransky, P. 2005. Urwald relict species – Saproxylic beetles indicating structural qualities and habitat tradition. Waldoekologie online (2010) . Vol. 2, pp: 106 - 113.

Müller, J. 2006. Critical values of dead wood for saproxylic beetles in beech forests. 4th Symposium and Workshop on the Conservation of Saproxylic Beetles. 27th to 29th of June, 2006. Vivoin. Sarthe. France.

Muona, Jyrki 2002. Family 56. Eucnemidae Eschscholtz 1829, pp: 152-157. In: Arnett, R.H., Jr., Thomas, M.C., S ., Skelley, P.E. & Howard, Frank, J. (Eds.). American Beetles, Volume 2: Polyphagha: Scarabaeoidea through Curculionoidea. CRC Press, Boca Raton, USA. 861 pages.

Murphy, D. H. 1990. The recorgnition of some insects associated with mangroves in Thailand. Mangrove Ecosystem occasional paper 7, UNDP/UNESCO, New Delhi, pp: 15-23.

Mylarstoreonline (date: 2009) available online (2010) in: www.mylarstoreonline.com/TotalBlack-OutMylar25ft.html

Mynhardt, Glené & Wenzel, John W. 2010. Phylogenetic analysis of the myrmecophilous *Cremastocheilus* Knoch (Coleoptera, Scarabaeidae, Cetoniinae), based on external adult morphology. ZooKeys. Vol.34, pp: 129–140. Available online (2010) in: http://pensoftonline.net/zookeys/index.php/journal/article/download/262/358

Myrtle, J. Ashmole and Ashmote, N. P. 1 988. Arthropod communities supported by biological fallout on recent lava flows in the Canary Islands. Entomologica scand. Supplement. Vol. 32, pp: 67-88.

Nagel, Peter 1979. The Classification of Carabidae en: On the evolution of the behavior in Carabid Beetles. Miscellaneous Papers. Agricultural University Wageningen the Netherlands. N° 18. edited by P.J. Den Boer, H. U. Thiele and F. Weber, pp: 1-9.

Nagel, Peter 1997. New fossil paussids from Dominican amber with notes on the phylogenetic systematics of the paussine complex (Coleoptera: Carabidae). Systematic Entomology. Vol. 22, pp: 345 – 362.

Nalini, Nadkarni and Nathaniel, T. Wheelwright 2000. Monteverde Ecology and conservation of a tropical cloud forest. Oxford University Press.573 pags.

Nápoles, J. Romero; S. A. Rosales; A. E. Martínez y H. M. González 1997. Lista de Platypodidae y Scolytidae de

Mexico (Insecta: Coleoptera). Acta Zool. Mex. (n. s.). Vol. 70, pp: 35 - 63.

Nathan, Lord, P. 2009. Bothrideridae. Cocoon-forming beetles. Version 22 September 2009. http://tolweb.org/Bothrideridae/9165/2009.09.22 in The Tree of Life Web Project, http://tolweb.org/

Navajas, E. 1950. Manifestaçoes de predatismo em Scarabaeidae do Brasil e alguns dados bionómicos de *Canthon virens* (Mannh.) (Col., Scarabaeidae). Ciéncia e Cultura. Vol.2. Fasc.4, pp: 284-285.

Navarrete-Heredia, J. L. 1996 Coleopteros micetócolos de *Basidiomycetes* de San José de los Laureles, Morelos, México. M. Sc. Thesis, Universidad Nacional Autónoma de México. 179 Pages.

Navarrete-Heredia, J. L. 1997. Coleópteros asociados a Basidiomycetes. In: González-Soriano E., Dirzo R. & Vogt R.C. (Eds), Historia Natural de Los Tuxtlas. Instituto de Biología e Instituto de Ecología, UNAM, México D.F., pp: 381-382.

Navarrete-Heredia, J. L. 2001. Beetles associated with Atta and Acromymex ants (Hymenoptera: Formicidae: Attini). Transactions of the American Entomological Society 127:381–429.

Neboiss, Arturs 1960. On the family Cupedidae, Coleoptera. Proceedings of the Royal Society of Victoria 72, pp: 12–20 pls 4–5.

Neboiss, Arturs 1984. Reclassification of *Cupes* Fabricius (s.lat.), with descriptions of new genera and species (Cupedidae: Coleoptera). Systematic Entomology. Vol. 9. N°4, pp: 443-477.

Nel, A. and C. H. Scholtz. 1990. Comparative morphology of the mouthparts of adult Scarabaeoidea (Coleoptera). Entomology Memoires of the Republic of South Africa Department of Agricultural Development. Vol. 80, pp: 1 - 84.

Nelson, G. H. & C. L. Bellamy. 1991. A revision and phylogenetic re-evaluation of the family Schizopodidae (Coleoptera, Buprestidae). Journal of Natural History. Vol. 25, pp: 985 - 1026.

Neumann Peter; W. Ritter 2004. A scientific note on the association of *Cychramus luteus* (Coleoptera: Nitidulidae) with honeybee (*Apis mellifera*) colonies. INRA/DIB-AG-IB/ EDP Sciences, Apidologie. Vol. 35, pp: 665 – 666. Available online (2010) in: http://www.apidologie.org/index.php?option=com_article&access=standard&Itemid=129&url=/articles/apido/pdf/2004/07/M4034.pdf

New, T. R. 1998. Invertebrate surveys for conservation. Oxford University Press, Oxford.

Newton, A. F. JR. 1975. the systematic position of *Glypholoma* Jeannel, with a new synonymy (Coleoptera: Silphidae, Staphylinidae). Psyche, pp: 53- 58.

Newton, A. F. 1984. Mycophagy in Staphylinoidea, p. 302-353, In Wheeler Q, and Blackwell M (eds.) Fungus-insect Relationships; Perspectives in Ecology and Evolution. New York: Columbia Univ Press.

Newton Jr, A. F. 1998. Phylogenetic problems, current classification and generic catalog of the subfamily Leptodirinae. In Phylogeny and Evolution of Subterranean and Endogean Cholevidae (= Leiodidae, Cholevinae), edited by P.M.Giachino & S.B.Peck, Turin: Museo Regionale di Scienze Naturali

Newton, A. F., M. K. Thayer, J. S. Ashe, and D. S. Chandler. 2001. Staphylinidae [pp. 272- 418]. In: American beefles, Vol. 1 (R. H. Amett, Jr. and M. C. Thomas, editors). CRS Press, Boca Raton, xv + 443 Pages.

Newton, A. F. & Margaret K. Thayer (2004-2005). Catalog of higher taxa, genera, and subgenera of Staphyliniformia. Catalog of austral species of Staphylinoidea. Bibliography of catalog references. Field Museum of Natural History. 19 Pages: http://www.fieldmuseum.org/peet_staph

Newton, A.F. 2005. 10.2. Sphaeritidae Shuckard, 1839. Pp. 183-186 in: Beutel, R.G. & Leschen, R.A.B. (volume eds.) Coleoptera, Beetles. Volume 1: Morphology and systematics (Archostemata, Adephaga, Myxophaga, Polyphaga partim). In: Kristensen, N.P. & Beutel, R.G. (eds.) Handbook of zoology. A natural history of the phyla of the animal kingdom. Volume IV. Arthropoda: Insecta. Part 38. Berlin, New York: Walter de Gruyter.

Newton, A.F. 2005. 10.3. Synteliidae Lewis, 1882, pp: 187-190 in: Beutel, R.G. & Leschen, R.A.B. (volume eds.) Coleoptera, Beetles. Volume 1: Morphology and systematics (Archostemata, Adephaga, Myxophaga, Polyphaga partim). In: Kristensen, N.P. & Beutel, R.G. (eds.) Handbook of zoology. A natural history of the phyla of the animal kingdom. Volume IV. Arthropoda: Insecta. Part 38. Berlin, New York: Walter de Gruyter.

Newton, A.F. 2006. *Habrocerus capillaricornis* (Gravenhorst) (Coleoptera: Staphylinidae) discovered in South Africa : short communication. Sabinet Online Department of Zoology, Field Museum of Natural History, Chicago, Vol. 14. Issue 1, pp: 193 -194: http://www.sabinet.co.za/abstracts/ento/ento_v14_n1_a21.xml

Newton, Adrian C. 2007. Forest Ecology and Conservation. Techniques in Ecology and Conservation Series. Oxford University press. Sutherland, William J. (edit.). 470 pags.

Nicholas, G. 1960. Checklist of macroscopic troglobitic organisms of the United States. American Midland Naturalist. Vol. 64, pp: 123 – 60.

Nielsen, E. S. & Mound, L. A. 2000. Global diversity of insects: the problems of estimating numbers. Pp. 213–222. In: Raven, P.H. (Ed.), Nature and Human Society. The Quest for a Sustainable World. National Academy Press, Washington D.C. 644 pages.

Nieto, Ana and Keith N.A. Alexander 2010. European Red List of Saproxylic Beetles. IUCN Species Programme. Nieto, A. and Alexander, K.N.A. 2010. European Red List of Saproxylic Beetles. Luxembourg: Publications Office of the European Union. 44 pgs.

Nikitsky, N. B. 1974. Morphology of the larvae and mode of life of Nemosoma (Col., Trogossitidae)-Predators of bark beetles in the NW Caucasus. Zoologicheskii Zhurnal. Vol. 53, pp: 563–568.

Никитский, Я. Б. 1986. Жесткокрылые подсемейств Monotoiminae и Thioninae (Coleóptera, Rhizophagidae) дальнего востока ссср. 1986, том LXV, вып. 11. 8 Pages.

Nikitsky, N. B. 1992. The Trictenotomidae family - trictenotomids. In: Opredelitel' nasekomykh Dal'nego Vostoka SSSR. Handbook of insects of the Far East of the USSR. Vol. 3, part 2. St Petersburg: 493 Pages (in Russian).

Nikitsky, N. B. & Egorov, A. B. 1992. Fam. Ischaliidae, stat. n. – false fire-red beetles. Pp. 497–498 in: Ler, P. A. (ed.) Keys to the Insecta of the Far Eastern U.S.S.R. in Six Volumes. Vol. 3. Coleoptera or Beetles, Part 2. Nauka, St. Petersburg (in Russian).

Nikitsky, N. B. 1996. Fam. Ischaliidae – False Firered Beetles. Pp. 427–428 in: Ler, P. A. (ed.) Keys to the Insecta of the Far Eastern U.S.S.R. in Six Volumes. Vol. 3. Coleoptera or Beetles, Part 3. Dal'nauka, ladivostok (in Russian).

Nikitsky, N. B.1998. Generic Classification of the Beetle Family Tetratomidae (Coleoptera, Tenebrionoidea) of the World, with Description of New Taxa. 80 pp. Pensoft (Series Faunistica No. 9), Sofi a, Moscow.

Nilsson, A. N. 2001: Dytiscidae. World Catalogue of Insects. Vol. 3, pp: 1 - 395.

Nilsson, A. N. 2004. World Catalogue of Dytiscidae. corrections and additions, 2 (Coleoptera: Dytiscidae). Koleopterologische Rundschau. Vol.74, pp: 157 - 174.

Nilsson, A.N. & Fery H. 2006: World Catalogue of Dytiscidae. corrections and additions, 3 (Coleoptera: Dytiscidae). Koleopterologische Rundschau. Vol. 76, pp: 55 - 74.

Nikitsky, N. B. 2004 a. The beetles of the subfamily Tetratominae Billberg, 1820 (Coleoptera, Tetratomidae) of the world fauna. – Byulleten Moskovskogo Obshestva Ispytatelei Prirody, Otdel Biologicheskii. 109 Fasc.2, pp: 25–36 (in Russian).

Nikitsky, N. B. 2004 b. The beetles of the subfamily Piseninae Miyatake, 1960 (Coleoptera, Tetratomidae) of the world fauna. – Byulleten Moskovskogo Obshestva Ispytatelei Prirody, Otdel Biologicheskii 109. Fasc.4, pp: 3–7 (in Russian).

Nodoka, Hayashi 1969. On the larvae of some species of small families of Cucujoidea in Japan (Coleoptera). Insecta Matsumurana, Supplement 7, pp: 1- 16. Available online (2010) in: http://eprints.lib.hokudai.ac.jp/dspace/bitstream/2115/22233/1/7_P1-9.pdf

Norment, J. C. 1987. A Comparison of Three Methods for Measuring Arthropod Abundance in Tundra Habitats and its lmplications in Avian Ecology. Northwest Science. Vol. 6l No. 3. T 9 l, pp: 1- 8. Available in: www.research.wsulibs.wsu.edu:8443/jspui/bitstream/2376/1765/1/v61%20p191%20Norment.PDF

Noyes, J. S. 1989. A study of five methods of sampling Hymenoptera (Insecta) in a tropical rainforest, with special reference to the Parasitica. *Journal of Natural History. Vol.* 23, pp: 285–298.

Oakland, B. 1996. A comparison of three methods of trapping saproxylic beetles. Eur. Journal Entomologie.Vol. 93, pp: 195 - 209.

Oana, Moldovan, T. & Rajka, Géza 2007. Historical Biogeographyof subterranean beetles – "Plato's Cave" or scientific evidence?. Zgodovinska Biogeografija podzemeljskih hroščev – »Platonova Jama« ali znanstveni dokaz?. TIME in KARST, POSTOJNA, pp: 77-86. Available online (2010) in: http://carsologica.zrc-sazu.si/downloads/361/moldovan8.pdf

Oberprieler, R. G. Marvaldi, A.E. & Anderson, R. S. 2007. Weevils, weevils, weevils everywhere.Zootaxa 1668, pp: 491–520. Available online (2010) in: http://fossilinsects.net/pdfs/oberprieler_etal_2007_Zootaxa_WeevilsEverywhere.pdf

Ocampo F. C. 2002 A new species of *Hybochaetodus* from Peru. The Coleopterists Bulletin. Vol. 56. Fasc. 3, pp: 446-450.

Ocampo F. C. 2002 Hybosorids of the United States and Expanding distribution of the introduced species *Hybosorus illegeri* Annals of the entomological Society of America. Columbus 95(3):316-322.

Ocampo F. C. 2002 *Procoilodes adrastus* Ocampo, a new genus and species of Miocene Hybosorid in amber from the Dominican Republic. The Coleopterists Bulletin. Vol. 56. Fasc.1, pp: 123-126.

Ocampo F. C. 2006 Phylogenetic analysis of the scarab family Hybosoridae and monographic revision of the New World subfamily Anaidinae 3. Phylogenetic analysis of the subfamily Anaidinae Bulletin of the University of Nebraska State Museum 19:13-177.

OECD 2006. Safety assessment of transgenic organisms: OECD consensus documents, pag. 288. Volumen 1, pp: 1- 821.

Ohara, M. and Nakane, T. 1986. On the genus *Onthophilus* from Japan. Insecta Mastumurana (N.S.). Vol.35, pp: 1-15.

Öhara, M. 1994. A revision of the superfamily Histeroidea of Japan (Coleoptera). - Insecta Matsumurama (N.S.). Vol.51, pp: 1-283.

Ohara, M. 1994. A revision of the superfamily Histeroidea of Japan. Insecta Mastumurana (N.S.) Vol. 51, pp: 1-283.

Ôhira, Hitoo and Makihara Hiroshi 2007. A new *Hemiops*, genus new to Japan, species of Elateridae (Coleoptera) from Okinawa Island of the Ryukyu Islands, Japan. Bulletin of FFPRI), Vol.6, N°.4 (N°.405), pp: 227 - 230

O'Keefe, Sean T. O. 2002 Family 61. Phengodidae, pp: 181=186, in: Arnett, R.H., Jr., Thomas, M.C., S ., Skelley, P.E. & Howard, Frank, J. (Eds.). American Beetles, Volume 2: Polyphagha: Scarabaeoidea through Curculionoidea. CRC Press, Boca Raton, USA. 861 pages.

Økland, B. 1996. A comparison of three methods of trapping saproxylic beetles. Eur. J. Entomol. 93: 195-209.

Oldroyd, H. 1958. Collecting, preserving and studying insects. The Macmillan Company, New York, 327p.

Oliff, A. S. 1883. Description of two larvae and new genera and species of Clavicorn Coleoptera, and a synopsis of the genus *Helota*, MacLeay. Cistula Entomol. Vol. 3, pp: 49 - 61.

Oliva, Adriana 1981. El genus *Derallus* Sharp en la Argentina (Coleoptera, Hydrophilidae). Revista de la Sociedad Entomológica Argentina. Vol. 40. Fasc. 1 – 4, pp: 285 - 296.

Oliva, Adriana 1987. El género *Hemiosus* en los Andes meridionales y la Patagonia (Coleopt., Hydrophilidae). Revista de la Sociedad Entomológica Argentina. Vol.44. Fasc. 3-4. 1987 (1985), pp: 377 – 381.

Oliva, Adriana 1994. A revision of the genus *Hemiosus* Sharp, 1882 in South America (Coleoptera: Hydrophilidae). Bulletin et Annales de la Societé Royale Belge d' Entomologie. Vol.130. Fasc. (7-9), pp: 267 – 303.

Oliva, Adriana 2002. Especies de Berosinae (Coleoptera: Hydrophilidae) de Paraguay. Bol. Mus. Nac. Hist. Nat. Parag. Vol. 14. Fasc.: 1-2, pp: 30 – 35. Available online (2010) in: www.faunaparaguay.com/bmnhnpy2003oliva.pdf

Oliva, Adriana 2007. Distribución de Berosini (Coleoptera: Hydrophilidae) en América del Sur. Rev. Soc. Entomol. Argent. Vol. 66. Fasc. 3-4, pp: 47 – 56.

Oman, P. W. 1946. Collection and preservation of insects. U.S. Department of Agriculture, Milscellaneous Publication No. 601, 42p.

Opitz, W. 2002. Family 73. Cleridae. p. 267-280. *In*: R. H. Arnett, Jr., M. C. Thomas, P. E. Skelley, and Howard, J. Frank (eds.). American Beetles, Vol. 2. CRC Press; Boca Raton, FL. 861 pags.

Osella, G. & A. M. Zuppa. 2006. Otiorhynchus (Podonebistus) gasparoi n. sp., a blind weevil from Greece (Coleoptera, Curculionidae, Entiminae, Otiorhynchini). Revue Suisse de Zoologie. Vol. 113, pp: 67 – 75.

Osorio, M. and Cerda M. 1984. Icerya purchasi M. y otros

organismos asociados a *Ulex europaeus* L. y *Cytisus monspessulanus* L, .dos por el control químico. Bosque. Universidad Austral de Chile Facultad de ciencias forestales. Vol. 2. Fasc. 5, pp: 110- 114.

Otto, R. L. and Young, D. K. 1998.Description of the Larva of *Schizophilus subrufus* (Randall) (Coleoptera: Eucnemidae: Pseudomeninae), with Notes on Its Natural History. The Coleopterists Bulletin, Vol. 52, No. 4, pp: 306-312.

Owen, J. A. 1989. An emergence trap for insects breeding in dead wood. Br. J. Ent. Nat. Hist. 2: 65-67.

Owen, J. A. 1992. Experience with an emergence trap for insects breeding in dead wood. Br. J. Ent. Nat. Hist. 5: 17-20.

Ozanne, Claire M.P. 2005. Techniques and methods for sampling canopy insects. Chapter 7. Ppp: 146- 165 in: Leather et al 2005. Insect sampling in forest ecosystems. Blackwell, Malden, MA.

Özdikmen, Hüseyin and Semra, Turgut 2008. The Megalopodidae and Orsodacnidae of Turkey (Coleoptera: Chrysomeloidea) with zoogeographical remarks and a new record, *Zeugophora scutellaris* Suffrian, 1840. Mun. Entom. Zool. Vol. 3, No. 1, pp: 287– 290.

Paarmann, W., and Stork, N. E. 1987. Seasonality of ground beetles (Coleoptera: Carabidae) in the rain forests of N. Sulawesi (Indonesia). Insect Science and its Application. Vol. 8, pp: 483–487.

Pace, R. 1976. An exceptional endogeous beetle: Crowsoniella relicta n. gen. n. sp. of Archostemata Tetraphaleridae from central Italy. Boll. Mus. Civ. Stor. Nat. Verona. Vol. 2[1975] pp: 445–458.

Pagola Carte S. 2006a. Aiako Harria Parke Naturaleko Daphne cneorumen entomokomunitatea. Hurbilketa taxonomiko eta ekologikoa. Informe técnico para la Diputación Foral de Gipuzkoa.

Pagola Carte S. 2006b. Inventario y seguimiento de la entomofauna del hayedo de Oieleku (Oiartzun, Parque Natural de Aiako Harria) – Campaña 2006. Informe técnic para la Diputación Foral de Gipuzkoa.

Pagola, Carte S. 2007b. Inventario y seguimiento de la entomofauna del hayedo de Oieleku (Oiartzun, Parque Natural de Aiako Harria) – Campaña 2007. Informe técnico para la Diputación Foral de Gipuzkoa.

Pagola, C. S. 2007. Inventario y seguimiento de la entomofauna del hayedo de Oieleku (Oiartzun, Parque Natural de Aiako Harria) Campaña 2007.Gipuzkoako Entomologia Elkartea, Asoc. Gipuzkoana de Entom. Informe técnico para la Diputación Foral de Gipuzkoa. 108 Pages.

Pakaluk, J. 1984. Natural history and evolution of *Lycoperda ferrugiena* (Coleoptera: Endomychidae) with descriptions of immature stages. Proc. of the Entomological Society of Washington. Vol. 86. Fasc. 2. Pp312-325.

Pakaluk, J. 1987. Revision of the North American Rhymbomicrus Casey (Coleoptera, Endomychidae). Annals of the Entomological Society of America. Vol. 80, pp: 456-461.

Pakaluk, J and Slipinski, S. Adam 1993. A new genus and two new species of Neotropical Rhizophagidae (Coleoptera) from *Trigona* (Hymenoptera: Apidae) nests, with a review of rhizophagid biology. The Coleopterists Bulletin. Vol. 47. N° 4, pp: 349-358.

Pakaluk, J. & Ślipiński, S. A. 1995. Acritosomatinae, a new subfamily of Endomychidae for Acritosoma new genus based upon two new species from Mexico and Peru (Coleoptera, Cucujoidea). Genus. 6. Fasc.3-4, pp: 327-335.

Pakaluk, James; Ślipiński, S. A. and Crowson, R. A. 1995. Biology, phylogeny, and classification of Coleoptera: papers celebrating the 80th birthday of Roy A. Crowson, Vol.1. Edit: Muzeum i Instytut Zoologii PAN. 1092 pages.

Panshin, I. V. 1968. On the Theoretical Concept of Self-Extinction of Bark Beetles (O Teoreticheskoi Kontseptsii Samoudusheniya Koroedov). National Science Foundation, Washington, D. C. Special Foreign Currency Science Information Program. 2 Pages.

Parsons, C.T. 1943. A revision of nearctic Nitidulidae (Coleoptera). Bulletin of the Museum of Comparative Zoology, 92, pp: 121-278.

Paul, J. Johnson 1996/1997. South Dakota's High Plains Tiger (*Amblycheila cylindriformis* (Say)). Published in South Dakota Conservation Digest (March/April). 4 Pages. http://nathist.sdstate.edu/SMIRCOL/amblycheila.htm

Paulian, R. 1984. Les Orphnidae Americains (Coleopteres, Scarabaeoidea). Annals Société Entomologique de France (N.S.). Vol.20. Fasc.1, pp: 65 - 92.

PAULIAN, R. and J.-P. LUMARET. 1982. La larve des Orphnidae. Bulletin de la Société Entomologique de France 87: 263-272.

Paula, H. Valdujo, Cristiano Nogueira, Marcio Martins 2002. Ecology of *Bothrops neuwiedi pauloensis* (Serpentes: Viperidae: Crotalinae) in the Brazilian Cerrado. Journal of Herpetology Vol. 36. Fasc. 2, pp: 169 – 176.

Paulian, R. 1959. Un nouveau coléoptère scarabéide tipo de Madagascar d'une nouvelle sous-famille. Naturaliste Malgache. Vol. 10. Pp :39 - 46.

Paulian, R., 1980. Revision taxonomique des Liparochrus Erichson, genre Australo-Melanesien de Coleopteres Hybosoridae, Scarabaeioidea. Annales de la Société Entomologique de France (n.s.). Vol. 16, pp: 389–433.

Paulsen, M. J. 2007. Nomenclatural changes in the Nearctic Ochodaeinae and description of two new genera (Coleoptera: Scarabaeoidea: Ochodaeidae). Insecta Mundi. 0021, pp: 1-13.

Paulus, H. F. 1969. *Pseudochelonarium* (*Neochelonarium*) *kalimantanense* n. sp. aus Borneo, mit Bemerkungen zum System der Chelonariidae (Col. Dryopoidea). Zeitschrift der Arbeitsgemeinschaft der Österreichischen Entomologen.Vol. 21, pp: 105 -109.

Paulus, H. F. 1972. Die systematische und phylogenetische Stellung der Karumiidae, mit einer Beschreibung von *Escalerina serraticornis* n. sp. aus S-Persien (Ins: Coleoptera: Cantharoidea). Senckenbergiana Biologica. 53, pp: 37–54.

Payne, N. M. 1931. Food requirements for the pupation of two coleopterous larvae, *Synchroa punctata* Newn. and *Dendroides canadensis* LeC. (Melandryidae, Pyrochroidae). Entomological News. Vol.42, pp: 13-15.

Payne, J. A., and E. W. King. 1970. Coleoptera associated with pig carrion. Entomologist's Monthly Magazine. Vol.105, pp: 224-232.

Peacock, E. R. 1978: Notes on the subgenera of European Rhizophagidae (Coleoptera). Entomologica Scandinavica. Vol. 9, pp: 236 - 239.

Pearson, D. L. 1988. Biology of Tiger Beetles. Annual Review of Entomology. 33, pp: 123-147.

Pearson, D. L., C. B. Knisley and C. J. Kazilek. 2007. A field guide to the tiger beetles of the United States and Canada: Identification, natural history, and distribution of the Cicindelidae. Oxford Univ. Press, NY, 288 pages. Volume 11, N° 4.- pp: 1- 227.

Pechuman, L. L. 1937. An annotated list of the insects

found in the bark and wood of *Ulmus americana* L. in New York State. Bulletin of the Brooklyn Entomological Society. Vol. 32, pp: 8-21.

Peck Stewart B. 1975. A review of the distribution and habitats of North Amaerican Brathinus (Coleoptera; Staphylinidae; Omaliinae). Psyche. (March 17), pp: 59 – 66.

Peck Stewart B. and A. E. Davies. 1980. Collecting small beetles with large-area "window" traps. Coleopterists Bulletin. Vol. 34, pp: 237-239.

Peck Stewart B.; Kukalova-Peck & C. Bordón. 1989. Beetles (Coleoptera) of an Oil-Bird Cave: Cueva Del Guácharo, Venezuela. The Coleopterists Bulletin. Vol. 43. No. 2, pp: 151 – 156.

Peck Stewart B. 1992. A synopsis of the invertebrate cave fauna of Jamaica. National Speleological Society Bulletin. Vol. 54, pp: 37 - 60.

Peck Stewart B. and Cook J. 1992. Use of "car nets" to sample flying micro-Coleoptera. The Canadian Entomologist. Vol. 124, pp: 745-9.

Peck Stewart B. and M. C. Thomas 1998. A distributional checklist of the beetles (Coleoptera) of Florida. Arthropods of Florida and Neighboring Land Areas 16. Florida Department of Agriculture and Consumer Services, Gainesville. 180 pags.

Peck Stewart B. 1998. Phylogeny and evolution of subterranean and endogean Cholevinae (=Leiodidae, Cholevinae): an introduction. In Phylogeny and Evolution of Subterranean and Endogean Cholevidae (=Leiodidae Cholevinae), edited by P.M. Giachino & S.B.Peck, Turin: Museo Regionale di Scienze Naturali

Peck Stewart B. 1998. A summary of diversity and distribution of the obligate caveinhabiting faunas of the United States and Canada. Journal of Cave and Karst Studies. Vol. 60, pp: 18 – 26.

Peck Stewart B.; Amador, E. Ruiz-Baliú and Gabriel F. Garcés González 1998. The Cave-inhabiting Beetles of Cuba (Insecta: Coleoptera): Diversity, Distribution and Ecology. Journal of Cave and Karst Studies. Vol. 60. Fasc. 3, pp: 156 - 166.

Peck Stewart B. 2001. Review of carrion beetles of Australia and New Guinea (Coleoptera: Silphidae). Australian Journal of Entomology. Vol. 40, pp: 93 – 101.

Peck Stewart B. 2001. Silphidae Latreille, 1807. In: Arnett, R. H. Jr. and Thomas, M. C. (Eds.) American Beetles. Volume 1. CRC Press, Boca Raton, Florida. Pp. 268-71.

Peck Stewart B. 2001a: 18. Agyrtidae C. G. Thomson, 1859, pp: 247-249. In: Arnett R. H., Jr. & Thomas M.C. (eds): American beetles, Volume 1: Archostemata, Myxophaga, Adephaga, Polyphaga: Staphyliniformia. CRC Press, Boca Raton, London, New York, Washington, xi + 443 Pages.

Peck Stewart B. and Joyce Cook 2002. Systematics, distributions, and bionomics of the small carrion beetles (Coleoptera: Leiodidae: Cholevinae: Cholevini) of North America. The Canadian Entomologist. N° 134, pp: 723 – 787.

Peck Stewart B. 2005. The Beetles of the Galápagos Islands, Ecuador: Evolution, Ecology, and Diversity (Insecta: Coleoptera). A publication of the National Research Council of Canada Monograph Publishing Program. 312 pgs.

Peck Stewart B. 2006. The beetle fauna of Dominica, Lesser Antilles (Insecta: Coleoptera): Diversity and distribution. Insecta Mundi, Vol. 20, N° 3-4, pp: 165-207. Available online (2010) in: http://www.cerambycoidea.com/titles/peck2006.pdf

Peck Stewart B. 2010. *Derolathrus cavernicolus* n. sp., a Beetle Family New for North America (Coleoptera: Jacobsoniidae). Annals of the Entomological Society of America. Vol. 103. N° 1, pp: 1-6.

Pérez-Bilbao, A.; Benetti, C. J. & Garrido, J. 2009. Nuevas aportaciones al conocimiento de La familia Hydrochidae (Coleoptera) en Galicia (noroeste de España). Boletín Sociedad Entomológica Aragonesa, N° 45, pp: 241–244. Available online (2010) in: http://webs.uvigo.es/jgarrido/documentos/67-2009_Perez-Bilbao_et_al_Hydrochidae.pdf

Perkins, P. D. 1980. Aquatic beetles of the family Hydraenidae in the Western Hemisphere: Classification, Biogeography and Inferred Phylogeny (Insecta: Coleoptera). Quaestiones Entomologicae. Vol. 16, pp: 1-554.

Perkins, P. D. 2001. Hydraenidae. pp. 228-232 in Arnett, R.H., Jr., and M. C. Thomas. (eds.). American Beetles. Volume 1. Archostemata, Myxophaga, Adephaga, Polyphaga: Staphyliniformia. CRC Press LLC, Boca Raton, 443 pp.

Peter K. L. Ng and N. Sivasothi (editors) 2001. Volume 1: The Ecosystem and Plant Diversity and Volume 2: Animal Diversity Authors: Kelvin K. P. Lim, Dennis H. Murphy, T. Morgany, N. Sivasothi, Peter K. L. Ng, B. C. Soong, Hugh T. W. Tan, K. S. Tan & T. K. Tan BP. A Guide to Mangroves of Singapore. Guide to Nature Series published by the Singapore Science Centre, sponsored by British Petroleum Raffles Museum of Biodiversity Research, The National University of Singapore & The Singapore Science Centre.

Philips, J. R. 1979. Decomposition of raptor pellets. Raptor Research. Vol.13, pp: 102-1 1.

Philips, T. Keith 2001 A record of Micromalthus debilis (Coleoptera: Micromalthidae) from Central America and a discussion of its distribution. Scientific Notes. Florida Entomologist. Vol.84. Fasc.1 pp: 159-160.

Philips, T. Keith and Karen L. Bell 2009. Ptinidae Latreille, 1802, pp: 217-225. in: Handbook of Zoology, Bd. 4/2 - Arthropoda 2. Hälfte Insecta Coleoptera, Beetles, Pt. 39: Vol. 2: Morphology and Systematics (Elateroidea, Bostrichiformia, Cucujiformia partim).

Pinto, J. D.; Bologna, M. A. and Bouseman, J. K. 1996. First-instar larvae, courtship and oviposition in Eletica: amend-ing the definition of the Meloidae (Coleoptera: Tenebrionoidea). Systematic Entomology. Vol.21, pp: 63-74.

Pinto, J. D. & M. A. Bologna. 1999. The New Word genera of Meloidae (Coleoptera): a key and synopsis. J. Nat. Hist. Vol. 33, pp: 569-620.

Griffiths, Picker M, C, Weaving A 2002. Field guide to insects of South Africa. Struik, Cape Town S Afr.

Pollock, D. A. 1991. Natural history, classification, reconstructed phylogeny, and geographic history of *Pytho* Latreille (Coleoptera: Heteromera: Pythidae). Memoirs of the Entomological Society of Canada. Vol. 154, pp: 1 – 104.

Pollock, D. A. 1995. Classification, reconstructed phylogeny and geographical history of genera of Pilipalpinae (Coleoptera: Tenebrionoidea: Pyrochroidae). Invertebrate Taxonomy. Vol. 9. N° 4-6, pp: 563-708.

Pollock, D. A. & Normark B.B. 2002. The life cycle of *Micromalthus debilis* LeConte (Coleoptera: Archostemata: Micromalthidae): historical review and evolutionary perspective. J. Zool. Syst. Evol. Res. Vol. 40. PP: 105 – 112.

Pollock, D. A. 2002. 113. BORIDAE C. G. Thomson 1859,

pp: 534-536, in: Arnett, R. H. Jr.; M. C. Thomas, P. E. Skelley and Howard, J. Frank (Eds.). 2002. American Beetles. Vol. 2: Polyphaga: Scarabaeoidea through Curculionoidea. CRC Press, Boca Raton. XVI + 861 Pages.

Pollock, D. A., and M.A. Ivie. 2004. Description of the larva of *Phaeogala rufa* Abdullah (Coleoptera: Mycteridae: Lacconotinae), with nomenclatural notes for the genus *Phaeogala*. Proceedings of the Entomological Society of Washington. Vol. 106. Fasc.3, pp: 697 - 704.

Pond Action 1994. The national pond survey. 1989-1993. Interim Report to the World Wide Fund for Nature. Pond Action. 64 pages.

Portevin, G. 1929 Leiodidae, Clambidae. Coleopt. Catal. Vol. 8. Fasc.103, pp: 1-32.

Portevin, G. 1937 - Liodides nouveaux des collections du Muséum. Rev. Fr. Ent., Vol. 4, pp: 31-36, 12 figs.

Poulson, T. L.; Culver, D.C. 1968. Diversity in terrestrial cave communities. Ecology. Vol. 50. Fasc. 1, pp: 153 - 158.

Price, Michele B. 2002. Family 78. Smicripidae Horn 1879, pp: 316-318. In: Arnett, R.H., Jr., Thomas, M.C., S ., Skelley, P.E. & Howard, Frank, J. (Eds.). American Beetles, Volume 2: Polyphagha: Scarabaeoidea through Curculionoidea. CRC Press, Boca Raton, USA. 861 pages.

Price, M. B., and D. K. Young. 2007. An annotated checklist of Wisconsin handsome fungus beetles (Coleoptera: Endomychidae). The Great Lakes Entomologist 40, pp: 177-188.

Prieto, Fernando Piloña & Pérez, Javier Valcárcel 2002. Catálogo de los Silphidae y Agyrtidae (Coleoptera) de la Península Ibérica e Islas Baleares. Bol. S.E.A.N°30, pp: 1—32. Available online (2010) in: http://www.sea-entomologia.org/PDF/BOLETIN_30/B30-001-001.pdf

Pringle, J. A. 1938. A contribution to the knowledge of *Micromalthus debilis* LeC. (Coleoptera). Trans. R. Entomol. Soc. London. Vol. 87, pp: 271 – 287.

Proshchalykin, M. Yu. 2008. The bees of family Colletidae (Hymenoptera, Apoidea) of Transbaikalia. Far Eastern Entomologist. N° 187, pp: 1 – 9.

Quinn, Joyce A. 2008. Arctic and alpine Biomes. Greenwood Guides to Biomes of the World. 228 pages.

Raffray 1873. Revue et magasin de Zoologie. Vol. 36, 3ª Série, 1. Page 376

Rakovitza, E. G. 1907. Essai sur les problémes biospéologiques. Arch. zool. exp. et gen. Vol. 6, pp: 371 - 488.

Ratcliffe, B. C. 1980. New species of Coprini (Coleoptera: Scarabaeidae: Scarabaeinae) taken from the pellage of three toed sloths (*Bradypus tridactylus* L.) (Edentata: Bradypodidae) in Central Amazonia with a brief commentary on scarab-sloth relationships. The Coleopterists Bulletin. Vol. 34. Fasc. 4, pp: 337–350.

Ratcliffe, B. C. 1984. A review of the Penichrolucaninae with analyses of phylogeny and biogeography, and description of a second New World species from the Amazon Basin (Coleoptera: Lucanidae). Quaestiones entomologicae. Vol. 20, pp: 60-87.

Ratti, E. 2007. I coleotteri silvanidi in Italia (Coleoptera Cucujoidea Silvanidae). Boll. Mus. civ. St. natur. Venezia. Vol. 58, pp: 83-137.

Reddell, J. R. 1994. The cave fauna of Texas with special reference to the Eastern Edwards Plateau. Pp. 31–50, in The Caves and Karst of Texas (W. R. Elliott and G. Veni, eds.). National Speleological Society, Huntsville, Alabama. 252 pgs.

Reddell, J. R., and G. Veni. 1996. Biology of the Chiquibul cave system, Belize and Guatemala. Journal of Cave and Karst Studies. 58, pp: 131-138.

Reichardt, A. N. 1941. Sem.Sphaeritidae i Histeridae (chast 1). In: Fauna SSSR. Nasekoye zhestkokrylye. V.3. xiii+419 pags. Moskva-Leningrad.

Reichard, T. H. 1973. A critical study of the suborder Myxophaga, with a taxonomic revision of the Brazilian Torridincolidae and Hydroscaphidae (Coleoptera). Arquivos de Zoologia (São Paulo). Vol.24, pp: 73-162.

Reichardt, T. H. 1977. A synopsis of the genera of Neotropical Carabidae (Insecta: Coleoptera). Quaest. Entomol. Vol. 13, pp: 346–493.

Reichensperger, A. 1938 - Südamerikanische Paussiden (Co1.) und einige Vorbemerkungen. Rev. Entom. Vol. 8, pp: 68-79, 18 figs., 1 est.

Reichensperger, A. 1949 - Beitrag zur Kenntnis der amerikanischen Paussiden. Rev. Entom. Vol.20, pp: 89-92, 4 figs.

Reid, C. A. M. 2000. Spilopyrinae Chapuis: a new subfamily in the Chrysomelidae and its systematic placement (Coleoptera). Invertebr. Taxon. Vol.14. PP: 837 - 862.

Rheinheimer, J. 2004. Illustrierter Katalog und Bibliographie der Anthribidae der Welt (Insecta: Coleoptera). Mitteilungen des Entomologischen Vereins Stuttgart, 39. Pp; 3–242.

Reitter, E. 1876. Revision der *Philothermus*-Arten. Deutsche Ent. Zeitschr. Vol. 20, pp: 301 - 303.

Reling, D. & Taylor, R. A. J. 1984. A collapsible tow net used for sampling arthropods by airplane. Journal of Economic Entomology. Vol. 77 pp: 1615–1617.

Rembialkowska, E. 1982. Energy balance of the developmental period of Geotrupes stercorosus Scriba (Scarabaeidae, Coleoptera). Ekol. Polska. 30, pp: 393-427.

Ren, Dong And Tan, Jingjing 2006. A new cupedid genus (Coleoptera: Archostemata: Cupedidae) from Jehol Biota of Western Liaoning, China. Anna L E S Zoologi C I (Warszawa). Vol.36.Fasc.3, pp: 457-464.

Reyes-Castillo, P. 1970. Coleoptera: Passalidae; morfología y división en grandes Grupos; genera americanos. Folia Entomologica Mexicana. Vol. 20 Fasc.22, pp: 1 - 240.

Reyes-Castillo, P. and G. halffter 1984. La estructura Social de los Passalidae (Coleoptera: Lamellicornia). Folia Entomologica Mexicana. Vol. 61, pp: 49 - 72.

Ribera, I. & Foster, G.N. 1992. Uso de coleópteros acuáticos como indicadores biológicos (Coleoptera). Elytron. Vol. 6, pp: 61 - 75.

Ribera, I. & Hernando, C. 1999. Taxonomic revision of the genus Platypelochares Champion (Coleoptera: Limnichidae). Koleopterologische Rundschau. Vol. 69, pp: 103–110.

Ribera, I., Garth, N. Foster& Alfred P. Vogler. 2003. Does habitat use explain large scale species richness patterns of aquatic beetles in Europe?. Ecography. Vol. 26, pp: 145 - 152.

Ribera, R. G. Beutel, M. Balke, and A. P. Vogler, 2002. Discovery of Aspidytidae, a new family of aquatic Coleoptera. Proc.The Royal Society. London, pp: 1-6.

Rich, Catherine and Longcore, Travis 2006. Ecological Consequences *of* Artificial Night Lighting. Island Press. 478 pags.

Richard, E. W. 1983. A field Guide to the Beetles of North America. The Peterson Field Guide Series. 29. Houghton Mefflin Company Boston. 368 Pages. 12 color plates.

Richards, P. W. 1952. The tropical rain forest: an ecological study. Cambridge University Press, Cambridge, England. 450 pages.

Rieske, L. K. & Raffa, K. F. 1993. Potential use of baited pitfall traps in monitoring pine root weevil, Hylobius picivorus, and Hylobius radicis (Coleoptera: Curculionidae) populations and infestation levels. Forest Entomology. Vol. 86, pp: 475–485.

Rifkind, J. 1997. *Enoclerus (Enoclerus) insidiosus* (Gorham) (Coleoptera: Cleridae): another likely mimic of *Camponotus sericeiventris* Guerin (Hymenoptera: Formicidae). Coleopterists Bulletin. Vol. 51, pp: 298-302.

Ritcher, Paul O. 1958. Biology of Scarabaeidae. Annual Reviews.Entomol. Vol. 3, pp: 311-334.

Ritcher, P. O. 1966. White Grubs and Their Allies: A Study of North American Scarabaeoid Larvae. Oregon State University Press, Corvallis. 219 pp.

Riveros, Fernando (sf) El Gran Chaco. Documento en línea del Departamento de Agricultura de la FAO.Available online (2010) in: http://www.fao.org/ag/AGP/AGPC/doc/Bulletin/GranChaco.htm

Roberts, A. W. R. 1958. «On the taxonomy of Erotylidae (Coleoptera), with special reference to the morphological characters of the larvae. II» Transactions of the Royal Entomological Society of London. Vol. 110, pp: 245 - 285.

Roberts, H. 1969. A note on the Nigerian species of the genus *Sosylus* Erichson (Col. Colydiidae), parasites and predators of ambrosia beetles. Journal of Natural History. Vol. 3, pp: 85 - 91.

Robertson, J. A.; J. V. McHugh y Whiting, M. F. 2004. «A molecular phylogenetic analysis of the pleasing fungus beetles (Coleoptera: Erotylidae): evolution of colour patterns, gregariousness and mycophagy», en Systematic Entomology. Vol. 29, N° 2, pp: 173-?

Rocío García, C., Patricia Chacón De Ulloa 2005. Rove beetles (Coleoptera Staphyiinidae) in dry forest fragments of the Cauca river valley. Rev. Colomb. Entomol. Vol. 31. N° 1, pp: 43-50.

Roig - Juñent, S. and Flores, G. E., 2001. Historia biogeográfica de las áreas áridas de América del Sur austral. In: Morrone, J.J., Llorente Bousquets, J. (Eds.), Introducción a la Biogeografía en Latinoamérica: Teorías, Conceptos, Métodos y Aplicaciones. Las Prensas de Ciencias, Facultad de Ciencias, UNAM, México, D.F. Pp. 257–266. Available in: http://tux.uis.edu.co/labsist/docencia/20070_2009/informes/informe_II_biogeografia/2041379-20070.pdf

Roig-Juñent, S. & Domínguez, M. C. 2001. Diversity of the family Carabidae (Coleoptera) in Chile. Revista Chilena de Historia Natural.. Vol.74 N°. 3.

Roig-Juñent; M. C. Domínguez; Flores, G. E.; Mattoni, C. 2006. Biogeographic History of South American arid lands: A view from its arthropods using TASS analysis. Journal of Arid Environments. Vol. 66, pp: 404 – 420. Available in: http://research.amnh.org/users/lorenzo/PDF/Roig.2006.JAE.TASS.pdf

Roig-Juñent, Sergio and Lagos, S. 2008. Karumiinae in Biodiversidad de Artrópodoa argentinos. Edit. Sociedad Entomologica Argentina, pp: 559-562. Available online (2010) in: http://www.conicet.gov.ar/scp/vista_resumen.php?produccion=327320&id=2272&keywords=arid

Rojas, P. E. y Gallardo, V. R. 2004. Manual de insectos asociados a maderas en la zona sur de Chile. Servicio Agricola y Ganadero Division Proteccion Agricola Proyecto Vigilancia y Control de Plagas Forestales. Gobierno de Chile, Ministerio de Agricultura S. A. G, pp: 1 - 64. Available online (2010) in: http://www.sag.gob.cl/common/asp/pagAtachadorVisualizador.asp?argCryptedData=GP1TkTXdhRJAS2Wp3v88hCvF%2BQ3IGjCv&argModo=&argOrigen=BD&argFlagYaGrabados=&argArchivoId=3843

Romero, L. S.; Ochoa, P. R.; Iturrondobeitia, J. C.; Lafuente, A. G. 2007. Los escolítidos de las coníferas del País Vasco. Guía práctica para su identificación y control). 1ª ed. Vitoria-Gasteiz: Eusko Jaurlaritzaren Argitalpen Zerbitzu Nagusia = Servicio Central de Publicaciones del Gobierno Vasco. Colección Lur. N° 11. 189 Pages.

Rosado- Hernández, M. C. y C. Deloya 2002. Interacción entre trogidos (Coleoptera: Trogidae) y tortugas marinas (Reptilia: Cheloniidae) en el Pacífico mexicano. Acta Zoológica mexicana (n.s). Vol. 87, pp: 29 – 46.

Rosenbauer, Tom 1999. The Orvis Guide to Reading Trout Streams (orvis Guides). Publisher: Lyons Press (Paperback). 176 pages.

Ross, H. Arnett & J. R. Arnett. 1973. The Beetles of the United States. American Entomological Institute, pp. 747-751.

Ross, H. Arnett and Thomas, Michael Charles 2001. American Beetles: Volume 1. CRC Press. 464 Pages.

Rücker, W. H. 1992. Latridiidae. In: Lohse, G. A. & Lucht, W. H. (eds.). Die Käfer Mitteleuropas, Supplementband mit Katalogteil. 2. Goecke & Evers, Krefeld, pp: 139 - 160.

Rücker, W. 2003. Holoparamecus gabrielae sp. nov., eine neue Merophysiinae aus México (Coleoptera, Endomychidae). Mitteilungsblatt für Systematik und Taxonomie der Latridiidae. 1. Page 10.

Rücker, W. 2008. Latridiidae of the world (Latridiidae, Merophysiidae, Dasyceridae). Available online (2010) : http://www.latridiidae.de

Rundel, P. W. 1978. Ecological relationships of desert fog zone lichens. The Bryologist. Vol. 81. Fasc. 2, pp: 277 - 93. Available in: http://www.ucm.es/BUCM/tesis/far/ucm-t%2026155.pdf

Rutanen, T. and J. Muona 1982. Coleoptera collected with a car net in Finland. Notulae Entomologicae. Vol. 62. P. 72.

Ruzicka, J. & Schneider, J. 1995. A new species of *Apteroloma* from the Far East, and new faunistic records on Palaearctic Agyrtidae (Coleoptera). Entomological Problems. Vol 26, pp: 111-115.

Růžička, Jan & Jan Schneider 2003. Interesting distributional records of Agyrtidae and Silphidae (Coleoptera) from the Palaearctic and Oriental regions. Klapalekiana. Vol. 39, pp: 307 - 311.

Roig-Juñent, S; Domínguez, M. C.; Flores, G.E.; Mattoni, C. 2006. Biogeographic History of South American arid lands: A view from its arthropods using TASS analysis. Journal of Arid Environments. Vol. 66, pp: 404–420.

SA2000 (The 27 Committees of Systematics Agenda 2000). 1994. Systematics Agenda 2000: Charting the Biosphere:Technical Report, pp: 1- 34.

Sabrosky, C. W. 1934. Notes on the larvae and larval habits of *Isohydnocera curtipennis* (Newman) (Coleoptera: Cleridae). Journal of the Kansas Entomological Society. Vol. 7, pp: 65-68.

Sackmann, Paula 2006. Effects of temporal variation and trapping methods on the efficiency of Coleoptera sampling in Loma del Medio Nature Reserve, El Bolsón, Río Negro. Laboratorio Ecotono CRUB-Universidad

Nacional del Comahue/CONICET. Rev. Soc. Entomol. Argent. Vol. 65. Fasc. 3- 4, pp: 35 – 50.

Sáiz, F.; Solervicens, J. & Ojeda, P. 1989. Coleópteros del Parque Nacional La Campana y Chile Central. Ediciones Universitarias de Valparaíso, Universidad Católica de Valparaíso, Chile. 124 pag.

Sakchoowong, W.; Jaitrong, W.; Ogata; Nomura, K. and Chanpaisaeng, J. 2008. Diversity of Soil-Litter Insects: Comparison of the Pselaphine Beetles(Coleoptera: Staphylinidae: Pselaphinae) and the Ground Ants (Hymenoptera: Formicidae). Thai Journal of Agricultural Science. N°41. Fasc.1-2, pp: 11-18. Available online (2010) in: http://www.thaiagj.org/files/41-1/02-TJ-AGR-0208-04.pdf

Salas-Araiza, M. D. y Boradonenko, A. 2006. Insectos Asociados al Amaranto *Amaranthus hypocondriacus* L. (Amaranthaceae) en Irapuato, Guanajuato, México. Acta Universitaria, vol. 16, número 001. Pp. 50-55

Sánchez, B. D.; Macias S. J. E.; Nepatami, M. R.; Cortés L. J. 2008. Respuesta Kairomonal de coleópteros asociados a *Dendroctonus frontalis* y dos species de *Ips* (Coleoptera: Curculionidae) en bosques de Chiapas, México. Revista Mexicana de Biodiversidad. Vol. 79. N° 001. Pp. 175-183.

Sarmiento, Fausto O. 2001. Diccionario de Ecología de paisajes, conservación y desarrollo sustentable para Latinoamérica. Editorial Abya-Yala. Ecuador.361 pages.

Sasaji, H. 1978. On the larva of a predaceous endomychid, *Saula japonica* Gorham (Coleoptera). Kontyû, Vol. 46. Fasc.1, pp: 24 – 28.

Satô, M. 1966. The Limnichid beetles of Japan. Transactions of the Shikoku Entomological Society. Vol. 9, pp: 55–62.

Satô, M. 1982. Discovery of Torridincolidae (Coleoptera) in Japan. Annotationes Zoologicae Japonenses 55, pp: 276-283.

Sawada, H. and Wiesner, J. 2000. The Tiger Beetles of Paraguay (Insecta: Coleoptera: Cicindelidae). 67th Contribution towards the knowledge of Cicindelidae. Coleoptera 4, pp: 285-325.

Schabel, Hans G. 2006. Forest Entomology in East Africa Forest Insects of Tanzania. Springer Science & Business Media.387 pags.

Schaefer, L. 1949. Les Buprestides de France. En: Miscellanea Entomologica. Tableaux analytiques des Coléoptères de la faune franco-rhénane. Famille LVI. Editions scientifiques du cabinet entomologique E. Le Moult. Paris. 511 pags. XXV.Planches.

Sharp, D. 1876. On the Colydiidœ of New Zealand. Ánnals and Magazine of Natural History. Art. LI, pp: 890 - 401.

Schawaller, W. 1991. *Prostomidae* (Coleoptera) of Hymalaya mit einem Beitrag zur Larvenmorpholoigie. Stuttgs. Beitr. Naturk. (A) Vol. 461, pp: 1 – 17.

Schawaller, W. 1992. The genus *Prostomis* (Coleoptera: Prostomidae) Sudost Asien. Rev. Suisse Zool. Vol. 99, pp: 255 – 262.

Schawaller, W. 1993. The genus *Prostomis* (Coleoptera: Prostomidae) in Australia and adjacent regions. Stuttgs. Beitr. Naturk. (A) Vol. 489, pp: 1 – 12.

Schawaller, W. 1994. A new species of *Prostomis* Latreille from New Caledonia. Doriana. Vol. 6, pp: 1 – 3.

Schigel, D. S. 2008. Collecting and rearing fungivorous Coleoptera. Rev. Écol. (Terre Vie). Vol.63, pp: 7 – 12.

Schmitz, Helmut; Anke, Schmitz; Stefan, Trenner & Horst, Bleckmann 2002. A new type of insect infrared organ of low thermal mass. Naturwissenschaften. Vol. 89, pp: 226 - 229.

Scholtz, C. H. and Evans, A. V. 1987. A revision of the African Ochodaeidae (Coleoptera: Scarabaeoidea). Journal of the Entomological Society of Southern Africa. Vol. 50, pp: 399-426.

Scholtz, C. H., D. D'Hotman; Evans, A. V. and Nel, A. 1988. Phylogeny and systematics of the Ochodaeidae (Insecta: Coleoptera: Scarabaeoidea). Journal of the Entomological Society of South Africa. Vol. 51, pp: 207-240.

Scholtz, C. H. and D. J. Browne. 1996. Polyphyly in the Geotrupidae (Coleoptera: Scarabaeoidea): a case for a new family. Journal of Natural History 30, pp: 597-614.

Schunger, L.; Beutel, R. G. and Britz, R. 2003. Morphology of inmature stages of *Prostomis mandibularis* (Coleoptera: Tenebrionoidea: Prostomidae). Eur. Journ. Entomol. Vol. 100, pp: 357 – 370.

Schuster, J. C.; Cano, E. y Cardona, C. 2000. Un método sencillo para priorizar la conservación de los bosques nubosos de Guatemala, usando Passalidae (Coleoptera) como organismos indicadores.Acta Zoologica Mexicana (Nueva Serie).N° 080, pp: 187-209.

Scott, A. C. 1938. Paedogenesis in the Coleoptera. Z. Morph. Ökol. Tiere 33, pp: 633 – 653.

Scott, H. 1920. Notes on the biology of some inquilines and parasites in a nest of *Bombus derhamellicus* Kirby, with a description of the larva and the pupa of Eouraea depressa Illig. (= aestiva auct.). Transactions of the Entomological Society of London, 1920, pp: 99-127.

Scullen, H. A., and Wold, J. L. 1969. Biology of wasps of the tribe Cercerini, with a list of Coleoptera used as prey. Ann. Ent. Soc. Amer. Vol. 62, pp: 209-214.

Sean, T. O'Keefe 2002. "Phengodidae", in Ross H. Arnett, Jr. and Michael C. Thomas, American Beetles (CRC Press). Vol. 2.

Seeno, T. N. & Wilcox, J. A. 1982. Leaf beetle genera (Coleoptera, Chrysomelidae), in Eichlin, T. D. & Papp, C. S. (eds.) Entomography, An annual review for biosystematics. Sacramento, California: Entomography Publications. Vol.1, pp: 1-221.

Seevers, C. H. & H. S. Dybas 1943 - A synopsis of the Limulodidae (Coleoptera). A new family proposed for myrmecophiles of the subfamilies Limulodinae (Ptiliidae) and Cephaloplectinae (Staphylinidae). Ann. Ent. Soc. Amer. Vol. 36, pp: 546-586, 5 ests. (41 figs.).

Selander, R. B. 1981a. Evidence for a third type of larval prey in blister beetles (Coleoptera: Meloidae). J. Kans. Entomol. Soc. Vol.54, pp: 757-783.

Selander, Richard B. and Fasulo, T. R. 2003. Blister Beetles (Insecta: Coleoptera: Meloidae). EENY-166 (originally published as DPI Entomology Circular 268), pp: 1-9. Available online (2010) in: http://entomology.ifas.ufl.edu/creatures

Sen, Gupta T. and Crowson, R. A. 1966. A new family of cucujoid beetles, based on six Australian and one New Zealand genera. Annals and Magazine of Natural History. Vol.13.pp: 61 – 85.

Sen, Gupta T. 1968. Review of the genera of the tribe Loberini Coleoptera: Languriidae). Proceedings of the Royal Entomological Society. N° 303. 27 pages. Available online (2010) in: http://www.archive.org/stream/breviora296325harv/breviora296325harv_djvu.txt

Sen, Gupta T. & Crowson, R. A. 1969. On a new family of clavicornia (Coleoptera) and a new genus of Languriidae. Proceedings of the Royal Entomological Society,

London. Vol. 38, pp: 125–131.

SenGupta, T. and R. A. Crowson. 1969a. Further observations on the family Boganiidae with definition of two new families Cavognathidae and Phloeostichidae. Journal of Natural History 3, pp: 371-590.

Sen, Gupta T. and Crowson, R. A. 1971. «A review of the classification of the family Languriidae (Coleoptera: Clavicornia) and the place of Languriidae in the natural system of Clavicornia», in Memoirs of the Zoological Survey of India. Vol. 15, N° 2, pp: 1-42

Sharp, D. 1882a. On aquatic carnivorous Coleoptera or Dytiscidae. Scientific Transactions of the Royal Dublin Society Vol. 2, Fasc. 2, pp: 179 – 1003.

Shavrin, A. V. 2007. The Rove Beetles (Coleoptera, Staphylinidae) of Pribaikalsky National Park. (Рябцев В. В. Труд Прибайкальского национального парка. Выпуск 2.). Availabale online (2010) in: http://www.zin.ru/animalia/coleoptera/pdf/Shavrin_2007.pdf

Shepard, W. D. 2002. Family 45. Lutrochidae Kasap and Crowson 1975, pp: 123-124; Family 46. Limnichidae Erichson 1846, pp: 125-126; Family 48. Psephenidae Lacordaire 1854, pp: 133-134. Family 51. Eulichadidae Crowson 1973 pp: 142-143. In: Arnett, R.H., Jr., Thomas, M.C., S ., Skelley, P.E. & Howard, Frank, J. (Eds.). American Beetles, Volume 2: Polyphagha: Scarabaeoidea through Curculionoidea. CRC Press, Boca Raton, USA. 861 pages.

Shepard, W. D.; Roughley, R. E. and Porras, Wendy. 2005. The natural history of *Lepicerus inaequalis* MOtschulsky (Coleoptera: Myxophaga: Lepiceridae) in Costa Rica, and additional morphological descriptions. Folia Entomol. Méx. Vol. 44 (Supl. 1), pp: 97 - 105.

Shepard, W. D. & Aguilar, J. C. 2010. Estudio Preliminar de las Familias de escarabajos acuáticos Dryopidae, Elmidae, Lutrochidae y Psephenidae conocidos de Paraguay (Coleoptera: Byrrhoidea). Boletin del Museo Nacional de Historia Natural de Paraguay. Vol. 16. Fasc.1, pp: 30 - 42: Available online (2010) in: http://www.4shared.com/document/y5JlX-O ... 2010.html

Shockley, F. W.; Tomaszewska, K. W. & Mchugh, J.V. 2009. An annotated checklist of the handsome fungus beetles of the world (Coleoptera: Cucujoidea: Endomychidae). Zootaxa. N° 1999. Magnolia Press. 113 Pages. Introduction available (in part) online (2010) in: http://www.mapress.com/zootaxa/2009/f/z01999p113f.pdf

Shockley, Floyd W. 2009. Endomychidae. Handsome fungus beetles. Version 31 March 2009. http://tolweb.org/Endomychidae/9169/2009.03.31 in The Tree of Life Web Project, available online (2010) in: http://tolweb.org/

Shockley, Floyd W. and Cline, Andrew R. 2009. Discolomatidae Horn 1878. Discolomatidbeetles, Discolomid beetles, Mexican hat beetles. Version 21 September 2009.

Shockley, F. W.;Tomaszewskay, K. W. and McHugh, J. V. 2009. Review of the Natural History of the Handsome Fungus Beetles (Coleoptera: Cucujoidea: Endomychidae). Center for Systematic Entomology, Gainesville, Florida. Insecta Mundi. N° 0072, pp: 1-24. Available online (2010) in: http://digitalcommons.unl.edu/cgi/viewcontent.cgi?article=1599&context=insectamundi

Short, A. E. Z.: Hydrophiloidea.org Facilitating research on water scavenger beetles: available online (2010) in: www.hydrophiloidea.org/home.html.

Short, A. E. Z. & Hebauer, F. 2006. World Catalogue of Hydrophiloidea. Additions and corrections. I (1999-2005). Koleopterplogische Rundschau. Vol. 76, pp: 315 - 359.

Shubeck, P. P. 1970. Silphidae attraction to carrion-baited air cans versus carrion-baited ground cans. The Coleopterists Bulletin. Vol. 24, pp: 66-70.

Shubeck, P. P. 1971. Diel periodicities of certain carrion beetles (Coleoptera: Silphidae). The Coleopterists Bulletin. Vol. 25, pp: 41-46.

Shubeck, P. P. 1976. Carrion beetle responses to poikilotherm and homoiotherm carrion (Coleoptera: Silphidae). Entomological News. Vol. 87, pp: 265-269.

Siitonen, J. 1994. Decaying wood and saproxylic Coleoptera in two old spruce forests: a comparison based on two sampling methods. Ann. Zool. Fennici 31: 89-95.

Sikes, D. S., and Raithel, C. J. 2002. A review of hypotheses of decline of the endangered American burying beetle (Silphidae: Nicrophorus americanus Olivier). Journal of Insect Conservation. Vol. 6, pp: 103-113.

Sikes, Derek S. 2005. Silphidae Latreille, 1807. in: Handbook of Zoology, Hexapoda (Kristensen, N. P. & Beutel, R. G. eds.), Coleoptera I (Beutel, R. G. & Leschen, R. A. B. eds.). De Gruyter, Berlin, pp: 288 – 296.

Silvey, J. R. G. 1935. An investigation of the burrowing inner-beach insects of some fresh water lakes. Papers of the Michigan Academy of Science. Arts and Letters, 21, pp: 655-696.

Silvia I. R.; James, F. P. and Cantliffe, D. J. 2004. Sap Beetle (Coleoptera: Nitidulidae) Management in Strawberries. University of Florida IFAS extension pp: 1- 4.

Similä, M.; Kouki, J.; Martikainen, P. 2003. Saproxylic beetles in managed and seminatural Scots pine forests: quality of dead wood matters. Forest Ecolog. Manag. Vol. 174, pp: 365 – 381.

Simon, H. R. 1996. Die artenreichsten Insekten-Ordnungen. Teil 2: Anzahl rezenter Arten mit Anmerkungen zur ökologischen Signifikanz. Entomologische Zeitschrift. 106, pp: 5–29.

Simmons, P. and G. W. Ellington. 1925. The ham beetle, *Necrobia rufipes* DeGeer. Journal of Agricultural Research. Vol. 30, pp: 845-863.

Siitonen, J. 2001. Forest management, coarse woody debris and saproxylic organisms: Fennoscandian boreal forests as an example. Ecological Bulletins. Vol. 49, pp: 11–41.

Singh, Surendra 2006. Geography For The Upsc Civil Services Preliminary Examination. Tata McGraw-Hill (edit.). 900 pages.

Skelley, Paul E. 1999. Pleasing Fungus Beetles, Pseudischyrus, Tritoma, Megalodacne, Ischyrus spp. (Insecta: Coleoptera: Erotylidae). University of Florida IFAS Extension. DPI Entomology Circular No. 313, pp: 1-3.

Slawomir Mazur and Masahiro Ohara 2003. A Revision Of the subfamily Saprininae from Thailand (Coleoptera: Histeridae). Hokkaido University. New Series 60, pp: 1-30.

Slipinski, S. A. 1998a. Revisión y filogenia de Protocucujidae (Coleoptera, Cucujoidea).Annales Zoologici (Warszawa). Vol. 48, pp: 273 - 296.

Slipinski, S. A., and S. Mazur. 1999. *Epuraeosoma*, a new genus of Histerinae and phylogeny of the family Histeridae. Annales Zoologici (Warszawa). Vol.49, pp: 209-230.

Slipinski, A. 2007. Silvanidae. En: Audisio P (Ed.). Fauna Europaea: Coleoptera. Fauna Europaea, versión 1.3 in: http://www.faunaeur.org.

Ślipiński, A. 2007. Australian ladybird beetles (Coleoptera: Coccinellidae). Their biology and classification. Australian Biological Resources Study, Canberra.

Ślipiński, A.; Tomaszewska, W.; Lawrence, J.F. 2009. Phylogeny and classification of Corylophidae (Coleoptera: Cucujoidea) with descriptions of new genera and larvae. Systematic entomology. Vol.34, pp: 409-433

Sloof, R. and Marbs, E. N. 1965. Mosquitoes (culicadae) biting a fish Periophthalmidae). Journ. Medic. Entom. Vol. 2. Pag: 16.

Smith, Andrew and Génier Francois 2001. Revision of the genus Holocephalus (Coleoptera: Scarabaeidae: Scarabaeinae: Coprini). The canadian Entomologist. Vol.133, pp: 777-791.

Solís, Ángel 2001. Métodos y técnicas de recolecta para Coleópteros Scarabaeoideos. Documento del Instituto de Biodiversidad de Costa Rica, Sto Domingo, pp: 1 – 13; Available online (2010) : http://inbio.ac.cr/papers/meto-col-scarabaeoidea/metoscar.pdf.

Sømme, L., R.; Davidson, L. ; Onore, G. 1996. Adaptations of insects at high altitudes of Chimborazo, Ecuador. European Journal of Entomology. Vol. 93, pp: 313 – 318.

Soto, A.; Orengo, L.; Estrela, A. 2002. Estudio de poblaciones de insectos escolítidos (Coleoptera: Scolytidae) en las masas de *Pinus halepensis* Miller del Parque Natural del Montgó (Alicante). Bull. San. Veg. Plagas, 28: 445-456.

Southwood, T. R. E. 1987. Ecologycal Methods with particular reference to the study of insect populations. Chapman & Hall. 524 Pags.

Southwood, T. R. E. and Henderson, P. A. 2000. *Ecological methods*. 3ª ed. Blackwell Science, Oxford. 596 Pags.

Spangler, P. J. 1980. Chelonariid larvae, aquatic or not?. The Coleopterist's Bulletin. Vol. 34. Fasc.1, pp: 105 - 114.

Spangler, J. P. 1981. Coleoptera. En: Hurlbert, R. and Santos. N. edit. Aquatic Biota of tropical South America, Part 1; Arthropoda. San Diego State University, San Diego, California. XII, pp: 129- 220.

Spangler, P. J. 1982. Aquatic biota of México, Central American and the West Indies. En: Humbert, S. H. & A. Villalobos –Figueroa (eds.), Coleoptera, San Diego State University, San Diego, California, pp: 328-323.

Spangler, P. J. 1983. Immature stages and biology of Tetraglossa palpalis (Coleoptera: Ptilodactylidae).Entomol. News. Vol. 94. Fasc. 5, pp: 161 - 175.

Spangler, P. J. 1985. A new genus and species of riffle beetle, *Neblinagena prima*, from the Venezuelan tepui, Cerro de la Neblina (Coleoptera: Elmidae: Larinae). Proc. Entomol. Soc. Washington. Vol. 87. Fasc. 3, pp: 538 - 544.

Spangler, P. J. 1985. A new species of the aquatic beetle genus *Dryopomorphus* from Borneo (Coleoptera: Elmidae: Larinae). Proc. Biol. Soc. Washington. Vol. 98. Fasc. 2, pp: 416 - 421.

Spangler, P. J. 1985. A survey of aquatic insects of Ecuador. Nat. Geogr. Soc. 1985, pp: 611 - 614.

Spangler, P. J. 1991. Chelonariidae (Dryopoidea). In: Immature Insects. (ed. F. Stehr). Kendall/Hunt Publishing Co., Dubuque. Vol. 2. Pp.394 - 395.

Spangler, P. J. 1991. A durable, lightweight net and a manual aspirator for collecting aquatic organisms. The Pan-Pacific Entomologist. 1981; 57: 245-250.

Spangler, P. J. 1991. *Guaranius carlosi*, a new Genus and species of dryopid beetle from Paraguay (Coleoptera; Dryopidae). The Coleopterists Bulletin, Vol. 45. Fasc.2, pp: 147- 151.

Spangler, P. and Warren, E. Steiner Jr. 2005. A new aquatic beetle family, Meruidae, from Venezuela (Coleoptera: Adephaga). Systematic Entomology. Vol.10, pp: 1-19.

Speight, M. C. D. 1989. Saproxylic invertebrates and their conservation. Nature and Environment Series 46, Council of Europe, Strasbourg

Speight, M. C. D. 1989. Saproxylic Invertebrates and their Conservation. Council of Europe. Strasbourg, France. 79 Pages.

Speight, Martin R. 2005. Sampling insects from trees: shoots, stems, and trunks. Chapter 5, Ppo:77-105. In: Leather, S. R.: Lawton, J. H. and Likens, G. E. (edits.) 2005. Insect sampling in forest ecosystems. Blackwell, Malden, MA.. 315 pags.

Spence, J. R. and Niemela, J. K. 1994. Sampling carabid assemblages with pitfall traps: the madness and the method. The Canadian Entomologist. Vol. 126, pp: 881–894.

Sperry, Arthur P. and Banks, Nathan 2007. Fauna of the caves of Yucatan. Carnegie Institution of Washington. Número 491 de Carnegie Institution of Washington publication. 304 pags.

Spilman, T. J. 1967. The heteromerous intertidal beetles (Coleoptera: Salpingidae: Aegialitinae). Pacific Insects. Vol. 9, pp: 1 - 21. Available online (2010) in: http://hbs.bishopmuseum.org/pi/pdf/9(1)-1.pdf

Spilman, T. J. 1968. Two new species of *Niptus* from North American caves (Coleoptera: Ptinidae). The Southwestern Naturalist. Vol. 13. Fasc. 2, pp: 193-200.

Spilman, T. 1971. Bredin-Archbold. Smithsonian Biological Survey of Dominica: Bostrichidae, Inopeplidae, Lagriidae, Lyctidae, Lymexylonidae, Melandryidae, Monommidae, Rhipiceridae, and Rhipiphoridae (Goleoptera). Smithsonian Contr. to Zoology.. N° 70, pp: 1- 10.

Spinelli, Gustavo R.; Muzón, Javier y Massaferro, Julieta (editores) 2009. Sección especial: Biodiversidad de Insectos de la Patagonia. Dedicado a la memoria de Don Mario Gentili. Rev. Soc. Entomol. Argentina. Vol. 68 Fasc.1- 2, pp: 143-144.

Springate, N.D. & Basset, Y. 1996. Diel activity of arboreal arthropods associated with Papua New Guinea trees. Journal of Natural History. Vol. 30, pp: 101–112.

St. George, R. A. 1931. The larva of *Boros unicolor* Say and the systematic position of the family Boridae Herbst. Proceedings of the Entomological Society of Washington. Vol.33, pp: 103-113.

Staines, C. L. 2002. The New World tribes and genera of hispines (Coleoptera: Chrysomelidae: Cassidinae). Proc. Entomol. Soc. Wash. Vol. 104. Fasc. 3, pp: 721 - 784.

Stan Melania 2006. New contributions to the knowledge of the Rove Beetle fauna (Coleoptera: Staphylinidae) of Maramure (Romania). Travaux du Muséum National d'Histoire Naturelle «Grigore Antipa». Vol. XLIX, pp: 173 – 184.

Stanger-Hall, Kathrin F.; Lloyd, James E. & Hillis, David M. 2007. Phylogeny of North American fireflies (Coleoptera: Lampyridae): Implications for the evolution of light signals. Mol. Phylogenet. Evol. Vol.45. Fasc.1 pp: 33-49.

Stebbing, Edward Percy 1914. Indian forest insects of economic importance. Coleoptera. Publisher: London Eyre & Spottiswoode. 648 pags.

Stebnicka, Zdzislawa 1999. Lomanoxia Martínez, 1951, and a new tribe Lomanoxiini with notes on comparative morphology (Coleoptera: Scarabaeoidea: Aphodiinae). Acta Zool. cracov. Vol.42. Fasc.2, pp: 279-286.

Stedman, Michael 2006. Small Hive Beetle (SHB): Aethina tumida Murray (Coleoptera: Nitidulidae). Government of South Australia. Primary Industries and Resources SA, pp: 1-15.

Steffan, A. W. 1964: Torridincolidae, coleopterorum nova familia e regione aethiopica. Entomologische Zeitschrift (Stuttgart) 74, pp: 193-200.

Steffan A. W. 1973: Torridincolidarum tres novae species e regione aethiopica (Coleoptera: Sphaerioidea). Revue de Zoologie et de Botanique Africaines 87, pp: 633-646.

Stehr, F. W.(ed.) 1991a. Immature Insects. Vol. 2, Kendall Hunt Publishing Company, Dubuque, Iowa, 975 pp.

Steiner, W. E. Jr. 2004. The first records of *Bycrea Villosa* Pascoe (Coleoptera: Tenebrionidae) in The United States, Central America and Colombia and notes on its Association with Leaf-Cutting Ants. The Coleopterists Bulletin. Vol. 58. Fasc.3, pp: 329–334.

Steiner, W. E. Jr. 2005. Notes and proposed studies on the darkling beetles (Coleoptera: Tenebrionidae) of the Bahamian region. Pp. 121-137 in Buckner, S. D. & McGrath, T. S. eds., Proceedings of the Tenth Symposium on the Natural History of the Bahamas. Gerace Research Center, San Salvador, Bahamas.

Steiner, W. E. Jr. 2009. The Helopini (Coleoptera: Tenebrionidae) of Virginia, pp: 331-339, in: S. M. Roble and J. C. Mitchell (eds.). 2009. A Lifetime of Contributions to Myriapodology and the Natural History of Virginia: A Festschrift in Honor of Richard L. Hoffman's 80th Birthday. Virginia Museum of Natural History Special Publication No. 16, Martinsville, VA.

Stephens, A. Stephanie 2003. "An Overview of the Coleopteran Family Histeridae and its significance to Forensic Entomology.".web Page: www.beetlelady.com/?page_id=7

Stephenson, Gary 1999. Vehicle impacts on the biota of sandy beaches and coastal dunes. A review from a New Zealand perspective. Science for conservation 121. Published by Department of Conservation Wellington, New Zealand, pp: 1-48. Available online (2010) in: http://www.doc.govt.nz/upload/documents/science-and-technical/Sfc121.pdf

Stern, A. C., et al. 1984. Fundamentals of Air Pollution. 2da. Edición. Nueva York: Academic Press.

Stewart, Joseph A. & Wright, A. F. 1995. A new inexpensive suction apparatus for sampling arthropods in grassland. Ecological Entomology. Vol. 20, pp: 98–102.

Stewart, Joseph A. 2007. Riparian Vegetation of the Grande Ronde River. Back ground papers, pp: 1-21.

Steyskal, G.C., W.L. Murphy, and E.M. Hoover. Editors. 1986. Insects and mites: techniques for collection and preservation. United States Department of Agriculture. Agricultural Research Service. Miscellaneous Publication, # 1443. 103 pp.

Strand, Larry L. 1999. Integrated pest management for stone fruits. ANR Publications. 264 pgs.

Stribling, J. B. and Seymour, R. L. 1988. Evidence of Mycophagy in Ptilodactylidae (Coleoptera: Dryopoidea) with Notes on Phylogenetic Implications The Coleopterists Bulletin. Vol. 42. N° 2, pp: 152 – 154.

Strohecker, H. F. 1953. Coleoptera Fam. Endomychidae. In, Wytsman P. (ed.), Genera Insectorum. Desmet-Verneuil, Brussels, fasc. 210. 145 Pages.

Strong, D. R. J. 1982. Harmonious coexistence of hispine beetles on *Heliconia* in experimental and natural communities. Ecology 63, pp: 1039–1049.

Stuntz, S., C. Ziegler, U. Simons & G. Zotz 2002. Diversity and structure of the arthropod fauna within three canopy epiphyte species in central Panama. Journal of Tropical Ecology. Vol.18, pp: 161-176.

Stylianos Chatzimanolis; James, S. Ashe And Rodney, S. Hanley 2004. Diurnal/nocturnal activity of Rove Beetles (Coleóptera: Staphylinidae) on Barro Colorado Island, Panama assayed by flight intercept trap. The Coleopterists Bulletin. Vol. 58. Fasc. 4, pp: 569 - 577.

Sugihara, Y. 1938. An observation on the inter-tidal rock-dwelling beetle, *Aegialitis stejnegeri sugiharai* Kono, in the Kuriles. Ent. World. Vol. 6. Fasc. 46, pp: 6-12, 5 figs. [In Japanese.]

Sung-Oui, Suh and Meredith, Blackwell and Joseph, V. McHugh 2001. Beetles and their yeast endosymbionts from basidiocarp habitats Available online (2010) in: http://lsb380.plbio.lsu.edu/beetlebellyfolder/beetlebelly.methods

Stuntz, S. 2001. The influence of epiphytes on arthropods in the tropical forest canopy. Ph.D. Dissertation. Julius–Maximilians–Universität Würzburg; Wurzburg. 111 p.

Sustek, Z. 1981. Mrchozroutovití, Ceskoslovesnka, Coleoptera, Silphidae. Zprávy Ceskoslovenské Spolecnosti Entomologické pri CSAV. Klice k Urcováni Hmyzu. Vol. 2, pp: 1 - 46.

Sustek, Z. 1981. Mrchožroutovití brouci Československa (Coleoptera, Silphidae). Zprávy Čs. Společ. Entomol. ČSAV, Klíče k určování hmyzu. Vol. 2, pp: 1–46.

Sutherland, W. J. (ed.). 1996. Ecological census techniques: a handbook. Cambridge University Press, Cambridge. 336 Pages.

Sutherland, Andrew M. and Parrella, Michael P. 2009. Mycophagy in Coccinellidae: Review and synthesis. Volume 51, Issue 2. Trophic Ecology of the Coccinellidae, pp: 284-293.

Sutton, S. L. & Hudson, P. J. 1980. The vertical distribution of small flying insects in the lowland rainforest of Zaire. Zoological Journal of the Linnean Society. Vol. 68, pp: 111 - 123.

Sutton, S. L. 1983. The spatial distribution of flying insects in tropical rainforests. (In) Sutton, S.L., Whitmore, T.C. & Chadwick, A.C. (eds.), Tropical rainforest: ecology and management. Blackwell, Oxford, pp: 77 - 91.

Tan, Jing-Jing and and Dong, Ren 2006. *Ovatocupes*: A New cupedid genus (Coleoptera: Archostemata: Cupedidae) from the Jehol Biota (Late Jurassic) of Western Liaoning, China. Entomological News. Vol. 117, N° 2l, pp: 223-232.

Taylor, R. and Doran, N. 2004. Identification of species and functional groups that give early warning of major environmental change (Indicator 1.2c). Part D: Use of terrestrial invertebrates as indicators of the ecological sustainability of forest management under the Montreal Process. Forest & Wood Products Research & Development Corporation. Victoria, Australia.

Telnov, D. 1999. Zoogeographie der Trictenotomidae Blanchard, 1845 (Coleoptera: Heteromera). 7. DPU zinatniskas konferences rakstu krajums, A9, pp: 95 - 97.

Telnov, D.; Gailis, J.; Kalnins, M.; Napolov, A.; Piterans, U.; Vilks, K. and Whitehead, P. F. 2005. Contributions to the knowledge of Latvian Coleoptera. 4. - Latvijas Entomologs. Vol. 42, pp: 18-47.

Tero, Toivanen 2007. Short-term Effects of Forest Restorationon Beetle Diversity. Jyväskyla studies in Biological and Envieronmental Science. 175. 33 pags.

Tero, Toivanen and Kotiaho, J. S. 2004. The effect of prescribed burning and decaying wood on the diversity of beetles. – Transactions of the Faculty of Forestry, Estonian Agricultural University. 37. Pp : 92–96.

Thangam, T. S. and Kathiresan, K. 1993. The mosquito consposition and seasonal distribution of *Culex quinquefasciatus* in a coastal town of wouth India. Tropical Biomedicine. Vol. 10, pp: 175-177.

Thayer, M. K. and Lawrence, J. F. 2002. Family 98. Ciidae (Leach in Samouelle 1819, pp: 403-412. in: Arnett, R. H. Jr. And Thomas, M. C., eds. American Beetles. vol. I. CRC Press, Florida, pp: 233-246.

Thayer, M. K. and Newton, A. F.. 2004. PEET: Monography, Phylogeny, and Historical Biogeography of Austral Staphylinidae (Coleoptera) available online (2010) in: www.fm1.fieldmuseum.org/aa/Files/thayer/Collecting-details.pdf

Thayer, M. K. 2009. Biology and phylogenetic relationships of *Neophonus bruchi*, an anomalous south Andean staphylinid (Coleoptera). Systematic Entomology. The Royal Entomological Society. Vol. 12 Issue 3, Pages: 389 – 404.

Thérond, J. 1931. Un piège naturel: l'Arum dracunculus et description d'une variété indédite de Saprinus furvus Er. - Miscellanea Entomologica. 33, pp: 59-60.

Thibaud, J. M. & Massoud, Z. 1973. Essai de classification des Insectes Collemboles "cavernicoles" européens. C. R. Acad. Sc. Paris. N° 277, pp: 2137 - 2140.

Thibaud, J. M. & Vannier, G. 1986. Caracterisation biologique et ecophysiologique des Insectes Collemboles cavernicoles. En: DALLAI, R. (editor). 2nd International Seminar on Apterygota. University of Siena, Italy, pp: 129-137.

Thiele, H. U. 1977. Carabid beetles in their evironments. Zoophysiology and Ecology 10. SpringerVerlag, Berlin.

Thomas, J. B. and H. Herdy 1961. A note on the life history of *Cimberis elongatus* (LeC.) (Coleoptera: Anthribidae). Canadian Entomologist. Vol. 93, pp: 406 - 408.

Thomas, M. C. Family 80. Silvanidae, pp: 322-326; Family 81. Passandridae Erichson 1845, pp: 327- 328; Family 82. Cucujidae Latreille 1802, pp: 329-330; Family 83. Laemophloeidae Ganglbauer 1899, pp: 331-334. in: Arnett, R.H., Jr., Thomas, M.C., S ., Skelley, P.E. & Howard, Frank, J. (Eds.). American Beetles, Volume 2: Polyphagha: Scarabaeoidea through Curculionoidea. CRC Press, Boca Raton, USA. 861 pages.

Thornton, I. W. B. 1957. Faunal succession in umbels of *Cyperus papyrus* L. on the Upper White Nile.Proceedings of the Royal Entomological Society of London (A) 32, pp: 119-131.

Thornhill, E. W. 1978. A motorised insect sampler. Pest Articles and News summaries. Vol. 24, pp: 205–207.

Throne, J. E., and L. D. Cline. 1994. Seasonal ßight activity and seasonal abundance of selected stored-product Coleoptera around grain storages in South Carolina. J. Agric. Entomol. Vol.11, pp: 321-338.

Tiemann, D. 1967. Observation on the natural history of the Eastern banded glow-worm *Zarhipis integripennis*. Proceedings of the California Academy of Sciences. Vol.35, pp: 235-264.

Tiemann, D. 1970. Nature's toy train, the railroad worm. National Geographic. Vol.138, pp: 58-67.

Tillyard, R. J. Angus & Robertson Ltd, 1926. Insects of Australia and New Zealand. Sydney. Pàg. 243.

Tishechkin, Alexey, K. 2005. Phylogenetic Revision of the Genus *Mesynodites* Reichartd (Coleoptera: Histeridae: Hetaeriinae). A dissertation submitted to the Graduate Faculty of the Louisiana State University and Agricultural and Mechanical College in partial fulfillment of the requirements for the degree of Doctor of Philosophy In The Department of Entomology. 201 pages. Available online (2010) in:
http://etd.lsu.edu/docs/available/etd-02232005-134727/unrestricted/Tishechkin_dis.pdf

Tishechkin, Alexey K. & Caterino, Michael S. 2007. Description of the first Chlamydopsinae (Coleoptera: Histeridae) from the Philippines. Zootaxa. N° 1527, pp: 39–44.

Tishechkin, Alexey K. 2009. Discovery of Chlamydopsinae (Insecta, Coleoptera, Histeridae) in Vanuatu with the description of eight new species from Espiritu Santo Island. ZOOSYSTEMA. 31. Fasc. 3, pp: 661- 690.

Tod, M. E.. 1973. Notes on beetle predators of molluscs. Entomologist. Vol. 106, pp: 196- 201.

Tomaszewska, W, Slipinski, S. A. 1995. A review of the family Hobartiidae (Coleoptera: Cucujoidea). Genus. 6, pp: 303-325.

Tomaszewska, K. W. 2000. Morphology, phylogeny and classification of adult Endomychidae (Coleoptera: Cucujoidea). Annales Zoologici (Warsaw) 50, pp: 449-558. Available online (2010) in: http://www.miiz.waw.pl/pliki/pracownicy/tomaszewska/Tomaszewskaphylogenyof Endo2000.pdf

Tomaszewska, Wioletta K. 2002. Two new species of *Cyclotoma* MULSANT from India and Burma (Coleoptera: Endomychidae). Genus Vol. 13. Fasc.2, pp: 203-210.

Tomaszewska, K. W. and Slipinski, A. 2002. A review of the genus Periptyctus Blackburn (Coleoptera: Corylophidae). Annales Zoologici. Vol. 52, pp: 499-532.

Tomaszewska, K. W. 2005. Phylogeny and generic classification of the subfamily Lycoperdininae with a re-analysis of the family Endomychidae (Coleoptera, Cucujoidea). Annales Zoologici (Warszawa), 55. suppl.1, pp: 1–172.

Tomaszewska, K. W.; Ślipiński, A. 2008. Review of the genus *Myrabolia* Reitter (Coleoptera: Cucujoidea: Myraboliidae). Australian journal of entomology. Vol. 47, pp: 32 - 50.

Toskina, I. N. 1993. New species of Anobiidae (Coleoptera) from Paraguay. Russian Entomological Journal. Vol. 2. Fasc. 1, pp: 23 - 34.

Toskina, I. N. 2000. New wood-boring beetles (Coleoptera: Anobiidae) from Paraguay. Russian Entomol. J. Vol.9. N°3, pp: 199 – 240.

Toth, M.; Klein, M. G. and Imrei, Z. 2003. Field Screening for Attractants of Scarab (Coleoptera: Scarabaeidae) Pests in Hungary. Acta Phytopathologica Hungarica. Vol. 38. Fasc. 3/4, pp: 323-331.

Toumanoff C. 1939. Les Ennemies des Abeilles, Imprimerie d'Extrême-Orient, Hanoi, North Vietnam.

Towwnsend, C. H. 1863. Some notes on Michigan Insects, principallly Coleoptera, affecting forest trees. The Canadian Entomologist. author: Entomol Society Of Canada (1863-1871). page 204.

Triplehorn, C. A. 1990. Review of the genus *Corticeus* (Coleoptera: Tenebrionidae) of America North of Mexico. Annals of the Entomological Society of America. Vol. 83, pp: 287-306.

Troukens, Willy 2006. Schijnsnuitkevers, vuurkevers, snoerhalskevers en waaierkevers aan de Eastrand van Brussel (Coleoptera: Pythidae, Pyrochroidae, Anthicidae & Rhipiphoridae). Phegea. Vol. 34. Fasc. 3. Page 99

Tschinkel, W. R. 1992. Brood raiding in the fire ant, Solenopsis invicta (Hymenoptera: Formicidae): laboratory and field observations. Annals of the Entomological Society of America. Vol. 85, pp: 638 - 646.

UK BAP Grouped Action Plan for river shingle beetles (September 2003): http://www.ukbap.org.uk

Ulke, H. 1902. List of beetles of the District of Columbia. Proc. U. S. Natl. Mus. Vol. 25, pp: 1 - 57.

Uiyshen, M. D. and Hanula, James, L. 2004. Diversity and Seasonal Activity of Carrion Beetles (Coleoptera: Silphidae) in Northeastern Georgia. Note. Journ.Entomol. Sci. Vol. 39. Fasc. 3, pp: 460 - 463.

Ulyshen, M. D.; Hanula, J. L. and Horn, S. 2005. Using Malaise traps to sample ground beetles (Coleoptera: Carabidae). Can. Entomol. Vol. 137, pp: 251–256.

Ulyshen, M. D., Horn, S. and Hanula, J. L. 2007. Burying beetles (Coleoptera: Silphidae) in the forest canopy: The unusual case of *Nicrophorus pustulatus* Herschel. The Coleopterists Bulletin Vol. 61, pp: 121-123.

Ulyshen, M. D. and Hanula, James L. 2009. Habitat associations of saproxylic beetles in the southeastern United States: A comparison of forest types, tree species and wood postures Forest Ecology and Management. Vol. 257, pp: 653–664.

U. S. Fish and Wildlife Service 2010. Petition to list six sand dune beetles under the U.S. endangered species act. Petitioner: Wild Earth Guardians, pp: 1- 24.

Usis, J. D. and MacLean, D. B. 1998. The ground beetles (Coleoptera: Carabidae) of Stillfork Swamp Nature Preserve, Carroll County, Ohio. Ohio Journal of Science. Vol. 98, pp: 66–68.

Vadim Viviani 1996. Occurrence of Aggressive Mimicry In A Brazilian *Bicellonycha* Firefly (Lampyridae; Photurinae). Fireflyer Companion. Vol. 1. N° 2. Pag:22. Available online in: entnemdept.ufl.edu/lloyd/firefly/ffcomp1-2.pdf

Valladares, L. F.; Vega, F. J.; Mazé, R. A.; Régil, J. A. and García-Criado, F. 2002. Biodiversidad de los macroinvertebrados acuáticos del Parque Natural de Valderejo (Álava): Complicaciones en conservación. Boletìn de la Asociaciòn española de Entomologìa. Vol. 26. Fasc. 3-4, pp: 37-55.

Van, Dam A. R. and Van, Dam M. H. 2008. Impact of off-road vehicle use on dune endemic Coleoptera. Annals Entomological Soc'y America. Vol. 101, Fasc. 2, pp: 411-417.

Vanin, S. A. & Costa, C. 2001. Description of immature stages of *Claudiella ingens* Reichardt & Vanin, 1976 and comparative notes on other Torridincolidae (Coleoptera, Torridincolidae). Aquatic Insects 23: 1-10.

Varga, L. 1928. Ein interessanter Biotop der Biocönose von Wasserorganismen. Biologisches Zentralblatt. Vol 48, pp: 143–162.

Vasily, V. Grebennikov & Alfred, F. 2008. Newton Minute larvae of Leptotyphlinae (Coleop.: Staphylinidae): description of three genera with discussion on the monophyly and phylogenetic position of the subfamily as inferred from larval morphology. Zootaxa. N° 1817, pp: 49 – 58.

Vaurie, P. 1948. A review of the North American Languriidae.Bulletin of the American Museum of Natural History. Vol. 92, pp: 119 - 155.

Vaz-de-Mello, F. Z. 2007. Revision and phylogeny of the dung beetle genus *Zonocopris* Arrow 1932 (Coleoptera: Scarabaeidae: Scarabaeinae), a phoretic of land snails. Ann. Zool. Entom. Fr. (n.a.) Vol.43. Fasc.2. Pp231-239.

Vaz-De-Mello, F. Z. 2008. Synopsis of the new subtribe Scatimina (Coleoptera: Scarabaeidae: Scarabaeinae: Ateuchini), with descriptions of twelve new genera and review of *Genieridium*, new genus. Zootaxa, pp: 1-75.

Vázquez, X. A. 1993. Coleoptera Oedemeridae, Pyrochroidae, Pythidae, Mycteridae. En Fauna Ibérica, vol. 5. Ramos M. A. et al. (eds.). Museo Nacional de Ciencias Naturales, Consejo Superior de Investigaciones Científicas, Madrid, pp: 1-181.

Vázquez, X. A. 1995. Faunística y mapas de distribución de los Oedemeridae (Coleoptera) de la fauna Ibérica y Balear. Zapateri, Revista aragonesa de entomología. Vol. 5, pp: 35-49.

Vega, F. E. and Blackwell, M. Editors 2005. Insect–Fungal Associations: Ecology and Evolution. Oxford University Press. 333 Pages.

Vidal, Sarmiento J. A. 1969. El sistema reproductor masculino y organos de copulation en Cupesidae (Coleoptera – Archostemata). Rev. Soc. Entomol. Argentina. Vol. 31, pp: 43–48.

Vinson, J. 1995. The genus *Cybocephalus* Erichson in the Mascarene Islands (Coleoptera: Nitidulidae). Proceedings of the Royal Entomological Society of London. Series B, Taxonomy. Vol. 28. N° 1-2, pp: 7-18.

Viviani, V. 2001. Fireflies (Coleoptera: Lampyridae) from Southeastern Brazil: Habitats, Life History, and Bioluminescence. Entomological Society of America. Vol. 94. N° 1, pp: 129 – 145.

Vladimír Malohlava & Ladislav Bocak 2009. A revision of Paratelius Kazantsev, 1992 (Coleoptera: Lycidae). Zootaxa N° 2306, pp: 44–50.

Vondel, B. J. van 2005. Haliplidae, pp. 20-86. In: Nilsson A.N. & Vondel B.J. van. 2005. Amphizoidae, Aspidytidae, Haliplidae, Noteridae and Paelobiidaer (Coleoptera, Adephaga). - In World Catalogue of Insects. 7, pp: 1-171.

Vondel, B. J. van 2007. World Catalogue of Haliplidae - corrections and additions 1 (Coleoptera: Haliplidae). - Koleopterologische Rundschau 77, pp: 89-96.

Vrålstad, T.; Holst-Jensen, A. and Schumacher, T. 1998. The postfire discomycete Geopyxis carbonaria (Ascomycota) is a biotrophic root associate with Norway spruce (Picea abies) in nature. Molecular Ecology. Vol. 7, pp: 609 – 616.

Vulcano, M. A. and Pereira, F. S. 1961. A subfamilia Penichrolucaninae (Col. Lucanidae) representada em América. Studia entomologia. Vol.4, pp: 471-480.

Vulcano, M. A. & Pereira, F. S. 1975. Cupesidae (Coleoptera). Arquivos do Instituto Biologico, São Paulo. Vol. 42, pp: 31 – 68.

Walden Pond. Wikipedia, the free encyclopedia 2010. Available online (2010) in: http://en.wikipedia.org/wiki/Walden_Pond (last access: 13. VIII. 2010).

Walker, T. W. 1955. "The Ecology of Tussock Grasslands: Discussion" Proc. NZ Ecol. Soc 3:7 "One fifth of New Zealand carries tussock or bunch grass vegetation related to the so called steppes of the world "

Walker, E. D. & Merritt, R. W. 1988. The significance of leaf detritus to mosquito (Diptera: Culicidae) productivity from treeholes. *Environmental Entomology. Vol.* 17, pp: 199–206.

Walters, Martin 2008. Chinese wildlife: a visitor's guide Bradt Guides Bradt Travel Guide Chinese Wildlife Series Bradt travel guides. Edit. Bradt Travel Guides. 208 pp. ISBN: 1841622206, 9781841622200

Wanat, Jerzy S. 2008. Nanophyes brevis Boheman, 1845 (Coleoptera: Curculionoidea: Nanophyidae) in Poland. Polish Journal of Entomology. Vol.77, pp: 183- 189.

Ward, D. and Seely, M. K. 1996a. Competition and habitat selection in Namib Desert tenebrionid beetles. Evolutionary Ecology. Vol. 10, pp: 341- 59.

Ward, David 2009. The Biology of Deserts. Oxford Uni-

versity Press Inc., New York. 352 Pages.
Ward, R. N.; Ernst, C. Ebert and Kenneth, E. Ward 2007. Occurrence of Clover Stem Borer, Languria mozardi (Coleoptera: Languriidae), on Canola: a new host record. Florida Entomologist. Vol. 90. Fasc. 4, pp: 732 – 737.
Warner, William B. 1997. Don't Catch That-or- Disease Risk Reduction & Rodent Inquiline Collecting. SCARABS. Occasional Issue. N° 13, pp: 1-8.
Watt, J. C. 1967a. The families Perimylopidae and Dacoderidae (Coleoptera, Heteromera). Proc. R. ent. Soc. Lond. Vol.36, pp: 109-18.
Watt, J. C. 1969. Keys to Genera and some Species of New Zealand Lathridiidae (Coleoptera). Entomology Division, D.S.I.R., Nelson.Vol. 204, pp: 49- 67.
Watt, J. C. 1974. Chalcodryidae: a new family of heteromerous beetles. (Coleoptera: Tenebrionoidea). Journal of the Royal Society of N.Z. Vol. 4. Fasc. 1, pp: 19 – 38.
Watt, J.C. 1980. Zeonidicola (Coleoptera: Cavognathidae) - beetles inhabiting birds' nests. Journal of the Royal Society of New Zealand. Vol. 10, pp: 331-339. Available (in part) online (2010) in: http://books.google.com.py/books?id=jLQ4AAAAIAAJ&pg=PA332&dq=Taphropiestes&hl=es&ei=zHVITLHxF8OC8gbYhJXZDg&sa=X&oi=book_result&ct=result&resnum=1&ved=0CCkQ6AEwAA#v=onepage&q=Taphropiestes&f=false
Watt, J. Charles 1983. Beetles (Coleoptera) of Auckland. TANE 29, pp: 31-50. Available online (2010) in: http://www.thebookshelf.auckland.ac.nz/docs/Tane/Tane-29/4%20Beetles%20(Coleoptera)%20of% 20Auckland.pdf
Webb, G. A. 1986. A breeding aggregation of Chauliognathus pulchellus Macleay (Coleoptera: Cantharidae). Victorian Naturalist. Vol.103. Fasc. 4, pp: 114 – 116
Wegrzynowicz P. 2000. Catalogue of the Helotidae (Coleoptera: Cucujoidea). Zool. Med. Leiden. Vol. 73, pp: 391 - 411. Available online (2010) in: http://www.repository.naturalis.nl/document/44306
Węgrzynowicz, Piotr 2002. «Morphology, phylogeny and classification of the family Erotylidae based on adult characters (Coleoptera: Cucujoidea)», en Genus. Vol.13. N° 4, pp: 435-504.
Weinreich, E. 1968. Über den Klebfangapparat der Imagines von Stenus Latr. (Coleopt., Staphylinidae) mit einem Beitrag zur Kenntnis der Jugendstadien dieser Gattung. Zeitschrift für Morphologie der Tiere. Vol. 62, pp: 162 - 210.
Weise, J. 1897. Biologische Mitteilungen. Deutsche Entomologische Zeitschrift 1897, pp: 389-395.
Weiss, H. B. and East, E. 1921. Additional fungous insects and their hosts. Proceedings of the Biological Society of Washington, Vol. 34, pp: 59 - 62.
Werner, F. G. 1974. A review of the chilean Anthicidae (Coleoptera). Revista Chilena de Entomología. Vol. 8, pp: 27-34.
Werner, F. G. and Chandler 1995. Anthicidae (Insecta: Coleoptera). Fauna of New Zealand Ko te Aitanga Pepeke o Aotearoa. N° 34, pp: 5- 61.
Eastcott, R. L. 1976. Observations on the biology and ethology of Lichnante rathvoni LeConte (Coleoptera: Scarabaeidae) with emphasis on mating. University of Idaho, Department of Entomology Anniversary Publication, No. 11: 85-90.
Eastwood, J. O. 1855. Description of a new genus of Coleopterous insects inhabiting the interior of ants nests, in Brazil. Trans. Ent. Soc, London, (n.s.) Vol. 3, pp: 90-94, est. 8. 21 figs,
Wetistein, W; Schmid, B. 1999. Conservation of arthropod diversity in montane Wctlands: effect of altitude, habitat quality and habitat fragmentation on butterflies and grasshoppers. Journal of Applied Ecology. Vol. 36, pp: 363 - 373.
Wheeler, Quentin and Blackwell, Meredith (edits) 1984. Fungus-insect relationships: perspectives in ecology and evolution. Columbia University Press Hardcover, 514 pages. Available (in part) online in: http://books.google.com.py/books?id=McUWyL0wsh8C&printsec=frontcover&dq=Fungus-insect+relationships:+perspectives+in+ecology+and+evolution&hl=es&ei=yWFITJ_tBYG78gbU2NjSDg&sa=X&oi=book_result&ct=book-thumbnail&resnum=1&ved=0CCoQ6wEwAA#v=onepage&q&f=false Wheeler, Quentin 1984. Notes on Host Associations and Habitats of Dasyceridae (Coleoptera) in the Southern Appalachian Mountains The Coleopterists Bulletin. Vol. 38. Fasc. 3, pp: 227- 231.
Wheeler Quentin D. and Hoebeke, R. E. 1984. A review of mycophagy in the Eucinetoidea (Coleoptera), with notes on an association of the eucinetid beetle, Eucinetus oviformis, with a Coniophoraceae fungus (Basidiomycetes: Aphyllophorales). Proceedings of the Entomological Society of Washington, Vol. 86, pp: 274 - 277.
Wheeler, Quentin D. and McHugh, Joseph D. 1987. A portable and convertible «Moczarski/Tullgren» extractor for fungus and litter Coleoptera. The Coleopterists Bullettin. Vol. 41. Fasc. 1, pp: 9-12. Available online (2010) in: http://www.bio-nica.info/biblioteca/Wheeler1987-WinklerSac.pdf
Wheeler, Q. D. 1991. Lymexylidae, pp: 446 - 447. In: F. W. Stehr, ed., Immature Insects. Volume 2. Kendall/Hunt. Dubuque, IA.
White, R. E. 1962. The Anobiidae of Ohio. – Bulletin of the Ohio Biological Survey (new series) 1:1–58.
White, D. S. and D. E. Jennings. 1973. A rearing technique for various aquatic Coleoptera. Annals of the Entomological Society of America 66, pp: 1174-1175.
White, R. E. 1983. A Field Guide to the Beetles of North America. Boston, MA: Houghton Mifflin, pp: 1-172.
White, D. S. & Brigham, W. U. 1996. An Introduction to the Aquatic Insects of North America. In: Merrit, R. W. & K. W. Cummins. (eds.), Coleoptera. Third Edition. Kendall/Hunt Publishing Co., Dubuque, Lowa. pp. 399-473.
White, Richard E. and Peterson, Roger Tory 1998. A Field Guide to the Beetles: Of North America. Houghton Mifflin Harcourt. 384 pages.
Whiteman, Charles D. 2000. Mountain meteorology: fundamentals and applications. Mountain meteorology: fundamentals and applications. Oxford University Press US. 355 pages.
Whiting, M. F.; Carpenter, C. J.; Wheeler, Q. D. and Wheeler, W. C. 1997. The Stresiptera Problem: Phylogeny of the Holometabolous Insect Orders Inferred from 18S and 28S Ribosomal DNA Sequences and Morphology, Society of Systematic Biologists. Vol. 46. Fasc. 1, pp: 1- 68.
Wiggins, G. B.; Mackay, R. J. & Smith, I. M. 1980. Evolutionary and ecological strategies of animals in annual temporary pools. Arch. Hydrobiol., Suppl. 58, pp: 97-206.
Wikars, L. -O. 1992. Skogsbränder och insekter. Entomologisk Tidskr. 113 (In Swedish with english summary), pp: 1-11.

Wikars, L.-O. 1997. Effects of fire and ecology of fire-adapted insects. Comprehensive Summaries of Uppsala Dissertations from the Faculty of Science and Technology 272. Acta Universitas Upsaliensis, Uppsala.

Wildermuth, V. L., and Gates, F. H. 1920. Clover stem-borer as an alfalfa pest. Bull. U. S. Dept Agr. N°. 889, pp: 1 - 25. Figs. 1-6, pl. 1.

Williams. R. N.; Blackmer, J. L.; Richmond, D. S. and Sean Ellis M. 1992. Nitidulidae (Coleoptera) Diversity in Three Natural Preserves in Portage County, Ohio. Ohio Journal Scienc. Vo. 92. Fasc. 4, pp: 82 - 87,

Williams, G.A., 1993. Hidden rainforests: subtropical rainforests and their invertebrate biodiversity. Sydney: New South Wales University Press and The Australian Museum.

Williams, D. D. 1996. Environmental constraints in temporary fresh waters and their consequences for the insect fauna. J. N. Am. Benthol. Soc. Vol. 15. Fasc. 4, pp: 634 - 650.

Williams, D. D. 1997. Temporary ponds and their invertebrate communities. Aquat. Conserv. Vol. 7. Fasc. 2, pp: 105 - 117.

Williams, Geoff 2002. A Taxonomic and Biogeographic Review of the Invertebrates of the Central Eastern Rainforest Reserves of Australia (CERRA) World Heritage Area, and Adjacent Regions.Technical Reports of the Australian Museum. N°16, pp: 1-206.

Wilson, E. O.; Eisner, T. and Valentine, B. D. 1954. The beetle genus *Paralimulodes* Bruch In North America, with notes on morphology and behavior (Coleoptera Limulodidae). Psyche.

Wilson, E. 1992. La Diversidad de la Vida. Grupo Gribaldo-Mandadori. Barcelona. 410 pages.

Willemstein, S. C. 1987. An evolutionary basis for pollination ecology. Leiden botanical series. Vol. 10 de 425 páginas.

Wink, M.; Mikes, Z. and Rheinheimer 1997. Phylogenetic Relationships in Weevils (Coleoptera: Curculionoidea) Inferred from Nucleotide Sequences of Mitochondrial 16S rDNA. Naturwissenschaften. Vol. 84, pp: 318-321.

Wittmer, W. 1958. Insects of Micronesia Coleoptera: Lampyridae, Cantharidae, Malachiidae, Prionoceridae. Vol. 16, No.2, 1958 pp: 67 – 74.

Wladimir Gatica 2009. El secreto de los Pozones. Ríos y Senderos.com. Pesca con mosca / outdoors. Chile. Available online (2010) in: http://www.riosysenderos.com/baul/pozonessecreto.htm (Last access: 22. V. 2010).

Wojcik, D. P. 1980. Fire ant myrmecophiles: behavior of Myrmecosaurus ferrugineus Bruch (Coleoptera: Staphylinidae) with comments on its abundance. Sociobiology. Vol. 5, pp: 63 - 68.

Wolfgang, H. Rücker 1998. *Cartodere (Aridius) delamarei* Dajoz, 1962 eine neue Adventivart für Europa (Coleoptera: Latridiidae). Mitt. internat. entomol. Ver. Band 22. Heft 3/4, pp: 179-180.

Rücker, W. H. 2003e. *Holoparamecus gabrielae* sp. nov., eine neue Merophysiinae aus Mexico (Coleoptera: Endomychidae) – Mitteilungsblatt für Systematik und Taxonomie der Latridiidae ISSN 1613-2440, Heft 1, Seite 10.

Wood, S. L. 1986. A Reclassification of the Genera of Scolytidae (Coleoptera). Great basin naturalist memoirs 10, Brigham Young Univ., Provo, Utah. 126 Pages.

Woodruff, R. E. 1973. The scarab beetles of Florida (Coleoptera: Scarabaeidae). Part. I. The Laparosticti (subfamilies: Scarabaeinae, Aphodiinae, Hybosorinae, Ochodaeinae, Geotrupinae, Acanthocerinae). Arthropods of Florida and Neighboring Land Areas, 8: 1-220.

Wooldridge, D. P. 1986. A catalog of the Coleoptera of America North of Mexico. Family: Limnichidae. Agriculture Handbook, United States Dep. Agric., Washington, pp: 1-8.

Worland, R.; Block, W.; and Rothery, P. 1992. Survival of sub-zero temperatures by two South Georgian beetles (Col., Perimylopidae).Polar Biology. Volume 11, Number 8, pp: 607-613.

Yanega, D. & R. A. B. Lescmer. 1994. Beetles associated with bee nests (Hymenoptera: Apidae) in Chiapas, México, with descriptions of the immatures stages of Vanomis balteatus Wernner (Coleoptera: Aderidae, Eudomychidae, Meloidae). Coleopterists Bull.Vol 48, pp: 355-360.

Yanli, Yue; Yunyun, Zhao & Dong, Ren 2009. *Glabrimycetoporus amoenus*, a new tachyporine genus and species of Mesozoic Staphylinidae (Coleoptera) from Liaoning, China. Zootaxa. N° 2225, pp: 63 – 68.

Yanoviak, S. P. 1999a. Community ecology of water-filled tree holes in Panama. PhD thesis, University of Oklahoma, Norman.

Yanoviak, S. P. 1999b. Community structure in water-filled tree holes of Panama: effects of hole height and size. *Selbyana. Vol.* 20, pp: 106–115.

Yanoviak, S.; Nadkarni, N. & Gering, J. 2003. Arthropods in epiphytes: a diversity component that is not effectively sampled by canopy fogging. Biodiversity and Conservation. Vol.12, pp: 731 - 741.

Yasuda, K. 1996. Attractiveness of pitfall traps to East-Indian sweet potato weevil, Euscepes postfasciatus (Fairmaire) (Coleoptera: Curculionidae). Japanese Journal of Applied Entomology and Zoology. Vol. 40, pp: 97–102.

Yearian, W. C., & Warren, L. O. 1964. Insects of pine cones in Arkansas. Journal of the Kansas Entomological Society. Vol. 37. Faac. 3, pp: 259-264.

Yee, M.; Yuan, Z. -Q.; Mohammed, C. 2001. Not just waste wood: decaying logs as key habitats in Tasmania's wet sclerophyll Eucalyptus obliqua production forests: the ecology of large and small logs compared. Tasforests 13: 119-128.

Yee, Marie; Grove, Simon; Alastair MM.Richardson and Caroline L. Mohammed 2006. Brown rot in inner heartwood: Why large logs support characteristic saproxylic beetle Assemblages of conservation concern, pp: 42-56. In: Grove, Simon J.; Hanula, James L., eds. 2006. Insect biodiversity and dead wood: proceedings of a symposium for the 22nd International Congress of Entomology. Gen. Tech. Rep. SRS-93. Asheville, NC: U.S. Department of Agriculture Forest Service, Southern Research Station. 109 p.

Ye-Jun, Zhang & Hong-Zhang, Zhou 2007. Taxonomy of the tribe Paromalini Reitter (Coleoptera: Histeridae, Dendrophilinae) from China. Zootaxa 1544, pp: 1–40.

Yélamos, T. 1992. Data on Afrotropical Histeridae (Coleoptera), with description of *Hister sindarae* n. sp. MIsc. Zool. Vol. 16, pp: 37-43.

Yélamos, T. 1993. Un nuevo *Atholus* Thomas, 1859 de Marruecos (Coleoptera, Histeridae). Nouvelle Revue d'Entomologie (N.S.). 10, pp: 145-151.

Yélamos, T. 2002. Coleoptera, Histeridae. En: Ramos et al. (Eds.). Fauna Ibérica, vol.17. Museo Nacional de Ciencias Naturales, CSIC. Madrid.

Yoshitomi, H. 2005. Systematic revision of the family Scirtidae of Japan, with phylogeny, morphology and bionomics (Insecta: Coleoptera, Scirtoidea). Japanese Journal of systematic Entomology, Monographic series, (3),

212 pages., Matsuyama.

Younan, E. G.; Hain, F. P. 1982. Evaluation of five trap designs for sampling insects associated with severed pines. Can. Ent. 114: 789-796.

Young, D. K. 1985. Description of the Larva of *Ischalia vancouverensis* Harrington (Coleoptera: Anthicidae: Ischaliinae), with Observations on the Systematic Position of the Genus. The Coleopterists Bulletin. Vol. 39, N° 3, pp: 201- 206.

Young, D. K. 1989. Notes on the bionomics of *Xenomycetes morrisoni* Horn (Coleoptera: Endomychidae), another cantharidin-orienting fungus beetle. Pan-Pacific Entomologist. Vol. 65, pp: 447-448.

Young, D. K. 1991. Prostomidae. Pp. 515-516. In: F. W. Stehr, ed., Immature insects, Volume 2. Kendall/Hunt. Dubuque, IA.

Young, D. K. 1991. Scraptiidae (Tenebrionoidea). Pp. 555-556. In: F. W. Stehr, ed. Immature Insects. Vol. 2. Kendall/ Hunt. Dubuque, IA. xvi + 975 pages.

Young, D. K. 1991a. Boridae (Tenebrionoidea). In: Stehr, F.W. (Ed.). Immature Insects. Volume 2 Kendall/Hunt, Dubuque, Iowa, pp: 537– 539.

Young, D. K. 1991b. Salpingidae (Tenebrionoidea). In: F. W. Stehr, ed. Immature Insects.Vol. 2, pp: 549- 551. Kendall/Hunt. Dubuque, I A, xvi + 975 Pages.

Young, D. K. 1991c. Inopeplidae (Tenebrionoidea). In: F. W. Stehr, ed. Immature Insects. Vol. 2, pp: 551-552. Kendall/Hunt. Dubuque, IA, xvi + 975 Pages.

Young, D. K.; Katovich, K. and Schwengel, M. 1996. The larval habitat of *Lecontia discicollis* (LeConte) (Boridae). Mola. Vol. 6, pp: 2-3.

Young, D. K. 2002. Family 36. Clambidae Jacquelin du Val 1857, pp. 85–86; Family 52. Callirhipidae Emden 1924, pp: 144-145; Family 53. Artematopodidae Lacordaire 1857, pp: 146-147; Family 54. Brachypsectridae Bøving and Craighead 1931, pp: 148-149; Family 71. Lymexylidae, pp: 261-262; Family 97. Archeocrypticidae Kaszab 1964, pp: 401-402; Family 107. Prostomidae Horn 1888, pp: 510-511; Family 108. Synchroidae Horn 1888, pp: 512-513; Family 110. Stenotrachelidae, pp: 520-521; Family. 115. Pyrochroidae Latreille 1807, pp: 540-543; In: Arnett, R.H., Jr., Thomas, M.C., S ., Skelley, P.E. & Howard, Frank, J. (Eds.). American Beetles, Volume 2: Polyphagha: Scarabaeoidea through Curculionoidea. CRC Press, Boca Raton, USA. 861 pages.

Young, Mark 2005. Insects in flight. Chapter 6, pp: 116-143, in: Leather, S. R.; Watt, A. D. Sampling theory and practice. In: Leather, S. R. (edit.) Insect sampling in forest ecosystems: 1-15. Blackwell, Malden, MA.

Zahradník, J. 1990. Guía de los coleópteros de España y de Europa. Omega. Barcelona.

Zahradník, Petr 2006. World catalogue of the family Endecatomidae (Coleoptera: Bostrichoidea). Studies and reports of District Museum Prague-East Taxonomical Series. Vol. 2. Fasc.1/2, pp: 142 – 144.

Zajac, A. & Zajac, M. (eds) 2001. Distribution Atlas of vascular Plants in Poland. Laboratory of Computer Chorology, Institute of Botany, Jagiellonian University and Fundation of Jagiellonian University, Cracow, xii + 716 pages.

Zaragoza-Caballero, S. 2008. Nueva species de *Pseudotelegeusis* (Coleoptera: Telegeusidae) del estado de Colima, México. Revista Mexicana de Biodiversidad. Vol. 79: 369- 372.

Zeck, E. H. 1919. Swarming of the soldier beetle, *Telephorus pulchellus*. Australian Zoologist. Vol.1, pp: 204 – 206.

Zhou, H. Z. and T.-H. Lou. 2001. On the genus *Onthophilus* Leach from China. Coleopterists Bulletin. N° 55, pp: 507-514.

Zhou, H. Z. and Yu, X. D. 2003. Rediscovery of the family Synteliidae (Coleoptera: Histeroidea) and two new species from China. The Coleopterists Bulletin. Vol. 57. Fasc.3, pp: 265– 273.

ZIKAN, J. F. 1939. *Amblyopinus henseli* Kolbe, um coleoptero da familia Staphylinidae que parasita mammiferos. Revista Entom. Vol.10, pp: 219-226, 3 figs.

Zimmerman, E. C. 1994a. Australian Weevils (Coleoptera: Curculionoidea). Volume I — Orthoceri, Anthribidae to Attelabidae. The Primitive Weevils. CSIRO, Australia. 741 pags.

Zimmerman, E. C. 1994b. Australian weevils. Volume II. Brentidae, Eurhynchidae, Apionidae. Melbourne: CSIRO Publishing.

Impreso y Encuadernado en
mastergraf srl
Gral. Pagola 1823 - CP 11800 - Tel.: 2203 4760*
Montevideo - Uruguay
E-mail: mastergraf@netgate.com.uy

Depósito Legal 353.856 - Comisión del Papel
Edición Amparada al Decreto 218/96